Grand Unification Theory
The Theory of Everything

Contents

1 Grand Unified Theory   1
- 1.1 History 1
- 1.2 Motivation 1
- 1.3 Unification of matter particles 2
  - 1.3.1 SU(5) 2
  - 1.3.2 SO(10) 4
  - 1.3.3 SU(8) 5
  - 1.3.4 O(16) 5
  - 1.3.5 Symplectic Groups and Quaternion Representations 5
  - 1.3.6 E8 and Octonion Representations 5
  - 1.3.7 Beyond Lie Groups 6
- 1.4 Unification of forces and the role of supersymmetry 6
- 1.5 Neutrino masses 6
- 1.6 Proposed theories 6
- 1.7 Ingredients 7
- 1.8 Current status 7
- 1.9 See also 7
- 1.10 Notes 8
- 1.11 References 8
  - 1.11.1 Further reading 8
- 1.12 External links 8

2 Theory of everything   9
- 2.1 Historical antecedents 10
  - 2.1.1 From ancient Greece to Einstein 10
  - 2.1.2 Twentieth century and the nuclear interactions 11
- 2.2 Modern physics 11
  - 2.2.1 Conventional sequence of theories 11
  - 2.2.2 String theory and M-theory 12
  - 2.2.3 Loop quantum gravity 12
  - 2.2.4 Other attempts 13
  - 2.2.5 Present status 13
- 2.3 Theory of everything and philosophy 13

- 2.4 Arguments against a theory of everything 13
  - 2.4.1 Gödel's incompleteness theorem 13
  - 2.4.2 Fundamental limits in accuracy 14
  - 2.4.3 Lack of fundamental laws 14
  - 2.4.4 Impossibility of being "of everything" 14
  - 2.4.5 Infinite number of onion layers 15
  - 2.4.6 Impossibility of calculation 15
- 2.5 See also 15
- 2.6 References 16
  - 2.6.1 Footnotes 16
  - 2.6.2 Bibliography 17
- 2.7 External links 17

3 Technicolor (physics) 18

- 3.1 Introduction 18
- 3.2 Early technicolor 19
- 3.3 Extended technicolor 20
- 3.4 Walking technicolor 21
  - 3.4.1 Top quark mass 22
- 3.5 Minimal Walking Models 22
- 3.6 Technicolor on the lattice 22
- 3.7 Technicolor phenomenology 22
  - 3.7.1 Precision electroweak tests 23
  - 3.7.2 Hadron collider phenomenology 24
  - 3.7.3 Dark matter 24
- 3.8 See also 25
- 3.9 References 25

4 Kaluza–Klein theory 30

- 4.1 The Kaluza Hypothesis 30
- 4.2 Field Equations from the Kaluza Hypothesis 31
- 4.3 Equations of Motion from the Kaluza Hypothesis 32
- 4.4 Kaluza's Hypothesis for the Matter Stress-Energy Tensor 34
- 4.5 The Quantum Interpretation of Klein 34
- 4.6 Quantum Field Theory Interpretation 34
- 4.7 Group Theory Interpretation 34
- 4.8 Space-time-matter theory 36
- 4.9 Geometric interpretation 36
  - 4.9.1 The Einstein equations 36
  - 4.9.2 The Maxwell equations 37
  - 4.9.3 The Kaluza–Klein geometry 37
  - 4.9.4 Generalizations 38

- 4.10 Empirical tests 38
- 4.11 See also 38
- 4.12 Notes 39
- 4.13 References 40
- 4.14 Further reading 40

5 String theory — 41
- 5.1 Overview 41
  - 5.1.1 Strings 42
  - 5.1.2 Branes 42
  - 5.1.3 Dualities 43
  - 5.1.4 Extra dimensions 44
- 5.2 Testability and experimental predictions 45
  - 5.2.1 String harmonics 45
  - 5.2.2 Cosmology 45
  - 5.2.3 Supersymmetry 46
- 5.3 AdS/CFT correspondence 46
  - 5.3.1 Examples of the correspondence 46
  - 5.3.2 Applications to quantum chromodynamics 46
  - 5.3.3 Applications to condensed matter physics 47
- 5.4 Connections to mathematics 47
  - 5.4.1 Mirror symmetry 47
  - 5.4.2 Vertex operator algebras 47
- 5.5 History 48
  - 5.5.1 Early results 48
  - 5.5.2 First superstring revolution 49
  - 5.5.3 Second superstring revolution 50
- 5.6 Criticisms 50
  - 5.6.1 High energies 50
  - 5.6.2 Number of solutions 50
  - 5.6.3 Background independence 51
- 5.7 See also 51
- 5.8 References 51
- 5.9 Further reading 54
  - 5.9.1 Popular books 54
  - 5.9.2 Textbooks 54
  - 5.9.3 Online material 55
- 5.10 External links 55

6 Superfluid vacuum theory — 61
- 6.1 History 61
- 6.2 Relation to other concepts and theories 61

	6.2.1	Lorentz and Galilean symmetries	62
	6.2.2	Relativistic quantum field theory	62
	6.2.3	Curved space-time	62
	6.2.4	Cosmological constant	63
	6.2.5	Gravitational waves and gravitons	63
	6.2.6	Mass generation and Higgs boson	63
6.3	Logarithmic BEC vacuum theory	64
6.4	See also	64
6.5	Notes	64
6.6	References	65

7 Supersymmetry   66

- 7.1 History . . . 66
- 7.2 Motivations . . . 67
- 7.3 Applications . . . 67
  - 7.3.1 Extension of possible symmetry groups . . . 68
  - 7.3.2 The Supersymmetric Standard Model . . . 68
  - 7.3.3 Supersymmetric quantum mechanics . . . 69
  - 7.3.4 Supersymmetry: Applications to condensed matter physics . . . 70
  - 7.3.5 Supersymmetry in optics . . . 70
  - 7.3.6 Mathematics . . . 70
- 7.4 General supersymmetry . . . 70
  - 7.4.1 Extended supersymmetry . . . 70
  - 7.4.2 Supersymmetry in alternate numbers of dimensions . . . 71
- 7.5 Supersymmetry as a quantum group . . . 71
- 7.6 Supersymmetry in quantum gravity . . . 71
- 7.7 Falsifiability . . . 72
- 7.8 Current status . . . 72
- 7.9 See also . . . 73
- 7.10 References . . . 73
- 7.11 Further reading . . . 75
  - 7.11.1 Theoretical introductions, free and online . . . 75
  - 7.11.2 Monographs . . . 75
  - 7.11.3 On experiments . . . 76
- 7.12 External links . . . 76

8 Quantum gravity   77

- 8.1 Overview . . . 77
  - 8.1.1 Effective field theories . . . 78
  - 8.1.2 Quantum gravity theory for the highest energy scales . . . 78
- 8.2 Quantum mechanics and general relativity . . . 79
  - 8.2.1 The graviton . . . 79

	8.2.2	The dilaton	79
	8.2.3	Nonrenormalizability of gravity	79
	8.2.4	QG as an effective field theory	80
	8.2.5	Spacetime background dependence	80
	8.2.6	Semi-classical quantum gravity	81
	8.2.7	Points of tension	81
8.3	Candidate theories		81
	8.3.1	String theory	82
	8.3.2	Loop quantum gravity	82
	8.3.3	Other approaches	83
8.4	Weinberg–Witten theorem		83
8.5	Experimental tests		84
8.6	See also		84
8.7	References		84
8.8	Further reading		86

9 Simple group — 92

9.1	Examples		92
	9.1.1	Finite simple groups	92
	9.1.2	Infinite simple groups	92
9.2	Classification		92
	9.2.1	Finite simple groups	93
9.3	Structure of finite simple groups		93
9.4	History for finite simple groups		93
	9.4.1	Construction	93
	9.4.2	Classification	94
9.5	Tests for nonsimplicity		94
9.6	See also		94
9.7	References		94
	9.7.1	Notes	95
	9.7.2	Textbooks	95
	9.7.3	Papers	95
9.8	External links		95

10 Georgi–Glashow model — 96

10.1	Breaking SU(5)		97
10.2	Minimal supersymmetric SU(5)		98
	10.2.1	Spacetime	98
	10.2.2	Spatial symmetry	98
	10.2.3	Gauge symmetry group	98
	10.2.4	Global internal symmetry	98
	10.2.5	Matter parity	98

- 10.2.6 Vector superfields 98
- 10.2.7 Chiral superfields 98
- 10.2.8 Superpotential 98
- 10.2.9 Vacua 99
- 10.3 Lee Smolin's view of SU(5) 100
- 10.4 Popular culture 100
- 10.5 References 100

11 Simple Lie group   101
- 11.1 Comments on the definition 101
- 11.2 Method of classification 102
- 11.3 Real forms 102
- 11.4 Relationship of simple Lie algebras to groups 102
- 11.5 Classification by Dynkin diagram 102
- 11.6 Infinite series 102
  - 11.6.1 A series 103
  - 11.6.2 B series 103
  - 11.6.3 C series 103
  - 11.6.4 D series 103
- 11.7 Exceptional cases 103
- 11.8 Simply laced groups 103
- 11.9 See also 103
- 11.10 References 104

12 Group representation   105
- 12.1 Branches of group representation theory 105
- 12.2 Definitions 106
- 12.3 Examples 107
- 12.4 Reducibility 107
- 12.5 Generalizations 107
  - 12.5.1 Set-theoretical representations 107
  - 12.5.2 Representations in other categories 108
- 12.6 See also 108
- 12.7 References 108

13 SO(10) (physics)   109
- 13.1 Important subgroups 110
- 13.2 Spontaneous symmetry breaking 111
- 13.3 The electroweak Higgs and the doublet-triplet splitting problem 111
- 13.4 Matter 111
- 13.5 Proton decay 112
- 13.6 See also 112

13.7 Notes 112

14 Lie superalgebra   114

    14.1 Definition 114

         14.1.1 Distinction from graded Lie algebra 114

         14.1.2 Even and odd parts 114

         14.1.3 Involution 115

    14.2 Examples 115

    14.3 Classification 115

    14.4 Classification of infinite-dimensional simple linearly compact Lie superalgebras 116

    14.5 Category-theoretic definition 116

    14.6 See also 117

    14.7 References 118

    14.8 External links 118

15 Yang–Mills theory   119

    15.1 History and theoretical description 119

    15.2 Mathematical overview 120

    15.3 Quantization of Yang–Mills theory 121

    15.4 Propagators 123

    15.5 Beta function and running coupling 125

    15.6 Open problems 125

    15.7 See also 126

    15.8 References 126

    15.9 Further reading 128

    15.10 External links 128

16 Minimal Supersymmetric Standard Model   129

    16.1 Theoretical motivations 129

         16.1.1 Naturalness 130

         16.1.2 Gauge-coupling unification 130

         16.1.3 Dark matter 131

    16.2 Predictions of the MSSM regarding hadron colliders 131

         16.2.1 Neutralinos 132

         16.2.2 Charginos 132

         16.2.3 Squarks 132

         16.2.4 Gluinos 133

         16.2.5 Sleptons 133

    16.3 MSSM fields 133

         16.3.1 MSSM superfields 133

         16.3.2 MSSM Higgs Mass 134

    16.4 The MSSM Lagrangian 134

16.4.1 Soft Susy breaking	. . . .	135

- 16.5 Problems with the MSSM 135
- 16.6 Theories of supersymmetry breaking 136
  - 16.6.1 Gravity-mediated supersymmetry breaking 136
  - 16.6.2 Gauge-mediated supersymmetry breaking (GMSB) 136
  - 16.6.3 Anomaly-mediated supersymmetry breaking (AMSB) 136
- 16.7 Phenomenological MSSM (pMSSM) 136
- 16.8 See also 137
- 16.9 References 137
- 16.10 External links 138

17 Grand unification energy — 139
- 17.1 See also 139
- 17.2 References 139

18 Hierarchy problem — 140
- 18.1 Technical definition 140
- 18.2 The Higgs mass 140
- 18.3 Theoretical solutions 141
  - 18.3.1 Supersymmetric solution 141
  - 18.3.2 Conformal solution 142
  - 18.3.3 Solution via extra dimensions 142
- 18.4 The cosmological constant 143
- 18.5 See also 143
- 18.6 References 143

19 Left–right symmetry — 145
- 19.1 Particle Physics 145

20 Trinification — 147
- 20.1 References 148

21 SU(6) (physics) — 149
- 21.1 References 149

22 E6 (mathematics) — 150
- 22.1 Real and complex forms 150
- 22.2 E_6 as an algebraic group 151
- 22.3 Algebra 151
  - 22.3.1 Dynkin diagram 151
  - 22.3.2 Roots of E_6 151
  - 22.3.3 Weyl group 154
  - 22.3.4 Cartan matrix 155
- 22.4 Important subalgebras and representations 155

CONTENTS

- 22.5 E6 polytope 155
- 22.6 Chevalley and Steinberg groups of type E_6 and $^2E_6$ 155
- 22.7 Importance in physics 156
- 22.8 See also 157
- 22.9 References 158

23 331 model — **159**
- 23.1 References 159
- 23.2 See also 159

24 Chiral color — **160**
- 24.1 References 160

25 Flipped SU(5) — **161**
- 25.1 The Model 161
- 25.2 Comparison with the standard SU(5) 162
- 25.3 spacetime 163
- 25.4 spatial symmetry 163
- 25.5 gauge symmetry group 163
- 25.6 global internal symmetry 163
- 25.7 vector superfields 163
- 25.8 chiral superfields 163
- 25.9 Superpotential 163
- 25.10 References 164

26 Pati–Salam model — **165**
- 26.1 Core theory 165
- 26.2 Differences from the SU(5) unification 166
- 26.3 Minimal supersymmetric Pati–Salam 166
  - 26.3.1 Spacetime 166
  - 26.3.2 Spatial symmetry 166
  - 26.3.3 Gauge symmetry group 166
  - 26.3.4 Global internal symmetry 166
  - 26.3.5 Vector superfields 166
  - 26.3.6 Chiral superfields 167
  - 26.3.7 Superpotential 167
  - 26.3.8 Left-right extension 167
- 26.4 Sources 167
- 26.5 References 167
- 26.6 External links 167

27 Flipped SO(10) — **168**
- 27.1 References 169

28 Little Higgs — 170
28.1 References — 170

29 Preon — 171
29.1 Background — 171
29.2 Motivations — 172
29.3 History — 172
29.4 Rishon model — 173
29.5 Criticisms — 173
29.5.1 The mass paradox — 173
29.5.2 Constraints — 174
29.6 Conflicts with observed physics — 174
29.7 Popular culture — 174
29.8 See also — 174
29.9 Notes — 174
29.10 Further reading — 175

30 M-theory — 176
30.1 Background — 176
30.1.1 Quantum gravity and strings — 176
30.1.2 Number of dimensions — 177
30.1.3 Dualities — 178
30.1.4 Supersymmetry — 179
30.1.5 Branes — 180
30.2 History and development — 180
30.2.1 Early work on supergravity — 180
30.2.2 Relationships between string theories — 180
30.2.3 Membranes and fivebranes — 182
30.2.4 Second superstring revolution — 183
30.2.5 Origin of the term — 184
30.3 AdS/CFT correspondence — 184
30.3.1 Overview — 184
30.3.2 6D (2,0) superconformal field theory — 185
30.3.3 ABJM superconformal field theory — 187
30.4 Phenomenology — 187
30.4.1 Overview — 187
30.4.2 Compactification on G_2 manifolds — 188
30.4.3 Heterotic M-theory — 189
30.5 Notes — 189
30.6 References — 191
30.7 Further reading — 193
30.8 External links — 193

31 Loop quantum gravity — 194

- 31.1 History — 194
- 31.2 General covariance and background independence — 195
- 31.3 Constraints and their Poisson Bracket Algebra — 196
  - 31.3.1 The constraints of classical canonical general relativity — 196
  - 31.3.2 The Poisson bracket algebra — 197
  - 31.3.3 Dirac observables — 198
- 31.4 Quantization of the constraints - the equations of Quantum General Relativity — 198
  - 31.4.1 Pre-history and Ashtekar new variables — 198
  - 31.4.2 Quantum constraints as the equations of quantum general relativity — 199
  - 31.4.3 Introduction of the loop representation — 199
  - 31.4.4 Geometric operators, the need for intersecting Wilson loops and spin network states — 201
  - 31.4.5 Real variables, modern analysis and LQG — 202
  - 31.4.6 Solving the quantum constraints — 204
- 31.5 Spin foams — 205
  - 31.5.1 Spin foam derived from the Hamiltonian constraint operator — 205
  - 31.5.2 Spin foams from BF theory — 206
  - 31.5.3 Modern formulation of spin foams — 207
  - 31.5.4 Spin foam derived from the Master constraint operator — 207
  - 31.5.5 Spin foams from consistent discretisations — 207
- 31.6 The semi-classical limit — 207
  - 31.6.1 What is the semiclassical limit? — 207
  - 31.6.2 Why might LQG not have general relativity as its semiclassical limit? — 207
  - 31.6.3 Difficulties checking the semiclassical limit of LQG — 208
  - 31.6.4 Progress in demonstrating LQG has the correct semiclassical limit — 208
- 31.7 Improved dynamics and the Master constraint — 209
  - 31.7.1 The Master constraint — 209
  - 31.7.2 Testing the Master constraint — 210
  - 31.7.3 Applications of the Master constraint — 210
  - 31.7.4 Spin foam from the Master constraint — 210
  - 31.7.5 Algebraic quantum gravity — 210
- 31.8 Physical applications of LQG — 211
  - 31.8.1 Black hole entropy — 211
  - 31.8.2 Loop quantum cosmology — 213
  - 31.8.3 Loop Quantum Gravity phenomenology — 213
  - 31.8.4 Background independent scattering amplitudes — 213
  - 31.8.5 planck stars — 214
- 31.9 Gravitons, string theory, super symmetry, extra dimensions in LQG — 214
- 31.10 LQG and related research programs — 214
- 31.11 Problems and comparisons with alternative approaches — 215
- 31.12 See also — 216

- 31.13 Notes 216
- 31.14 References 219
- 31.15 External links 221

32 Causal dynamical triangulation — 222
- 32.1 Introduction 222
- 32.2 Derivation 222
- 32.3 Advantages and Disadvantages 223
- 32.4 Related theories 223
- 32.5 See also 223
- 32.6 References 224
- 32.7 External links 224

33 Lie algebra — 225
- 33.1 Definitions 225
  - 33.1.1 Generators and dimension 226
  - 33.1.2 Homomorphisms, subalgebras, and ideals 226
  - 33.1.3 Direct sum and semidirect product 226
- 33.2 Properties 226
  - 33.2.1 Admits an enveloping algebra 227
  - 33.2.2 Representation 227
- 33.3 Examples 227
  - 33.3.1 Vector spaces 227
  - 33.3.2 Subspaces 227
  - 33.3.3 Real matrix groups 228
  - 33.3.4 Three dimensions 228
  - 33.3.5 Infinite dimensions 228
- 33.4 Structure theory and classification 229
  - 33.4.1 Abelian, nilpotent, and solvable 229
  - 33.4.2 Simple and semisimple 229
  - 33.4.3 Cartan's criterion 230
  - 33.4.4 Classification 230
- 33.5 Relation to Lie groups 230
- 33.6 Category theoretic definition 231
- 33.7 See also 232
- 33.8 Notes 232
- 33.9 References 232
- 33.10 External links 233

34 Lie group — 234
- 34.1 Overview 234
- 34.2 Definitions and examples 235

CONTENTS

- 34.2.1 First examples 235
- 34.2.2 Related concepts 236
- 34.3 More examples of Lie groups 236
  - 34.3.1 Examples with a specific number of dimensions 236
  - 34.3.2 Examples with n dimensions 237
  - 34.3.3 Constructions 237
  - 34.3.4 Related notions 238
- 34.4 Early history 238
- 34.5 The concept of a Lie group, and possibilities of classification 239
- 34.6 Properties 239
- 34.7 Types of Lie groups and structure theory 239
- 34.8 The Lie algebra associated with a Lie group 240
- 34.9 Homomorphisms and isomorphisms 241
- 34.10 The exponential map 242
- 34.11 Infinite-dimensional Lie groups 244
- 34.12 See also 244
- 34.13 References 245
- 34.14 Notes 245
- 34.15 References 245

35 Heterotic string theory 247
- 35.1 String duality 247
- 35.2 References 247

36 Topological defect 248
- 36.1 Cosmology 248
  - 36.1.1 Symmetry breakdown 248
  - 36.1.2 Types of topological defects 249
  - 36.1.3 Observation 249
- 36.2 Condensed matter 249
  - 36.2.1 Classification 249
  - 36.2.2 Stable defects 251
- 36.3 Images 251
- 36.4 See also 251
- 36.5 References 252
- 36.6 External links 252

37 Magnetic monopole 253
- 37.1 Historical background 254
  - 37.1.1 Pre-twentieth century 254
  - 37.1.2 Twentieth century 254
- 37.2 Poles and magnetism in ordinary matter 254

- 37.3 Maxwell's equations 255
  - 37.3.1 In Gaussian cgs units 256
  - 37.3.2 In SI units 256
  - 37.3.3 Tensor formulation 256
  - 37.3.4 Duality transformation 257
- 37.4 Dirac's quantization 257
- 37.5 Topological interpretation 258
  - 37.5.1 Dirac string 258
  - 37.5.2 Grand unified theories 259
  - 37.5.3 String theory 260
  - 37.5.4 Mathematical formulation 260
- 37.6 Grand unified theories 261
- 37.7 Searches for magnetic monopoles 262
- 37.8 "Monopoles" in condensed-matter systems 262
- 37.9 Further descriptions in particle physics 263
- 37.10 See also 264
- 37.11 Notes 265
- 37.12 References 268
- 37.13 External links 268

38 Cosmic string — 269
- 38.1 Theories containing cosmic strings 269
- 38.2 Dimensions 269
- 38.3 Gravitation 270
  - 38.3.1 Negative Mass Cosmic String 270
- 38.4 Observational evidence 270
- 38.5 String theory and cosmic strings 271
- 38.6 See also 271
- 38.7 References 272
- 38.8 External links 272

39 Domain wall (string theory) — 274
- 39.1 String theory 274
- 39.2 See also 274
- 39.3 References 274

40 Inflation (cosmology) — 275
- 40.1 Overview 276
  - 40.1.1 Space expands 277
  - 40.1.2 Few inhomogeneities remain 278
  - 40.1.3 Key requirement 278
  - 40.1.4 Reheating 278

40.2 Motivations 278
    40.2.1 Horizon problem 279
    40.2.2 Flatness problem 279
    40.2.3 Magnetic-monopole problem 279

40.3 History 280
    40.3.1 Precursors 280
    40.3.2 Early inflationary models 280
    40.3.3 Slow-roll inflation 281
    40.3.4 Effects of asymmetries 281

40.4 Observational status 282

40.5 Theoretical status 283
    40.5.1 Fine-tuning problem 283
    40.5.2 Eternal inflation 283
    40.5.3 Initial conditions 284
    40.5.4 Hybrid inflation 284
    40.5.5 Inflation and string cosmology 285
    40.5.6 Inflation and loop quantum gravity 285
    40.5.7 Inflation and generalized uncertainty principle (GUP) 285

40.6 Alternatives to inflation 285

40.7 Criticisms 286

40.8 See also 286

40.9 Notes 287

40.10 References 292

40.11 External links 293

41 Doublet–triplet splitting problem   **294**

41.1 Doublet–triplet splitting and the μ-problem 294
    41.1.1 Dimopoulos–Wilczek mechanism 294

41.2 Higgs representations in Grand Unified Theories 295

41.3 Proton decay 295

41.4 References 295

42 Quantum chromodynamics   **297**

42.1 Terminology 297

42.2 History 298

42.3 Theory 299
    42.3.1 Some definitions 299
    42.3.2 Additional remarks: duality 299
    42.3.3 Symmetry groups 299
    42.3.4 Lagrangian 300
    42.3.5 Fields 300
    42.3.6 Dynamics 301

	42.3.7 Area law and confinement 301
	42.4 Methods 302

- 42.3.7 Area law and confinement . . . 301
- 42.4 Methods . . . 302
  - 42.4.1 Perturbative QCD . . . 302
  - 42.4.2 Lattice QCD . . . 302
  - 42.4.3 1/N expansion . . . 302
  - 42.4.4 Effective theories . . . 302
  - 42.4.5 QCD sum rules . . . 302
  - 42.4.6 Nambu–Jona-Lasinio model . . . 303
- 42.5 Experimental tests . . . 303
- 42.6 Cross-relations to solid state physics . . . 303
- 42.7 See also . . . 304
- 42.8 References . . . 304
- 42.9 Further reading . . . 305
- 42.10 External links . . . 305

43 Classical unified field theories — 307

- 43.1 Overview . . . 307
- 43.2 Early work . . . 307
- 43.3 Differential geometry and field theory . . . 307
- 43.4 Weyl's infinitesimal geometry . . . 308
- 43.5 Kaluza's fifth dimension . . . 308
- 43.6 Eddington's affine geometry . . . 308
- 43.7 Einstein's geometric approaches . . . 308
- 43.8 Schrödinger's pure-affine theory . . . 309
- 43.9 Later work . . . 309
- 43.10 See also . . . 310
- 43.11 References . . . 310

44 Riemannian geometry — 311

- 44.1 Introduction . . . 311
- 44.2 Classical theorems in Riemannian geometry . . . 312
  - 44.2.1 General theorems . . . 312
  - 44.2.2 Geometry in large . . . 313
- 44.3 See also . . . 314
- 44.4 Literature . . . 314
- 44.5 References . . . 315
- 44.6 External links . . . 315

45 Affine connection — 316

- 45.1 Motivation and history . . . 317
  - 45.1.1 Motivation from surface theory . . . 317
  - 45.1.2 Motivation from tensor calculus . . . 318

- 45.1.3 Approaches 318
- 45.2 Formal definition as a differential operator 318
  - 45.2.1 Elementary properties 319
- 45.3 Parallel transport for affine connections 319
- 45.4 Formal definition on the frame bundle 320
- 45.5 Affine connections as Cartan connections 321
  - 45.5.1 Explanations and historical intuition 321
  - 45.5.2 Affine space as the flat model geometry 322
  - 45.5.3 General affine geometries: formal definitions 323
- 45.6 Further properties 324
  - 45.6.1 Curvature and torsion 324
  - 45.6.2 The Levi-Civita connection 325
  - 45.6.3 Geodesics 325
  - 45.6.4 Development 326
- 45.7 Surface theory revisited 326
  - 45.7.1 Example: the unit sphere in Euclidean space 326
- 45.8 See also 327
- 45.9 Notes 327
- 45.10 References 328
  - 45.10.1 Primary historical references 328
  - 45.10.2 Secondary references 328

46 De Sitter universe   330
- 46.1 Mathematical expression 330
- 46.2 Potential for the Universe 331
- 46.3 Relative expansion 331
- 46.4 Modelling cosmic inflation 331
- 46.5 See also 331
- 46.6 References 331

47 Standard Model   332
- 47.1 Historical background 333
- 47.2 Overview 333
- 47.3 Particle content 333
  - 47.3.1 Fermions 333
  - 47.3.2 Gauge bosons 335
  - 47.3.3 Higgs boson 337
  - 47.3.4 Full particle count 337
- 47.4 Theoretical aspects 337
  - 47.4.1 Construction of the Standard Model Lagrangian 337
- 47.5 Tests and predictions 339
- 47.6 Challenges 339

- 47.7 See also 340
- 47.8 Notes and references 340
- 47.9 References 340
- 47.10 Further reading 342
- 47.11 External links 343

48 Gauge theory   344

- 48.1 History and importance 344
- 48.2 Description 345
  - 48.2.1 Global and local symmetries 345
  - 48.2.2 Example of global symmetry 346
  - 48.2.3 Use of fiber bundles to describe local symmetries 346
  - 48.2.4 Gauge fields 346
  - 48.2.5 Physical experiments 347
  - 48.2.6 Continuum theories 347
  - 48.2.7 Quantum field theories 348
- 48.3 Classical gauge theory 348
  - 48.3.1 Classical electromagnetism 348
  - 48.3.2 An example: Scalar $O(n)$ gauge theory 348
  - 48.3.3 The Yang–Mills Lagrangian for the gauge field 351
  - 48.3.4 An example: Electrodynamics 351
- 48.4 Mathematical formalism 352
- 48.5 Quantization of gauge theories 353
  - 48.5.1 Methods and aims 353
  - 48.5.2 Anomalies 353
- 48.6 Pure gauge 354
- 48.7 See also 354
- 48.8 References 354
- 48.9 Bibliography 354
- 48.10 External links 355

49 Quantum field theory   356

- 49.1 History 357
  - 49.1.1 Foundations 357
  - 49.1.2 Gauge theory 357
  - 49.1.3 Grand synthesis 357
- 49.2 Principles 357
  - 49.2.1 Classical and quantum fields 357
  - 49.2.2 Single- and many-particle quantum mechanics 358
  - 49.2.3 Second quantization 359
  - 49.2.4 Dynamics 362
  - 49.2.5 Implications 362

- 49.2.6 Axiomatic approaches 363
- 49.3 Associated phenomena 364
  - 49.3.1 Renormalization 364
  - 49.3.2 Haag's theorem 365
  - 49.3.3 Gauge freedom 365
  - 49.3.4 Multivalued gauge transformations 365
  - 49.3.5 Supersymmetry 366
- 49.4 See also 366
- 49.5 Notes 367
- 49.6 References 367
- 49.7 Further reading 367
- 49.8 External links 369
- 49.9 Text and image sources, contributors, and licenses 370
  - 49.9.1 Text 370
  - 49.9.2 Images 379
  - 49.9.3 Content license 383

Chapter 1

Grand Unified Theory

For the album, see Grand Unification (album).

A **Grand Unified Theory** (**GUT**) is a model in particle physics in which at high energy, the three gauge interactions of the Standard Model which define the electromagnetic, weak, and strong interactions or forces, are merged into one single force. This unified interaction is characterized by one larger gauge symmetry and thus several force carriers, but one unified coupling constant. If Grand Unification is realized in nature, there is the possibility of a grand unification epoch in the early universe in which the fundamental forces are not yet distinct.

Models that do not unify all interactions using one simple Lie group as the gauge symmetry, but do so using semisimple groups, can exhibit similar properties and are sometimes referred to as Grand Unified Theories as well.

Unifying gravity with the other three interactions would provide a theory of everything (TOE), rather than a GUT. Nevertheless, GUTs are often seen as an intermediate step towards a TOE.

Because their masses are predicted to be just a few orders of magnitude below the Planck scale, at the GUT scale, well beyond the reach of foreseen particle colliders experiments, novel particles predicted by GUT models cannot be observed directly. Instead, effects of grand unification might be detected through indirect observations such as proton decay, electric dipole moments of elementary particles, or the properties of neutrinos.[1] Some grand unified theories predict the existence of magnetic monopoles.

As of 2012, all GUT models which aim to be completely realistic are quite complicated, even compared to the Standard Model, because they need to introduce additional fields and interactions, or even additional dimensions of space. The main reason for this complexity lies in the difficulty of reproducing the observed fermion masses and mixing angles. Due to this difficulty, and due to the lack of any observed effect of grand unification so far, there is no generally accepted GUT model.

1.1 History

Historically, the first true GUT which was based on the simple Lie group SU(5), was proposed by Howard Georgi and Sheldon Glashow in 1974.[2] The Georgi–Glashow model was preceded by the Semisimple Lie algebra Pati–Salam model by Abdus Salam and Jogesh Pati,[3] who pioneered the idea to unify gauge interactions.

The acronym GUT was first coined in 1978 by CERN researchers John Ellis, Andrzej Buras, Mary K. Gaillard, and Dimitri Nanopoulos, however in the final version of their paper[4] they opted for the less anatomical *GUM* (Grand Unification Mass). Nanopoulos later that year was the first to use[5] the acronym in a paper.[6]

1.2 Motivation

The fact that the electric charges of electrons and protons seem to cancel each other exactly to extreme precision is essential for the existence of the macroscopic world as we know it, but this important property of elementary particles is not explained in the Standard Model of particle physics. While the description of strong and weak interactions within the Standard Model is based on gauge symmetries governed by the simple symmetry groups SU(3) and SU(2) which allow only discrete charges, the remaining component, the weak hypercharge interaction is described by an abelian symmetry U(1) which in principle allows for arbitrary charge assignments.[note 1] The observed charge quantization, namely the fact that all known elementary particles carry electric charges which appear to be exact multiples of 1/3 of the "elementary" charge, has led to the idea that hypercharge interactions and possibly the strong and weak interactions might be embedded in one Grand Unified interaction described by a single, larger simple symmetry group containing the Standard Model. This would automatically predict the quantized nature and values of all elementary particle charges. Since this also results in a prediction for the relative strengths of the fundamental interactions which we observe, in particular the weak mixing angle, Grand Unification ideally reduces the number of independent input parameters, but is also constrained by observations.

Grand Unification is reminiscent of the unification of electric and magnetic forces by Maxwell's theory of electromagnetism in the 19th century, but its physical implications and mathematical structure are qualitatively different.

1.3 Unification of matter particles

$$\begin{bmatrix}
0 & W & X_r & X_g & X_b & 0 & 0 & 0 & 0 & 0 & 0 & 0 & 0 & 0 & 0 \\
\overline{W} & 0 & Y_r & Y_g & Y_b & 0 & 0 & 0 & 0 & 0 & 0 & 0 & 0 & 0 & 0 \\
\overline{X}_r & \overline{Y}_r & 0 & g_{r\bar{g}} & g_{r\bar{b}} & 0 & 0 & 0 & 0 & 0 & 0 & 0 & 0 & 0 & 0 \\
\overline{X}_g & \overline{Y}_g & \bar{g}_{r\bar{g}} & 0 & g_{g\bar{b}} & 0 & 0 & 0 & 0 & 0 & 0 & 0 & 0 & 0 & 0 \\
\overline{X}_b & \overline{Y}_b & \bar{g}_{r\bar{b}} & \bar{g}_{g\bar{b}} & 0 & 0 & 0 & 0 & 0 & 0 & 0 & 0 & 0 & 0 & 0 \\
0 & 0 & 0 & 0 & 0 & 0 & Y_r & Y_g & Y_b & X_r & X_g & X_b & 0 & 0 & 0 \\
0 & 0 & 0 & 0 & 0 & \overline{Y}_r & 0 & g_{r\bar{g}} & g_{r\bar{b}} & W & 0 & 0 & 0 & X_b & X_g \\
0 & 0 & 0 & 0 & 0 & \overline{Y}_g & \bar{g}_{r\bar{g}} & 0 & g_{g\bar{b}} & 0 & W & 0 & X_b & 0 & X_r \\
0 & 0 & 0 & 0 & 0 & \overline{Y}_b & \bar{g}_{r\bar{b}} & \bar{g}_{g\bar{b}} & 0 & 0 & 0 & W & X_g & X_r & 0 \\
0 & 0 & 0 & 0 & 0 & \overline{X}_r & \overline{W} & 0 & 0 & 0 & g_{r\bar{g}} & g_{r\bar{b}} & 0 & Y_b & Y_g \\
0 & 0 & 0 & 0 & 0 & \overline{X}_g & 0 & \overline{W} & 0 & \bar{g}_{r\bar{g}} & 0 & g_{g\bar{b}} & Y_b & 0 & Y_r \\
0 & 0 & 0 & 0 & 0 & \overline{X}_b & 0 & 0 & \overline{W} & \bar{g}_{r\bar{b}} & \bar{g}_{g\bar{b}} & 0 & Y_g & Y_r & 0 \\
0 & 0 & 0 & 0 & 0 & 0 & 0 & \overline{X}_b & \overline{X}_g & 0 & \overline{Y}_b & \overline{Y}_g & 0 & g_{r\bar{g}} & g_{r\bar{b}} \\
0 & 0 & 0 & 0 & 0 & 0 & \overline{X}_b & 0 & \overline{X}_r & \overline{Y}_b & 0 & \overline{Y}_r & \bar{g}_{r\bar{g}} & 0 & g_{g\bar{b}} \\
0 & 0 & 0 & 0 & 0 & 0 & \overline{X}_g & \overline{X}_r & 0 & \overline{Y}_g & \overline{Y}_r & 0 & \bar{g}_{r\bar{b}} & \bar{g}_{g\bar{b}} & 0
\end{bmatrix}
\begin{bmatrix}
\nu_e \\ e \\ \bar{d}_r \\ \bar{d}_g \\ \bar{d}_b \\ \bar{e} \\ d_r \\ d_g \\ d_b \\ u_r \\ u_g \\ u_b \\ \bar{u}_r \\ \bar{u}_g \\ \bar{u}_b
\end{bmatrix}$$

Schematic representation of fermions and bosons in SU(5) GUT showing 5 + 10 split in the multiplets. Neutral bosons (photon, Z-boson, and neutral gluons) are not shown but occupy the diagonal entries of the matrix in complex superpositions

> For an elementary introduction to how Lie algebras are related to particle physics, see the article Particle physics and representation theory.

1.3.1 SU(5)

Main article: SU(5) (physics)

SU(5) is the simplest GUT. The smallest simple Lie group which contains the standard model, and upon which the first Grand Unified Theory was based, is

$$SU(5) \supset SU(3) \times SU(2) \times U(1)$$

1.3. UNIFICATION OF MATTER PARTICLES

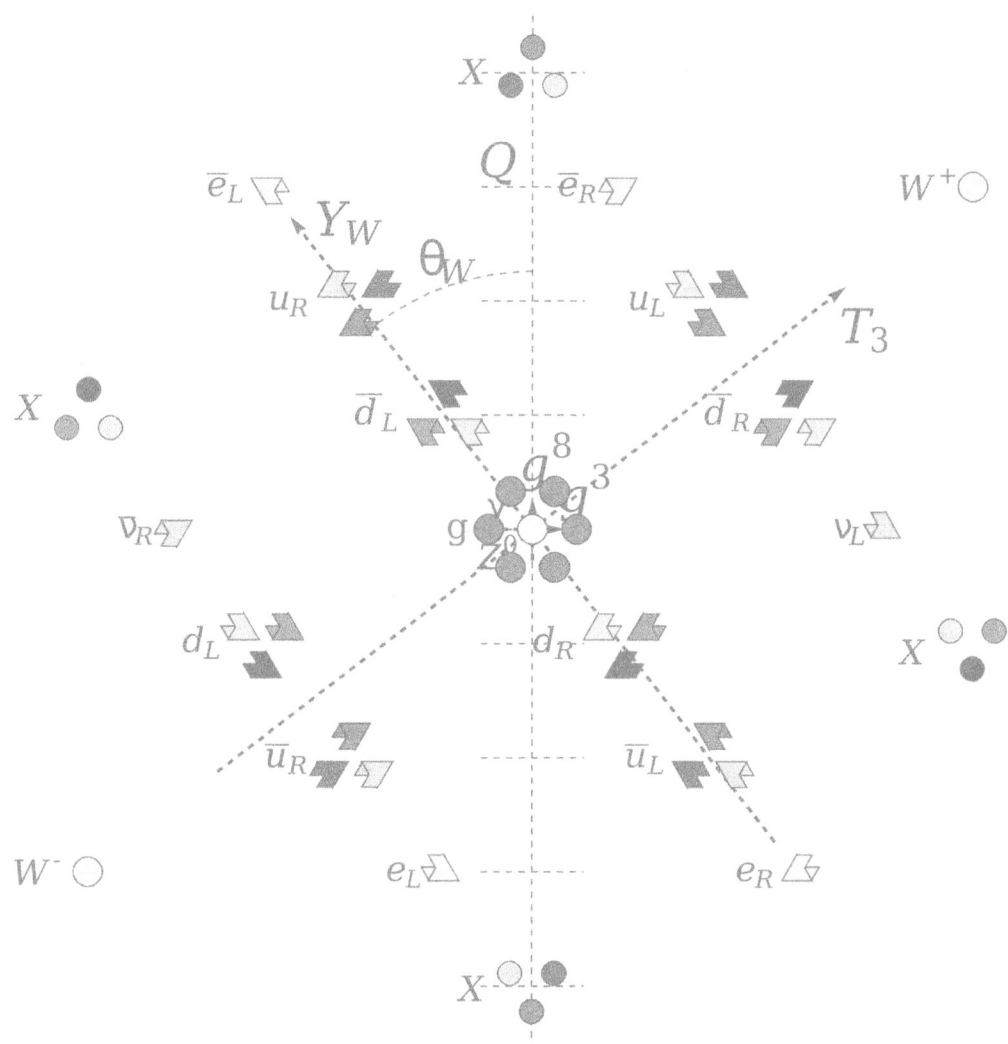

The pattern of weak isospins, weak hypercharges, and strong charges for particles in the SU(5) model, rotated by the predicted weak mixing angle, showing electric charge roughly along the vertical. In addition to Standard Model particles, the theory includes twelve colored X bosons, responsible for proton decay.

Such group symmetries allow the reinterpretation of several known particles as different states of a single particle field. However, it is not obvious that the simplest possible choices for the extended "Grand Unified" symmetry should yield the correct inventory of elementary particles. The fact that all currently known (2009) matter particles fit nicely into three copies of the smallest group representations of SU(5) and immediately carry the correct observed charges, is one of the first and most important reasons why people believe that a Grand Unified Theory might actually be realized in nature.

The two smallest irreducible representations of SU(5) are **5** and **10**. In the standard assignment, the **5** contains the charge conjugates of the right-handed down-type quark color triplet and a left-handed lepton isospin doublet, while the **10** contains the six up-type quark components, the left-handed down-type quark color triplet, and the right-handed electron. This scheme has to be replicated for each of the three known generations of matter. It is notable that the theory is anomaly free with this matter content.

The hypothetical right-handed neutrinos are not contained in any of these representations, which can explain their relative heaviness (see seesaw mechanism).

1.3.2 SO(10)

Main article: SO(10) (physics)
The next simple Lie group which contains the standard model is

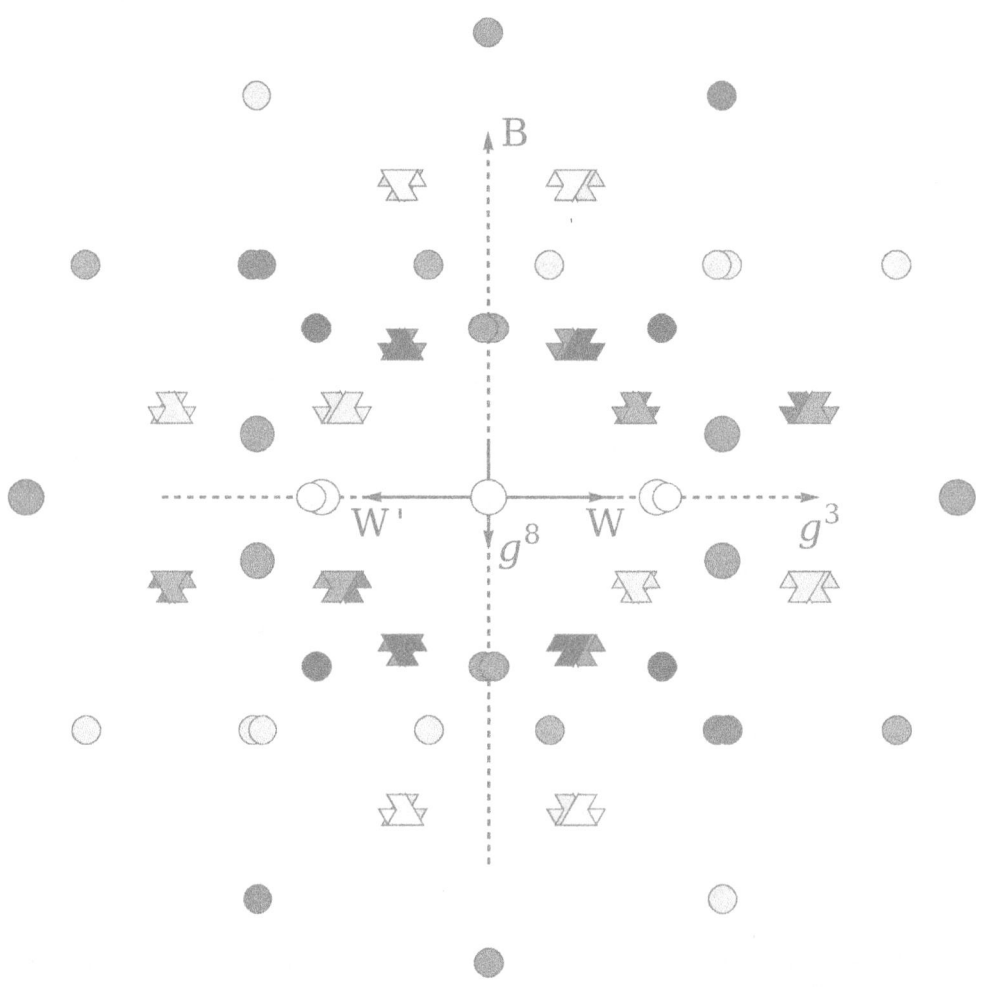

The pattern of weak isospin, W, weaker isospin, W', strong g3 and g8, and baryon minus lepton, B, charges for particles in the SO(10) Grand Unified Theory, rotated to show the embedding in E_6.

$SO(10) \supset SU(5) \supset SU(3) \times SU(2) \times U(1)$

Here, the unification of matter is even more complete, since the irreducible spinor representation **16** contains both the **5** and **10** of SU(5) and a right-handed neutrino, and thus the complete particle content of one generation of the extended standard model with neutrino masses. This is already the largest simple group which achieves the unification of matter in a scheme involving only the already known matter particles (apart from the Higgs sector).

Since different standard model fermions are grouped together in larger representations, GUTs specifically predict relations among the fermion masses, such as between the electron and the down quark, the muon and the strange quark, and the tau lepton and the bottom quark for SU(5) and SO(10). Some of these mass relations hold approximately, but most don't (see Georgi-Jarlskog mass relation).

The boson matrix for SO(10) is found by taking the 15×15 matrix from the **10 + 5** representation of SU(5) and adding an extra row and column for the right handed neutrino. The bosons are found by adding a partner to each of the 20 charged bosons (2 right-handed W bosons, 6 massive charged gluons and 12 X/Y type bosons) and adding an extra heavy neutral Z-boson to make 5 neutral bosons in total. The boson matrix will have a boson or its new partner in each row and column. These pairs combine to create the familiar 16D Dirac spinor matrices of SO(10).

1.3.3 SU(8)

Assuming 4 generations of fermions instead of 3 makes a total of **64** types of particles. These can be put into **64 = 8 + 56** representations of SU(8). This can be divided into SU(5) × SU(3)F × U(1) which is the SU(5) theory together with some heavy bosons which act on the generation number.

1.3.4 O(16)

Again assuming 4 generations of fermions, the **128** particles and anti-particles can be put into a single spinor representation of O(16).

1.3.5 Symplectic Groups and Quaternion Representations

Symplectic gauge groups could also be considered. For example Sp(8) has a representation in terms of 4 × 4 quaternion unitary matrices which has a **16** dimensional real representation and so might be considered as a candidate for a gauge group. Sp(8) has 32 charged bosons and 4 neutral bosons. It's subgroups include SU(4) so can at least contain the gluons and photon of SU(3) × U(1). Although it's probably not possible to have weak bosons acting on chiral fermions in this representation. A quaternion representation of the fermions might be:

$$\begin{bmatrix} e + i\bar{e} + jv + k\bar{v} \\ u_r + i\overline{u_r} + jd_r + k\overline{d_r} \\ u_g + i\overline{u_g} + jd_g + k\overline{d_g} \\ u_b + i\overline{u_b} + jd_b + k\overline{d_b} \end{bmatrix}_L$$

A further complication with quaternion representations of fermions is that there are two types of multiplication: left multiplication and right multiplication which must be taken into account. It turns out that including left and right-handed 4 × 4 quaternion matrices is equivalent to including a single right-multiplication by a unit quaternion which adds an extra SU(2) and so has an extra neutral boson and two more charged bosons. Thus the group of left and right handed 4 × 4 quaternion matrcies is Sp(8) × SU(2) which does include the standard model bosons:

$$SU(4, H)_L \times H_R = Sp(8) \times SU(2) \supset SU(4) \times SU(2) \supset SU(3) \times SU(2) \times U(1)$$

If ψ is a quaternion valued spinor, A_μ^{ab} is quaternion hermitian 4 × 4 matrix coming from Sp(8) and B_μ is a pure imaginary quaternion (both of which are 4-vector bosons) then the interaction term is:

$$\overline{\psi^a}\gamma_\mu \left(A_\mu^{ab}\psi^b + \psi^a B_\mu \right)$$

1.3.6 E8 and Octonion Representations

It can be noted that a generation of 16 fermions can be put into the form of an Octonion with each element of the octonion being an 8-vector. If the 3 generations are then put in a 3x3 hermitian matrix with certain additions for the diagonal elements then these matrices form an exceptional (grassman-) Jordan algebra, which has the symmetry group of one of the exceptional Lie groups (F_4, E_6, E_7 or E_8) depending on the details.

$$\psi = \begin{bmatrix} a & e & \mu \\ \bar{e} & b & \tau \\ \bar{\mu} & \bar{\tau} & c \end{bmatrix}$$

$[\psi_A, \psi_B] \subset J_3(O)$

Because they are fermions the anti-commutators of the Jordan algebra become commutators. It is known that E_6 has subgroup O(10) and so is big enough to include the Standard Model. An E_8 gauge group, for example, would have 8 neutral bosons, 120 charged bosons and 120 charged anti-bosons. To account for the 248 fermions in the lowest multiplet of E_8, these would either have to include anti-particles (and so have Baryogenesis), have new undiscovered particles, or have gravity-like (Spin connection) bosons affecting elements of the particles spin direction. Each of these poses theoretical problems.

1.3.7 Beyond Lie Groups

Other structures have been suggested including Lie 3-algebras and Lie superalgebras. Neither of these fit with Yang–Mills theory. In particular Lie superalgebras would introduce bosons with the wrong statistics. Supersymmetry however does fit with Yang–Mills. For example N=4 Super Yang Mills Theory requires an SU(N) gauge group.

1.4 Unification of forces and the role of supersymmetry

The unification of forces is possible due to the energy scale dependence of force coupling parameters in quantum field theory called renormalization group running, which allows parameters with vastly different values at usual energies to converge to a single value at a much higher energy scale.[7]

The renormalization group running of the three gauge couplings in the Standard Model has been found to nearly, but not quite, meet at the same point if the hypercharge is normalized so that it is consistent with SU(5) or SO(10) GUTs, which are precisely the GUT groups which lead to a simple fermion unification. This is a significant result, as other Lie groups lead to different normalizations. However, if the supersymmetric extension MSSM is used instead of the Standard Model, the match becomes much more accurate. In this case, the coupling constants of the strong and electroweak interactions meet at the grand unification energy, also known as the GUT scale:

$\Lambda_{\text{GUT}} \approx 10^{16}$ GeV

It is commonly believed that this matching is unlikely to be a coincidence, and is often quoted as one of the main motivations to further investigate supersymmetric theories despite the fact that no supersymmetric partner particles have been experimentally observed (May 2014). Also, most model builders simply assume supersymmetry because it solves the hierarchy problem—i.e., it stabilizes the electroweak Higgs mass against radiative corrections.

1.5 Neutrino masses

Since Majorana masses of the right-handed neutrino are forbidden by SO(10) symmetry, SO(10) GUTs predict the Majorana masses of right-handed neutrinos to be close to the GUT scale where the symmetry is spontaneously broken in those models. In supersymmetric GUTs, this scale tends to be larger than would be desirable to obtain realistic masses of the light, mostly left-handed neutrinos (see neutrino oscillation) via the seesaw mechanism.

1.6 Proposed theories

Several such theories have been proposed, but none is currently universally accepted. An even more ambitious theory that includes *all* fundamental forces, including gravitation, is termed a theory of everything. Some common mainstream GUT models are:

Not quite GUTs:

Note: These models refer to Lie algebras not to Lie groups. The Lie group could be [SU(4) × SU(2) × SU(2)]/$\mathbf{Z}_2$, just to take a random example.

The most promising candidate is SO(10). (Minimal) SO(10) does not contain any exotic fermions (i.e. additional fermions besides the Standard Model fermions and the right-handed neutrino), and it unifies each generation into a single irreducible representation. A number of other GUT models are based upon subgroups of SO(10). They are the minimal left-right model, SU(5), flipped SU(5) and the Pati–Salam model. The GUT group E_6 contains SO(10), but models based upon it are significantly more complicated. The primary reason for studying E_6 models comes from $E_8 \times E_8$ heterotic string theory.

GUT models generically predict the existence of topological defects such as monopoles, cosmic strings, domain walls, and others. But none have been observed. Their absence is known as the monopole problem in cosmology. Most GUT models also predict proton decay, although not the Pati–Salam model; current experiments still haven't detected proton decay. This experimental limit on the proton's lifetime pretty much rules out minimal SU(5).

- Proton Decay. These graphics refer to the X bosons and Higgs bosons.
- Dimension 6 proton decay mediated by the *X* boson in SU(5) GUT
- Dimension 6 proton decay mediated by the *X* boson in flipped SU(5) GUT
- Dimension 6 proton decay mediated by the triplet Higgs and the anti-triplet Higgs in SU(5) GUT

Some GUT theories like SU(5) and SO(10) suffer from what is called the doublet-triplet problem. These theories predict that for each electroweak Higgs doublet, there is a corresponding colored Higgs triplet field with a very small mass (many orders of magnitude smaller than the GUT scale here). In theory, unifying quarks with leptons, the Higgs doublet would also be unified with a Higgs triplet. Such triplets have not been observed. They would also cause extremely rapid proton decay (far below current experimental limits) and prevent the gauge coupling strengths from running together in the renormalization group.

Most GUT models require a threefold replication of the matter fields. As such, they do not explain why there are three generations of fermions. Most GUT models also fail to explain the little hierarchy between the fermion masses for different generations.

1.7 Ingredients

A GUT model basically consists of a gauge group which is a compact Lie group, a connection form for that Lie group, a Yang–Mills action for that connection given by an invariant symmetric bilinear form over its Lie algebra (which is specified by a coupling constant for each factor), a Higgs sector consisting of a number of scalar fields taking on values within real/complex representations of the Lie group and chiral Weyl fermions taking on values within a complex rep of the Lie group. The Lie group contains the Standard Model group and the Higgs fields acquire VEVs leading to a spontaneous symmetry breaking to the Standard Model. The Weyl fermions represent matter.

1.8 Current status

As of 2012, there is still no hard evidence that nature is described by a Grand Unified Theory. Moreover, since we have no idea which Higgs particle has been observed, the smaller electroweak unification is still pending.[8] The discovery of neutrino oscillations indicates that the Standard Model is incomplete and has led to renewed interest toward certain GUT such as SO(10). One of the few possible experimental tests of certain GUT is proton decay and also fermion masses. There are a few more special tests for supersymmetric GUT.

The gauge coupling strengths of QCD, the weak interaction and hypercharge seem to meet at a common length scale called the GUT scale and equal approximately to 10^{16} GeV, which is slightly suggestive. This interesting numerical observation is called the **gauge coupling unification**, and it works particularly well if one assumes the existence of superpartners of the Standard Model particles. Still it is possible to achieve the same by postulating, for instance, that ordinary (non supersymmetric) SO(10) models break with an intermediate gauge scale, such as the one of Pati–Salam group

1.9 See also

- Paradigm shift

- Classical unified field theories
- X and Y bosons
- B − L quantum number

1.10 Notes

[1] There are however certain constraints on the choice of particle charges from theoretical consistency, in particular anomaly cancellation.

1.11 References

[1] Ross, G. (1984). *Grand Unified Theories*. Westview Press. ISBN 978-0-8053-6968-7.

[2] Georgi, H.; Glashow, S.L. (1974). "Unity of All Elementary Particle Forces". *Physical Review Letters* **32**: 438–441. Bibcode:1974PhRvL..32..438G. doi:10.1103/PhysRevLett.32.438.

[3] Pati, J.; Salam, A. (1974). "Lepton Number as the Fourth Color". *Physical Review D* **10**: 275–289. Bibcode:1974PhRvD..10..275P. doi:10.1103/PhysRevD.10.275.

[4] Buras, A.J.; Ellis, J.; Gaillard, M.K.; Nanopoulos, D.V. (1978). "Aspects of the grand unification of strong, weak and electromagnetic interactions". *Nuclear Physics B* **135** (1): 66–92. Bibcode:1978NuPhB.135...66B. doi:10.1016/0550-3213(78)90214-6. Retrieved 2011-03-21.

[5] Nanopoulos, D.V. (1979). "Protons Are Not Forever". *Orbis Scientiae* **1**: 91. Harvard Preprint HUTP-78/A062.

[6] Ellis, J. (2002). "Physics gets physical". *Nature* **415** (6875): 957. Bibcode:2002Natur.415..957E. doi:10.1038/415957b.

[7] Ross, G. (1984). *Grand Unified Theories*. Westview Press. ISBN 978-0-8053-6968-7.

[8] Hawking, S.W. (1996). *A Brief History of Time: The Updated and Expanded Edition. (2nd ed.)*. Bantam Books. p. XXX. ISBN 0-553-38016-8.

1.11.1 Further reading

- Stephen Hawking, A Brief History of Time, includes a brief popular overview.
- DR. Chaim Tejman http://www.grandunifiedtheory.org.il/

1.12 External links

- The Algebra of Grand Unified Theories
- Scholarpedia: Grand Unification

Chapter 2

Theory of everything

This article is about the physical concept. For other uses, see Theory of everything (disambiguation).

A **theory of everything** (ToE) or **final theory**, **ultimate theory**, or **master theory** refers to the hypothetical presence of a single, all-encompassing, coherent theoretical framework of physics that fully explains and links together all physical aspects of the universe.[1] ToE is one of the major unsolved problems in physics. Over the past few centuries, two theoretical frameworks have been developed that, as a whole, most closely resemble a ToE. The two theories upon which all modern physics rests are General Relativity (GR) and Quantum Field Theory (QFT). GR is a theoretical framework that only focuses on the force of gravity for understanding the universe in regions of both large-scale and high-mass: stars, galaxies, clusters of galaxies, etc. On the other hand, QFT is a theoretical framework that only focuses on three non-gravitational forces for understanding the universe in regions of both small scale and low mass: sub-atomic particles, atoms, molecules, etc. QFT successfully implemented the Standard Model and unified the interactions (so-called Grand Unified Theory) between the three non-gravitational forces: weak, strong, and electromagnetic force.

Through years of research, physicists have experimentally confirmed with tremendous accuracy virtually every prediction made by these two theories when in their appropriate domains of applicability. In accordance with their findings, scientists also learned that GR and QFT, as they are currently formulated, are mutually incompatible - they cannot both be right. Since the usual domains of applicability of GR and QFT are so different, most situations require that only one of the two theories be used. As it turns out, this incompatibility between GR and QFT is only an apparent issue in regions of extremely small-scale and high-mass, such as those that exist within a black hole or during the beginning stages of the universe (i.e., the moment immediately following the Big Bang). To resolve this conflict, a theoretical framework revealing a deeper underlying reality, unifying gravity with the other three interactions, must be discovered to harmoniously integrate the realms of GR and QFT into a seamless whole: a single theory that, in principle, is capable of describing all phenomena. In pursuit of this goal, quantum gravity has recently become an area of active research.

Over the past few decades, a single explanatory framework, called "string theory", has emerged that may turn out to be the ultimate theory of the universe. Many physicists believe that, at the beginning of the universe (up to 10^{-43} seconds after the Big Bang), the four fundamental forces were once a single fundamental force. Unlike most (if not all) other theories, string theory may be on its way to successfully incorporating each of the four fundamental forces into a unified whole. According to string theory, every particle in the universe, at its most microscopic level (Planck length), consists of varying combinations of vibrating strings (or strands) with preferred patterns of vibration. String theory claims that it is through these specific oscillatory patterns of strings that a particle of unique mass and force charge is created (that is to say, the electron is a type of string that vibrates one way, while the up-quark is a type of string vibrating another way, and so forth).

Initially, the term *theory of everything* was used with an ironic connotation to refer to various overgeneralized theories. For example, a grandfather of Ijon Tichy — a character from a cycle of Stanisław Lem's science fiction stories of the 1960s — was known to work on the "General Theory of Everything". Physicist John Ellis[2] claims to have introduced the term into the technical literature in an article in *Nature* in 1986.[3] Over time, the term stuck in popularizations of theoretical physics research.

2.1 Historical antecedents

2.1.1 From ancient Greece to Einstein

Archimedes was possibly the first scientist that is known to have described nature with axioms (or principles) and then deduce new results from them.[4] He thus tried to describe "everything" starting from a few axioms. Any "theory of everything" is similarly expected to be based on axioms and to deduce all observable phenomena from them.

The concept of 'atom', introduced by Democritus, unified all phenomena observed in nature as the motion of atoms. In ancient Greek times philosophers speculated that the apparent diversity of observed phenomena was due to a single type of interaction, namely the collisions of atoms. Following atomism, the mechanical philosophy of the 17th century posited that all forces could be ultimately reduced to contact forces between the atoms, then imagined as tiny solid particles.[5]

In the late 17th century, Isaac Newton's description of the long-distance force of gravity implied that not all forces in nature result from things coming into contact. Newton's work in his *Principia* dealt with this in a further example of unification, in this case unifying Galileo's work on terrestrial gravity, Kepler's laws of planetary motion and the phenomenon of tides by explaining these apparent actions at a distance under one single law: the law of universal gravitation.

In 1814, building on these results, Laplace famously suggested that a sufficiently powerful intellect could, if it knew the position and velocity of every particle at a given time, along with the laws of nature, calculate the position of any particle at any other time:

> An intellect which at a certain moment would know all forces that set nature in motion, and all positions of all items of which nature is composed, if this intellect were also vast enough to submit these data to analysis, it would embrace in a single formula the movements of the greatest bodies of the universe and those of the tiniest atom; for such an intellect nothing would be uncertain and the future just like the past would be present before its eyes.
> — *Essai philosophique sur les probabilités*, Introduction. 1814

Laplace thus envisaged a combination of gravitation and mechanics as a theory of everything. Modern quantum mechanics implies that uncertainty is inescapable, and thus that Laplace's vision needs to be amended: a theory of everything must include gravitation and quantum mechanics.

In 1820, Hans Christian Ørsted discovered a connection between electricity and magnetism, triggering decades of work that culminated in 1865, in James Clerk Maxwell's theory of electromagnetism. During the 19th and early 20th centuries, it gradually became apparent that many common examples of forces – contact forces, elasticity, viscosity, friction, and pressure – result from electrical interactions between the smallest particles of matter.

In his experiments of 1849–50, Michael Faraday was the first to search for a unification of gravity with electricity and magnetism.[6] However, he found no connection.

In 1900, David Hilbert published a famous list of mathematical problems. In Hilbert's sixth problem, he challenged researchers to find an axiomatic basis to all of physics. In this problem he thus asked for what today would be called a theory of everything.

In the late 1920s, the new quantum mechanics showed that the chemical bonds between atoms were examples of (quantum) electrical forces, justifying Dirac's boast that "the underlying physical laws necessary for the mathematical theory of a large part of physics and the whole of chemistry are thus completely known".[7]

After 1915, when Albert Einstein published the theory of gravity (general relativity), the search for a unified field theory combining gravity with electromagnetism began with a renewed interest. In Einstein's day, the strong and the weak forces had not yet been discovered, yet, he found the potential existence of two other distinct forces - gravity and electromagnetism- far more alluring. This launched his thirty-year voyage in search of the so-called "unified field theory" that he hoped would show that these two forces are really manifestations of one grand underlying principle. During these last few decades of his life, this quixotic quest isolated Einstein from the mainstream of physics. Understandably, the mainstream was instead far more excited about the newly emerging framework of quantum mechanics. Einstein wrote to a friend in the early 1940s, "I have become a lonely old chap who is mainly known because he doesn't wear socks and who is exhibited as a curiosity on special occasions." Prominent contributors were Gunnar Nordström, Hermann Weyl, Arthur Eddington, Theodor Kaluza, Oskar Klein, and most notably, Albert

Einstein and his collaborators. Einstein intensely searched for, but ultimately failed to find, a unifying theory.[8] (But see:Einstein–Maxwell–Dirac equations.) More than a half a century later, Einstein's dream of discovering a unified theory has become the Holy Grail of modern physics.

2.1.2 Twentieth century and the nuclear interactions

In the twentieth century, the search for a unifying theory was interrupted by the discovery of the strong and weak nuclear forces (or interactions), which differ both from gravity and from electromagnetism. A further hurdle was the acceptance that in a ToE, quantum mechanics had to be incorporated from the start, rather than emerging as a consequence of a deterministic unified theory, as Einstein had hoped.

Gravity and electromagnetism could always peacefully coexist as entries in a list of classical forces, but for many years it seemed that gravity could not even be incorporated into the quantum framework, let alone unified with the other fundamental forces. For this reason, work on unification, for much of the twentieth century, focused on understanding the three "quantum" forces: electromagnetism and the weak and strong forces. The first two were combined in 1967–68 by Sheldon Glashow, Steven Weinberg, and Abdus Salam into the "electroweak" force.[9] Electroweak unification is a broken symmetry: the electromagnetic and weak forces appear distinct at low energies because the particles carrying the weak force, the W and Z bosons, have non-zero masses of 80.4 GeV/c^2 and 91.2 GeV/c^2, whereas the photon, which carries the electromagnetic force, is massless. At higher energies Ws and Zs can be created easily and the unified nature of the force becomes apparent.

While the strong and electroweak forces peacefully coexist in the Standard Model of particle physics, they remain distinct. So far, the quest for a theory of everything is thus unsuccessful on two points: neither a unification of the strong and electroweak forces – which Laplace would have called `contact forces' – has been achieved, nor has a unification of these forces with gravitation been achieved.

2.2 Modern physics

2.2.1 Conventional sequence of theories

A Theory of Everything would unify all the fundamental interactions of nature: gravitation, strong interaction, weak interaction, and electromagnetism. Because the weak interaction can transform elementary particles from one kind into another, the ToE should also yield a deep understanding of the various different kinds of possible particles. The usual assumed path of theories is given in the following graph, where each unification step leads one level up:

In this graph, electroweak unification occurs at around 100 GeV, grand unification is predicted to occur at 10^{16} GeV, and unification of the GUT force with gravity is expected at the Planck energy, roughly 10^{19} GeV.

Several Grand Unified Theories (GUTs) have been proposed to unify electromagnetism and the weak and strong forces. Grand unification would imply the existence of an electronuclear force; it is expected to set in at energies of the order of 10^{16} GeV, far greater than could be reached by any possible Earth-based particle accelerator. Although the simplest GUTs have been experimentally ruled out, the general idea, especially when linked with supersymmetry, remains a favorite candidate in the theoretical physics community. Supersymmetric GUTs seem plausible not only for their theoretical "beauty", but because they naturally produce large quantities of dark matter, and because the inflationary force may be related to GUT physics (although it does not seem to form an inevitable part of the theory). Yet GUTs are clearly not the final answer; both the current standard model and all proposed GUTs are quantum field theories which require the problematic technique of renormalization to yield sensible answers. This is usually regarded as a sign that these are only effective field theories, omitting crucial phenomena relevant only at very high energies.

The final step in the graph requires resolving the separation between quantum mechanics and gravitation, often equated with general relativity. Numerous researchers concentrate their efforts on this specific step; nevertheless, no accepted theory of quantum gravity – and thus no accepted theory of everything – has emerged yet. It is usually assumed that the ToE will also solve the remaining problems of GUTs.

In addition to explaining the forces listed in the graph, a ToE may also explain the status of at least two candidate forces suggested by modern cosmology: an inflationary force and dark energy. Furthermore, cosmological experiments also suggest the existence of dark matter, supposedly composed of fundamental particles outside the scheme of the standard model. However, the existence of these forces and particles has not been proven yet.

2.2.2 String theory and M-theory

Since the 1990s, many physicists believe that 11-dimensional M-theory, which is described in some limits by one of the five perturbative superstring theories, and in another by the maximally-supersymmetric 11-dimensional supergravity, is the theory of everything. However, there is no widespread consensus on this issue.

A surprising property of string/M-theory is that extra dimensions are required for the theory's consistency. In this regard, string theory can be seen as building on the insights of the Kaluza–Klein theory, in which it was realized that applying general relativity to a five-dimensional universe (with one of them small and curled up) looks from the four-dimensional perspective like the usual general relativity together with Maxwell's electrodynamics. This lent credence to the idea of unifying gauge and gravity interactions, and to extra dimensions, but didn't address the detailed experimental requirements. Another important property of string theory is its supersymmetry, which together with extra dimensions are the two main proposals for resolving the hierarchy problem of the standard model, which is (roughly) the question of why gravity is so much weaker than any other force. The extra-dimensional solution involves allowing gravity to propagate into the other dimensions while keeping other forces confined to a four-dimensional spacetime, an idea that has been realized with explicit stringy mechanisms.[10]

Research into string theory has been encouraged by a variety of theoretical and experimental factors. On the experimental side, the particle content of the standard model supplemented with neutrino masses fits into a spinor representation of SO(10), a subgroup of E8 that routinely emerges in string theory, such as in heterotic string theory[11] or (sometimes equivalently) in F-theory.[12][13] String theory has mechanisms that may explain why fermions come in three hierarchical generations, and explain the mixing rates between quark generations.[14] On the theoretical side, it has begun to address some of the key questions in quantum gravity, such as resolving the black hole information paradox, counting the correct entropy of black holes[15][16] and allowing for topology-changing processes.[17][18][19] It has also led to many insights in pure mathematics and in ordinary, strongly-coupled gauge theory due to the Gauge/String duality.

In the late 1990s, it was noted that one major hurdle in this endeavor is that the number of possible four-dimensional universes is incredibly large. The small, "curled up" extra dimensions can be compactified in an enormous number of different ways (one estimate is 10^{500}) each of which leads to different properties for the low-energy particles and forces. This array of models is known as the string theory landscape.

One proposed solution is that many or all of these possibilities are realised in one or another of a huge number of universes, but that only a small number of them are habitable, and hence the fundamental constants of the universe are ultimately the result of the anthropic principle rather than dictated by theory. This has led to criticism of string theory,[20] arguing that it cannot make useful (i.e., original, falsifiable, and verifiable) predictions and regarding it as a pseudoscience. Others disagree,[21] and string theory remains an extremely active topic of investigation in theoretical physics.

2.2.3 Loop quantum gravity

Current research on loop quantum gravity may eventually play a fundamental role in a ToE, but that is not its primary aim.[22] Also loop quantum gravity introduces a lower bound on the possible length scales.

There have been recent claims that loop quantum gravity may be able to reproduce features resembling the Standard Model. So far only the first generation of fermions (leptons and quarks) with correct parity properties have been modelled by Sundance Bilson-Thompson using preons constituted of braids of spacetime as the building blocks.[23] However, there is no derivation of the Lagrangian that would describe the interactions of such particles, nor is it possible to show that such particles are fermions, nor that the gauge groups or interactions of the Standard Model are realised. Utilization of quantum computing concepts made it possible to demonstrate that the particles are able to survive quantum fluctuations.[24]

This model leads to an interpretation of electric and colour charge as topological quantities (electric as number and chirality of twists carried on the individual ribbons and colour as variants of such twisting for fixed electric charge).

Bilson-Thompson's original paper suggested that the higher-generation fermions could be represented by more complicated braidings, although explicit constructions of these structures were not given. The electric charge, colour, and parity properties of such fermions would arise in the same way as for the first generation. The model was expressly generalized for an infinite number of generations and for the weak force bosons (but not for photons or gluons) in a 2008 paper by Bilson-Thompson, Hackett, Kauffman and Smolin.[25]

2.2.4 Other attempts

Any ToE must include general relativity and the Standard Model of particle physics.

A recent and very prolific attempt is called Causal Sets. As some of the approaches mentioned above, its direct goal isn't necessarily to achieve a ToE but primarily a working theory of quantum gravity, which might eventually include the standard model and become a candidate for a ToE. Its founding principle is that spacetime is fundamentally discrete and that the spacetime events are related by a partial order. This partial order has the physical meaning of the causality relations between relative past and future distinguishing spacetime events.

Outside the previously mentioned attempts there is Garrett Lisi's E8 proposal. This theory provides an attempt of identifying general relativity and the standard model within the Lie group E8. The theory doesn't provide a novel quantization procedure and the author suggests its quantization might follow the Loop Quantum Gravity approach above mentioned.[26]

Christoph Schiller's Strand Model attempts to account for the gauge symmetry of the Standard Model of particle physics, $U(1) \times SU(2) \times SU(3)$, with the three Reidemeister moves of knot theory by equating each elementary particle to a different tangle of one, two, or three strands (selectively a long prime knot or unknotted curve, a rational tangle, or a braided tangle respectively).

2.2.5 Present status

At present, there is no candidate theory of everything that includes the standard model of particle physics and general relativity. For example, no candidate theory is able to calculate the fine structure constant or the mass of the electron. Most particle physicists expect that the outcome of the ongoing experiments – the search for new particles at the large particle accelerators and for dark matter – are needed in order to provide further input for a ToE.

2.3 Theory of everything and philosophy

Main article: Theory of everything (philosophy)

The philosophical implications of a physical ToE are frequently debated. For example, if philosophical physicalism is true, a physical ToE will coincide with a philosophical theory of everything.

The "system building" style of metaphysics attempts to answer *all* the important questions in a coherent way, providing a complete picture of the world. Plato and Aristotle could be said to have created early examples of comprehensive systems. In the early modern period (17th and 18th centuries), the system-building *scope* of philosophy is often linked to the rationalist *method* of philosophy, which is the technique of deducing the nature of the world by pure *a priori* reason. Examples from the early modern period include the Leibniz's Monadology, Descarte's Dualism, and Spinoza's Monism. Hegel's Absolute idealism and Whitehead's Process philosophy were later systems.

2.4 Arguments against a theory of everything

In parallel to the intense search for a ToE, various scholars have seriously debated the possibility of its discovery.

2.4.1 Gödel's incompleteness theorem

A number of scholars claim that Gödel's incompleteness theorem suggests that any attempt to construct a ToE is bound to fail. Gödel's theorem, informally stated, asserts that any formal theory expressive enough for elementary arithmetical facts to be expressed and strong enough for them to be proved is either inconsistent (both a statement and its denial can be derived from its axioms) or incomplete, in the sense that there is a true statement that can't be derived in the formal theory.

Stanley Jaki, in his 1966 book *The Relevance of Physics*, pointed out that, because any "theory of everything" will certainly be a consistent non-trivial mathematical theory, it must be incomplete. He claims that this dooms searches for a deterministic theory of everything.[27] In a later reflection, Jaki states that it is wrong to say that a final theory is impossible, but rather that "when it is on hand one cannot know rigorously that it is a final theory."[28]

Freeman Dyson has stated that

Stephen Hawking was originally a believer in the Theory of Everything but, after considering Gödel's Theorem, concluded that one was not obtainable.

Jürgen Schmidhuber (1997) has argued against this view; he points out that Gödel's theorems are irrelevant for computable physics.[29] In 2000, Schmidhuber explicitly constructed limit-computable, deterministic universes whose pseudo-randomness based on undecidable, Gödel-like halting problems is extremely hard to detect but does not at all prevent formal ToEs describable by very few bits of information.[30]

Related critique was offered by Solomon Feferman,[31] among others. Douglas S. Robertson offers Conway's game of life as an example:[32] The underlying rules are simple and complete, but there are formally undecidable questions about the game's behaviors. Analogously, it may (or may not) be possible to completely state the underlying rules of physics with a finite number of well-defined laws, but there is little doubt that there are questions about the behavior of physical systems which are formally undecidable on the basis of those underlying laws.

Since most physicists would consider the statement of the underlying rules to suffice as the definition of a "theory of everything", most physicists argue that Gödel's Theorem does *not* mean that a ToE cannot exist. On the other hand, the scholars invoking Gödel's Theorem appear, at least in some cases, to be referring not to the underlying rules, but to the understandability of the behavior of all physical systems, as when Hawking mentions arranging blocks into rectangles, turning the computation of prime numbers into a physical question.[33] This definitional discrepancy may explain some of the disagreement among researchers.

2.4.2 Fundamental limits in accuracy

No physical theory to date is believed to be precisely accurate. Instead, physics has proceeded by a series of "successive approximations" allowing more and more accurate predictions over a wider and wider range of phenomena. Some physicists believe that it is therefore a mistake to confuse theoretical models with the true nature of reality, and hold that the series of approximations will never terminate in the "truth". Einstein himself expressed this view on occasions.[34] Following this view, we may reasonably hope for *a* theory of everything which self-consistently incorporates all currently known forces, but we should not expect it to be the final answer.

On the other hand it is often claimed that, despite the apparently ever-increasing complexity of the mathematics of each new theory, in a deep sense associated with their underlying gauge symmetry and the number of fundamental physical constants, the theories are becoming simpler. If this is the case, the process of simplification cannot continue indefinitely.

2.4.3 Lack of fundamental laws

There is a philosophical debate within the physics community as to whether a theory of everything deserves to be called *the* fundamental law of the universe.[35] One view is the hard reductionist position that the ToE is the fundamental law and that all other theories that apply within the universe are a consequence of the ToE. Another view is that emergent laws, which govern the behavior of complex systems, should be seen as equally fundamental. Examples of emergent laws are the second law of thermodynamics and the theory of natural selection. The advocates of emergence argue that emergent laws, especially those describing complex or living systems are independent of the low-level, microscopic laws. In this view, emergent laws are as fundamental as a ToE.

The debates do not make the point at issue clear. Possibly the only issue at stake is the right to apply the high-status term "fundamental" to the respective subjects of research. A well-known one took place between Steven Weinberg and Philip Anderson

2.4.4 Impossibility of being "of everything"

Although the name "theory of everything" suggests the determinism of Laplace's quotation, this gives a very misleading impression. Determinism is frustrated by the probabilistic nature of quantum mechanical predictions, by the extreme sensitivity to initial conditions that leads to mathematical chaos, by the limitations due to event horizons, and by the extreme mathematical difficulty of applying the theory. Thus, although the current standard model of particle physics "in principle" predicts almost all known non-gravitational phenomena, in practice only a few quantitative results have been derived from the full theory (e.g., the masses of some of the simplest hadrons), and these results (especially the particle masses which are most relevant for low-energy physics) are less accurate than existing experimental measurements. The ToE would almost certainly be even harder to apply for the prediction of experimental results, and thus might be of limited use.

A motive for seeking a ToE, apart from the pure intellectual satisfaction of completing a centuries-long quest, is that prior examples of unification have predicted new phenomena, some of which (e.g., electrical generators) have proved of great practical importance. And like in these prior examples of unification, the ToE would probably allow us to confidently define the domain of validity and residual error of low-energy approximations to the full theory.

2.4.5 Infinite number of onion layers

Lee Smolin regularly argues that the layers of nature may be like the layers of an onion, and that the number of layers might be infinite. This would imply an infinite sequence of physical theories.

The argument is not universally accepted, because it is not obvious that infinity is a concept that applies to the foundations of nature.

2.4.6 Impossibility of calculation

Weinberg[36] points out that calculating the precise motion of an actual projectile in the Earth's atmosphere is impossible. So how can we know we have an adequate theory for describing the motion of projectiles? Weinberg suggests that we know *principles* (Newton's laws of motion and gravitation) that work "well enough" for simple examples, like the motion of planets in empty space. These principles have worked so well on simple examples that we can be reasonably confident they will work for more complex examples. For example, although general relativity includes equations that do not have exact solutions, it is widely accepted as a valid theory because all of its equations with exact solutions have been experimentally verified. Likewise, a ToE must work for a wide range of simple examples in such a way that we can be reasonably confident it will work for every situation in physics.

2.5 See also

- Absolute (philosophy)
- An Exceptionally Simple Theory of Everything
- Attractor
- Beyond the standard model
- Brownian motion
- Chaos theory
- Electroweak interaction
- Holographic principle
- Mathematical universe hypothesis
- Multiverse
- Standard Model (mathematical formulation)
- Zero-energy universe
- The Theory Of Everything - a feature film about Prof.Stephen Hawking and his first wife Jane Hawking.
- Timeline of the Big Bang
- Chronology of the universe
- Big Bang

2.6 References

2.6.1 Footnotes

[1] Weinberg (1993)

[2] Ellis, John (2002). "Physics gets physical (correspondence)". *Nature* **415** (6875): 957. Bibcode:2002Natur.415..957E. doi:10.1038/415957b.

[3] Ellis, John (1986). "The Superstring: Theory of Everything, or of Nothing?". *Nature* **323** (6089): 595–598. Bibcode:1986Natur.323..595E. doi:10.1038/323595a0.

[4] Rorres, Chris (2009). "ARCHIMEDES AND THE QUEST FOR THE THEORY OF EVERYTHING".

[5] Shapin, Steven (1996). *The Scientific Revolution.* University of Chicago Press. ISBN 0-226-75021-3.

[6] Faraday, M. (1850). "Experimental Researches in Electricity. Twenty-Fourth Series. On the Possible Relation of Gravity to Electricity". *Abstracts of the Papers Communicated to the Royal Society of London* **5**: 994–995. doi:10.1098/rspl.1843.0267.

[7] Dirac, P.A.M. (1929). "Quantum mechanics of many-electron systems". *Proceedings of the Royal Society of London A* **123** (792): 714. Bibcode:1929RSPSA.123..714D. doi:10.1098/rspa.1929.0094.

[8] Pais (1982), Ch. 17.

[9] Weinberg (1993), Ch. 5

[10] Holloway, M (2005). "The Beauty of Branes". *Scientific American* (Scientific American) **293** (4): 38. Bibcode:2005SciAm.293d..38H. doi:10.1038/scientificamerican1005-38. PMID 16196251. Retrieved August 13, 2012.

[11] Nilles, Hans Peter; Ramos-Sánchez, Saúl; Ratz, Michael; Vaudrevange, Patrick K. S. (2008). "From strings to the MSSM". *The European Physical Journal C* **59** (2): 249. arXiv:0806.3905. Bibcode:2009EPJC...59..249N. doi:10.1140/epjc/s10052-008-0740-1.

[12] Beasley, Chris; Heckman, Jonathan J; Vafa, Cumrun (2009). "GUTs and exceptional branes in F-theory — I". *Journal of High Energy Physics* **2009**: 058. arXiv:0802.3391. Bibcode:2009JHEP...01..058B. doi:10.1088/1126-6708/2009/01/058.

[13] Donagi, Ron and Wijnholt, Martijn (2008) Model Building with F-Theory

[14] Heckman, Jonathan J. and Vafa, Cumrun (2008) Flavor Hierarchy From F-theory

[15] Strominger, Andrew; Vafa, Cumrun (1996). "Microscopic origin of the Bekenstein-Hawking entropy". *Physics Letters B* **379**: 99. arXiv:hep-th/9601029. Bibcode:1996PhLB..379...99S. doi:10.1016/0370-2693(96)00345-0.

[16] Horowitz, Gary (1996) The Origin of Black Hole Entropy in String Theory

[17] Greene, Brian R.; Morrison, David R.; Strominger, Andrew (1995). "Black hole condensation and the unification of string vacua". *Nuclear Physics B* **451**: 109. arXiv:hep-th/9504145. Bibcode:1995NuPhB.451..109G. doi:10.1016/0550-3213(95)00371-X.

[18] Aspinwall, Paul S.; Greene, Brian R.; Morrison, David R. (1994). "Calabi-Yau moduli space, mirror manifolds and space-time topology change in string theory". *Nuclear Physics B* **416** (2): 414. arXiv:hep-th/9309097. Bibcode:1994NuPhB.416..414A. doi:10.1016/0550-3213(94)90321-2.

[19] Adams, Allan; Liu, Xiao; McGreevy, John; Saltman, Alex; Silverstein, Eva (2005). "Things fall apart: Topology change from winding tachyons". *Journal of High Energy Physics* **2005** (10): 033. arXiv:hep-th/0502021. Bibcode:2005JHEP...10..033A. doi:10.1088/1126-6708/2005/10/033.

[20] Smolin, Lee (2006). *The Trouble With Physics: The Rise of String Theory, the Fall of a Science, and What Comes Next.* Houghton Mifflin. ISBN 978-0-618-55105-7.

[21] Duff, M. J. (2011). "String and M-Theory: Answering the Critics". *Foundations of Physics* **43**: 182. arXiv:1112.0788. Bibcode:2013FoPh...43..182D. doi:10.1007/s10701-011-9618-4.

[22] Potter, Franklin (15 February 2005). "Leptons And Quarks In A Discrete Spacetime". *Frank Potter's Science Gems.* Retrieved 2009-12-01.

[23] Bilson-Thompson, Sundance O.; Markopoulou, Fotini; Smolin, Lee (2007). "Quantum gravity and the standard model". *Classical and Quantum Gravity* **24** (16): 3975–3994. arXiv:hep-th/0603022. Bibcode:2007CQGra..24.3975B. doi:10.1088/0264-9381/24/16/002.

[24] Castelvecchi, Davide; Valerie Jamieson (August 12, 2006). "You are made of space-time". *New Scientist* (2564).

[25] Sundance Bilson-Thompson; Jonathan Hackett; Lou Kauffman; Lee Smolin (2008). "Particle Identifications from Symmetries of Braided Ribbon Network Invariants". arXiv:0804.0037 [hep-th].

[26] A. G. Lisi (2007). "An Exceptionally Simple Theory of Everything". arXiv:0711.0770 [hep-th].

[27] Jaki, S.L. (1966). *The Relevance of Physics*. Chicago Press. pp. 127–130.

[28] Stanley L. Jaki (2004) "A Late Awakening to Gödel in Physics", pp. 8–9.

[29] Schmidhuber, Jürgen (1997). *A Computer Scientist's View of Life, the Universe, and Everything. Lecture Notes in Computer Science*. Springer. pp. 201–208. doi:10.1007/BFb0052071. ISBN 978-3-540-63746-2.

[30] Schmidhuber, Jürgen (2002). "Hierarchies of generalized Kolmogorov complexities and nonenumerable universal measures computable in the limit". arXiv:quant-ph/0011122.

[31] Feferman, Solomon (17 November 2006). "The nature and significance of Gödel's incompleteness theorems". Institute for Advanced Study. Retrieved 2009-01-12.

[32] Robertson, Douglas S. (2007). "Goedel's Theorem, the Theory of Everything, and the Future of Science and Mathematics". *Complexity* **5** (5): 22–27. doi:10.1002/1099-0526(200005/06)5:5<22::AID-CPLX4>3.0.CO;2-0.

[33] Hawking, Stephen (20 July 2002). "Gödel and the end of physics". Retrieved 2009-12-01.

[34] Einstein, letter to Felix Klein, 1917. (On determinism and approximations.) Quoted in Pais (1982), Ch. 17.

[35] Weinberg (1993), Ch 2.

[36] Weinberg (1993) p. 5

2.6.2 Bibliography

- Pais, Abraham (1982) *Subtle is the Lord...: The Science and the Life of Albert Einstein* (Oxford University Press, Oxford, . Ch. 17, ISBN 0-19-853907-X

- Weinberg, Steven (1993) *Dreams of a Final Theory: The Search for the Fundamental Laws of Nature*, Hutchinson Radius, London, ISBN 0-09-177395-4

2.7 External links

- The Elegant Universe — a *Nova* episode about the search for the theory of everything and string theory.

- 'Theory of Everything' Freeview video by the Vega Science Trust and the BBC/OU.

- Are we getting closer, or is a final theory of matter and the universe impossible? John Ellis (physicist), Frank Close, and Nicholas Maxwell debate in The Theory of Everything

- Leading physicist Laura Mersini-Houghton, cosmologist George Francis Rayner Ellis and Oxford University philosopher David Wallace ask whether dark matter or infinite parallel universes explain why the world exists, or are such questions fundamentally insoluble in Why The World Exists

Chapter 3

Technicolor (physics)

Technicolor theories are models of physics beyond the standard model that address electroweak gauge symmetry breaking, the mechanism through which W and Z bosons acquire masses. Early technicolor theories were modelled on quantum chromodynamics (QCD), the "color" theory of the strong nuclear force, which inspired their name.

Instead of introducing elementary Higgs bosons to explain observed phenomena, technicolor models hide electroweak symmetry and generate masses for the W and Z bosons through the dynamics of new gauge interactions. Although asymptotically free at very high energies, these interactions must become strong and confining (and hence unobservable) at lower energies that have been experimentally probed. This dynamical approach is natural and avoids issues of Quantum triviality and the hierarchy problem of the Standard Model.[1]

In order to produce quark and lepton masses, technicolor has to be "extended" by additional gauge interactions. Particularly when modelled on QCD, extended technicolor is challenged by experimental constraints on flavor-changing neutral current and precision electroweak measurements. It is not known what is the extended technicolor dynamics.

Much technicolor research focuses on exploring strongly interacting gauge theories other than QCD, in order to evade some of these challenges. A particularly active framework is "walking" technicolor, which exhibits nearly conformal behavior caused by an infrared fixed point with strength just above that necessary for spontaneous chiral symmetry breaking. Whether walking can occur and lead to agreement with precision electroweak measurements is being studied through non-perturbative lattice simulations.[2]

Experiments at the Large Hadron Collider are expected to discover the mechanism responsible for electroweak symmetry breaking, and will be critical for determining whether the technicolor framework provides the correct description of nature. In 2012 these experiments declared the discovery of a Higgs-like boson with mass approximately 125 GeV/c^2;[3][4][5] such a particle is not generically predicted by technicolor models, but can be accommodated by them.

3.1 Introduction

The mechanism for the breaking of electroweak gauge symmetry in the Standard Model of elementary particle interactions remains unknown. The breaking must be spontaneous, meaning that the underlying theory manifests the symmetry exactly (the gauge-boson fields are massless in the equations of motion), but the solutions (the ground state and the excited states) do not. In particular, the physical *W* and *Z* gauge bosons become massive. This phenomenon, in which the *W* and *Z* bosons also acquire an extra polarization state, is called the "Higgs mechanism". Despite the precise agreement of the electroweak theory with experiment at energies accessible so far, the necessary ingredients for the symmetry breaking remain hidden, yet to be revealed at higher energies.

The simplest mechanism of electroweak symmetry breaking introduces a single complex field and predicts the existence of the Higgs boson. Typically, the Higgs boson is "unnatural" in the sense that quantum mechanical fluctuations produce corrections to its mass that lift it to such high values that it cannot play the role for which it was introduced. Unless the Standard Model breaks down at energies less than a few TeV, the Higgs mass can be kept small only by a delicate fine-tuning of parameters.

Technicolor avoids this problem by hypothesizing a new gauge interaction coupled to new massless fermions. This interaction is asymptotically free at very high energies and becomes strong and confining as the energy decreases to the electroweak scale of 246 GeV. These strong forces spontaneously break the massless fermions' chiral symmetries, some of which are weakly gauged as part of the Standard Model. This is the dynamical version of the Higgs mechanism. The electroweak gauge symmetry is thus broken, producing masses for the W and Z bosons.

The new strong interaction leads to a host of new composite, short-lived particles at energies accessible at the Large Hadron Collider (LHC). This framework is natural because there are no elementary Higgs bosons and, hence, no fine-tuning of parameters. Quark and lepton masses also break the electroweak gauge symmetries, so they, too, must arise spontaneously. A mechanism for incorporating this feature is known as extended technicolor. Technicolor and extended technicolor face a number of phenomenological challenges, in particular issues of flavor-changing neutral currents, precision electroweak tests, and the top quark mass. Technicolor models also do not generically predict Higgs-like bosons as light as 125 GeV/c^2; such a particle was discovered by experiments at the Large Hadron Collider in 2012.[3][4][5] Some of these issues can be addressed with a class of theories known as walking technicolor.

3.2 Early technicolor

Technicolor is the name given to the theory of electroweak symmetry breaking by new strong gauge-interactions whose characteristic energy scale ΛTC is the weak scale itself, ΛTC ≅ FEW ≡ 246 GeV. The guiding principle of technicolor is "naturalness": basic physical phenomena should not require fine-tuning of the parameters in the Lagrangian that describes them. What constitutes fine-tuning is to some extent a subjective matter, but a theory with elementary scalar particles typically is very finely tuned (unless it is supersymmetric). The quadratic divergence in the scalar's mass requires adjustments of a part in $\mathcal{O}\left(\frac{M_{bare}^2}{M_{physical}^2}\right)$, where M_{bare} is the cutoff of the theory, the energy scale at which the theory changes in some essential way. In the standard electroweak model with $M_{bare} \sim 10^{15}$ GeV (the grand-unification mass scale), and with the Higgs boson mass $M_{physical}$ = 100–500 GeV, the mass is tuned to at least a part in 10^{25}.

By contrast, a natural theory of electroweak symmetry breaking is an asymptotically free gauge theory with fermions as the only matter fields. The technicolor gauge group GTC is often assumed to be SU(NTC). Based on analogy with quantum chromodynamics (QCD), it is assumed that there are one or more doublets of massless Dirac "technifermions" transforming vectorially under the same complex representation of GTC, T_iL,R = (U_i,D_i)L,R, i = 1,2, ..., N_f/2. Thus, there is a chiral symmetry of these fermions, e.g., $SU(N_f)$L ⊗ $SU(N_f)$R, if they all transform according the same complex representation of GTC. Continuing the analogy with QCD, the running gauge coupling αTC(μ) triggers spontaneous chiral symmetry breaking, the technifermions acquire a dynamical mass, and a number of massless Goldstone bosons result. If the technifermions transform under $[SU(2) \otimes U(1)]$EW as left-handed doublets and right-handed singlets, three linear combinations of these Goldstone bosons couple to three of the electroweak gauge currents.

In 1973 Jackiw and Johnson[6] and Cornwall and Norton[7] studied the possibility that a (non-vectorial) gauge interaction of fermions can break itself; i.e., is strong enough to form a Goldstone boson coupled to the gauge current. Using Abelian gauge models, they showed that, *if* such a Goldstone boson is formed, it is "eaten" by the Higgs mechanism, becoming the longitudinal component of the now massive gauge boson. Technically, the polarization function Π(p^2) appearing in the gauge boson propagator, Δμν = (pμ pν/p^2 - gμν)/[p^2(1 - g^2 Π(p^2))] develops a pole at p^2 = 0 with residue F^2, the square of the Goldstone boson's decay constant, and the gauge boson acquires mass $M \cong g\,F$. In 1973, Weinstein[8] showed that composite Goldstone bosons whose constituent fermions transform in the "standard" way under $SU(2) \otimes U(1)$ generate the weak boson masses

(1) $\quad M_{W^\pm} = \frac{1}{2} g F_{EW}$ and $\quad M_Z = \frac{1}{2}\sqrt{g^2 + g'^2} F_{EW} \equiv \frac{M_W}{\cos\theta_W}$.

This standard-model relation is achieved with elementary Higgs bosons in electroweak doublets; it is verified experimentally to better than 1%. Here, g and g' are $SU(2)$ and $U(1)$ gauge couplings and tanθW = g'/g defines the weak mixing angle.

The important idea of a *new* strong gauge interaction of massless fermions at the electroweak scale FEW driving the spontaneous breakdown of its global chiral symmetry, of which an $SU(2) \otimes U(1)$ subgroup is weakly gauged, was first proposed in 1979 by S. Weinberg[9] and L. Susskind.[10] This "technicolor" mechanism is natural in that no fine-tuning of parameters is necessary.

3.3 Extended technicolor

Elementary Higgs bosons perform another important task. In the Standard Model, quarks and leptons are necessarily massless because they transform under $SU(2) \otimes U(1)$ as left-handed doublets and right-handed singlets. The Higgs doublet couples to these fermions. When it develops its vacuum expectation value, it transmits this electroweak breaking to the quarks and leptons, giving them their observed masses. (In general, electroweak-eigenstate fermions are not mass eigenstates, so this process also induces the mixing matrices observed in charged-current weak interactions.)

In technicolor, something else must generate the quark and lepton masses. The only natural possibility, one avoiding the introduction of elementary scalars, is to enlarge G_{TC} to allow technifermions to couple to quarks and leptons. This coupling is induced by gauge bosons of the enlarged group. The picture, then, is that there is a large "extended technicolor" (ETC) gauge group $G_{ETC} \supset G_{TC}$ in which technifermions, quarks, and leptons live in the same representations. At one or more high scales Λ_{ETC}, G_{ETC} is broken down to G_{TC}, and quarks and leptons emerge as the TC-singlet fermions. When $\alpha_{TC}(\mu)$ becomes strong at scale $\Lambda_{TC} \cong F_{EW}$, the fermionic condensate $\langle \bar{T}T \rangle_{TC} \cong 4\pi F_{EW}^3$ forms. (The condensate is the vacuum expectation value of the technifermion bilinear $\bar{T}T$. The estimate here is based on naive dimensional analysis of the quark condensate in QCD, expected to be correct as an order of magnitude.) Then, the transitions $q_L(\text{or } \ell_L) \to T_L \to T_R \to q_R(\text{or } \ell_R)$ can proceed through the technifermion's dynamical mass by the emission and reabsorption of ETC bosons whose masses $M_{ETC} \cong g_{ETC} \Lambda_{ETC}$ are much greater than Λ_{TC}. The quarks and leptons develop masses given approximately by

$$(2) \quad m_{q,\ell}(M_{ETC}) \cong \frac{g_{ETC}^2 \langle \bar{T}T \rangle_{ETC}}{M_{ETC}^2} \cong \frac{4\pi F_{EW}^3}{\Lambda_{ETC}^2}.$$

Here, $\langle \bar{T}T \rangle_{ETC}$ is the technifermion condensate renormalized at the ETC boson mass scale,

$$(3) \quad \langle \bar{T}T \rangle_{ETC} = \exp\left(\int_{\Lambda_{TC}}^{M_{ETC}} \frac{d\mu}{\mu} \gamma_m(\mu)\right) \langle \bar{T}T \rangle_{TC},$$

where $\gamma_m(\mu)$ is the anomalous dimension of the technifermion bilinear $\bar{T}T$ at the scale μ. The second estimate in Eq. (2) depends on the assumption that, as happens in QCD, $\alpha_{TC}(\mu)$ becomes weak not far above Λ_{TC}, so that the anomalous dimension γ_m of $\bar{T}T$ is small there. Extended technicolor was introduced in 1979 by Dimopoulos and Susskind,[11] and by Eichten and Lane.[12] For a quark of mass $m_q \cong 1$ GeV, and with $\Lambda_{TC} \cong 246$ GeV, one estimates $\Lambda_{ETC} \cong 15$ TeV. Therefore, assuming that $g_{ETC}^2 \gtrsim 1$, M_{ETC} will be at least this large.

In addition to the ETC proposal for quark and lepton masses, Eichten and Lane observed that the size of the ETC representations required to generate all quark and lepton masses suggests that there will be more than one electroweak doublet of technifermions.[12] If so, there will be more (spontaneously broken) chiral symmetries and therefore more Goldstone bosons than are eaten by the Higgs mechanism. These must acquire mass by virtue of the fact that the extra chiral symmetries are also explicitly broken, by the standard-model interactions and the ETC interactions. These "pseudo-Goldstone bosons" are called technipions, πT. An application of Dashen's theorem[13] gives for the ETC contribution to their mass

$$(4) \quad F_{EW}^2 M_{\pi T}^2 \cong \frac{g_{ETC}^2 \langle \bar{T}T\bar{T}T \rangle_{ETC}}{M_{ETC}^2} \cong \frac{16\pi^2 F_{EW}^6}{\Lambda_{ETC}^2}.$$

The second approximation in Eq. (4) assumes that $\langle \bar{T}T\bar{T}T \rangle_{ETC} \cong \langle \bar{T}T \rangle_{ETC}^2$. For $F_{EW} \cong \Lambda_{TC} \cong 246$ GeV and $\Lambda_{ETC} \cong 15$ TeV, this contribution to $M_{\pi T}$ is about 50 GeV. Since ETC interactions generate $m_{q,\ell}$ and the coupling of technipions to quark and lepton pairs, one expects the couplings to be Higgs-like; i.e., roughly proportional to the masses of the quarks and leptons. This means that technipions are expected to decay to the heaviest $\bar{q}q$ and $\bar{\ell}\ell$ pairs allowed.

Perhaps the most important restriction on the ETC framework for quark mass generation is that ETC interactions are likely to induce flavor-changing neutral current processes such as $\mu \to e\gamma$, $K_L \to \mu e$, and $|\Delta S| = 2$ and $|\Delta B| = 2$ interactions that induce $K^0 \leftrightarrow \bar{K}^0$ and $B^0 \leftrightarrow \bar{B}^0$ mixing.[12] The reason is that the algebra of the ETC currents involved in $m_{q,\ell}$ generation imply $\bar{q}q'$ and $\bar{\ell}\ell'$ ETC currents which, when written in terms of fermion mass eigenstates, have no reason to conserve flavor. The strongest constraint comes from requiring that ETC interactions mediating K–$\bar{K}$ mixing contribute less than the Standard Model. This implies an effective Λ_{ETC} greater than 1000 TeV. The actual Λ_{ETC} may be reduced somewhat if CKM-like mixing angle factors are present. If these interactions are CP-violating, as they well may be, the constraint from the ε-parameter is that the effective $\Lambda_{ETC} > 10^4$ TeV. Such huge ETC mass scales imply tiny quark and lepton masses and ETC contributions to $M_{\pi T}$ of at most a few GeV, in conflict with LEP searches for πT at the Z^0.

Extended technicolor is a very ambitious proposal, requiring that quark and lepton masses and mixing angles arise from experimentally accessible interactions. *If* there exists a successful model, it would not only predict the masses and mixings of quarks and leptons (and technipions), it would explain why there are three families of each: they

3.4 Walking technicolor

Since quark and lepton masses are proportional to the bilinear technifermion condensate divided by the ETC mass scale squared, their tiny values can be avoided if the condensate is enhanced above the weak-αTC estimate in Eq. (2), $\langle \bar{T}T \rangle_{ETC} \cong \langle \bar{T}T \rangle_{TC} \cong 4\pi F_{EW}^3$.

During the 1980s, several dynamical mechanisms were advanced to do this. In 1981 Holdom suggested that, if the αTC(μ) evolves to a nontrivial fixed point in the ultraviolet, with a large positive anomalous dimension γ_m for $\bar{T}T$, realistic quark and lepton masses could arise with ΛETC large enough to suppress ETC-induced K--$\bar{K}$ mixing.[14] However, no example of a nontrivial ultraviolet fixed point in a four-dimensional gauge theory has been constructed. In 1985 Holdom analyzed a technicolor theory in which a "slowly varying" αTC(μ) was envisioned.[15] His focus was to separate the chiral breaking and confinement scales, but he also noted that such a theory could enhance $\langle \bar{T}T \rangle_{ETC}$ and thus allow the ETC scale to be raised. In 1986 Akiba and Yanagida also considered enhancing quark and lepton masses, by simply assuming that αTC is constant and strong all the way up to the ETC scale.[16] In the same year Yamawaki, Bando and Matumoto again imagined an ultraviolet fixed point in a non-asymptotically free theory to enhance the technifermion condensate.[17]

In 1986 Appelquist, Karabali and Wijewardhana discussed the enhancement of fermion masses in an asymptotically free technicolor theory with a slowly running, or "walking", gauge coupling.[18] The slowness arose from the screening effect of a large number of technifermions, with the analysis carried out through two-loop perturbation theory. In 1987 Appelquist and Wijewardhana explored this walking scenario further.[19] They took the analysis to three loops, noted that the walking can lead to a power law enhancement of the technifermion condensate, and estimated the resultant quark, lepton, and technipion masses. The condensate enhancement arises because the associated technifermion mass decreases slowly, roughly linearly, as a function of its renormalization scale. This corresponds to the condensate anomalous dimension γ_m in Eq. (3) approaching unity (see below).[20]

In the 1990s, the idea emerged more clearly that walking is naturally described by asymptotically free gauge theories dominated in the infrared by an approximate fixed point. Unlike the speculative proposal of ultraviolet fixed points, fixed points in the infrared are known to exist in asymptotically free theories, arising at two loops in the beta function providing that the fermion count N_f is large enough. This has been known since the first two-loop computation in 1974 by Caswell.[21] If N_f is close to the value $\hat{N}_f$ at which asymptotic freedom is lost, the resultant infrared fixed point is weak, of parametric order $\hat{N}_f - N_f$, and reliably accessible in perturbation theory. This weak-coupling limit was explored by Banks and Zaks in 1982.[22]

The fixed-point coupling αIR becomes stronger as N_f is reduced from $\hat{N}_f$. Below some critical value N_{fc} the coupling becomes strong enough ($> \alpha_\chi SB$) to break spontaneously the massless technifermions' chiral symmetry. Since the analysis must typically go beyond two-loop perturbation theory, the definition of the running coupling αTC(μ), its fixed point value αIR, and the strength α_χ SB necessary for chiral symmetry breaking depend on the particular renormalization scheme adopted. For $0 < (\alpha_{IR} - \alpha_{\chi SB})/\alpha_{IR} \ll 1$; i.e., for N_f just below N_{fc}, the evolution of αTC(μ) is governed by the infrared fixed point and it will evolve slowly (walk) for a range of momenta above the breaking scale ΛTC. To overcome the M_{ETC}^2-suppression of the masses of first and second generation quarks involved in K--$\bar{K}$ mixing, this range must extend almost to their ETC scale, of $\mathcal{O}(10^3 \text{ TeV})$. Cohen and Georgi argued that $\gamma_m = 1$ is the signal of spontaneous chiral symmetry breaking, i.e., that $\gamma_m(\alpha_\chi SB) = 1$.[20] Therefore, in the walking-αTC region, $\gamma_m \cong 1$ and, from Eqs. (2) and (3), the light quark masses are enhanced approximately by METC/ΛTC.

The idea that αTC(μ) walks for a large range of momenta when αIR lies just above α_χ SB was suggested by Lane and Ramana.[23] They made an explicit model, discussed the walking that ensued, and used it in their discussion of walking technicolor phenomenology at hadron colliders. This idea was developed in some detail by Appelquist, Terning and Wijewardhana.[24] Combining a perturbative computation of the infrared fixed point with an approximation of α_χ SB based on the Schwinger-Dyson equation, they estimated the critical value N_{fc} and explored the resultant electroweak physics. Since the 1990s, most discussions of walking technicolor are in the framework of theories assumed to be dominated in the infrared by an approximate fixed point. Various models have been explored, some with the technifermions in the fundamental representation of the gauge group and some employing higher representations.[25][26][27]

The possibility that the technicolor condensate can be enhanced beyond that discussed in the walking literature, has also been considered recently by Luty and Okui under the name "conformal technicolor".[28] They envision an infrared stable fixed point, but with a very large anomalous dimension for the operator $\bar{T}T$. It remains to be seen whether this can be realized, for example, in the class of theories currently being examined using lattice techniques.

3.4.1 Top quark mass

The walking enhancement described above may be insufficient to generate the measured top quark mass, even for an ETC scale as low as a few TeV. However, this problem could be addressed if the effective four-technifermion coupling resulting from ETC gauge boson exchange is strong and tuned just above a critical value.[29] The analysis of this strong-ETC possibility is that of a Nambu–Jona–Lasinio model with an additional (technicolor) gauge interaction. The technifermion masses are small compared to the ETC scale (the cutoff on the effective theory), but nearly constant out to this scale, leading to a large top quark mass. No fully realistic ETC theory for all quark masses has yet been developed incorporating these ideas. A related study was carried out by Miransky and Yamawaki.[30] A problem with this approach is that it involves some degree of parameter fine-tuning, in conflict with technicolor's guiding principle of naturalness.

Finally, it should be noted that there is a large body of closely related work in which ETC does not generate m_t. These are the top quark condensate,[31] topcolor and top-color-assisted technicolor models,[32] in which new strong interactions are ascribed to the top quark and other third-generation fermions. As with the strong-ETC scenario described above, all these proposals involve a considerable degree of fine-tuning of gauge couplings.

3.5 Minimal Walking Models

In 2004 Francesco Sannino and Kimmo Tuominen proposed technicolor models with technifermions in higher-dimensional representations of the technicolor gauge group.[26] They argued that these more "minimal" models required fewer flavors of technifermions in order to exhibit walking behavior, making it easier to pass precision electroweak tests.

For example, SU(2) and SU(3) gauge theories may exhibit walking with as few as two Dirac flavors of fermions in the adjoint or two-index symmetric representation. In contrast, at least eight flavors of fermions in the fundamental representation of SU(3) (and possibly SU(2) as well) are required to reach the near-conformal regime.[27]

These results continue to be investigated by various methods, including lattice simulations discussed below, which have confirmed the near-conformal dynamics of these minimal walking models. The first comprehensive effective Lagrangian for minimal walking models, featuring a light composite Higgs, spin-one states, tree-level unitarity, and consistency with phenomenological constraints was constructed in 2007 by Foadi, Frandsen, Ryttov and Sannino.[33]

3.6 Technicolor on the lattice

Lattice gauge theory is a non-perturbative method applicable to strongly interacting technicolor theories, allowing first-principles exploration of walking and conformal dynamics. In 2007, Catterall and Sannino used lattice gauge theory to study $SU(2)$ gauge theories with two flavors of Dirac fermions in the symmetric representation,[34] finding evidence of conformality that has been confirmed by subsequent studies.[35]

As of 2010, the situation for $SU(3)$ gauge theory with fermions in the fundamental representation is not as clear-cut. In 2007, Appelquist, Fleming and Neil reported evidence that a non-trivial infrared fixed point develops in such theories when there are twelve flavors, but not when there are eight.[36] While some subsequent studies confirmed these results, others reported different conclusions, depending on the lattice methods used, and there is not yet consensus.[37]

Further lattice studies exploring these issues, as well as considering the consequences of these theories for precision electroweak measurements, are underway by several research groups.[38]

3.7 Technicolor phenomenology

Any framework for physics beyond the Standard Model must conform with precision measurements of the electroweak parameters. Its consequences for physics at existing and future high-energy hadron colliders, and for the dark matter

3.7.1 Precision electroweak tests

In 1990, the phenomenological parameters S, T, and U were introduced by Peskin and Takeuchi to quantify contributions to electroweak radiative corrections from physics beyond the Standard Model.[39] They have a simple relation to the parameters of the electroweak chiral Lagrangian.[40][41] The Peskin-Takeuchi analysis was based on the general formalism for weak radiative corrections developed by Kennedy, Lynn, Peskin and Stuart,[42] and alternate formulations also exist.[43]

The S, T, and U-parameters describe corrections to the electroweak gauge boson propagators from physics Beyond the Standard Model. They can be written in terms of polarization functions of electroweak currents and their spectral representation as follows:

$$
\begin{align}
(5) \quad S &= 16\pi \frac{d}{dq^2} \left[\Pi_{33}^{\text{new}}(q^2) - \Pi_{3Q}^{\text{new}}(q^2) \right]_{q^2=0} \\
&= 4\pi \int \frac{dm^2}{m^4} \left[\sigma_V^3(m^2) - \sigma_A^3(m^2) \right]^{\text{new}} ;
\end{align}
$$

$$
\begin{align}
(6) \quad T &= \frac{16\pi}{M_Z^2 \sin^2 2\theta_W} \left[\Pi_{11}^{\text{new}}(0) - \Pi_{33}^{\text{new}}(0) \right] \\
&= \frac{4\pi}{M_Z^2 \sin^2 2\theta_W} \int_0^\infty \frac{dm^2}{m^2} \left[\sigma_V^1(m^2) + \sigma_A^1(m^2) - \sigma_V^3(m^2) - \sigma_A^3(m^2) \right]^{\text{new}} ,
\end{align}
$$

where only new, beyond-standard-model physics is included. The quantities are calculated relative to a minimal Standard Model with some chosen reference mass of the Higgs boson, taken to range from the experimental lower bound of 117 GeV to 1000 GeV where its width becomes very large.[44] For these parameters to describe the dominant corrections to the Standard Model, the mass scale of the new physics must be much greater than M_W and M_Z, and the coupling of quarks and leptons to the new particles must be suppressed relative to their coupling to the gauge bosons. This is the case with technicolor, so long as the lightest technivector mesons, ρT and aT, are heavier than 200–300 GeV. The S-parameter is sensitive to all new physics at the TeV scale, while T is a measure of weak-isospin breaking effects. The U-parameter is generally not useful; most new-physics theories, including technicolor theories, give negligible contributions to it.

The S and T-parameters are determined by global fit to experimental data including Z-pole data from LEP at CERN, top quark and W-mass measurements at Fermilab, and measured levels of atomic parity violation. The resultant bounds on these parameters are given in the Review of Particle Properties.[44] Assuming $U = 0$, the S and T parameters are small and, in fact, consistent with zero:

$$
(7) \quad \begin{aligned} S &= -0.04 \pm 0.09 \, (-0.07), \\ T &= 0.02 \pm 0.09 \, (+0.09), \end{aligned}
$$

where the central value corresponds to a Higgs mass of 117 GeV and the correction to the central value when the Higgs mass is increased to 300 GeV is given in parentheses. These values place tight restrictions on beyond-standard-model theories—when the relevant corrections can be reliably computed.

The S parameter estimated in QCD-like technicolor theories is significantly greater than the experimentally allowed value.[39][43] The computation was done assuming that the spectral integral for S is dominated by the lightest ρT and aT resonances, or by scaling effective Lagrangian parameters from QCD. In walking technicolor, however, the physics at the TeV scale and beyond must be quite different from that of QCD-like theories. In particular, the vector and axial-vector spectral functions cannot be dominated by just the lowest-lying resonances.[45] It is unknown whether higher energy contributions to $\sigma_{V,A}^3$ are a tower of identifiable ρT and aT states or a smooth continuum. It has been conjectured that ρT and aT partners could be more nearly degenerate in walking theories (approximate parity doubling), reducing their contribution to S.[46] Lattice calculations are underway or planned to test these ideas and obtain reliable estimates of S in walking theories.[2][47]

The restriction on the T-parameter poses a problem for the generation of the top-quark mass in the ETC framework. The enhancement from walking can allow the associated ETC scale to be as large as a few TeV,[24] but—since the ETC interactions must be strongly weak-isospin breaking to allow for the large top-bottom mass splitting—the contribution to the T parameter,[48] as well as the rate for the decay $Z^0 \to \bar{b}b$,[49] could be too large.

3.7.2 Hadron collider phenomenology

Early studies generally assumed the existence of just one electroweak doublet of technifermions, or of one technifamily including one doublet each of color-triplet techniquarks and color-singlet technileptons (four electroweak doublets in total).[50] The number ND of electroweak doublets determines the decay constant F needed to produce the correct electroweak scale, as $F = F_{EW}/\sqrt{ND} = 246$ GeV/$\sqrt{ND}$. In the minimal, one-doublet model, three Goldstone bosons (technipions, πT) have decay constant $F = F_{EW} = 246$ GeV and are eaten by the electroweak gauge bosons. The most accessible collider signal is the production through $\bar{q}q$ annihilation in a hadron collider of spin-one $\rho_T^{\pm,0}$, and their subsequent decay into a pair of longitudinally polarized weak bosons, $W_L^{\pm} Z_L^0$ and $W_L^+ W_L^-$. At an expected mass of 1.5–2.0 TeV and width of 300–400 GeV, such ρT's would be difficult to discover at the LHC. A one-family model has a large number of physical technipions, with $F = F_{EW}/\sqrt{4} = 123$ GeV.[51] There is a collection of correspondingly lower-mass color-singlet and octet technivectors decaying into technipion pairs. The πT's are expected to decay to the heaviest possible quark and lepton pairs. Despite their lower masses, the ρT's are wider than in the minimal model and the backgrounds to the πT decays are likely to be insurmountable at a hadron collider.

This picture changed with the advent of walking technicolor. A walking gauge coupling occurs if $\alpha_\chi SB$ lies just below the IR fixed point value αIR, which requires either a large number of electroweak doublets in the fundamental representation of the gauge group, e.g., or a few doublets in higher-dimensional TC representations.[25][52] In the latter case, the constraints on ETC representations generally imply other technifermions in the fundamental representation as well.[12][23] In either case, there are technipions πT with decay constant $F \ll F_{EW}$. This implies $\Lambda_{TC} \ll F_{EW}$ so that the lightest technivectors accessible at the LHC—ρT, ωT, aT (with $I^G \, J^{PC} = 1^+ \, 1^{--}, \, 0^- \, 1^{--}, \, 1^- \, 1^{++}$)—have masses well below a TeV. The class of theories with many technifermions and thus $F \ll F_{EW}$ is called low-scale technicolor.[53]

A second consequence of walking technicolor concerns the decays of the spin-one technihadrons. Since technipion masses $M_{\pi_T}^2 \propto \langle \bar{T}T\bar{T}T \rangle_{M_{ETC}}$ (see Eq. (4)), walking enhances them much more than it does other technihadron masses. Thus, it is very likely that the lightest $M_\rho T < 2 M \pi T$ and that the two and three-πT decay channels of the light technivectors are closed.[25] This further implies that these technivectors are very narrow. Their most probable two-body channels are $W_L^{\pm,0} \pi_T$, WL WL, $\gamma \pi T$ and γ WL. The coupling of the lightest technivectors to WL is proportional to F/F_{EW}.[54] Thus, all their decay rates are suppressed by powers of $(F/F_{EW})^2 \ll 1$ or the fine-structure constant, giving total widths of a few GeV (for ρT) to a few tenths of a GeV (for ωT and T).

A more speculative consequence of walking technicolor is motivated by consideration of its contribution to the S-parameter. As noted above, the usual assumptions made to estimate STC are invalid in a walking theory. In particular, the spectral integrals used to evaluate STC cannot be dominated by just the lowest-lying ρT and aT and, if STC is to be small, the masses and weak-current couplings of the ρT and aT could be more nearly equal than they are in QCD.

Low-scale technicolor phenomenology, including the possibility of a more parity-doubled spectrum, has been developed into a set of rules and decay amplitudes.[54] An April 2011 announcement of an excess in jet pairs produced in association with a W boson measured at the Tevatron[55] has been interpreted by Eichten, Lane and Martin as a possible signal of the technipion of low-scale technicolor.[56]

The general scheme of low-scale technicolor makes little sense if the limit on M_{ρ_T} is pushed past about 700 GeV. The LHC should be able to discover it or rule it out. Searches there involving decays to technipions and thence to heavy quark jets are hampered by backgrounds from $\bar{t}t$ production; its rate is 100 times larger than that at the Tevatron. Consequently, the discovery of low-scale technicolor at the LHC relies on all-leptonic final-state channels with favorable signal-to-background ratios: $\rho_T^{\pm} \to W_L^{\pm} Z_L^0$, $a_T^{\pm} \to \gamma W_L^{\pm}$ and $\omega_T \to \gamma Z_L^0$.[57]

3.7.3 Dark matter

Technicolor theories naturally contain dark matter candidates. Almost certainly, models can be built in which the lowest-lying technibaryon, a technicolor-singlet bound state of technifermions, is stable enough to survive the evolution of the universe.[44][58] If the technicolor theory is low-scale ($F \ll F_{EW}$), the baryon's mass should be no more than 1–2 TeV. If not, it could be much heavier. The technibaryon must be electrically neutral and satisfy constraints on its abundance. Given the limits on spin-independent dark-matter-nucleon cross sections from dark-matter search experiments ($\lesssim 10^{-42}$ cm^2 for the masses of interest[59]), it may have to be electroweak neutral (weak isospin $I = 0$) as well. These considerations suggest that the "old" technicolor dark matter candidates may be difficult to produce at the LHC.

A different class of technicolor dark matter candidates light enough to be accessible at the LHC was introduced by Francesco Sannino and his collaborators.[60] These states are pseudo Goldstone bosons possessing a global charge that makes them stable against decay.

3.8 See also

- Higgsless model
- Topcolor
- Top quark condensate

3.9 References

[1] For a recent introductions to and reviews of technicolor, see:
Christopher T. Hill and Elizabeth H. Simmons (2003). "Strong Dynamics and Electroweak Symmetry Breaking". *Physics Reports* **381** (4-6): 235–402. arXiv:hep-ph/0203079. Bibcode:2003PhR...381..235H. doi:10.1016/S0370-1573(03)00140-6.
Kenneth Lane (2002). "Two Lectures on Technicolor". l'Ecole de GIF at LAPP, Annecy-le-Vieux, France. arXiv:hep-ph/0202255.
Robert Shrock (2007). "Some Recent Results on Models of Dynamical Electroweak Symmetry Breaking". In M. Tanabashi, M. Harada, and K. Yamawaki. "Nagoya 2006: The Origin of Mass and Strong Coupling Gauge Theories". International Workshop on Strongly Coupled Gauge Theories. pp. 227–241. arXiv:hep-ph/0703050.
Adam Martin (2008). "Technicolor Signals at the LHC". The 46th Course at the International School of Subnuclear Physics: Predicted and Totally Unexpected in the Energy Frontier Opened by LHC. arXiv:0812.1841.
Francesco Sannino (2009). "Conformal Dynamics for TeV Physics and Cosmology". *Acta Physica Polonica* **B40**: 3533–3745. arXiv:0911.0931. Bibcode:2009arXiv0911.0931S.

[2] George Fleming (2008). "Strong Interactions for the LHC". *Proceedings of Science*. LATTICE 2008: 21. arXiv:0812.2035. Bibcode:2008arXiv0812.2035F.

[3] "CERN experiments observe particle consistent with long-sought Higgs boson". CERN press release. 4 July 2012. Retrieved 4 July 2012.

[4] Taylor, Lucas (4 July 2012). "Observation of a New Particle with a Mass of 125 GeV". *CMS Public Web site*. CERN.

[5] "Latest Results from ATLAS Higgs Search". ATLAS. 4 July 2012. Retrieved 4 July 2012.

[6] R. Jackiw and K. Johnson (1973). "Dynamical Model of Spontaneously Broken Gauge Symmetries". *Physical Review* **D8** (8): 2386–2398. Bibcode:1973PhRvD...8.2386J. doi:10.1103/PhysRevD.8.2386.

[7] John M. Cornwall and Richard E. Norton (1973). "Spontaneous Symmetry Breaking Without Scalar Mesons". *Physical Review* **D8** (10): 3338–3346. Bibcode:1973PhRvD...8.3338C. doi:10.1103/PhysRevD.8.3338.

[8] Marvin Weinstein (1973). "Conserved Currents, Their Commutators, and the Symmetry Structure of Renormalizable Theories of Electromagnetic, Weak, and Strong Interactions". *Physical Review* **D8** (8): 2511–2524. Bibcode:1973PhRvD...8.2511W. doi:10.1103/PhysRevD.8.2511.

[9] Steven Weinberg (1976). "Implications of dynamical symmetry breaking". *Physical Review* **D13** (4): 974–996. Bibcode:1976PhRvD..13..974W. doi:10.1103/PhysRevD.13.974.
S. Weinberg (1979). "Implications of dynamical symmetry breaking: An addendum". *Physical Review* **D19** (4): 1277–1280. Bibcode:1979PhRvD..19.1277W. doi:10.1103/PhysRevD.19.1277.

[10] Leonard Susskind (1979). "Dynamics of spontaneous symmetry breaking in the Weinberg-Salam theory". *Physical Review* **D20** (10): 2619–2625. Bibcode:1979PhRvD..20.2619S. doi:10.1103/PhysRevD.20.2619.

[11] Savas Dimopoulos and Leonard Susskind (1979). "Mass without scalars". *Nuclear Physics* **B155** (1): 237–252. Bibcode:1979NuPhB.155..237D. doi:10.1016/0550-3213(79)90364-X.

[12] Estia Eichten and Kenneth Lane (1980). "Dynamical breaking of weak interaction symmetries". *Physics Letters* **B90** (1-2): 125–130. Bibcode:1980PhLB...90..125E. doi:10.1016/0370-2693(80)90065-9.

[13] Roger Dashen (1969). "Chiral SU(3)⊗SU(3) as a Symmetry of the Strong Interactions". *Physical Review* **183** (5): 1245–1260. Bibcode:1969PhRv..183.1245D. doi:10.1103/PhysRev.183.1245.
Roger Dashen (1971). "Some Features of Chiral Symmetry Breaking". *Physical Review* **D3** (8): 1879–1889. Bibcode:1971PhRvD...3.1879D. doi:10.1103/PhysRevD.3.1879.

[14] Bob Holdom (1981). "Raising the sideways scale". *Physical Review* **D24** (5): 1441–1444. Bibcode:1981PhRvD..24.1441H. doi:10.1103/PhysRevD.24.1441.

[15] Bob Holdom (1985). "Techniodor". *Physics Letters* **B150** (4): 301–305. Bibcode:1985PhLB..150..301H. doi:10.1016/0370-2693(85)91015-9.

[16] T. Akiba and T. Yanagida (1986). "Hierarchic chiral condensate". *Physics Letters* **B169** (4): 432–435. Bibcode:1986PhLB..169..432A. doi:10.1016/0370-2693(86)90385-0.

[17] Koichi Yamawaki, Masako Bando, and Ken-iti Matumoto (1986). "Scale-Invariant Hypercolor Model and a Dilaton". *Physical Review Letters* **56** (13): 1335–1338. Bibcode:1986PhRvL..56.1335Y. doi:10.1103/PhysRevLett.56.1335. PMID 10032641.

[18] Thomas Appelquist, Dimitra Karabali, and L. C. R. Wijewardhana (1986). "Chiral Hierarchies and Flavor-Changing Neutral Currents in Hypercolor". *Physical Review Letters* **57** (8): 957–960. Bibcode:1986PhRvL..57..957A. doi:10.1103/PhysRevLett.57.957. PMID 10034209.

[19] Thomas Appelquist and L. C. R. Wijewardhana (1987). "Chiral hierarchies from slowly running couplings in technicolor theories". *Physical Review* **D36** (2): 568–580. Bibcode:1987PhRvD..36..568A. doi:10.1103/PhysRevD.36.568.

[20] Andrew Cohen and Howard Georgi (1989). "Walking beyond the rainbow". *Nuclear Physics* **B314** (1): 7–24. Bibcode:1989NuPhB.314....7 doi:10.1016/0550-3213(89)90109-0.

[21] William E. Caswell (1974). "Asymptotic Behavior of Non-Abelian Gauge Theories to Two-Loop Order". *Physical Review Letters* **33** (4): 244–246. Bibcode:1974PhRvL..33..244C. doi:10.1103/PhysRevLett.33.244.

[22] T. Banks and A. Zaks (1982). "On the phase structure of vector-like gauge theories with massless fermions". *Nuclear Physics* **B196** (2): 189–204. Bibcode:1982NuPhB.196..189B. doi:10.1016/0550-3213(82)90035-9.

[23] Kenneth Lane and M. V. Ramana (1991). "Walking technicolor signatures at hadron colliders". *Physical Review* **D44** (9): 2678–2700. Bibcode:1991PhRvD..44.2678L. doi:10.1103/PhysRevD.44.2678.

[24] Thomas Appelquist, John Terning and L. C. R. Wijewardhana (1997). "Postmodern Technicolor". *Physical Review Letters* **79** (15): 2767–2770. arXiv:hep-ph/9706238. Bibcode:1997PhRvL..79.2767A. doi:10.1103/PhysRevLett.79.2767.

[25] Kenneth Lane and Estia Eichten (1989). "Two-scale technicolor". *Physics Letters* **B222** (2): 274–280. Bibcode:1989PhLB..222..274L. doi:10.1016/0370-2693(89)91265-3.

[26] Francesco Sannino and Kimmo Tuominen (2005). "Orientifold theory dynamics and symmetry breaking". *Physical Review* **D71** (5): 051901. arXiv:hep-ph/0405209. Bibcode:2005PhRvD..71e1901S. doi:10.1103/PhysRevD.71.051901.

[27] Dennis D. Dietrich, Francesco Sannino and Kimmo Tuominen (2005). "Light composite Higgs boson from higher representations versus electroweak precision measurements: Predictions for CERN LHC". *Physical Review* **D72** (5): 055001. arXiv:hep-ph/0505059. Bibcode:2005PhRvD..72e5001D. doi:10.1103/PhysRevD.72.055001.
Dennis D. Dietrich, Francesco Sannino and Kimmo Tuominen (2006). "Light composite Higgs and precision electroweak measurements on the Z resonance: An update". *Physical Review* **D73** (3): 037701. arXiv:hep-ph/0510217. Bibcode:2006PhRvD..73c7701D. doi:10.1103/PhysRevD.73.037701.
Dennis D. Dietrich and Francesco Sannino (2007). "Conformal window of SU(N) gauge theories with fermions in higher dimensional representations". *Physical Review* **D75** (8): 085018. arXiv:hep-ph/0611341. Bibcode:2007PhRvD..75h5018D. doi:10.1103/PhysRevD.75.085018.
Thomas A. Ryttov and Francesco Sannino (2007). "Conformal windows of SU(N) gauge theories, higher dimensional representations, and the size of the unparticle world". *Physical Review* **D76** (10): 105004. arXiv:0707.3166. Bibcode:2007PhRvD..76j5004R. doi:10.1103/PhysRevD.76.105004.
Thomas A. Ryttov and Francesco Sannino (2008). "Supersymmetry inspired QCD beta function". *Physical Review* **D78** (6): 065001. arXiv:0711.3745. Bibcode:2008PhRvD..78f5001R. doi:10.1103/PhysRevD.78.065001.

[28] Markus A. Luty and Takemichi Okui (2006). "Conformal technicolor". *Journal of High Energy Physics* **0609** (09): 070. arXiv:hep-ph/0409274. Bibcode:2006JHEP...09..070L. doi:10.1088/1126-6708/2006/09/070.
Markus A. Luty (2009). "Strong conformal dynamics at the LHC and on the lattice". *Journal of High Energy Physics* **0904** (04): 050. arXiv:0806.1235. Bibcode:2009JHEP...04..050L. doi:10.1088/1126-6708/2009/04/050.
Jared A. Evans, Jamison Galloway, Markus A. Luty and Ruggero Altair Tacchi (2010). "Minimal conformal technicolor and precision electroweak tests". *Journal of High Energy Physics* **1010** (10): 086. arXiv:1001.1361. Bibcode:2010JHEP...10..086E. doi:10.1007/JHEP10(2010)086.

[29] Thomas Appelquist, T. Takeuchi, Martin Einhorn and L. C. R. Wijewardhana (1989). "Higher mass scales and mass hierarchies". *Physics Letters* **B220** (1-2): 223–228. Bibcode:1989PhLB..220..223A. doi:10.1016/0370-2693(89)90041-5.

[30] V. A. Miransky and K. Yamawaki (1989). "On Gauge Theories with Additional Four Fermion Interaction". *Modern Physics Letters* **A4** (2): 129–135. Bibcode:1989MPLA....4..129M. doi:10.1142/S0217732389000186.

[31] Y. Nambu (1989). "BCS mechanism, quasi supersymmetry, and fermion masses". In Z. Adjduk, S. Pokorski, and A. Trautman. "Proceedings of the Kazimierz 1988 Conference on New Theories in Physics". XI International Symposium on Elementary Particle Physics. pp. 406–415.
V. A. Miransky, Masaharu Tanabashi and Koichi Yamawaki (1989). "Is the t Quark Responsible for the Mass of W and Z Bosons?". *Modern Physics Letters* **A4** (11): 1043–1053. Bibcode:1989MPLA....4.1043M. doi:10.1142/S0217732389001210.
V. A. Miransky, Masaharu Tanabashi and Koichi Yamawaki (1989). "Dynamical electroweak symmetry breaking with large anomalous dimension and t quark condensate". *Physics Letters* **B221** (2): 177–183. Bibcode:1989PhLB..221..177M. doi:10.1016/0370-2693(89)91494-9.
William A. Bardeen, Christopher T. Hill, and Manfred Lindner (1990). "Minimal dynamical symmetry breaking of the standard model". *Physical Review* **D41** (5): 1647–1660. Bibcode:1990PhRvD..41.1647B. doi:10.1103/PhysRevD.41.1647.

[32] Christopher T. Hill (1991). "Topcolor: top quark condensation in a gauge extension of the standard model". *Physics Letters* **B266** (3-4): 419–424. Bibcode:1991PhLB..266..419H. doi:10.1016/0370-2693(91)91061-Y.
Christopher T. Hill (1995). "Topcolor assisted technicolor". *Physics Letters* **B345** (4): 483–489. arXiv:hep-ph/9411426. Bibcode:1995PhLB..345..483H. doi:10.1016/0370-2693(94)01660-5.

[33] Roshan Foadi, Mads T. Frandsen, Thomas A. Ryttov and Francesco Sannino (2007). "Minimal walking technicolor: Setup for collider physics". *Physical Review* **D76** (5): 055005. arXiv:0707.1696. Bibcode:2007PhRvD..76e5005F. doi:10.1103/PhysRevD.76.055005.

[34] Simon Catterall and Francesco Sannino (2007). "Minimal Walking on the Lattice". *Physical Review* **D76** (3): 034504. arXiv:0705.1664. Bibcode:2007PhRvD..76c4504C. doi:10.1103/PhysRevD.76.034504.

[35] Simon Catterall, Joel Giedt, Francesco Sannino and Joe Schneible (2008). "Phase diagram of SU(2) with 2 flavors of dynamical adjoint quarks". *Journal of High Energy Physics* **0811** (11): 009. arXiv:0807.0792. Bibcode:2008JHEP...11..009C. doi:10.1088/1126-6708/2008/11/009.
Ari J. Hietanen, Kari Rummukainen and Kimmo Tuominen (2009). "Evolution of the coupling constant in SU(2) lattice gauge theory with two adjoint fermions". *Physical Review* **D80** (9): 094504. arXiv:0904.0864. Bibcode:2009PhRvD..80i4504H. doi:10.1103/PhysRevD.80.094504.

[36] Thomas Appelquist, George T. Fleming and Ethan T. Neil (2008). "Lattice Study of the Conformal Window in QCD-like Theories". *Physical Review Letters* **100** (17): 171607. arXiv:0712.0609. Bibcode:2008PhRvL.100q1607A. doi:10.1103/PhysRevLett.100.171(PMID 18518277.

[37] Albert Deuzeman, Maria Paola Lombardo and Elisabetta Pallante (2008). "The physics of eight flavours". *Physics Letters* **B670** (1): 41–48. arXiv:0804.2905. Bibcode:2008PhLB..670...41D. doi:10.1016/j.physletb.2008.10.039.
Thomas Appelquist, George T. Fleming and Ethan T. Neil (2009). "Lattice study of conformal behavior in SU(3) Yang-Mills theories". *Physical Review* **D79** (7): 076010. arXiv:0901.3766. Bibcode:2009PhRvD..79g6010A. doi:10.1103/PhysRevD.79.076010.
Erek Bilgici *et al.* (2009). "New scheme for the running coupling constant in gauge theories using Wilson loops". *Physical Review* **D80** (3): 034507. arXiv:0902.3768. Bibcode:2009PhRvD..80c4507B. doi:10.1103/PhysRevD.80.034507.
Xiao-Yong Jin and Robert D. Mawhinney (2009). "Lattice QCD with 8 and 12 degenerate quark flavors". *Proceedings of Science*. LAT2009: 049.
Zoltan Fodor, Kieran Holland, Julius Kuti, Daniel Nogradi and Chris Schroeder (2009). "Chiral symmetry breaking in nearly conformal gauge theories". *Proceedings of Science*. LAT2009: 058. arXiv:0911.2463. Bibcode:2009arXiv0911.2463F.
Anna Hasenfratz (2010). "Conformal or Walking? Monte Carlo renormalization group studies of SU(3) gauge models with fundamental fermions". *Physical Review* **D82** (1): 014506. arXiv:1004.1004. Bibcode:2010PhRvD..82a4506H. doi:10.1103/PhysRevD.82.014506.

[38] Thomas DeGrand, Yigal Shamir and Benjamin Svetitsky (2009). "Phase structure of SU(3) gauge theory with two flavors of symmetric-representation fermions". *Physical Review* **D79** (3): 034501. arXiv:0812.1427. Bibcode:2009PhRvD..79c4501D. doi:10.1103/PhysRevD.79.034501.
Thomas Appelquist *et al.* (2009). "Toward TeV Conformality". *Physical Review Letters* **104** (7): 071601. arXiv:0910.2224. Bibcode:2010PhRvL.104g1601A. doi:10.1103/PhysRevLett.104.071601. PMID 20366870.

[39] Michael E. Peskin and Tatsu Takeuchi (1990). "New constraint on a strongly interacting Higgs sector". *Physical Review Letters* **65** (8): 964–967. Bibcode:1990PhRvL..65..964P. doi:10.1103/PhysRevLett.65.964. PMID 10043071.
Michael E. Peskin and Tatsu Takeuchi (1992). "Estimation of oblique electroweak corrections". *Physical Review* **D46** (1): 381–409. Bibcode:1992PhRvD..46..381P. doi:10.1103/PhysRevD.46.381.

[40] Thomas Appelquist and Claude Bernard (1980). "Strongly interacting Higgs bosons". *Physical Review* **D22** (1): 200–213. Bibcode:1980PhRvD..22..200A. doi:10.1103/PhysRevD.22.200.

[41] Anthony C. Longhitano (1980). "Heavy Higgs bosons in the Weinberg-Salam model". *Physical Review* **D22** (5): 1166–1175. Bibcode:1980PhRvD..22.1166L. doi:10.1103/PhysRevD.22.1166.
Anthony C. Longhitano (1981). "Low-energy impact of a heavy Higgs boson sector". *Nuclear Physics* **B188** (1): 118–154. Bibcode:1981NuPhB.188..118L. doi:10.1016/0550-3213(81)90109-7.

[42] B. W. Lynn, Michael Edward Peskin, and R. G. Stuart (1985). "Radiative Corrections in SU(2) x U(1): LEP / SLC". In Bryan W. Lynn and Claudio Verzegnassi. "Tests of electroweak theories: polarized processes and other phenomena". Second Conference on Tests of Electroweak Theories, Trieste, Italy, 10–12 June 1985. p. 213.
D. C. Kennedy and B. W. Lynn (1989). "Electroweak radiative corrections with an effective lagrangian: Four-fermions processes". *Nuclear Physics* **B322** (1): 1–54. Bibcode:1989NuPhB.322....1K. doi:10.1016/0550-3213(89)90483-5.

[43] Mitchell Golden and Lisa Randall (1991). "Radiative corrections to electroweak parameters in technicolor theories". *Nuclear Physics* **B361** (1): 3–23. Bibcode:1991NuPhB.361....3G. doi:10.1016/0550-3213(91)90614-4.
B. Holdom and J. Terning (1990). "Large corrections to electroweak parameters in technicolor theories". *Physics Letters* **B247** (1): 88–92. Bibcode:1990PhLB..247...88H. doi:10.1016/0370-2693(90)91054-F.
G. Altarelli, R. Barbieri, and S. Jadach (1992). "Toward a model-independent analysis of electroweak data". *Nuclear Physics* **B369** (1-2): 3–32. Bibcode:1992NuPhB.369....3A. doi:10.1016/0550-3213(92)90376-M.

[44] Particle Data Group (C. Amsler *et al.*) (2008). "Review of Particle Physics". *Physics Letters* **B667** (1-5): 1. Bibcode:2008PhLB..667....1P. doi:10.1016/j.physletb.2008.07.018.

[45] Kenneth Lane (1994). "An introduction to technicolor". In K. T. Mahantappa. "Boulder 1993 Proceedings: The building blocks of creation". Theoretical Advanced Study Institute (TASI 93) in Elementary Particle Physics: The Building Blocks of Creation - From Microfermis to Megaparsecs, Boulder, Colorado, 6 June - 2 July 1993. pp. 381–408. arXiv:hep-ph/9401324.
Kenneth Lane (1995). "Technicolor and precision tests of the electroweak interactions". In P. J. Bussey and I. G. Knowles. "High energy physics: Proceedings". 27th International Conference on High Energy Physics (ICHEP), Glasgow, Scotland, 20–27 July 1994 **II**. p. 543. arXiv:hep-ph/9409304.

[46] Thomas Appelquist and Francesco Sannino (1999). "Physical spectrum of conformal SU(N) gauge theories". *Physical Review* **D59** (6): 067702. arXiv:hep-ph/9806409. Bibcode:1999PhRvD..59f7702A. doi:10.1103/PhysRevD.59.067702.
Johannes Hirn and Verónica Sanz (2006). "Negative S Parameter from Holographic Technicolor". *Physical Review Letters* **97** (12): 121803. arXiv:hep-ph/0606086. Bibcode:2006PhRvL..97l1803H. doi:10.1103/PhysRevLett.97.121803. PMID 17025952.
R. Casalbuoni, D. Dominici, A. Deandrea, R. Gatto, S. De Curtis, and M. Grazzini (1996). "Low energy strong electroweak sector with decoupling". *Physical Review* **D53** (9): 5201–5221. arXiv:hep-ph/9510431. Bibcode:1996PhRvD..53.5201C. doi:10.1103/PhysRevD.53.5201.

[47] Lattice Strong Dynamics Collaboration.

[48] Thomas Appelquist, Mark J. Bowick, Eugene Cohler, and Avi I. Hauser (1985). "Breaking of isospin symmetry in theories with a dynamical Higgs mechanism". *Physical Review* **D31** (7): 1676–1684. Bibcode:1985PhRvD..31.1676A. doi:10.1103/PhysRevD.31.1676.
R. S. Chivukula, B. A. Dobrescu, and J. Terning (1995). "Isospin breaking and fine-tuning in top-color assisted technicolor". *Physics Letters* **B353** (2-3): 289–284. arXiv:hep-ph/9503203. Bibcode:1995PhLB..353..289C. doi:10.1016/0370-2693(95)00569-7.

[49] R. Sekhar Chivukula, Stephen B. Selipsky, and Elizabeth H. Simmons (1992). "Nonoblique effects in the Zbb⁻ vertex from extended technicolor dynamics". *Physical Review Letters* **69** (4): 575–577. arXiv:hep-ph/9204214. Bibcode:1992PhRvL..69..575C. doi:10.1103/PhysRevLett.69.575. PMID 10046976.
Elizabeth H. Simmons, R.S. Chivukula, and J. Terning (1996). "Testing extended technicolor with R(b)". *Progress of Theoretical Physics Supplement* **123**: 87–96. arXiv:hep-ph/9509392. Bibcode:1996PThPS.123...87S. doi:10.1143/PTPS.123.87.

[50] E. Eichten, I. Hinchliffe, K. Lane, and C. Quigg (1984). "Supercollider physics". *Reviews of Modern Physics* **56** (4): 579–707. Bibcode:1984RvMP...56..579E. doi:10.1103/RevModPhys.56.579.
E. Eichten, I. Hinchliffe, K. Lane, and C. Quigg (1986). "Erratum: Supercollider physics". *Reviews of Modern Physics* **58** (4): 1065–1073. Bibcode:1986RvMP...58.1065E. doi:10.1103/RevModPhys.58.1065.

[51] E. Farhi and L. Susskind (1979). "Grand unified theory with heavy color". *Physical Review* **D20** (12): 3404–3411. Bibcode:1979PhRvD..20.3404F. doi:10.1103/PhysRevD.20.3404.

[52] Dennis D. Dietrich, Francesco Sannino, and Kimmo Tuominen (2005). "Light composite Higgs boson from higher representations versus electroweak precision measurements: Predictions for CERN LHC". *Physical Review* **D72** (5): 055001. arXiv:hep-ph/0505059. Bibcode:2005PhRvD..72e5001D. doi:10.1103/PhysRevD.72.055001.

[53] Kenneth Lane and Estia Eichten (1995). "Natural topcolor-assisted technicolor". *Physics Letters* **B352** (3-4): 382–387. arXiv:hep-ph/9503433. Bibcode:1995PhLB..352..382L. doi:10.1016/0370-2693(95)00482-Z.
Estia Eichten and Kenneth Lane (1996). "Low-scale technicolor at the Tevatron". *Physics Letters* **B388** (4): 803–807. arXiv:hep-ph/9607213. Bibcode:1996PhLB..388..803E. doi:10.1016/S0370-2693(96)01211-7.
Estia Eichten, Kenneth Lane, and John Womersley (1997). "Finding low-scale technicolor at hadron colliders". *Physics Letters* **B405** (3-4): 305–311. arXiv:hep-ph/9704455. Bibcode:1997PhLB..405..305E. doi:10.1016/S0370-2693(97)00637-0.

[54] Kenneth Lane (1999). "Technihadron production and decay in low-scale technicolor". *Physical Review* **D60** (7): 075007. arXiv:hep-ph/9903369. Bibcode:1999PhRvD..60g5007L. doi:10.1103/PhysRevD.60.075007.
Estia Eichten and Kenneth Lane (2008). "Low-scale technicolor at the Tevatron and LHC". *Physics Letters* **B669** (3-4): 235–238. arXiv:0706.2339. Bibcode:2008PhLB..669..235E. doi:10.1016/j.physletb.2008.09.047.

[55] CDF Collaboration (T. Aaltonen *et al.*) (2011). "Invariant Mass Distribution of Jet Pairs Produced in Association with a W boson in ppbar Collisions at sqrt(s) = 1.96 TeV". arXiv:1104.0699.

[56] Estia J. Eichten, Kenneth Lane and Adam Martin (2011). "Technicolor at the Tevatron". arXiv:1104.0976.

[57] Gustaaf H. Brooijmans *et al.* (New Physics Working Group) (2008). "New Physics at the LHC: A Les Houches Report". "Les Houches 2007: Physics at TeV Colliders". 5th Les Houches Workshop on Physics at TeV Colliders 11–29 June 2007, Les Houches, France. pp. 363–489. arXiv:0802.3715.

[58] S. Nussinov (1985). "Technocosmology — could a technibaryon excess provide a "natural" missing mass candidate?". *Physics Letters* **B165** (1-3): 55–58. Bibcode:1985PhLB..165...55N. doi:10.1016/0370-2693(85)90689-6.
R. S. Chivukula and Terry P. Walker (1990). "Technicolor cosmology". *Nuclear Physics* **B329** (2): 445–463. Bibcode:1990NuPhB.329..445C. doi:10.1016/0550-3213(90)90151-3.
John Bagnasco, Michael Dine and Scott Thomas (1994). "Detecting technibaryon dark matter". *Physics Letters* **B320** (1-2): 99–104. arXiv:hep-ph/9310290. Bibcode:1994PhLB..320...99B. doi:10.1016/0370-2693(94)90830-3.
Sven Bjarke Gudnason, Chris Kouvaris, and Francesco Sannino (2006). "Dark matter from new technicolor theories". *Physical Review* **D74** (9): 095008. arXiv:hep-ph/0608055. Bibcode:2006PhRvD..74i5008G. doi:10.1103/PhysRevD.74.095008.

[59] D. McKinsey, "Direct Dark Matter Detection Using Noble Liquids", 2009 Institute for Advanced Study Workshop on Current Trends in Dark Matter.

[60] Sven Bjarke Gudnason, Chris Kouvaris and Francesco Sannino (2006). "Towards working technicolor: Effective theories and dark matter". *Physical Review* **D73** (11): 115003. arXiv:hep-ph/0603014. Bibcode:2006PhRvD..73k5003G. doi:10.1103/PhysRevD.73.115003.
Sven Bjarke Gudnason, Chris Kouvaris and Francesco Sannino (2006). "Dark matter from new technicolor theories". *Physical Review* **D74** (9): 095008. arXiv:hep-ph/0608055. Bibcode:2006PhRvD..74i5008G. doi:10.1103/PhysRevD.74.095008.
Thomas A. Ryttov and Francesco Sannino (2008). "Ultraminimal technicolor and its dark matter technicolor interacting massive particles". *Physical Review* **D78** (11): 115010. arXiv:0809.0713. Bibcode:2008PhRvD..78k5010R. doi:10.1103/PhysRevD.78.115010.
Enrico Nardi, Francesco Sannino and Alessandro Strumia (2009). "Decaying Dark Matter can explain the e± excesses". *Journal of Cosmology and Astroparticle Physics* **0901** (01): 043. arXiv:0811.4153. Bibcode:2009JCAP...01..043N. doi:10.1088/1475-7516/2009/01/043.
Roshan Foadi, Mads T. Frandsen and Francesco Sannino (2009). "Technicolor dark matter". *Physical Review* **D80** (3): 037702. arXiv:0812.3406. Bibcode:2009PhRvD..80c7702F. doi:10.1103/PhysRevD.80.037702.
Mads T. Frandsen and Francesco Sannino (2010). "Isotriplet technicolor interacting massive particle as dark matter". *Physical Review* **D81** (9): 097704. arXiv:0911.1570. Bibcode:2010PhRvD..81i7704F. doi:10.1103/PhysRevD.81.097704.

Chapter 4

Kaluza–Klein theory

This article is about gravitation and electromagnetism. For the mathematical generalization of K theory, see KK-theory.

In physics, **Kaluza–Klein theory** (**KK theory**) is a unified field theory of gravitation and electromagnetism built around the idea of a fifth dimension beyond the usual 4 of space and time. The five-dimensional theory was developed in three steps. The original hypothesis came from Theodor Kaluza, who sent his results to Einstein in 1919,[1] and published them in 1921.[2] Kaluza's theory was a purely classical extension of general relativity to five dimensions. The 5-dimensional metric has 15 components. 10 components are identified with the 4-dimensional spacetime metric, 4 components with the electromagnetic vector potential, and one component with an unidentified scalar field sometimes called the "radion" or the "dilaton". Correspondingly, the 5-dimensional Einstein equations yield the 4-dimensional Einstein field equations, the Maxwell equations for the electromagnetic field, and an equation for the scalar field. Kaluza also introduced the hypothesis known as the "cylinder condition", that no component of the 5-dimensional metric depends on the fifth dimension. Without this assumption, the field equations of 5-dimensional relativity are enormously more complex. Standard 4-dimensional physics seems to manifest the cylinder condition. Kaluza also set the scalar field equal to a constant, in which case standard general relativity and electrodynamics are recovered identically.

In 1926, Oskar Klein gave Kaluza's classical 5-dimensional theory a quantum interpretation,[3][4] to accord with the then-recent discoveries of Heisenberg and Schroedinger. Klein introduced the hypothesis that the fifth dimension was curled up and microscopic, to explain the cylinder condition. Klein also calculated a scale for the fifth dimension based on the quantum of charge.

It wasn't until the 1940s that the classical theory was completed, and the full field equations including the scalar field were obtained by 3 independent research groups:[5] Thiry,[6][7][8] working in France on his dissertation under Lichnerowicz; Jordan, Ludwig, and Müller in Germany,[9][10][11][12][13] with critical input from Pauli and Fierz; and Scherrer [14][15][16] working alone in Switzerland. Jordan's work led to the famous scalar-tensor theory of Brans & Dicke;[17] Brans and Dicke were apparently unaware of Thiry or Scherrer.

4.1 The Kaluza Hypothesis

In his 1921 paper,[2] Kaluza established all the elements of the classical 5-dimensional theory: the metric, the field equations, the equations of motion, the stress-energy tensor, and the cylinder condition. The theory has no free parameters; it merely extends general relativity to five dimensions. One starts by hypothesizing a form of the 5-dimensional metric $\widetilde{g}_{ab}$, where roman indices span 5 dimensions. Let us also introduce the 4-dimensional spacetime metric $g_{\mu\nu}$, where Greek indices span the usual 4 dimensions of space and time; a 4-vector A^μ which will be identified with the electromagnetic vector potential; and a scalar field ϕ. Then decompose the 5D metric so that the 4D metric is framed by the electromagnetic vector potential, with the scalar field at the fifth diagonal. This can be visualized as:

$$\widetilde{g}_{ab} \equiv \begin{bmatrix} g_{\mu\nu} + \phi^2 A_\mu A_\nu & \phi^2 A_\mu \\ \phi^2 A_\nu & \phi^2 \end{bmatrix}.$$

More precisely, we can write

$$\widetilde{g}_{\mu\nu} \equiv g_{\mu\nu} + \phi^2 A_\mu A_\nu, \qquad \widetilde{g}_{5\nu} \equiv \widetilde{g}_{\nu 5} \equiv \phi^2 A_\nu, \qquad \widetilde{g}_{55} \equiv \phi^2$$

where the index 5 indicates the fifth coordinate by convention even though the first four coordinates are indexed with 0, 1, 2, and 3. The associated inverse metric is

$$\widetilde{g}^{ab} \equiv \begin{bmatrix} g^{\mu\nu} & -A^\mu \\ -A^\nu & g_{\alpha\beta}A^\alpha A^\beta + \frac{1}{\phi^2} \end{bmatrix}.$$

So far, this decomposition is quite general and all terms are dimensionless. Kaluza then applies the machinery of standard general relativity to this metric. The field equations are obtained from 5-dimensional Einstein equations, and the equations of motion are obtained from the 5-dimensional geodesic hypothesis. The resulting field equations provide both the equations of general relativity and of electrodynamics; the equations of motion provide the 4-dimensional geodesic equation and the Lorentz force law. And one finds that electric charge is identified with motion in the fifth dimension.

The hypothesis for the metric implies an invariant 5-dimensional length element ds:

$$ds^2 \equiv \widetilde{g}_{ab} dx^a dx^b = g_{\mu\nu} dx^\mu dx^\nu + \phi^2 (A_\nu dx^\nu + dx^5)^2$$

4.2 Field Equations from the Kaluza Hypothesis

The field equations of the 5-dimensional theory were never adequately provided by Kaluza or Klein, mainly regarding the scalar field. The full Kaluza field equations are generally attributed to Thiry,[7] who most famously obtained vacuum field equations, although Kaluza [2] originally provided a stress-energy tensor for his theory and Thiry included a stress-energy tensor in his thesis. But as described by Gonner,[5] several independent groups worked on the field equations in the 1940s and earlier. Thiry is perhaps best known only because an English translation was provided by Applequist, Chodos, & Freund in their review book.[18] Applequist et al. also provided an English translation of Kaluza's paper. There are no English translations of the Jordan papers.[9][10][12]

To obtain the 5D field equations, the 5D connections $\widetilde{\Gamma}^a_{bc}$ are calculated from the 5D metric $\widetilde{g}_{ab}$, and the 5D Ricci tensor $\widetilde{R}_{ab}$ is calculated from the 5D connections.

The classic results of Thiry and other authors presume the cylinder condition:

$$\frac{\partial \widetilde{g}_{ab}}{\partial x^5} = 0$$

Without this assumption, the field equations become much more complex, providing many more degrees of freedom that can be identified with various new fields. Paul Wesson and colleagues have pursued relaxation of the cylinder condition to gain extra terms that can be identified with the matter fields,[19] for which Kaluza [2] otherwise inserted a stress-energy tensor by hand.

It has been an objection to the original Kaluza hypothesis to invoke the fifth dimension only to negate its dynamics. But Thiry argued [5] that the interpretation of the Lorentz force law in terms of a 5-dimensional geodesic mitigates strongly for a fifth dimension irrespective of the cylinder condition. Most authors have therefore employed the cylinder condition in deriving the field equations. Furthermore, vacuum equations are typically assumed for which

$$\widetilde{R}_{ab} = 0$$

where

$$\widetilde{R}_{ab} \equiv \partial_c \widetilde{\Gamma}^c_{ab} - \partial_b \widetilde{\Gamma}^c_{ca} + \widetilde{\Gamma}^c_{cd} \widetilde{\Gamma}^d_{ab} - \widetilde{\Gamma}^c_{bd} \widetilde{\Gamma}^d_{ac}$$

and

$$\widetilde{\Gamma}^a_{bc} \equiv \frac{1}{2}\widetilde{g}^{ad}(\partial_b \widetilde{g}_{dc} + \partial_c \widetilde{g}_{db} - \partial_d \widetilde{g}_{bc})$$

The vacuum field equations obtained in this way by Thiry [7] and Jordan's group [9][10][12] are as follows.

The field equation for ϕ is obtained from

$$\widetilde{R}_{55} = 0 \Rightarrow \Box\phi = \frac{1}{4}\phi^3 F^{\alpha\beta} F_{\alpha\beta}$$

where $F_{\alpha\beta} \equiv \partial_\alpha A_\beta - \partial_\beta A_\alpha$, where $\Box \equiv g^{\mu\nu}\nabla_\mu\nabla_\nu$, and where ∇_μ is a standard, 4D covariant derivative. It shows that the electromagnetic field is a source for the scalar field. Note that the scalar field cannot be set to a constant without constraining the electromagnetic field. The earlier treatments by Kaluza and Klein did not have an adequate description of the scalar field, and did not realize the implied constraint on the electromagnetic field by assuming the scalar field to be constant.

The field equation for A^ν is obtained from

$$\widetilde{R}_{5\alpha} = 0 = \frac{1}{2}g^{\beta\mu}\nabla_\mu(\phi^3 F_{\alpha\beta})$$

It has the form of the vacuum Maxwell equations if the scalar field is constant.

The field equation for the 4D Ricci tensor $R_{\mu\nu}$ is obtained from

$$\widetilde{R}_{\mu\nu} - \frac{1}{2}\widetilde{g}_{\mu\nu}\widetilde{R} = 0 \Rightarrow R_{\mu\nu} - \frac{1}{2}g_{\mu\nu}R = \frac{1}{2}\phi^2\left(g^{\alpha\beta}F_{\mu\alpha}F_{\nu\beta} - \frac{1}{4}g_{\mu\nu}F_{\alpha\beta}F^{\alpha\beta}\right) + \frac{1}{\phi}(\nabla_\mu\nabla_\nu\phi - g_{\mu\nu}\Box\phi)$$

where R is the standard 4D Ricci scalar.

This equation shows the remarkable result, called the "Kaluza miracle", that the precise form for the electromagnetic stress-energy tensor emerges from the 5D vacuum equations as a source in the 4D equations: field from the vacuum. This relation allows the definitive identification of A^μ with the electromagnetic vector potential. Therefore the field needs to be rescaled with a conversion constant k such that $A^\mu \to kA^\mu$.

The relation above shows that we must have

$$\frac{k^2}{2} = \frac{8\pi G}{c^4}\frac{1}{\mu_0}$$

where G is the gravitational constant and μ_0 is the permeability of free space. In the Kaluza theory, the gravitational constant can be understood as an electromagnetic coupling constant in the metric. There is also a stress-energy tensor for the scalar field. The scalar field behaves like a variable gravitational constant, in terms of modulating the coupling of electromagnetic stress energy to spacetime curvature. The sign of ϕ^2 in the metric is fixed by correspondence with 4D theory so that electromagnetic energy densities are positive. This turns out to imply that the 5th coordinate is spacelike in its signature in the metric.

4.3 Equations of Motion from the Kaluza Hypothesis

The equations of motion are obtained from the 5-dimensional geodesic hypothesis [2] in terms of a 5-velocity $\widetilde{U}^a \equiv dx^a/ds$:

$$\widetilde{U}^b \widetilde{\nabla}_b \widetilde{U}^a = \frac{d\widetilde{U}^a}{ds} + \widetilde{\Gamma}^a_{bc}\widetilde{U}^b\widetilde{U}^c = 0$$

4.3. EQUATIONS OF MOTION FROM THE KALUZA HYPOTHESIS

This equation can be recast in several ways, and it has been studied in various forms by authors including Kaluza,[2] Pauli,[20] Gross & Perry,[21] Gegenberg & Kunstatter,[22] and Wesson & Ponce de Leon,[23] but it is instructive to convert it back to the usual 4-dimensional length element $c^2 d\tau^2 \equiv g_{\mu\nu} dx^\mu dx^\nu$, which is related to the 5-dimensional length element ds as given above:

$$ds^2 = c^2 d\tau^2 + \phi^2 (kA_\nu dx^\nu + dx^5)^2$$

Then the 5D geodesic equation can be written [24] for the spacetime components of the 4velocity, $U^\nu \equiv dx^\nu/d\tau$:

$$\frac{dU^\nu}{d\tau} + \widetilde{\Gamma}^\mu_{\alpha\beta} U^\alpha U^\beta + 2\widetilde{\Gamma}^\mu_{5\alpha} U^\alpha U^5 + \widetilde{\Gamma}^\mu_{55} (U^5)^2 + U^\mu \frac{d}{d\tau} \ln\left(\frac{cd\tau}{ds}\right) = 0$$

The term quadratic in U^ν provides the 4D geodesic equation plus some electromagnetic terms:

$$\widetilde{\Gamma}^\mu_{\alpha\beta} = \Gamma^\mu_{\alpha\beta} + \frac{1}{2} g^{\mu\nu} k^2 \phi^2 (A_\alpha F_{\beta\nu} + A_\beta F_{\alpha\nu} + A_\alpha A_\beta \partial_\nu \ln \phi^2)$$

The term linear in U^ν provides the Lorentz force law:

$$\widetilde{\Gamma}^\mu_{5\alpha} = \frac{1}{2} g^{\mu\nu} k \phi^2 (F_{\alpha\nu} - A_\alpha \partial_\nu \ln \phi^2)$$

This is another expression of the "Kaluza miracle". The same hypothesis for the 5D metric that provides electromagnetic stress-energy in the Einstein equations, also provides the Lorentz force law in the equation of motions along with the 4D geodesic equation. Yet correspondence with the Lorentz force law requires that we identify the component of 5-velocity along the 5th dimension with electric charge:

$$kU^5 = k\frac{dx^5}{d\tau} \to \frac{q}{mc}$$

where m is particle mass and q is particle electric charge. Thus, electric charge is understood as motion along the 5th dimension. The fact that the Lorentz force law could be understood as a geodesic in 5 dimensions was to Kaluza a primary motivation for considering the 5-dimensional hypothesis, even in the presence of the aesthetically-unpleasing cylinder condition.

Yet there is a problem: the term quadratic in U^5.

$$\widetilde{\Gamma}^\mu_{55} = -\frac{1}{2} g^{\mu\alpha} \partial_\alpha \phi^2$$

If there is no gradient in the scalar field, the term quadratic in U^5 vanishes. But otherwise the expression above implies

$$U^5 \sim c \frac{q/m}{G^{1/2}}$$

For elementary particles, $U^5 > 10^{20} c$. The term quadratic in U^5 should dominate the equation, perhaps in contradiction to experience. This was the main shortfall of the 5-dimensional theory as Kaluza saw it,[2] and he gives it some discussion in his original article.

The equation of motion for U^5 is particularly simple under the cylinder condition. Start with the alternate form of the geodesic equation, written for the covariant 5-velocity:

$$\frac{d\widetilde{U}_a}{ds} = \frac{1}{2} \widetilde{U}^b \widetilde{U}^c \frac{\partial \widetilde{g}_{bc}}{\partial x^a}$$

This means that under the cylinder condition, $\widetilde{U}_5$ is a constant of the 5-dimensional motion:

$$\widetilde{U}_5 = \widetilde{g}_{5a} \widetilde{U}^a = \phi^2 \frac{cd\tau}{ds} (kA_\nu U^\nu + U^5) = \text{constant}$$

4.4 Kaluza's Hypothesis for the Matter Stress-Energy Tensor

Kaluza [2] proposed a 5D matter stress tensor $\widetilde{T}_M^{ab}$ of the form

$$\widetilde{T}_M^{ab} = \rho \frac{dx^a}{ds} \frac{dx^b}{ds}$$

where ρ is a density and the length element ds is as defined above.

Then, the spacetime component gives a typical "dust" stress energy tensor:

$$\widetilde{T}_M^{\mu\nu} = \rho \frac{dx^\mu}{ds} \frac{dx^\nu}{ds}$$

The mixed component provides a 4-current source for the Maxwell equations:

$$\widetilde{T}_M^{5\mu} = \rho \frac{dx^\mu}{ds} \frac{dx^5}{ds} = \rho U^\mu \frac{q}{kmc}$$

Just as the 5-dimensional metric comprises the 4-D metric framed by the electromagnetic vector potential, the 5-dimensional stress-energy tensor comprises the 4-D stress-energy tensor framed by the vector 4-current.

4.5 The Quantum Interpretation of Klein

Kaluza's original hypothesis was purely classical and extended discoveries of general relativity. By the time of Klein's contribution, the discoveries of Heisenberg, Schroedinger, and de Broglie were receiving a lot of attention. Klein's Nature paper [4] suggested that the fifth dimension is closed and periodic, and that the identification of electric charge with motion in the fifth dimension be interpreted as standing waves of wavelength λ^5, much like the electrons around a nucleus in the Bohr model of the atom. The quantization of electric charge could then be nicely understood in terms of integer multiples of fifth-dimensional momentum. Combining the previous Kaluza result for U^5 in terms of electric charge, and a de Broglie relation for momentum $p^5 = h/\lambda^5$, Klein [4] obtained an expression for the 0th mode of such waves:

$$mU^5 = \frac{cq}{G^{1/2}} = \frac{h}{\lambda^5} \rightarrow \lambda^5 \sim \frac{hG^{1/2}}{cq}$$

where h is the Planck constant. Klein found $\lambda^5 \sim 10^{-30}$ cm, and thereby an explanation for the cylinder condition in this small value.

Klein's Zeitschrift fur Physik paper of the same year,[3] gave a more-detailed treatment that explicitly invoked the techniques of Schroedinger and de Broglie. It recapitulated much of the classical theory of Kaluza described above, and then departed into Klein's quantum interpretation. Klein solved a Schroedinger-like wave equation using an expansion in terms of fifth-dimensional waves resonating in the closed, compact fifth dimension.

4.6 Quantum Field Theory Interpretation

4.7 Group Theory Interpretation

A splitting of five-dimensional spacetime into the Einstein equations and Maxwell equations in four dimensions was first discovered by Gunnar Nordström in 1914, in the context of his theory of gravity, but subsequently forgotten. Kaluza published his derivation in 1921 as an attempt to unify electromagnetism with Einstein's general relativity.

In 1926, Oskar Klein proposed that the fourth spatial dimension is curled up in a circle of a very small radius, so that a particle moving a short distance along that axis would return to where it began. The distance a particle can travel

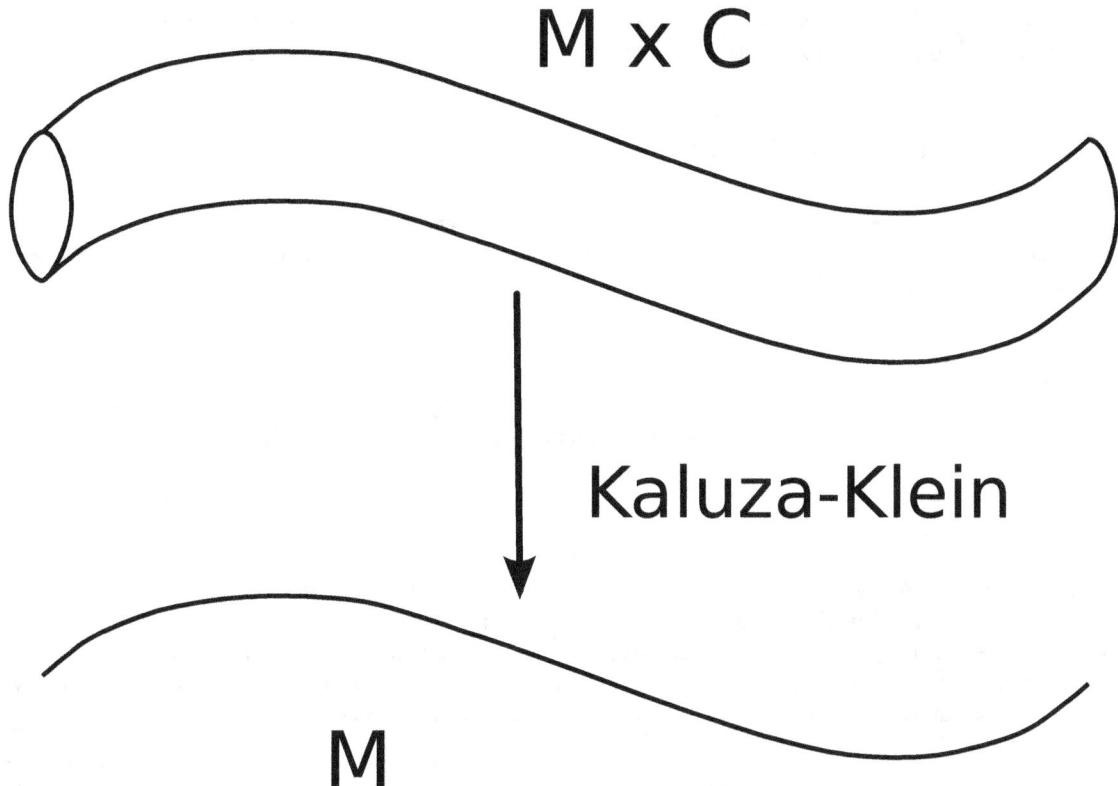

The space M × C is compactified over the compact set C, and after Kaluza–Klein decomposition we have an effective field theory over M.

before reaching its initial position is said to be the size of the dimension. This extra dimension is a compact set, and the phenomenon of having a space-time with compact dimensions is referred to as compactification.

In modern geometry, the extra fifth dimension can be understood to be the circle group U(1), as electromagnetism can essentially be formulated as a gauge theory on a fiber bundle, the circle bundle, with gauge group U(1). In Kaluza–Klein theory this group suggests that gauge symmetry is the symmetry of circular compact dimensions. Once this geometrical interpretation is understood, it is relatively straightforward to replace $U(1)$ by a general Lie group. Such generalizations are often called Yang–Mills theories. If a distinction is drawn, then it is that Yang–Mills theories occur on a flat space-time, whereas Kaluza–Klein treats the more general case of curved spacetime. The base space of Kaluza–Klein theory need not be four-dimensional space-time; it can be any (pseudo-)Riemannian manifold, or even a supersymmetric manifold or orbifold or even a noncommutative space.

As an approach to the unification of the forces, it is straightforward to apply the Kaluza–Klein theory in an attempt to unify gravity with the strong and electroweak forces by using the symmetry group of the Standard Model, SU(3) × SU(2) × U(1). However, an attempt to convert this interesting geometrical construction into a bona-fide model of reality flounders on a number of issues, including the fact that the fermions must be introduced in an artificial way (in nonsupersymmetric models). Nonetheless, KK remains an important touchstone in theoretical physics and is often embedded in more sophisticated theories. It is studied in its own right as an object of geometric interest in K-theory.

Even in the absence of a completely satisfying theoretical physics framework, the idea of exploring extra, compactified, dimensions is of considerable interest in the experimental physics and astrophysics communities. A variety of predictions, with real experimental consequences, can be made (in the case of large extra dimensions/warped models). For example, on the simplest of principles, one might expect to have standing waves in the extra compactified dimension(s). If a spatial extra dimension is of radius R, the invariant mass of such standing waves would be $Mn = nh/Rc$ with n an integer, h being Planck's constant and c the speed of light. This set of possible mass values is often called the **Kaluza–Klein tower**. Similarly, in Thermal quantum field theory a compactification of the euclidean time dimension leads to the Matsubara frequencies and thus to a discretized thermal energy spectrum.

Examples of experimental pursuits include work by the CDF collaboration, which has re-analyzed particle collider data for the signature of effects associated with large extra dimensions/warped models.

Brandenberger and Vafa have speculated that in the early universe, cosmic inflation causes three of the space dimensions to expand to cosmological size while the remaining dimensions of space remained microscopic.

4.8 Space-time-matter theory

One particular variant of Kaluza–Klein theory is **space-time-matter theory** or **induced matter theory**, chiefly promulgated by Paul Wesson and other members of the so-called Space-Time-Matter Consortium.[25] In this version of the theory, it is noted that solutions to the equation

$$\widetilde{R}_{ab} = 0$$

may be re-expressed so that in four dimensions, these solutions satisfy Einstein's equations

$$G_{\mu\nu} = 8\pi T_{\mu\nu}$$

with the precise form of the $T\mu\nu$ following from the Ricci-flat condition on the five-dimensional space. In other words, the cylinder condition of the previous development is dropped, and the stress-energy now comes from the derivatives of the 5D metric with respect to the fifth coordinate. Since the energy–momentum tensor is normally understood to be due to concentrations of matter in four-dimensional space, the above result is interpreted as saying that four-dimensional matter is induced from geometry in five-dimensional space.

In particular, the soliton solutions of $\widetilde{R}_{ab} = 0$ can be shown to contain the Friedmann–Lemaître–Robertson–Walker metric in both radiation-dominated (early universe) and matter-dominated (later universe) forms. The general equations can be shown to be sufficiently consistent with classical tests of general relativity to be acceptable on physical principles, while still leaving considerable freedom to also provide interesting cosmological models.

4.9 Geometric interpretation

The Kaluza–Klein theory is striking because it has a particularly elegant presentation in terms of geometry. In a certain sense, it looks just like ordinary gravity in free space, except that it is phrased in five dimensions instead of four.

4.9.1 The Einstein equations

The equations governing ordinary gravity in free space can be obtained from an action, by applying the variational principle to a certain action. Let M be a (pseudo-)Riemannian manifold, which may be taken as the spacetime of general relativity. If g is the metric on this manifold, one defines the action $S(g)$ as

$$S(g) = \int_M R(g) \text{vol}(g)$$

where $R(g)$ is the scalar curvature and $\text{vol}(g)$ is the volume element. By applying the variational principle to the action

$$\frac{\delta S(g)}{\delta g} = 0$$

one obtains precisely the Einstein equations for free space:

$$R_{ij} - \frac{1}{2} g_{ij} R = 0$$

Here, R_{ij} is the Ricci tensor.

4.9.2 The Maxwell equations

By contrast, the Maxwell equations describing electromagnetism can be understood to be the Hodge equations of a principal U(1)-bundle or circle bundle $\pi\colon P \to M$ with fiber U(1). That is, the electromagnetic field F is a harmonic 2-form in the space $\Omega^2(M)$ of differentiable 2-forms on the manifold M. In the absence of charges and currents, the free-field Maxwell equations are

$$\mathrm{d}F = 0 \text{ and } \mathrm{d}{*}F = 0.$$

where $*$ is the Hodge star.

4.9.3 The Kaluza–Klein geometry

To build the Kaluza–Klein theory, one picks an invariant metric on the circle $\mathbf{S}^1$ that is the fiber of the U(1)-bundle of electromagnetism. In this discussion, an *invariant metric* is simply one that is invariant under rotations of the circle. Suppose this metric gives the circle a total length of Λ. One then considers metrics $\widehat{g}$ on the bundle P that are consistent with both the fiber metric, and the metric on the underlying manifold M. The consistency conditions are:

- The projection of $\widehat{g}$ to the vertical subspace $\mathrm{Vert}_p P \subset T_p P$ needs to agree with metric on the fiber over a point in the manifold M.

- The projection of $\widehat{g}$ to the horizontal subspace $\mathrm{Hor}_p P \subset T_p P$ of the tangent space at point $p \in P$ must be isomorphic to the metric g on M at $\pi(p)$.

The Kaluza–Klein action for such a metric is given by

$$S(\widehat{g}) = \int_P R(\widehat{g}) \operatorname{vol}(\widehat{g})$$

The scalar curvature, written in components, then expands to

$$R(\widehat{g}) = \pi^* \left(R(g) - \frac{\Lambda^2}{2} |F|^2 \right)$$

where π^* is the pullback of the fiber bundle projection $\pi\colon P \to M$. The connection A on the fiber bundle is related to the electromagnetic field strength as

$$\pi^* F = \mathrm{d}A$$

That there always exists such a connection, even for fiber bundles of arbitrarily complex topology, is a result from homology and specifically, K-theory. Applying Fubini's theorem and integrating on the fiber, one gets

$$S(\widehat{g}) = \Lambda \int_M \left(R(g) - \frac{1}{\Lambda^2} |F|^2 \right) \operatorname{vol}(g)$$

Varying the action with respect to the component A, one regains the Maxwell equations. Applying the variational principle to the base metric g, one gets the Einstein equations

$$R_{ij} - \frac{1}{2} g_{ij} R = \frac{1}{\Lambda^2} T_{ij}$$

with the stress–energy tensor being given by

$$T^{ij} = F^{ik}F^{jl}g_{kl} - \frac{1}{4}g^{ij}|F|^2,$$

sometimes called the **Maxwell stress tensor**.

The original theory identifies Λ with the fiber metric g_{55}, and allows Λ to vary from fiber to fiber. In this case, the coupling between gravity and the electromagnetic field is not constant, but has its own dynamical field, the radion.

4.9.4 Generalizations

In the above, the size of the loop Λ acts as a coupling constant between the gravitational field and the electromagnetic field. If the base manifold is four-dimensional, the Kaluza–Klein manifold P is five-dimensional. The fifth dimension is a compact space, and is called the **compact dimension**. The technique of introducing compact dimensions to obtain a higher-dimensional manifold is referred to as compactification. Compactification does not produce group actions on chiral fermions except in very specific cases: the dimension of the total space must be 2 mod 8 and the G-index of the Dirac operator of the compact space must be nonzero.[26]

The above development generalizes in a more-or-less straightforward fashion to general principal G-bundles for some arbitrary Lie group G taking the place of U(1). In such a case, the theory is often referred to as a Yang–Mills theory, and is sometimes taken to be synonymous. If the underlying manifold is supersymmetric, the resulting theory is a super-symmetric Yang–Mills theory.

4.10 Empirical tests

Up to now, no experimental or observational signs of extra dimensions have been officially reported. Many theoretical search techniques for detecting Kaluza–Klein resonances have been proposed using the mass couplings of such resonances with the top quark, however until the Large Hadron Collider (LHC) reaches full operational power observation of such resonances are unlikely. An analysis of results from the LHC in December 2010 severely constrains theories with large extra dimensions.[27]

The observation of a Higgs-like boson at the LHC puts a brand new empirical test in the search for Kaluza–Klein resonances and supersymmetric particles. The loop Feynman diagrams that exist in the Higgs Interactions allow any particle with electric charge and mass to run in such a loop. Standard Model particles besides the top quark and W boson do not make big contributions to the cross-section observed in the H $\to$ γγ decay, but if there are new particles beyond the Standard Model, they could potentially change the ratio of the predicted Standard Model H $\to$ γγ cross-section to the experimentally observed cross-section. Hence a measurement of any dramatic change to the H $\to$ γγ cross section predicted by the Standard Model is crucial in probing the physics beyond it.

4.11 See also

- Classical theories of gravitation
- DGP model
- Randall–Sundrum model
- Supergravity
- Superstring theory
- String theory
- Quantum gravity

4.12 Notes

[1] Pais, Abraham (1982). *Subtle is the Lord ...: The Science and the Life of Albert Einstein*. Oxford: Oxford University Press. pp. 329–330.

[2] Kaluza, Theodor (1921). "Zum Unitätsproblem in der Physik". *Sitzungsber. Preuss. Akad. Wiss. Berlin. (Math. Phys.)*: 966–972.

[3] Klein, Oskar (1926). "Quantentheorie und fünfdimensionale Relativitätstheorie". *Zeitschrift für Physik A* **37** (12): 895–906. Bibcode:1926ZPhy...37..895K. doi:10.1007/BF01397481.

[4] Klein, Oskar (1926). *Nature* **118**: 516.

[5] Goenner, H. (2012). *General Relativity and Gravitation* **44**: 2077.

[6] Lichnerowicz, A.; Thiry, M.Y. (1947). *Compt. Rend. Acad. Sci. Paris* **224**: 529–531.

[7] Thiry, M.Y. (1948). *Compt. Rend. Acad. Sci. Paris* **226**: 216–218.

[8] Thiry, M.Y. (1948). *Compt. Rend. Acad. Sci. Paris* **226**: 1881–1882.

[9] Jordan, P. (1946). *Naturwiss.* **11**: 250–251.

[10] Jordan, P.; Müller, C. (1947). *Z Naturf.* **2a**: 1–2.

[11] Ludwig, G. (1947). *Z Naturf.* **2a**: 3–5.

[12] Jordan, P. (1948). *Astr. Nachr.* **276**: 193–208.

[13] Ludwig, G.; Müller, C. (1948). *Ann. Phys. Leipzig* **2(6)**: 76–84.

[14] Scherrer, W. (1941). *Helv. Phys. Acta.* 14(2): 130.

[15] Scherrer, W. (1949). *Helv. Phys. Acta* **22**: 537–551.

[16] Scherrer, W. (1949). *Helv. Phys. Acta* **23**: 547–555.

[17] Brans, C. H.; Dicke, R. H. (November 1, 1961). "Mach's Principle and a Relativistic Theory of Gravitation". *Physical Review* **124** (3): 925–935. Bibcode:1961PhRv..124..925B. doi:10.1103/PhysRev.124.925.

[18] Appelquist, Thomas; Chodos, Alan; Freund, Peter G. O. (1987). *Modern Kaluza–Klein Theories*. Menlo Park, Cal.: Addison–Wesley. ISBN 0-201-09829-6.

[19] Wesson, Paul S. (1999). *Space-Time-Matter, Modern Kaluza-Klein Theory*. Singapore: World Scientific. ISBN 981-02-3588-7.

[20] Pauli, Wolfgang (1958). *Theory of Relativity* (translated by George Field ed.). New York: Pergamon Press. pp. Supplement 23.

[21] Gross, D.J.; Perry, M.J. (1983). *Nucl. Phys. B* **226**: 29.

[22] Gegenberg, J.; Kunstatter, G. (1984). *Phys. Lett.* **106A**: 410.

[23] Wesson, P.S.; Ponce de Leon, J. (1995). *Astron. & Astrophys.* **294**: 1.

[24] Williams, L.L. (2012). "Physics of the Electromagnetic Control of Spacetime and Gravity". *Proceedings of 48th AIAA Joint Propulsion Conference*. AIAA 2012-3916. doi:10.2514/6.2012-3916.

[25] 5Dstm.org

[26] L. Castellani et al., Supergravity and superstrings, Vol 2, chapter V.11

[27] CMS Collaboration, "Search for Microscopic Black Hole Signatures at the Large Hadron Collider", http://arxiv.org/abs/1012.3375

4.13 References

- Nordström, Gunnar (1914). "Über die Möglichkeit, das elektromagnetische Feld und das Gravitationsfeld zu vereinigen". *Physikalische Zeitschrift* **15**: 504–506. OCLC 1762351.

- Kaluza, Theodor (1921). "Zum Unitätsproblem in der Physik". *Sitzungsber. Preuss. Akad. Wiss. Berlin. (Math. Phys.)*: 966–972. http://archive.org/details/sitzungsberichte1921preussi

- Klein, Oskar (1926). "Quantentheorie und fünfdimensionale Relativitätstheorie". *Zeitschrift für Physik A* **37** (12): 895–906. Bibcode:1926ZPhy...37..895K. doi:10.1007/BF01397481.

- Witten, Edward (1981). "Search for a realistic Kaluza–Klein theory". *Nuclear Physics B* **186** (3): 412–428. Bibcode:1981NuPhB.186..412W. doi:10.1016/0550-3213(81)90021-3.

- Appelquist, Thomas; Chodos, Alan; Freund, Peter G. O. (1987). *Modern Kaluza–Klein Theories*. Menlo Park, Cal.: Addison–Wesley. ISBN 0-201-09829-6. *(Includes reprints of the above articles as well as those of other important papers relating to Kaluza–Klein theory.)*

- Brandenberger, Robert; Vafa, Cumrun (1989). "Superstrings in the early universe". *Nuclear Physics B* **316** (2): 391–410. Bibcode:1989NuPhB.316..391B. doi:10.1016/0550-3213(89)90037-0.

- Duff, M. J. (1994). "Kaluza–Klein Theory in Perspective". In Lindström, Ulf (ed.). *Proceedings of the Symposium 'The Oskar Klein Centenary'*. Singapore: World Scientific. pp. 22–35. ISBN 981-02-2332-3.

- Overduin, J. M.; Wesson, P. S. (1997). "Kaluza–Klein Gravity". *Physics Reports* **283** (5): 303–378. arXiv:gr-qc/9805018. Bibcode:1997PhR...283..303O. doi:10.1016/S0370-1573(96)00046-4.

- Wesson, Paul S. (1999). *Space-Time-Matter, Modern Kaluza-Klein Theory*. Singapore: World Scientific. ISBN 981-02-3588-7.

- Wesson, Paul S. (2006). *Five-Dimensional Physics: Classical and Quantum Consequences of Kaluza-Klein Cosmology*. Singapore: World Scientific. ISBN 981-256-661-9.

- Coquereaux, R.; Esposito-Farese, G. (1990). "The Theory of Kaluza-Klein-Jordan-Thiry revisited". *Annales de l'I.H.P., Section A* **52**: 113–150.

4.14 Further reading

- Grøn, Øyvind; Hervik, Sigbjørn (2007). *Einstein's General Theory of Relativity*. New York: Springer. ISBN 978-0-387-69199-2.

- Kaku, Michio and Robert O'Keefe. *Hyperspace: A Scientific Odyssey Through Parallel Universes, Time Warps, and the Tenth Dimension*. New York: Oxford University Press, 1994. ISBN 0-19-286189-1

- The CDF Collaboration, *Search for Extra Dimensions using Missing Energy at CDF*, (2004) *(A simplified presentation of the search made for extra dimensions at the Collider Detector at Fermilab (CDF) particle physics facility.)*

- John M. Pierre, *SUPERSTRINGS! Extra Dimensions*, (2003).

- TeV scale gravity, mirror universe, and ... dinosaurs Article from Acta Physica Polonica B by Z.K. Silagadze.

- Chris Pope, *Lectures on Kaluza–Klein Theory*.

- Edward Witten (2014). "A Note On Einstein, Bergmann, and the Fifth Dimension", arXiv:1401.8048; pdf

Chapter 5

String theory

For a more accessible and less technical introduction to this topic, see Introduction to M-theory.

In physics, **string theory** is a theoretical framework in which the point-like particles of particle physics are replaced by one-dimensional objects called strings.[1] String theory aims to explain all types of observed elementary particles using quantum states of these strings. In addition to the particles postulated by the standard model of particle physics, string theory naturally incorporates gravity and so is a candidate for a theory of everything, a self-contained mathematical model that describes all fundamental forces and forms of matter. Besides this potential role, string theory is now widely used as a theoretical tool and has shed light on many aspects of quantum field theory and quantum gravity.[2]

The earliest version of string theory, bosonic string theory, incorporated only the class of particles known as bosons. It was then developed into superstring theory, which posits that a connection – a "supersymmetry" – exists between bosons and the class of particles called fermions. String theory requires the existence of extra spatial dimensions for its mathematical consistency. In realistic physical models constructed from string theory, these extra dimensions are typically compactified to extremely small scales.

String theory was first studied in the late 1960s[3] as a theory of the strong nuclear force before being abandoned in favor of the theory of quantum chromodynamics. Subsequently, it was realized that the very properties that made string theory unsuitable as a theory of nuclear physics made it a promising candidate for a quantum theory of gravity. Five consistent versions of string theory were developed until it was realized in the mid-1990s that they were different limits of a conjectured single 11-dimensional theory now known as M-theory.[4]

Many theoretical physicists, including Stephen Hawking, Edward Witten and Juan Maldacena, believe that string theory is a step towards the correct fundamental description of nature: it accommodates a consistent combination of quantum field theory and general relativity, agrees with insights in quantum gravity (such as the holographic principle and black hole thermodynamics) and has passed many non-trivial checks of its internal consistency. According to Hawking, "M-theory is the *only* candidate for a complete theory of the universe."[5] Other physicists, such as Richard Feynman,[6][7] Roger Penrose[8] and Sheldon Lee Glashow,[9] have criticized string theory for not providing novel experimental predictions at accessible energy scales.

5.1 Overview

The starting point for string theory is the idea that the point-like particles of elementary particle physics can also be modeled as one-dimensional objects called *strings*. According to string theory, strings can oscillate in many ways. On distance scales larger than the string radius, each oscillation mode gives rise to a different species of particle, with its mass, charge, and other properties determined by the string's dynamics. Splitting and recombination of strings correspond to particle emission and absorption, giving rise to the interactions between particles. An analogy for strings' modes of vibration is a guitar string's production of multiple distinct musical notes. In this analogy, different notes correspond to different particles.

In string theory, one of the modes of oscillation of the string corresponds to a massless, spin-2 particle. Such a particle is called a graviton since it mediates a force which has the properties of gravity. Since string theory is believed to be a mathematically consistent quantum mechanical theory, the existence of this graviton state implies that string theory is a theory of quantum gravity.

String theory includes both *open* strings, which have two distinct endpoints, and *closed* strings, which form a complete loop. The two types of string behave in slightly different ways, yielding different particle types. For example, all string theories have closed string graviton modes, but only open strings can correspond to the particles known as photons. Because the two ends of an open string can always meet and connect, forming a closed string, all string theories contain closed strings.

The earliest string model, the bosonic string, incorporated only the class of particles known as bosons. This model describes, at low enough energies, a quantum gravity theory, which also includes (if open strings are incorporated as well) gauge bosons such as the photon. However, this model has problems. What is most significant is that the theory has a fundamental instability, believed to result in the decay (at least partially) of spacetime itself. In addition, as the name implies, the spectrum of particles contains only bosons, particles which, like the photon, obey particular rules of behavior. Roughly speaking, bosons are the constituents of radiation, but not of matter, which is made of fermions. Investigating how a string theory may include fermions led to the invention of supersymmetry, a mathematical relation between bosons and fermions. String theories that include fermionic vibrations are now known as superstring theories; several kinds have been described, but all are now thought to be different limits of a theory called M-theory.

Since string theory incorporates all of the fundamental interactions, including gravity, many physicists hope that it fully describes our universe, making it a theory of everything. One of the goals of current research in string theory is to find a solution of the theory that is quantitatively identical with the standard model, with a small cosmological constant, containing dark matter and a plausible mechanism for cosmic inflation. It is not yet known whether string theory has such a solution, nor is it known how much freedom the theory allows to choose the details.

One of the challenges of string theory is that the full theory does not yet have a satisfactory definition in all circumstances. The scattering of strings is most straightforwardly defined using the techniques of perturbation theory, but it is not known in general how to define string theory nonperturbatively. It is also not clear as to whether there is any principle by which string theory selects its vacuum state, the spacetime configuration that determines the properties of our universe (see string theory landscape).

5.1.1 Strings

The motion of a point-like particle can be described by drawing a graph of its position with respect to time. The resulting picture depicts the worldline of the particle in spacetime. In an analogous way, one can draw a graph depicting the progress of a *string* as time passes. The string, which looks like a small line by itself, will sweep out a two-dimensional surface known as the worldsheet. The different string modes (giving rise to different particles, such as the photon or graviton) appear as waves on this surface.

A closed string looks like a small loop, so its worldsheet will look like a pipe. An open string looks like a segment with two endpoints, so its worldsheet will look like a strip. In a more mathematical language, these are both Riemann surfaces, the strip having a boundary and the pipe none.

Strings can join and split. This is reflected by the form of their worldsheet, or more precisely, by its topology. For example, if a closed string splits, its worldsheet will look like a single pipe splitting into two pipes. This topology is often referred to as a *pair of pants* (see drawing at right). If a closed string splits and its two parts later reconnect, its worldsheet will look like a single pipe splitting to two and then reconnecting, which also looks like a torus connected to two pipes (one representing the incoming string, and the other representing the outgoing one). An open string doing the same thing will have a worldsheet that looks like an annulus connected to two strips.

In quantum mechanics, one computes the probability for a point particle to propagate from one point to another by summing certain quantities called probability amplitudes. Each amplitude is associated with a different worldline of the particle. This process of summing amplitudes over all possible worldlines is called path integration. In string theory, one computes probabilities in a similar way, by summing quantities associated with the worldsheets joining an initial string configuration to a final configuration. It is in this sense that string theory extends quantum field theory, replacing point particles by strings. As in quantum field theory, the classical behavior of fields is determined by an action functional, which in string theory can be either the Nambu–Goto action or the Polyakov action.

5.1.2 Branes

Main articles: Brane and D-brane

In string theory and related theories such as supergravity theories, a *brane* is a physical object that generalizes the notion of a point particle to higher dimensions.[10] For example, a point particle can be viewed as a brane of dimension zero, while a string can be viewed as a brane of dimension one. It is also possible to consider higher-dimensional branes. In dimension p, these are called p-branes. The word brane comes from the word "membrane" which refers to a two-dimensional brane.

Branes are dynamical objects which can propagate through spacetime according to the rules of quantum mechanics. They have mass and can have other attributes such as charge. A p-brane sweeps out a ($p+1$)-dimensional volume in spacetime called its *worldvolume*. Physicists often study fields analogous to the electromagnetic field which live on the worldvolume of a brane.

In string theory, D-branes are an important class of branes that arise when one considers open strings. As an open string propagates through spacetime, its endpoints are required to lie on a D-brane. The letter "D" in D-brane refers to the fact that we impose a certain mathematical condition on the system known as the Dirichlet boundary condition. The study of D-branes in string theory has led to important results such as the AdS/CFT correspondence, which has shed light on many problems in quantum field theory.

Branes are also frequently studied from a purely mathematical point of view[11] since they are related to subjects such as homological mirror symmetry and noncommutative geometry. Mathematically, branes may be represented as objects of certain categories, such as the derived category of coherent sheaves on a Calabi–Yau manifold, or the Fukaya category.

5.1.3 Dualities

In physics, the term *duality* refers to a situation where two seemingly different physical systems turn out to be equivalent in a nontrivial way. If two theories are related by a duality, it means that one theory can be transformed in some way so that it ends up looking just like the other theory. The two theories are then said to be *dual* to one another under the transformation. Put differently, the two theories are mathematically different descriptions of the same phenomena.

In addition to providing a candidate for a theory of everything, string theory provides many examples of dualities between different physical theories and can therefore be used as a tool for understanding the relationships between these theories.[12]

S-, T-, and U-duality

Main articles: S-duality, T-duality and U-duality

These are dualities between string theories which relate seemingly different quantities. Large and small distance scales, as well as strong and weak coupling strengths, are quantities that have always marked very distinct limits of behavior of a physical system in both classical and quantum physics. But strings can obscure the difference between large and small, strong and weak, and this is how these five very different theories end up being related. T-duality relates the large and small distance scales between string theories, whereas S-duality relates strong and weak coupling strengths between string theories. U-duality links T-duality and S-duality.

M-theory

Main article: M-theory

Before the 1990s, string theorists believed there were five distinct superstring theories: type I, type IIA, type IIB, and the two flavors of heterotic string theory (SO(32) and $E_8 \times E_8$). The thinking was that out of these five candidate theories, only one was the actual correct theory of everything, and that theory was the one whose low energy limit, with ten spacetime dimensions compactified down to four, matched the physics observed in our world today. It is now believed that this picture was incorrect and that the five superstring theories are related to one another by the dualities described above. The existence of these dualities suggests that the five string theories are in fact special cases of a more fundamental theory called M-theory.[13]

5.1.4 Extra dimensions

Number of dimensions

An intriguing feature of string theory is that it predicts extra dimensions. In classical string theory the number of dimensions is not fixed by any consistency criterion. However, to make a consistent quantum theory, string theory is required to live in a spacetime of the so-called "critical dimension": we must have 26 spacetime dimensions for the bosonic string and 10 for the superstring. This is necessary to ensure the vanishing of the conformal anomaly of the worldsheet conformal field theory. Modern understanding indicates that there exist less trivial ways of satisfying this criterion. Cosmological solutions exist in a wider variety of dimensionalities, and these different dimensions are related by dynamical transitions. The dimensions are more precisely different values of the "effective central charge", a count of degrees of freedom that reduces to dimensionality in weakly curved regimes.[14][15]

One such theory is the 11-dimensional M-theory, which requires spacetime to have eleven dimensions,[16] as opposed to the usual three spatial dimensions and the fourth dimension of time. The original string theories from the 1980s describe special cases of M-theory where the eleventh dimension is a very small circle or a line, and if these formulations are considered as fundamental, then string theory requires ten dimensions. But the theory also describes universes like ours, with four observable spacetime dimensions, as well as universes with up to 10 flat space dimensions, and also cases where the position in some of the dimensions is described by a complex number rather than a real number. The notion of spacetime dimension is not fixed in string theory: it is best thought of as different in different circumstances.[17]

Nothing in Maxwell's theory of electromagnetism or Einstein's theory of relativity makes this kind of prediction; these theories require physicists to insert the number of dimensions manually and arbitrarily, and this number is fixed and independent of potential energy. String theory allows one to relate the number of dimensions to scalar potential energy. In technical terms, this happens because a gauge anomaly exists for every separate number of predicted dimensions, and the gauge anomaly can be counteracted by including nontrivial potential energy into equations to solve motion. Furthermore, the absence of potential energy in the "critical dimension" explains why flat spacetime solutions are possible.

This can be better understood by noting that a photon included in a consistent theory (technically, a particle carrying a force related to an unbroken gauge symmetry) must be massless. The mass of the photon that is predicted by string theory depends on the energy of the string mode that represents the photon. This energy includes a contribution from the Casimir effect, namely from quantum fluctuations in the string. The size of this contribution depends on the number of dimensions, since for a larger number of dimensions there are more possible fluctuations in the string position. Therefore, the photon in flat spacetime will be massless—and the theory consistent—only for a particular number of dimensions.[18] When the calculation is done, the critical dimensionality is not four as one may expect (three axes of space and one of time). The subset of X is equal to the relation of photon fluctuations in a linear dimension. Flat space string theories are 26-dimensional in the bosonic case, while superstring and M-theories turn out to involve 10 or 11 dimensions for flat solutions. In bosonic string theories, the 26 dimensions come from the Polyakov equation.[19] Starting from any dimension greater than four, it is necessary to consider how these are reduced to four-dimensional spacetime.

Compact dimensions

Two ways have been proposed to resolve this apparent contradiction. The first is to compactify the extra dimensions; i.e., the 6 or 7 extra dimensions are so small as to be undetectable by present-day experiments.

To retain a high degree of supersymmetry, these compactification spaces must be very special, as reflected in their holonomy. A 6-dimensional manifold must have SU(3) structure, a particular case (torsionless) of this being SU(3) holonomy, making it a Calabi–Yau space, and a 7-dimensional manifold must have G_2 structure, with G_2 holonomy again being a specific, simple, case. Such spaces have been studied in attempts to relate string theory to the 4-dimensional Standard Model, in part due to the computational simplicity afforded by the assumption of supersymmetry. More recently, progress has been made constructing more realistic compactifications without the degree of symmetry of Calabi–Yau or G2 manifolds.

A standard analogy for this is to consider multidimensional space as a garden hose. If the hose is viewed from sufficient distance, it appears to have only one dimension, its length. Indeed, think of a ball just small enough to enter the hose. Throwing such a ball inside the hose, the ball would move more or less in one dimension; in any experiment we make by throwing such balls in the hose, the only important movement will be one-dimensional, that is, along the hose. However, as one approaches the hose, one discovers that it contains a second dimension, its circumference. Thus, an ant crawling inside it would move in two dimensions (and a fly flying in it would move in three dimensions). This "extra dimension" is only visible within a relatively close range to the hose, or if one "throws in" small enough objects. Similarly, the extra compact dimensions are only "visible" at extremely small distances, or by experimenting with particles with extremely small wavelengths (of the order of the compact dimension's radius), which in quantum mechanics means very high energies (see wave–particle duality).

Brane-world scenario

Another possibility is that we are "stuck" in a 3+1 dimensional (three spatial dimensions plus one time dimension) subspace of the full universe. Properly localized matter and Yang–Mills gauge fields will typically exist if the sub-spacetime is an exceptional set of the larger universe.[20] These "exceptional sets" are ubiquitous in Calabi–Yau n-folds and may be described as subspaces without local deformations, akin to a crease in a sheet of paper or a crack in a crystal, the neighborhood of which is markedly different from the exceptional subspace itself. However, until the work of Randall and Sundrum,[21] it was not known that gravity can be properly localized to a sub-spacetime. In addition, spacetime may be stratified, containing strata of various dimensions, allowing us to inhabit the 3+1-dimensional stratum—such geometries occur naturally in Calabi–Yau compactifications.[22] Such sub-spacetimes are D-branes, hence such models are known as brane-world scenarios.

Effect of the hidden dimensions

In either case, gravity acting in the hidden dimensions affects other non-gravitational forces such as electromagnetism. In fact, Kaluza's early work demonstrated that general relativity in five dimensions actually predicts the existence of electromagnetism. However, because of the nature of Calabi–Yau manifolds, no new forces appear from the small dimensions, but their shape has a profound effect on how the forces between the strings appear in our four-dimensional universe. In principle, therefore, it is possible to deduce the nature of those extra dimensions by requiring consistency with the standard model, but this is not yet a practical possibility. It is also possible to extract information regarding the hidden dimensions by precision tests of gravity, but so far these have only put upper limitations on the size of such hidden dimensions.

5.2 Testability and experimental predictions

Although a great deal of recent work has focused on using string theory to construct realistic models of particle physics, several major difficulties complicate efforts to test models based on string theory. The most significant is the extremely small size of the Planck length, which is expected to be close to the string length (the characteristic size of a string, where strings become easily distinguishable from particles). Another issue is the huge number of metastable vacua of string theory, which might be sufficiently diverse to accommodate almost any phenomena we might observe at lower energies.

5.2.1 String harmonics

One unique prediction of string theory is the existence of *string harmonics*. At sufficiently high energies, the string-like nature of particles would become obvious. There should be heavier copies of all particles, corresponding to higher vibrational harmonics of the string. It is not clear how high these energies are. In most conventional string models, they would be close to the Planck energy, which is 10^{14} times higher than the energies accessible in the newest particle accelerator, the LHC, making this prediction impossible to test with any particle accelerator in the near future. However, in models with large extra dimensions they could potentially be produced at the LHC, or at energies not far above its reach.

5.2.2 Cosmology

String theory as currently understood makes a series of predictions for the structure of the universe at the largest scales. Many phases in string theory have very large, positive vacuum energy.[23] Regions of the universe that are in such a phase will inflate exponentially rapidly in a process known as eternal inflation. As such, the theory predicts that most of the universe is very rapidly expanding. However, these expanding phases are not stable, and can decay via the nucleation of bubbles of lower vacuum energy. Since our local region of the universe is not very rapidly expanding, string theory predicts we are inside such a bubble. The spatial curvature of the "universe" inside the bubbles that form by this process is negative, a testable prediction.[24] Moreover, other bubbles will eventually form in the parent vacuum outside the bubble and collide with it. These collisions lead to potentially observable imprints on cosmology.[25] However, it is possible that neither of these will be observed if the spatial curvature is too small and the collisions are too rare.

Under certain circumstances, fundamental strings produced at or near the end of inflation can be "stretched" to astronomical proportions. These cosmic strings could be observed in various ways, for instance by their gravitational lensing effects. However, certain field theories also predict cosmic strings arising from topological defects in the field configuration.[26]

5.2.3 Supersymmetry

Main article: Supersymmetry

If confirmed experimentally, supersymmetry is often considered circumstantial evidence, because most consistent string theories are space-time supersymmetric. As with other physical theories, the existence of space-time supersymmetry is a desired feature addressing various issues we encounter in non-supersymmetric theories, like in the Standard Model. However, the absence of supersymmetric particles at energies accessible to the LHC will not actually disprove string theory, since the energy scale at which supersymmetry is broken could be well above the accelerator's range. This would make supersymmetric particles too heavy to be produced in relatively lower energies. On the other hand, there are fully consistent non-supersymmetric string-theories that can also provide phenomenologically relevant predictions.

5.3 AdS/CFT correspondence

Main article: AdS/CFT correspondence

The anti-de Sitter/conformal field theory (AdS/CFT) correspondence is a relationship which says that string theory is in certain cases equivalent to a quantum field theory. More precisely, one considers string or M-theory on an anti-de Sitter background. This means that the geometry of spacetime is obtained by perturbing a certain solution of Einstein's equation in the vacuum. In this setting, it is possible to define a notion of "boundary" of spacetime. The AdS/CFT correspondence states that this boundary can be regarded as the "spacetime" for a quantum field theory, and this field theory is equivalent to the bulk gravitational theory in the sense that there is a "dictionary" for translating calculations in one theory into calculations in the other.

5.3.1 Examples of the correspondence

The most famous example of the AdS/CFT correspondence states that Type IIB string theory on the product $\mathbf{AdS}_5 \times \mathbf{S}^5$ is equivalent to $N = 4$ super Yang–Mills theory on the four-dimensional conformal boundary.[27][28][29][30] Another realization of the correspondence states that M-theory on $\mathbf{AdS}_4 \times \mathbf{S}^7$ is equivalent to the ABJM superconformal field theory in three dimensions.[31] Yet another realization states that M-theory on $\mathbf{AdS}_7 \times \mathbf{S}^4$ is equivalent to the so-called (2,0)-theory in six dimensions.[32]

5.3.2 Applications to quantum chromodynamics

Main article: AdS/QCD

Since it relates string theory to ordinary quantum field theory, the AdS/CFT correspondence can be used as a theoretical tool for doing calculations in quantum field theory. For example, the correspondence has been used to study the quark–gluon plasma, an exotic state of matter produced in particle accelerators.

The physics of the quark–gluon plasma is governed by quantum chromodynamics, the fundamental theory of the strong nuclear force, but this theory is mathematically intractable in problems involving the quark–gluon plasma. In order to understand certain properties of the quark–gluon plasma, theorists have therefore made use of the AdS/CFT correspondence. One version of this correspondence relates string theory to a certain supersymmetric gauge theory called $N = 4$ super Yang–Mills theory. The latter theory provides a good approximation to quantum chromodynamics. One can thus translate problems involving the quark–gluon plasma into problems in string theory which are more tractable. Using these methods, theorists have computed the shear viscosity of the quark–gluon plasma.[33] In 2008, these predictions were confirmed at the Relativistic Heavy Ion Collider at Brookhaven National Laboratory.[34]

5.3.3 Applications to condensed matter physics

In addition, string theory methods have been applied to problems in condensed matter physics. Certain condensed matter systems are difficult to understand using the usual methods of quantum field theory, and the AdS/CFT correspondence may allow physicists to better understand these systems by describing them in the language of string theory. Some success has been achieved in using string theory methods to describe the transition of a superfluid to an insulator.[35][36]

5.4 Connections to mathematics

In addition to influencing research in theoretical physics, string theory has stimulated a number of major developments in pure mathematics. Like many developing ideas in theoretical physics, string theory does not at present have a mathematically rigorous formulation in which all of its concepts can be defined precisely. As a result, physicists who study string theory are often guided by physical intuition to conjecture relationships between the seemingly different mathematical structures that are used to formalize different parts of the theory. These conjectures are later proved by mathematicians, and in this way, string theory has served as a source of new ideas in pure mathematics.[37]

5.4.1 Mirror symmetry

Main article: Mirror symmetry (string theory)

One of the ways in which string theory influenced mathematics was through the discovery of mirror symmetry. In string theory, the shape of the unobserved spatial dimensions is typically encoded in mathematical objects called Calabi–Yau manifolds. These are of interest in pure mathematics, and they can be used to construct realistic models of physics from string theory. In the late 1980s, it was noticed that given such a physical model, it is not possible to uniquely reconstruct a corresponding Calabi–Yau manifold. Instead, one finds that there are *two* Calabi–Yau manifolds that give rise to the same physics. These manifolds are said to be "mirror" to one another. The existence of this mirror symmetry relationship between different Calabi–Yau manifolds has significant mathematical consequences as it allows mathematicians to solve many problems in enumerative algebraic geometry. Today mathematicians are still working to develop a mathematical understanding of mirror symmetry based on physicists' intuition.[38]

5.4.2 Vertex operator algebras

Main articles: Vertex operator algebra and Monstrous moonshine

In addition to mirror symmetry, applications of string theory to pure mathematics include results in the theory of vertex operator algebras. For example, ideas from string theory were used by Richard Borcherds in 1992 to prove the monstrous moonshine conjecture relating the monster group (a construction arising in group theory, a branch of algebra) and modular functions (a class of functions which are important in number theory).[39]

5.5 History

Main article: History of string theory

5.5.1 Early results

Some of the structures reintroduced by string theory arose for the first time much earlier as part of the program of classical unification started by Albert Einstein. The first person to add a fifth dimension to a theory of gravity was Gunnar Nordström in 1914, who noted that gravity in five dimensions describes both gravity and electromagnetism in four. Nordström attempted to unify electromagnetism with his theory of gravitation, which was however superseded by Einstein's general relativity in 1919.[40] Thereafter, German mathematician Theodor Kaluza combined the fifth dimension with general relativity, and only Kaluza is usually credited with the idea.[40] In 1926, the Swedish physicist Oskar Klein gave a physical interpretation of the unobservable extra dimension—it is wrapped into a small circle. Einstein introduced a non-symmetric metric tensor, while much later Brans and Dicke added a scalar component to gravity. These ideas would be revived within string theory, where they are demanded by consistency conditions.

String theory was originally developed during the late 1960s and early 1970s as a never completely successful theory of hadrons, the subatomic particles like the proton and neutron that feel the strong interaction. In the 1960s, Geoffrey Chew and Steven Frautschi discovered that the mesons make families called Regge trajectories with masses related to spins in a way that was later understood by Yoichiro Nambu, Holger Bech Nielsen and Leonard Susskind to be the relationship expected from rotating strings. Chew advocated making a theory for the interactions of these trajectories that did not presume that they were composed of any fundamental particles, but would construct their interactions from self-consistency conditions on the S-matrix. The S-matrix approach was started by Werner Heisenberg in the 1940s as a way of constructing a theory that did not rely on the local notions of space and time, which Heisenberg believed break down at the nuclear scale. While the scale was off by many orders of magnitude, the approach he advocated was ideally suited for a theory of quantum gravity.

Working with experimental data, R. Dolen, D. Horn and C. Schmid[41] developed some sum rules for hadron exchange. When a particle and antiparticle scatter, virtual particles can be exchanged in two qualitatively different ways. In the s-channel, the two particles annihilate to make temporary intermediate states that fall apart into the final state particles. In the t-channel, the particles exchange intermediate states by emission and absorption. In field theory, the two contributions add together, one giving a continuous background contribution, the other giving peaks at certain energies. In the data, it was clear that the peaks were stealing from the background—the authors interpreted this as saying that the t-channel contribution was dual to the s-channel one, meaning both described the whole amplitude and included the other.

The result was widely advertised by Murray Gell-Mann, leading Gabriele Veneziano to construct a scattering amplitude that had the property of Dolen-Horn-Schmid duality, later renamed world-sheet duality. The amplitude needed poles where the particles appear, on straight line trajectories, and there is a special mathematical function whose poles are evenly spaced on half the real line— the Gamma function— which was widely used in Regge theory. By manipulating combinations of Gamma functions, Veneziano was able to find a consistent scattering amplitude with poles on straight lines, with mostly positive residues, which obeyed duality and had the appropriate Regge scaling at high energy. The amplitude could fit near-beam scattering data as well as other Regge type fits, and had a suggestive integral representation that could be used for generalization.

Over the next years, hundreds of physicists worked to complete the bootstrap program for this model, with many surprises. Veneziano himself discovered that for the scattering amplitude to describe the scattering of a particle that appears in the theory, an obvious self-consistency condition, the lightest particle must be a tachyon. Miguel Virasoro and Joel Shapiro found a different amplitude now understood to be that of closed strings, while Ziro Koba and Holger Nielsen generalized Veneziano's integral representation to multiparticle scattering. Veneziano and Sergio Fubini introduced an operator formalism for computing the scattering amplitudes that was a forerunner of world-sheet conformal theory, while Virasoro understood how to remove the poles with wrong-sign residues using a constraint on the states. Claud Lovelace calculated a loop amplitude, and noted that there is an inconsistency unless the dimension of the theory is 26. Charles Thorn, Peter Goddard and Richard Brower went on to prove that there are no wrong-sign propagating states in dimensions less than or equal to 26.

In 1969, Yoichiro Nambu, Holger Bech Nielsen, and Leonard Susskind recognized that the theory could be given a description in space and time in terms of strings. The scattering amplitudes were derived systematically from the action principle by Peter Goddard, Jeffrey Goldstone, Claudio Rebbi, and Charles Thorn, giving a space-time picture to the vertex operators introduced by Veneziano and Fubini and a geometrical interpretation to the Virasoro conditions.

In 1970, Pierre Ramond added fermions to the model, which led him to formulate a two-dimensional supersymmetry to cancel the wrong-sign states. John Schwarz and André Neveu added another sector to the fermi theory a short time later. In the fermion theories, the critical dimension was 10. Stanley Mandelstam formulated a world sheet conformal theory for both the bose and fermi case, giving a two-dimensional field theoretic path-integral to generate the operator formalism. Michio Kaku and Keiji Kikkawa gave a different formulation of the bosonic string, as a string field theory, with infinitely many particle types and with fields taking values not on points, but on loops and curves.

In 1974, Tamiaki Yoneya discovered that all the known string theories included a massless spin-two particle that obeyed the correct Ward identities to be a graviton. John Schwarz and Joel Scherk came to the same conclusion and made the bold leap to suggest that string theory was a theory of gravity, not a theory of hadrons. They reintroduced Kaluza–Klein theory as a way of making sense of the extra dimensions. At the same time, quantum chromodynamics was recognized as the correct theory of hadrons, shifting the attention of physicists and apparently leaving the bootstrap program in the dustbin of history.

String theory eventually made it out of the dustbin, but for the following decade all work on the theory was completely ignored. Still, the theory continued to develop at a steady pace thanks to the work of a handful of devotees. Ferdinando Gliozzi, Joel Scherk, and David Olive realized in 1976 that the original Ramond and Neveu Schwarz-strings were separately inconsistent and needed to be combined. The resulting theory did not have a tachyon, and was proven to have space-time supersymmetry by John Schwarz and Michael Green in 1981. The same year, Alexander Polyakov gave the theory a modern path integral formulation, and went on to develop conformal field theory extensively. In 1979, Daniel Friedan showed that the equations of motions of string theory, which are generalizations of the Einstein equations of General Relativity, emerge from the Renormalization group equations for the two-dimensional field theory. Schwarz and Green discovered T-duality, and constructed two superstring theories—IIA and IIB related by T-duality, and type I theories with open strings. The consistency conditions had been so strong, that the entire theory was nearly uniquely determined, with only a few discrete choices.

5.5.2 First superstring revolution

In the early 1980s, Edward Witten discovered that most theories of quantum gravity could not accommodate chiral fermions like the neutrino. This led him, in collaboration with Luis Alvarez-Gaumé to study violations of the conservation laws in gravity theories with anomalies, concluding that type I string theories were inconsistent. Green and Schwarz discovered a contribution to the anomaly that Witten and Alvarez-Gaumé had missed, which restricted the gauge group of the type I string theory to be SO(32). In coming to understand this calculation, Edward Witten became convinced that string theory was truly a consistent theory of gravity, and he became a high-profile advocate. Following Witten's lead, between 1984 and 1986, hundreds of physicists started to work in this field, and this is sometimes called the first superstring revolution.

During this period, David Gross, Jeffrey Harvey, Emil Martinec, and Ryan Rohm discovered heterotic strings. The gauge group of these closed strings was two copies of E8, and either copy could easily and naturally include the standard model. Philip Candelas, Gary Horowitz, Andrew Strominger and Edward Witten found that the Calabi–Yau manifolds are the compactifications that preserve a realistic amount of supersymmetry, while Lance Dixon and others worked out the physical properties of orbifolds, distinctive geometrical singularities allowed in string theory. Cumrun Vafa generalized T-duality from circles to arbitrary manifolds, creating the mathematical field of mirror symmetry. Daniel Friedan, Emil Martinec and Stephen Shenker further developed the covariant quantization of the superstring using conformal field theory techniques. David Gross and Vipul Periwal discovered that string perturbation theory was divergent. Stephen Shenker showed it diverged much faster than in field theory suggesting that new non-perturbative objects were missing.

In the 1990s, Joseph Polchinski discovered that the theory requires higher-dimensional objects, called D-branes and identified these with the black-hole solutions of supergravity. These were understood to be the new objects suggested by the perturbative divergences, and they opened up a new field with rich mathematical structure. It quickly became clear that D-branes and other p-branes, not just strings, formed the matter content of the string theories, and the physical interpretation of the strings and branes was revealed—they are a type of black hole. Leonard Susskind had incorporated the holographic principle of Gerardus 't Hooft into string theory, identifying the long highly excited string states with ordinary thermal black hole states. As suggested by 't Hooft, the fluctuations of the black hole horizon, the world-sheet or world-volume theory, describes not only the degrees of freedom of the black hole, but all nearby objects too.

5.5.3 Second superstring revolution

In 1995, at the annual conference of string theorists at the University of Southern California (USC), Edward Witten gave a speech on string theory that in essence united the five string theories that existed at the time, and giving birth to a new 11-dimensional theory called M-theory. M-theory was also foreshadowed in the work of Paul Townsend at approximately the same time. The flurry of activity that began at this time is sometimes called the second superstring revolution.

During this period, Tom Banks, Willy Fischler, Stephen Shenker and Leonard Susskind formulated matrix theory, a full holographic description of M-theory using IIA D0 branes.[42] This was the first definition of string theory that was fully non-perturbative and a concrete mathematical realization of the holographic principle. It is an example of a gauge-gravity duality and is now understood to be a special case of the AdS/CFT correspondence. Andrew Strominger and Cumrun Vafa calculated the entropy of certain configurations of D-branes and found agreement with the semi-classical answer for extreme charged black holes. Petr Hořava and Witten found the eleven-dimensional formulation of the heterotic string theories, showing that orbifolds solve the chirality problem. Witten noted that the effective description of the physics of D-branes at low energies is by a supersymmetric gauge theory, and found geometrical interpretations of mathematical structures in gauge theory that he and Nathan Seiberg had earlier discovered in terms of the location of the branes.

In 1997, Juan Maldacena noted that the low energy excitations of a theory near a black hole consist of objects close to the horizon, which for extreme charged black holes looks like an anti-de Sitter space. He noted that in this limit the gauge theory describes the string excitations near the branes. So he hypothesized that string theory on a near-horizon extreme-charged black-hole geometry, an anti-deSitter space times a sphere with flux, is equally well described by the low-energy limiting gauge theory, the $N = 4$ supersymmetric Yang–Mills theory. This hypothesis, which is called the AdS/CFT correspondence, was further developed by Steven Gubser, Igor Klebanov and Alexander Polyakov, and by Edward Witten, and it is now well-accepted. It is a concrete realization of the holographic principle, which has far-reaching implications for black holes, locality and information in physics, as well as the nature of the gravitational interaction. Through this relationship, string theory has been shown to be related to gauge theories like quantum chromodynamics and this has led to more quantitative understanding of the behavior of hadrons, bringing string theory back to its roots.

5.6 Criticisms

Some critics of string theory say that it is a failure as a theory of everything.[43][44][45][46][47][48] Notable critics include Peter Woit, Lee Smolin, Philip Warren Anderson,[49] Sheldon Glashow,[50] Lawrence Krauss,[51] Carlo Rovelli[52] and Bert Schroer.[53] Some common criticisms include:

1. Very high energies needed to test quantum gravity.
2. Lack of uniqueness of predictions due to the large number of solutions.
3. Lack of background independence.

5.6.1 High energies

It is widely believed that any theory of quantum gravity would require extremely high energies to probe directly, higher by orders of magnitude than those that current experiments such as the Large Hadron Collider[54] can attain. This is because strings themselves are expected to be only slightly larger than the Planck length, which is twenty orders of magnitude smaller than the radius of a proton, and high energies are required to probe small length scales. Generally speaking, quantum gravity is difficult to test because gravity is much weaker than the other forces, and because quantum effects are controlled by Planck's constant h, a very small quantity. As a result, the effects of quantum gravity are extremely weak.

5.6.2 Number of solutions

String theory as it is currently understood has a huge number of solutions, called string vacua,[23] and these vacua might be sufficiently diverse to accommodate almost any phenomena we might observe at lower energies.

The vacuum structure of the theory, called the string theory landscape (or the anthropic portion of string theory vacua), is not well understood. String theory contains an infinite number of distinct meta-stable vacua, and perhaps 10^{520} of these or more correspond to a universe roughly similar to ours—with four dimensions, a high planck scale, gauge groups, and chiral fermions. Each of these corresponds to a different possible universe, with a different collection of particles and forces.[23] What principle, if any, can be used to select among these vacua is an open issue. While there are no continuous parameters in the theory, there is a very large set of possible universes, which may be radically different from each other. It is also suggested that the landscape is surrounded by an even more vast swampland of consistent-looking semiclassical effective field theories, which are actually inconsistent.[55]

Some physicists believe this is a good thing, because it may allow a natural anthropic explanation of the observed values of physical constants, in particular the small value of the cosmological constant.[56][57] The argument is that most universes contain values for physical constants that do not lead to habitable universes (at least for humans), and so we happen to live in the "friendliest" universe. This principle is already employed to explain the existence of life on earth as the result of a life-friendly orbit around the medium-sized sun among an infinite number of possible orbits (as well as a relatively stable location in the galaxy).

5.6.3 Background independence

See also: Background independence

A separate and older criticism of string theory is that it is background-dependent—string theory describes perturbative expansions about fixed spacetime backgrounds which means that mathematical calculations in the theory rely on preselecting a background as a starting point. This is because, like many quantum field theories, much of string theory is still only formulated perturbatively, as a divergent series of approximations.

Although the theory, defined as a perturbative expansion on a fixed background, is not background independent, it has some features that suggest non-perturbative approaches would be background-independent—topology change is an established process in string theory, and the exchange of gravitons is equivalent to a change in the background. Since there are dynamic corrections to the background spacetime in the perturbative theory, one would expect spacetime to be dynamic in the nonperturbative theory as well since they would have to predict the same spacetime.

This criticism has been addressed to some extent by the AdS/CFT duality, which is believed to provide a full, non-perturbative definition of string theory in spacetimes with anti-de Sitter space asymptotics. Nevertheless, a non-perturbative definition of the theory in arbitrary spacetime backgrounds is still lacking. Some hope that M-theory, or a non-perturbative treatment of string theory (such as "background independent open string field theory") will have a background-independent formulation.

5.7 See also

- Conformal field theory
- Glossary of string theory
- List of string theory topics
- Loop quantum gravity
- Supergravity
- Supersymmetry

5.8 References

[1] Sean Carroll, Ph.D., Caltech, 2007, The Teaching Company, *Dark Matter, Dark Energy: The Dark Side of the Universe*, Guidebook Part 2 page 59, Accessed Oct. 7, 2013, "...The idea that the elementary constituents of matter are small loops of string rather than pointlike particles ... we think of string theory as a candidate theory of quantum gravity..."

[2] Klebanov, Igor and Maldacena, Juan (2009). "Solving Quantum Field Theories via Curved Spacetimes" (PDF). *Physics Today* **62**: 28. Bibcode:2009PhT....62a..28K. doi:10.1063/1.3074260. Retrieved May 2013.

[3] http://superstringtheory.com/history/history4.html

[4] Schwarz, John H. (1999). "From Superstrings to M Theory". *Physics Reports* **315**: 107. arXiv:hep-th/9807135. Bibcode:1999PhR...315..107 doi:10.1016/S0370-1573(99)00016-2.

[5] Hawking, Stephen (2010). *The Grand Design*. Bantam Books. ISBN 055338466X.

[6] Woit, Peter (2006). *Not Even Wrong: The Failure of String Theory and the Search for Unity in Physical Law*. London: Jonathan Cape: New York: Basic Books. p. 174. ISBN 0-465-09275-6.

[7] P.C.W. Davies and J. Brown (ed), *Superstrings, A Theory of Everything?*, Cambridge University Press, 1988 (ISBN 0-521-35741-1).

[8] Penrose, Roger (2005). *The Road to Reality: A Complete Guide to the Laws of the Universe*. Knopf. ISBN 0-679-45443-8.

[9] Sheldon Glashow. "NOVA – The elegant Universe". Pbs.org. Retrieved on 2012-07-11.

[10] Moore, Gregory (2005). "What is... a Brane?" (PDF). *Notices of the AMS* **52**: 214. Retrieved June 2013.

[11] Aspinwall, Paul; Bridgeland, Tom; Craw, Alastair; Douglas, Michael; Gross, Mark; Kapustin, Anton; Moore, Gregory; Segal, Graeme; Szendrői, Balázs; Wilson, P.M.H., eds. (2009). *Dirichlet Branes and Mirror Symmetry*. American Mathematical Society.

[12] Duality in string theory in *nLab*

[13] Witten, Edward (1995). "String theory dynamics in various dimensions". *Nuclear Physics B* **443** (1): 85–126. arXiv:hep-th/9503124. Bibcode:1995NuPhB.443...85W. doi:10.1016/0550-3213(95)00158-O.

[14] Hellerman, Simeon; Swanson, Ian (2007). "Dimension-changing exact solutions of string theory". *Journal of High Energy Physics* **2007** (9): 096. arXiv:hep-th/0612051v3. Bibcode:2007JHEP...09..096H. doi:10.1088/1126-6708/2007/09/096.

[15] Aharony, Ofer; Silverstein, Eva (2007). "Supercritical stability, transitions, and (pseudo)tachyons". *Physical Review D* **75** (4). arXiv:hep-th/0612031v2. Bibcode:2007PhRvD..75d6003A. doi:10.1103/PhysRevD.75.046003.

[16] Duff, M. J.; Liu, James T. and Minasian, R. (1995). "Eleven Dimensional Origin of String/String Duality: A One Loop Test". *Nuclear Physics B* **452**: 261. arXiv:hep-th/9506126v2. Bibcode:1995NuPhB.452..261D. doi:10.1016/0550-3213(95)00368-3.

[17] Polchinski, Joseph (1998). *String Theory*, Cambridge University Press ISBN 0521672295.

[18] The calculation of the number of dimensions can be circumvented by adding a degree of freedom, which compensates for the "missing" quantum fluctuations. However, this degree of freedom behaves similar to spacetime dimensions only in some aspects, and the produced theory is not Lorentz invariant, and has other characteristics that do not appear in nature. This is known as the *linear dilaton* or non-critical string.

[19] Botelho, Luiz C. L. and Botelho, Raimundo C. L. (1999) "Quantum Geometry of Bosonic Strings – Revisited". Centro Brasileiro de Pesquisas Físicas.

[20] Hübsch, T. (1997). "A Hitchhiker's Guide to Superstring Jump Gates and Other Worlds". *Nuclear Physics B – Proceedings Supplements* **52**: 347. Bibcode:1997NuPhS..52..347H. doi:10.1016/S0920-5632(96)00589-0.

[21] Randall, Lisa (1999). "An Alternative to Compactification". *Physical Review Letters* **83** (23): 4690. arXiv:hep-th/9906064. Bibcode:1999PhRvL..83.4690R. doi:10.1103/PhysRevLett.83.4690.

[22] Aspinwall, Paul S.; Greene, Brian R.; Morrison, David R. (1994). "Calabi-Yau moduli space, mirror manifolds and space-time topology change in string theory". *Nuclear Physics B* **416** (2): 414. arXiv:hep-th/9309097. Bibcode:1994NuPhB.416..414A. doi:10.1016/0550-3213(94)90321-2.

[23] Kachru, Shamit; Kallosh, Renata; Linde, Andrei; Trivedi, Sandip (2003). "De Sitter vacua in string theory". *Physical Review D* **68** (4). arXiv:hep-th/0301240. Bibcode:2003PhRvD..68d6005K. doi:10.1103/PhysRevD.68.046005.

[24] Freivogel, Ben; Kleban, Matthew; Martínez, María Rodríguez; Susskind, Leonard (2006). "Observational consequences of a landscape". *Journal of High Energy Physics* **2006** (3): 039. arXiv:hep-th/0505232. Bibcode:2006JHEP...03..039F. doi:10.1088/1126-6708/2006/03/039.

[25] Kleban, Matthew; Levi, Thomas S.; Sigurdson, Kris (2013). "Observing the multiverse with cosmic wakes". *Physical Review D* **87** (4). arXiv:1109.3473. Bibcode:2013PhRvD..87d1301K. doi:10.1103/PhysRevD.87.041301.

[26] Polchinski, Joseph (2004). "Introduction to Cosmic F- and D-Strings". arXiv:hep-th/0412244 [hep-th].

[27] Maldacena, J. *The Large N Limit of Superconformal Field Theories and Supergravity*, arXiv:hep-th/9711200

5.8. REFERENCES

[28] Gubser, S. S.; Klebanov, I. R. and Polyakov, A. M. (1998). "Gauge theory correlators from non-critical string theory". *Physics Letters* **B428**: 105–114. arXiv:hep-th/9802109. Bibcode:1998PhLB..428..105G. doi:10.1016/S0370-2693(98)00377-3.

[29] Edward Witten (1998). "Anti-de Sitter space and holography". *Advances in Theoretical and Mathematical Physics* **2**: 253–291. arXiv:hep-th/9802150. Bibcode:1998hep.th....2150W.

[30] Aharony, O.; S.S. Gubser, J. Maldacena, H. Ooguri, Y. Oz (2000). "Large N Field Theories, String Theory and Gravity". *Phys. Rept.* **323** (3–4): 183–386. arXiv:hep-th/9905111. Bibcode:1999PhR...323..183A. doi:10.1016/S0370-1573(99)00083-6.

[31] Aharony, Ofer; Bergman, Oren; Jafferis, Daniel Louis; Maldacena, Juan (2008). "$N = 6$ superconformal Chern-Simons-matter theories, M2-branes and their gravity duals". *Journal of High Energy Physics* **2008** (10): 091. arXiv:0806.1218. Bibcode:2008JHEP...10..091A. doi:10.1088/1126-6708/2008/10/091.

[32] 6d (2,0)-supersymmetric QFT in *nLab*

[33] Kovtun, P. K.; Son, Dam T.; Starinets, A. O. (2001). "Viscosity in strongly interacting quantum field theories from black hole physics". *Physical review letters* **94** (11).

[34] Luzum, Matthew; Romatschke, Paul (2008). "Conformal relativistic viscous hydrodynamics: Applications to RHIC results at sqrt [s_ {NN}]= 200 GeV". *Physical Review C* **78** (3). arXiv:0804.4015. Bibcode:2008PhRvC..78c4915L. doi:10.1103/PhysRevC.78.034915.

[35] Merali, Zeeya (2011). "Collaborative physics: string theory finds a bench mate". *Nature* **478** (7369): 302–304. Bibcode:2011Natur.478..302M. doi:10.1038/478302a. PMID 22012369.

[36] Sachdev, Subir (2013). "Strange and stringy". *Scientific American* **308** (44): 44. Bibcode:2012SciAm.308a..44S. doi:10.1038/scientificamerican 44.

[37] Deligne, Pierre; Etingof, Pavel; Freed, Daniel; Jeffery, Lisa; Kazhdan, David; Morgan, John; Morrison, David; Witten, Edward, eds. (1999). *Quantum Fields and Strings: A Course for Mathematicians* **1**. American Mathematical Society. p. 1. ISBN 0821820125.

[38] Hori, Kentaro; Katz, Sheldon; Klemm, Albrecht; Pandharipande, Rahul; Thomas, Richard; Vafa, Cumrun; Vakil, Ravi; Zaslow, Eric, eds. (2003). *Mirror Symmetry*. American Mathematical Society. ISBN 0821829556.

[39] Frenkel, Igor; Lepowsky, James; Meurman, Arne (1988). *Vertex operator algebras and the Monster*. Pure and Applied Mathematics **134**. Boston: Academic Press. ISBN 0-12-267065-5.

[40] http://www.hs.fi/tiede/Suomalaistutkija+kilpailii+Einsteinin+kanssa+ja+keksi+viidennen+ulottuvuuden/a1410754065152

[41] Dolen, R.; Horn, D.; Schmid, C. (1968). "Finite-Energy Sum Rules and Their Application to πN Charge Exchange". *Physical Review* **166** (5): 1768. Bibcode:1968PhRv..166.1768D. doi:10.1103/PhysRev.166.1768.

[42] Banks, T.; Fischler, W.; Shenker, S. H.; Susskind, L. (1997). "M theory as a matrix model: A conjecture". *Physical Review D* **55** (8): 5112. arXiv:hep-th/9610043v3. Bibcode:1997PhRvD..55.5112B. doi:10.1103/PhysRevD.55.5112.

[43] Woit, Peter Not Even Wrong. Math.columbia.edu. Retrieved on 2012-07-11.

[44] Smolin, Lee. The Trouble With Physics. Thetroublewithphysics.com. Retrieved on 2012-07-11.

[45] The n-Category Cafe. Golem.ph.utexas.edu (2007-02-25). Retrieved on 2012-07-11.

[46] John Baez weblog. Math.ucr.edu (2007-02-25). Retrieved on 2012-07-11.

[47] Woit, P. (Columbia University), *String theory: An Evaluation*, February 2001, arXiv:physics/0102051

[48] Woit, P. Is String Theory Testable? INFN Rome March 2007

[49] God (or Not), Physics and, of Course, Love: Scientists Take a Leap, New York Times, 4 January 2005: "String theory is the first science in hundreds of years to be pursued in pre-Baconian fashion, without any adequate experimental guidance"

[50] "there ain't no experiment that could be done nor is there any observation that could be made that would say, `You guys are wrong.' The theory is safe, permanently safe" NOVA interview

[51] Krauss, Lawrence (8 November 2005) Science and Religion Share Fascination in Things Unseen. *New York Times*: "String theory [is] yet to have any real successes in explaining or predicting anything measurable".

[52] Rovelli, Carlo (2003). "A Dialog on Quantum Gravity". *International Journal of Modern Physics D [Gravitation; Astrophysics and Cosmology]* **12** (9): 1509. arXiv:hep-th/0310077. Bibcode:2003IJMPD..12.1509R. doi:10.1142/S0218271803004304.

[53] Schroer, B. (2008) *String theory and the crisis of particle physics II or the ascent of metaphoric arguments*, arXiv:0805.1911

[54] Kiritsis, Elias (2007) *String Theory in a Nutshell*, Princeton University Press, ISBN 1400839335.

[55] Vafa, Cumrun (2005). "The String landscape and the swampland". arXiv:hep-th/0509212.

[56] Arkani-Hamed, N.; Dimopoulos, S. and Kachru, S. *Predictive Landscapes and New Physics at a TeV*, arXiv:hep-th/0501082, SLAC-PUB-10928, HUTP-05-A0001, SU-ITP-04-44, January 2005

[57] Susskind, L. *The Anthropic Landscape of String Theory*, arXiv:hep-th/0302219, February 2003

5.9 Further reading

5.9.1 Popular books

General

- Davies, Paul; Julian R. Brown (Eds.) (1992). *Superstrings: A Theory of Everything?*. Cambridge: Cambridge University Press. ISBN 0-521-43775-X.

- Greene, Brian (2003). *The Elegant Universe: Superstrings, Hidden Dimensions, and the Quest for the Ultimate Theory*. New York: W.W. Norton & Company. ISBN 0-393-05858-1.

- Greene, Brian (2004). *The Fabric of the Cosmos: Space, Time, and the Texture of Reality*. New York: Alfred A. Knopf. ISBN 0-375-41288-3.

- Kaku, Michio (1994). *Hyperspace: A Scientific Odyssey Through Parallel Universes, Time Warps, and the Tenth Dimension*. Oxford: Oxford University Press. ISBN 0-19-508514-0.

- Musser, George (2008). *The Complete Idiot's Guide to String Theory*. Indianapolis: Alpha. ISBN 978-1-59257-702-6.

- Randall, Lisa (2005). *Warped Passages: Unraveling the Mysteries of the Universe's Hidden Dimensions*. New York: Ecco Press. ISBN 0-06-053108-8.

- Susskind, Leonard (2006). *The Cosmic Landscape: String Theory and the Illusion of Intelligent Design*. New York: Hachette Book Group/Back Bay Books. ISBN 0-316-01333-1.

- Yau, Shing-Tung; Nadis, Steve (2010). *The Shape of Inner Space: String Theory and the Geometry of the Universe's Hidden Dimensions*. Basic Books. ISBN 978-0-465-02023-2.

Critical

- Penrose, Roger (2005). *The Road to Reality: A Complete Guide to the Laws of the Universe*. Knopf. ISBN 0-679-45443-8.

- Smolin, Lee (2006). *The Trouble with Physics: The Rise of String Theory, the Fall of a Science, and What Comes Next*. New York: Houghton Mifflin Co. ISBN 0-618-55105-0.

- Woit, Peter (2006). *Not Even Wrong: The Failure of String Theory And the Search for Unity in Physical Law*. London: Jonathan Cape &: New York: Basic Books. ISBN 978-0-465-09275-8.

5.9.2 Textbooks

For physicists

- Becker, Katrin, Becker, Melanie, and Schwarz, John (2007) *String Theory and M-Theory: A Modern Introduction* . Cambridge University Press. ISBN 0-521-86069-5

- Dine, Michael (2007) *Supersymmetry and String Theory: Beyond the Standard Model*. Cambridge University Press. ISBN 0-521-85841-0.

- Kiritsis, Elias (2007) *String Theory in a Nutshell*. Princeton University Press. ISBN 978-0-691-12230-4.
- Michael Green, John H. Schwarz and Edward Witten (1987) *Superstring theory*. Cambridge University Press.
  - *Vol. 1: Introduction*. ISBN 0-521-35752-7.
  - *Vol. 2: Loop amplitudes, anomalies and phenomenology*. ISBN 0-521-35753-5.
- Johnson, Clifford (2003). *D-branes*. Cambridge: Cambridge University Press. ISBN 0-521-80912-6.
- Polchinski, Joseph (1998) *String theory*. Cambridge University Press.
  - *Vol. 1: An Introduction to the Bosonic String*. ISBN 0-521-63303-6.
  - *Vol. 2: Superstring Theory and Beyond*. ISBN 0-521-63304-4.
- Szabo, Richard J. (2007) *An Introduction to String Theory and D-brane Dynamics*. Imperial College Press. ISBN 978-1-86094-427-7.
- Zwiebach, Barton (2004) *A First Course in String Theory*. Cambridge University Press. ISBN 0-521-83143-1.

For mathematicians

- Aspinwall, Paul; Bridgeland, Tom; Craw, Alastair; Douglas, Michael; Gross, Mark; Kapustin, Anton; Moore, Gregory; Segal, Graeme; Szendrői, Balázs; Wilson, P.M.H., eds. (2009). *Dirichlet Branes and Mirror Symmetry*. American Mathematical Society.
- Deligne, Pierre; Etingof, Pavel; Freed, Daniel; Jeffery, Lisa; Kazhdan, David; Morgan, John; Morrison, David; Witten, Edward, eds. (1999). *Quantum Fields and Strings: A Course for Mathematicians*. American Mathematical Society. ISBN 0821820125.
- Hori, Kentaro; Katz, Sheldon; Klemm, Albrecht; Pandharipande, Rahul; Thomas, Richard; Vafa, Cumrun; Vakil, Ravi; Zaslow, Eric, eds. (2003). *Mirror Symmetry*. American Mathematical Society. ISBN 0821829556.

5.9.3 Online material

- Klebanov, Igor and Maldacena, Juan (January 2009). "Solving Quantum Field Theories via Curved Spacetimes". *Physics Today*.
- Schwarz, John H. (2000). "Introduction to Superstring Theory". arXiv:hep-ex/0008017 [hep-ex].
- Witten, Edward (June 2002). "The Universe on a String" (PDF). *Astronomy Magazine*. Retrieved December 19, 2005.
- Witten, Edward (1998). "Duality, Spacetime and Quantum Mechanics". Kavli Institute for Theoretical Physics. Retrieved December 16, 2005.
- Woit, Peter (2002). "Is string theory even wrong?". *American Scientist*. Retrieved December 16, 2005.

5.10 External links

- Why String Theory—An introduction to string theory.
- Dialogue on the Foundations of String Theory at MathPages
- Superstrings! String Theory Home Page—Online tutorial
- A Layman's Guide to String Theory—An explanation for the layperson
- Not Even Wrong—A blog critical of string theory
- The Official String Theory Web Site

- *The Elegant Universe*—A three-hour miniseries with Brian Greene by *NOVA* (original PBS Broadcast Dates: October 28, 8–10 p.m. and November 4, 8–9 p.m., 2003). Various images, texts, videos and animations explaining string theory.

- Beyond String Theory—A project by a string physicist explaining aspects of string theory to a broad audience

- String Theory and M-Theory a serious but amusing lecture with not to complicated math and not too advanced physics, by Prof. Leonard Susskind at Stanford University

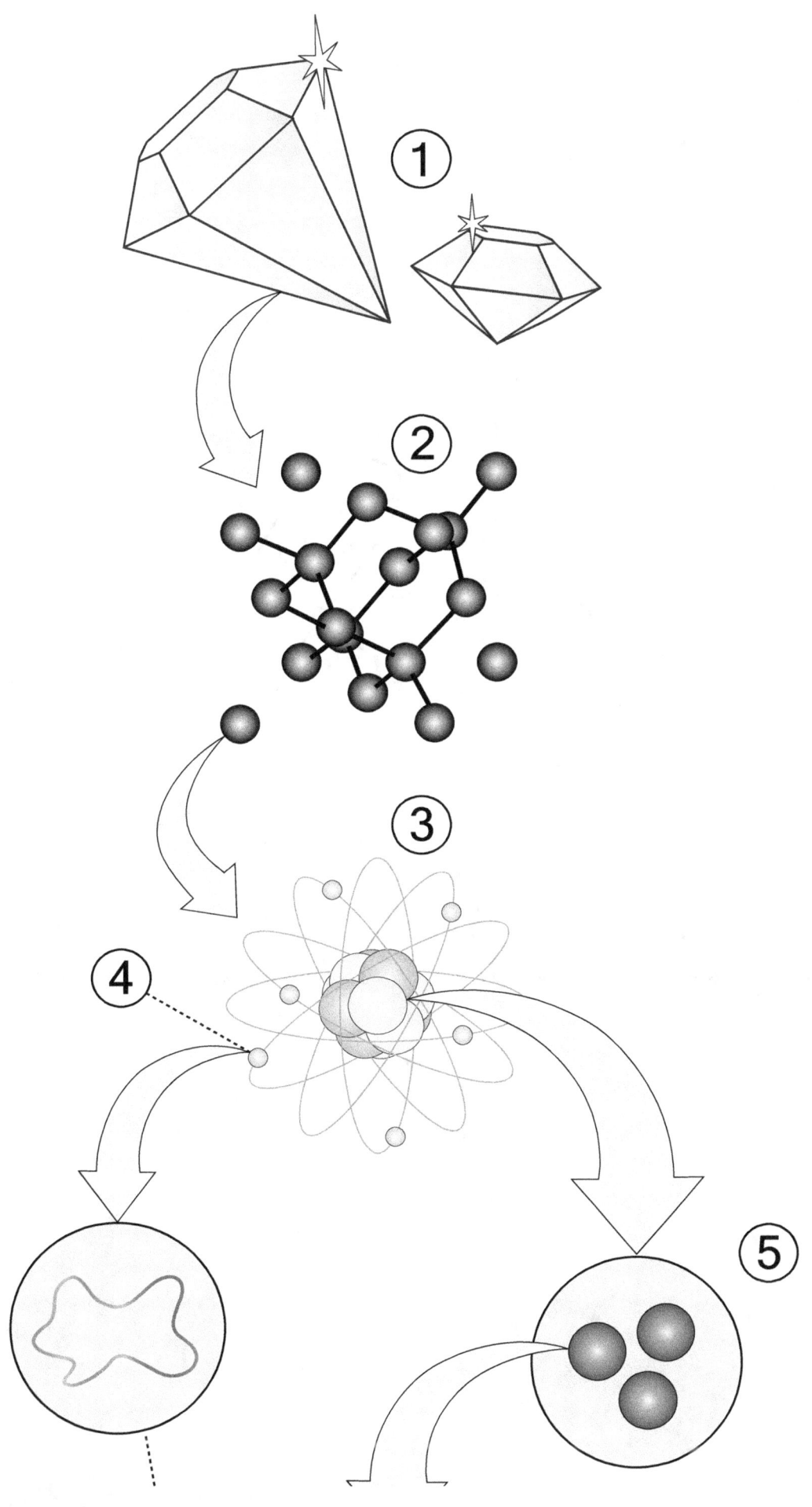

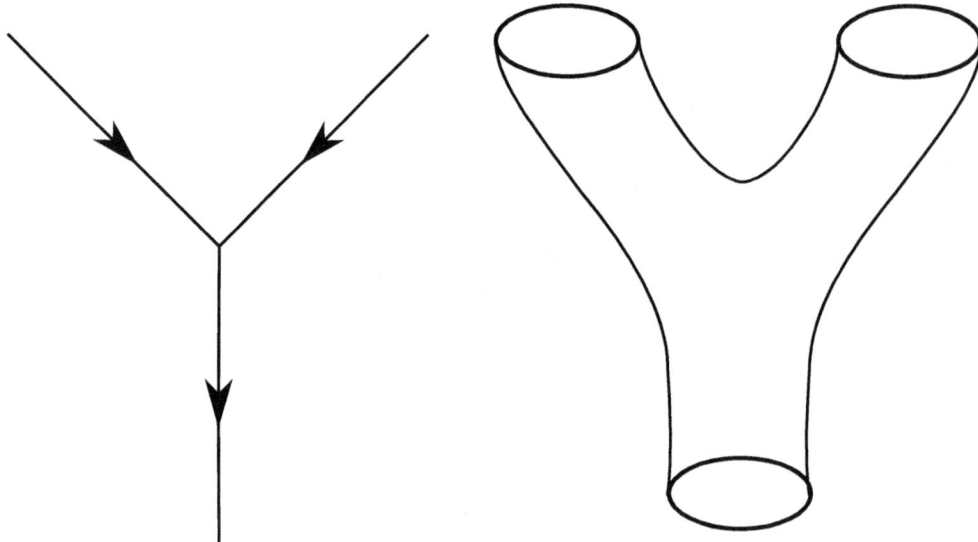

Interaction in the subatomic world: world lines of point-like particles in the Standard Model or a world sheet swept up by closed strings in string theory

Calabi–Yau manifold (3D projection)

Edward Witten

Chapter 6

Superfluid vacuum theory

Superfluid vacuum theory (SVT), sometimes known as the **BEC vacuum theory**, is an approach in theoretical physics and quantum mechanics where the fundamental physical vacuum (non-removable background) is viewed as superfluid or as a Bose–Einstein condensate (BEC).

The microscopic structure of this physical vacuum is currently unknown and is a subject of intensive studies in SVT. An ultimate goal of this approach is to develop scientific models that unify quantum mechanics (describing three of the four known fundamental interactions) with gravity, making SVT a candidate for the theory of quantum gravity and describing all known interactions in the Universe, at both microscopic and astronomic scales, as different manifestations of the same entity, superfluid vacuum.

6.1 History

The concept of a luminiferous aether as a medium sustaining electromagnetic waves was discarded after the advent of the special theory of relativity. The aether, as conceived in classical physics leads to several contradictions; in particular, aether having a definite velocity at each space-time point will exhibit a preferred direction. This conflicts with the relativistic requirement that all directions within a light cone are equivalent. However, as early as in 1951 P.A.M. Dirac published two papers where he pointed out that we should take into account quantum fluctuations in the flow of the aether.[1][2] His arguments involve the application of the uncertainty principle to the velocity of aether at any space-time point, implying that the velocity will not be a well-defined quantity. In fact, it will be distributed over various possible values. At best, one could represent the aether by a wave function representing the perfect vacuum state for which all aether velocities are equally probable. These works can be regarded as the birth point of the theory.

Inspired by the Dirac ideas, K. P. Sinha, C. Sivaram and E. C. G. Sudarshan published in 1975 a series of papers that suggested a new model for the aether according to which it is a superfluid state of fermion and anti-fermion pairs, describable by a macroscopic wave function.[3][4][5] They noted that particle-like small fluctuations of superfluid background obey the Lorentz symmetry, even if the superfluid itself is non-relativistic. Nevertheless, they decided to treat the superfluid as the relativistic matter - by putting it into the stress–energy tensor of the Einstein field equations. This did not allow them to describe the relativistic gravity as a small fluctuation of the superfluid vacuum, as subsequent authors have noted.

As an alternative to the better known string theories, a very different theory by Friedwardt Winterberg proposes instead, that the vacuum is a kind of superfluid plasma compound of positive and negative Planck masses, called a Planck mass plasma.[6][7]

Since then, several theories have been proposed within the SVT framework. They differ in how the structure and properties of the background superfluid must look like. In absence of observational data which would rule out some of them, these theories are being pursued independently.

6.2 Relation to other concepts and theories

6.2.1 Lorentz and Galilean symmetries

According to the approach, the background superfluid is assumed to be essentially non-relativistic whereas the Lorentz symmetry is not an exact symmetry of Nature but rather the approximate description valid only for small fluctuations. An observer who resides inside such vacuum and is capable of creating or measuring the small fluctuations would observe them as relativistic objects - unless their energy and momentum are sufficiently high to make the Lorentz-breaking corrections detectable.[8] If the energies and momenta are below the excitation threshold then the superfluid background behaves like the ideal fluid, therefore, the Michelson–Morley-type experiments would observe no drag force from such aether.[1][2]

Further, in the theory of relativity the Galilean symmetry (pertinent to our macroscopic non-relativistic world) arises as the approximate one - when particles' velocities are small compared to speed of light in vacuum. In SVT one does not need to go through Lorentz symmetry to obtain the Galilean one - the dispersion relations of most non-relativistic superfluids are known to obey the non-relativistic behavior at large momenta.[9][10][11]

To summarize, the fluctuations of vacuum superfluid behave like relativistic objects at "small"[nb 1] momenta (a.k.a. the "phononic limit")

$$E^2 \propto |\vec{p}|^2$$

and like non-relativistic ones

$$E \propto |\vec{p}|^2$$

at large momenta. The yet unknown nontrivial physics is believed to be located somewhere between these two regimes.

6.2.2 Relativistic quantum field theory

In the relativistic quantum field theory the physical vacuum is also assumed to be some sort of non-trivial medium to which one can associate certain energy. This is because the concept of absolutely empty space (or "mathematical vacuum") contradicts to the postulates of quantum mechanics. According to QFT, even in absence of real particles the background is always filled by pairs of creating and annihilating virtual particles. However, a direct attempt to describe such medium leads to the so-called ultraviolet divergences. In some QFT models, such as quantum electrodynamics, these problems can be "solved" using the renormalization technique, namely, replacing the diverging physical values by their experimentally measured values. In other theories, such as the quantum general relativity, this trick does not work, and reliable perturbation theory cannot be constructed.

According to SVT, this is because in the high-energy ("ultraviolet") regime the Lorentz symmetry starts failing so dependent theories cannot be regarded valid for all scales of energies and momenta. Correspondingly, while the Lorentz-symmetric quantum field models are obviously a good approximation below the vacuum-energy threshold, in its close vicinity the relativistic description becomes more and more "effective" and less and less natural since one will need to adjust the expressions for the covariant field-theoretical actions by hand.

6.2.3 Curved space-time

According to general relativity, gravitational interaction is described in terms of space-time curvature using the mathematical formalism of Riemannian geometry. This was supported by numerous experiments and observations in the regime of low energies. However, the attempts to quantize general relativity led to various severe problems, therefore, the microscopic structure of gravity is still ill-defined. There may be a fundamental reason for this—the degrees of freedom of general relativity are based on may be only approximate and effective. The question of whether general relativity is an effective theory has been raised for a long time.[12]

According to SVT, the curved space-time arises as the small-amplitude collective excitation mode of the non-relativistic background condensate.[8][13] The mathematical description of this is similar to fluid-gravity analogy which is being used also in the analog gravity models.[14] Thus, relativistic gravity is essentially a long-wavelength theory of the collective modes whose amplitude is small compared to the background one. Outside this requirement the curved-space description of gravity in terms of the Riemannian geometry becomes incomplete or ill-defined.

6.2.4 Cosmological constant

The notion of the cosmological constant makes sense in a relativistic theory only, therefore, within the SVT framework this constant can refer at most to the energy of small fluctuations of the vacuum above a background value but not to the energy of vacuum itself.[15] Thus, in SVT this constant does not have any fundamental physical meaning and the related problems, such as the vacuum catastrophe, simply do not occur in first place.

6.2.5 Gravitational waves and gravitons

According to general relativity, the conventional gravitational wave is:

1. the small fluctuation of curved spacetime which
2. has been separated from its source and propagates independently.

Superfluid vacuum theory brings into question the possibility that a relativistic object possessing both of these properties exists in nature.[13] Indeed, according to the approach, the curved spacetime itself is the small collective excitation of the superfluid background, therefore, the property (1) means that the graviton would be in fact the "small fluctuation of the small fluctuation", which does not look like a physically robust concept (as if somebody tried to introduce small fluctuations inside a phonon, for instance). As a result, it may be not just a coincidence that in general relativity the gravitational field alone has no well-defined stress–energy tensor, only the pseudotensor one.[16] Therefore, the property (2) cannot be completely justified in a theory with exact Lorentz symmetry which the general relativity is. Though, SVT does not *a priori* forbid an existence of the non-localized wave-like excitations of the superfluid background which might be responsible for the astrophysical phenomena which are currently being attributed to gravitational waves, such as the Hulse–Taylor binary. However, such excitations cannot be correctly described within the framework of a fully relativistic theory.

6.2.6 Mass generation and Higgs boson

The Higgs boson is the spin-0 particle that has been introduced in electroweak theory to give mass to the weak bosons. The origin of mass of the Higgs boson itself is not explained by electroweak theory. Instead, this mass is introduced as a free parameter by means of the Higgs potential, which thus makes it yet another free parameter of the Standard Model.[17] Within the framework of the Standard Model (or its extensions) the theoretical estimates of this parameter's value are possible only indirectly and results differ from each other significantly.[18] Thus, the usage of the Higgs boson (or any other elementary particle with predefined mass) alone is not the most fundamental solution of the mass generation problem but only its reformulation *ad infinitum*. Another known issue of the Glashow–Weinberg–Salam model is the wrong sign of mass term in the (unbroken) Higgs sector for energies above the symmetry-breaking scale.[nb 2]

While SVT does not explicitly forbid the existence of the electroweak Higgs particle, it has its own idea of the fundamental mass generation mechanism - elementary particles acquire mass due to the interaction with the vacuum condensate, similarly to the gap generation mechanism in superconductors or superfluids.[13][19] Although this idea is not entirely new, one could recall the relativistic Coleman-Weinberg approach,[20] SVT gives the meaning to the symmetry-breaking relativistic scalar field as describing small fluctuations of background superfluid which can be interpreted as an elementary particle only under certain conditions.[21] In general, one allows two scenarios to happen:

- Higgs boson exists: in this case SVT provides the mass generation mechanism which underlies the electroweak one and explains the origin of mass of the Higgs boson itself;
- Higgs boson does not exist: then the weak bosons acquire mass by directly interacting with the vacuum condensate.

Thus, the Higgs boson, even if it exists, would be a by-product of the fundamental mass generation phenomenon rather than its cause.[21]

Also, some versions of SVT favor a wave equation based on the logarithmic potential rather than on the quartic one. The former potential has not only the Mexican-hat shape, necessary for the spontaneous symmetry breaking, but also some other features which make it more suitable for the vacuum's description.

6.3 Logarithmic BEC vacuum theory

In this model the physical vacuum is conjectured to be strongly-correlated quantum Bose liquid whose ground-state wavefunction is described by the logarithmic Schrödinger equation. It was shown that the relativistic gravitational interaction arises as the small-amplitude collective excitation mode whereas relativistic elementary particles can be described by the particle-like modes in the limit of low energies and momenta.[19] The essential difference of this theory from others is that in the logarithmic superfluid the maximal velocity of fluctuations is constant in the leading (classical) order. This allows to fully recover the relativity postulates in the "phononic" (linearized) limit.[13]

The proposed theory has many observational consequences. They are based on the fact that at high energies and momenta the behavior of the particle-like modes eventually becomes distinct from the relativistic one - they can reach the speed of light limit at finite energy.[22] Among other predicted effects is the superluminal propagation and vacuum Cherenkov radiation.[23]

Theory advocates the mass generation mechanism which is supposed to replace or alter the electroweak Higgs one. It was shown that masses of elementary particles can arise as a result of interaction with the superfluid vacuum, similarly to the gap generation mechanism in superconductors.[13][19] For instance, the photon propagating in the average interstellar vacuum acquires a tiny mass which is estimated to be about 10^{-35} electronvolt. One can also derive an effective potential for the Higgs sector which is different from the one used in the Glashow–Weinberg–Salam model, yet it yields the mass generation and it is free of the imaginary-mass problem[nb 2] appearing in the conventional Higgs potential.[21]

6.4 See also

- Analog gravity
- Acoustic metric
- Bose–Einstein condensate
- Casimir vacuum
- Hawking radiation
- Induced gravity
- Planck scale
- Planck units
- Hořava–Lifshitz gravity
- Quantum gravity
- Quantum realm
- Macrocosm and microcosm
- Sonic black hole
- Vacuum energy

6.5 Notes

[1] The term "small" refers here to the linearized limit, in practice the values of these momenta may not be small at all.

[2] If one expands the Higgs potential then the coefficient at the quadratic term appears to be negative. This coefficient has a physical meaning of squared mass of a scalar particle.

6.6 References

[1] Dirac, P. A. M. (24 November 1951). "Is there an Æther?". *Letters to Nature* (Nature) **168** (4282): 906–907. Bibcode:1951Natur.168..906D. doi:10.1038/168906a0. Retrieved 16 October 2012.

[2] Dirac, P. A. M. (26 April 1952). "Is there an Æther?". *Nature* **169** (4304): 702–702. Bibcode:1952Natur.169..702D. doi:10.1038/169702b0.

[3] K. P. Sinha, C. Sivaram, E. C. G. Sudarshan, Found. Phys. 6, 65 (1976).

[4] K. P. Sinha, C. Sivaram, E. C. G. Sudarshan, Found. Phys. 6, 717 (1976).

[5] K. P. Sinha and E. C. G. Sudarshan, Found. Phys. 8, 823 (1978).

[6] Winterberg, Friedwardt (1988). "Substratum Approach to a Unified Theory of Elementary Particles". *Z.f. Naturforsch.- Physical Sciences.* **43a**.

[7] Winterberg, Friedwardt (2003). "Planck Mass Plasma Vacuum Conjecture". *Z. Naturforsch* **58a**: 231–267.

[8] G. E. Volovik, *The Universe in a helium droplet*, Int. Ser. Monogr. Phys. **117** (2003) 1-507.

[9] N. N. Bogoliubov, Izv. Acad. Nauk USSR 11, 77 (1947).

[10] N.N. Bogoliubov, J. Phys. 11, 23 (1947)

[11] V. L. Ginzburg, L. D. Landau, Zh. Eksp. Teor. Fiz. 20, 1064 (1950).

[12] A. D. Sakharov, Sov. Phys. Dokl. 12, 1040 (1968). This paper was reprinted in Gen. Rel. Grav. 32, 365 (2000) and commented in: M. Visser, Mod. Phys. Lett. A 17, 977 (2002).

[13] K. G. Zloshchastiev, *Spontaneous symmetry breaking and mass generation as built-in phenomena in logarithmic nonlinear quantum theory*, Acta Phys. Polon. B **42** (2011) 261-292 ArXiv:0912.4139.

[14] M. Novello, M. Visser, G. Volovik, *Artificial Black Holes*, World Scientific, River Edge, USA, 2002, p391.

[15] G.E. Volovik, Int. J. Mod. Phys. D15, 1987 (2006) ArXiv: gr-qc/0604062.

[16] L.D. Landau and E.M. Lifshitz, *The Classical Theory of Fields*, (1951), Pergamon Press, chapter 11.96.

[17] V. A. Bednyakov, N. D. Giokaris and A. V. Bednyakov, Phys. Part. Nucl. **39** (2008) 13-36 ArXiv:hep-ph/0703280.

[18] B. Schrempp and M. Wimmer, Prog. Part. Nucl. Phys. **37** (1996) 1-90 ArXiv:hep-ph/9606386.

[19] A. V. Avdeenkov and K. G. Zloshchastiev, *Quantum Bose liquids with logarithmic nonlinearity: Self-sustainability and emergence of spatial extent*, J. Phys. B: At. Mol. Opt. Phys. **44** (2011) 195303. ArXiv:1108.0847.

[20] S. R. Coleman and E. J. Weinberg, Phys. Rev. D7, 1888 (1973).

[21] V. Dzhunushaliev and K.G. Zloshchastiev (2013). "Singularity-free model of electric charge in physical vacuum: Non-zero spatial extent and mass generation". *Cent. Eur. J. Phys.* **11** (3): 325–335. arXiv:1204.6380. Bibcode:2013CEJPh..11..325D. doi:10.2478/s11534-012-0159-z.

[22] K. G. Zloshchastiev, *Logarithmic nonlinearity in theories of quantum gravity: Origin of time and observational consequences*, Grav. Cosmol. **16** (2010) 288-297 ArXiv:0906.4282.

[23] K. G. Zloshchastiev, *Vacuum Cherenkov effect in logarithmic nonlinear quantum theory*, Phys. Lett. A **375** (2011) 2305–2308 ArXiv:1003.0657.

Chapter 7

Supersymmetry

"SUSY" redirects here. For other uses, see Susy (disambiguation).

For the episode of the American TV series Angel, see Supersymmetry (Angel)

In particle physics, **supersymmetry** (**SUSY**) is a proposed extension of spacetime symmetry that relates two basic classes of elementary particles: bosons, which have an integer-valued spin, and fermions, which have a half-integer spin.[1] Each particle from one group is associated with a particle from the other, called its superpartner, whose spin differs by a half-integer. In a theory with perfectly unbroken supersymmetry, each pair of superpartners shares the same mass and internal quantum numbers besides spin - for example, a "selectron" (superpartner electron) would be a boson version of the electron, and would have the same mass energy and thus be equally easy to find in the lab. However, since no superpartners have been observed yet, supersymmetry must be a spontaneously broken symmetry if it exists. If supersymmetry is a true symmetry of nature, it would explain many mysterious features of particle physics and would help solve paradoxes such as the cosmological constant problem. The Minimal Supersymmetric Standard Model is one of the best studied candidates for physics beyond the Standard Model.

The failure of the Large Hadron Collider to find evidence for supersymmetry has led some physicists to suggest that the theory should be abandoned as a solution to such problems, as any superpartners that exist would now need to be too massive to solve the paradoxes anyway.[2] Experiments with the Large Hadron Collider also yielded extremely rare particle decay events which casts doubt on many versions of supersymmetry.[3]

Supersymmetry differs notably from currently known symmetries in that it establishes a symmetry between classical and quantum physics, which up to now has not been observed in any other domain. While any number of bosons can occupy the same quantum state, for fermions this is not possible because of the exclusion principle, which allows only one fermion in a given state. But when the occupation numbers become large, quantum physics approaches the classical limit. This means that while bosons also exist in classical physics, fermions do not. That makes it difficult to expect that bosons possess the same quantum numbers as fermions.[4] There is only indirect evidence for the existence of supersymmetry, primarily in the form of evidence for gauge coupling unification.[5] However this refers only to electroweak and strong interactions and does not provide the ultimate unification of all interactions, since it leaves gravitation untouched.

7.1 History

A supersymmetry relating mesons and baryons was first proposed, in the context of hadronic physics, by Hironari Miyazawa in 1966. This supersymmetry did not involve spacetime, that is it concerned internal symmetry, and was badly broken. His work was largely ignored at the time.[6][7][8][9]

J. L. Gervais and B. Sakita (in 1971),[10] Yu. A. Golfand and E. P. Likhtman (also in 1971), and D.V. Volkov and V.P. Akulov (in 1972),[11] independently rediscovered supersymmetry in the context of quantum field theory, a radically new type of symmetry of spacetime and fundamental fields, which establishes a relationship between elementary particles of different quantum nature, bosons and fermions, and unifies spacetime and internal symmetries of the microscopic world. Supersymmetry with a consistent Lie-algebraic graded structure on which the Gervais–Sakita rediscovery was based directly first arose in 1971[12] in the context of an early version of string theory by Pierre Ramond, John H. Schwarz and André Neveu.

Finally, J. Wess and B. Zumino (in 1974)[13] identified the characteristic renormalization features of four-dimensional supersymmetric field theories, which singled them out as remarkable QFTs, and they and Abdus Salam and their fellow researchers introduced early particle physics applications. The mathematical structure of supersymmetry (Graded Lie superalgebras) has subsequently been applied successfully to other areas of physics, in a variety of fields, ranging from nuclear physics,[14][15] critical phenomena,[16] quantum mechanics to statistical physics. It remains a vital part of many proposed theories of physics.

The first realistic supersymmetric version of the Standard Model was proposed in 1981 by Howard Georgi and Savas Dimopoulos and is called the Minimal Supersymmetric Standard Model or MSSM for short. It was proposed to solve the hierarchy problem and predicts superpartners with masses between 100 GeV and 1 TeV.

As of September 2011, no meaningful signs of the superpartners have been observed.[17][18] The Large Hadron Collider at CERN is producing the world's highest energy collisions and offers the best chance at discovering superparticles for the foreseeable future.

After the discovery of the Higgs particle in 2012, it was expected that supersymmetric particles would be found at CERN, but there has been still no evidence of them. The LHCb and CMS experiments at the LHC made the first definitive observation of a Strange B meson decaying into two muons, confirming a standard model prediction, but a blow for those hoping for signs of supersymmetry.[19] Neil Turok at Perimeter Institute concedes that theorists are disheartened at that situation, and that they are at a crossroad in theoretical (and particle) physics, calling it a deep crisis. He described the LHC results as "simple, yet extremely puzzling" and said "we have to get people to try to find the new principles that will explain the simplicity".[20]

7.2 Motivations

A central motivation for supersymmetry close to the TeV energy scale is the resolution of the hierarchy problem of the Standard Model. Without the extra supersymmetric particles, the Higgs boson mass is subject to quantum corrections which are so large as to naturally drive it close to the Planck mass barring its fine tuning to an extraordinarily tiny value. In the supersymmetric theory, on the other hand, these quantum corrections are canceled by those from the corresponding superpartners above the supersymmetry breaking scale, which becomes the new characteristic natural scale for the Higgs mass. Other attractive features of TeV-scale supersymmetry are the fact that it often provides a candidate dark matter particle at a mass scale consistent with thermal relic abundance calculations,[21][22] provides a natural mechanism for electroweak symmetry breaking and allows for the precise high-energy unification of the weak, the strong and electromagnetic interactions. Therefore, scenarios where supersymmetric partners appear with masses not much greater than 1 TeV are considered the most well-motivated by theorists.[23] These scenarios would imply that experimental traces of the superpartners should begin to emerge in high-energy collisions at the LHC relatively soon. As of September 2011, no meaningful signs of the superpartners have been observed,[17][18] which is beginning to significantly constrain the most popular incarnations of supersymmetry. However, the total parameter space of consistent supersymmetric extensions of the Standard Model is extremely diverse and can not be definitively ruled out at the LHC.

Supersymmetry is also motivated by solutions to several theoretical problems, for generally providing many desirable mathematical properties, and for ensuring sensible behavior at high energies. Supersymmetric quantum field theory is often much easier to analyze, as many more problems become exactly solvable. When supersymmetry is imposed as a *local* symmetry, Einstein's theory of general relativity is included automatically, and the result is said to be a theory of supergravity. It is also a necessary feature of the most popular candidate for a theory of everything, superstring theory.

Another theoretically appealing property of supersymmetry is that it offers the only "loophole" to the Coleman–Mandula theorem, which prohibits spacetime and internal symmetries from being combined in any nontrivial way, for quantum field theories like the Standard Model under very general assumptions. The Haag-Lopuszanski-Sohnius theorem demonstrates that supersymmetry is the only way spacetime and internal symmetries can be consistently combined.[24]

7.3 Applications

7.3.1 Extension of possible symmetry groups

One reason that physicists explored supersymmetry is because it offers an extension to the more familiar symmetries of quantum field theory. These symmetries are grouped into the Poincaré group and internal symmetries and the Coleman–Mandula theorem showed that under certain assumptions, the symmetries of the S-matrix must be a direct product of the Poincaré group with a compact internal symmetry group or if there is no mass gap, the conformal group with a compact internal symmetry group. In 1971 Golfand and Likhtman were the first to show that the Poincaré algebra can be extended through introduction of four anticommuting spinor generators (in four dimensions), which later became known as supercharges. In 1975 the Haag-Lopuszanski-Sohnius theorem analyzed all possible super-algebras in the general form, including those with an extended number of the supergenerators and central charges. This extended super-Poincaré algebra paved the way for obtaining a very large and important class of supersymmetric field theories.

The supersymmetry algebra

Main article: Supersymmetry algebra

Traditional symmetries in physics are generated by objects that transform under the tensor representations of the Poincaré group and internal symmetries. Supersymmetries, on the other hand, are generated by objects that transform under the spinor representations. According to the spin-statistics theorem, bosonic fields commute while fermionic fields anticommute. Combining the two kinds of fields into a single algebra requires the introduction of a $\mathbf{Z}_2$-grading under which the bosons are the even elements and the fermions are the odd elements. Such an algebra is called a Lie superalgebra.

The simplest supersymmetric extension of the Poincaré algebra is the Super-Poincaré algebra. Expressed in terms of two Weyl spinors, has the following anti-commutation relation:

$$\{Q_\alpha, \bar{Q}_{\dot\beta}\} = 2(\sigma^\mu)_{\alpha\dot\beta} P_\mu$$

and all other anti-commutation relations between the Qs and commutation relations between the Qs and Ps vanish. In the above expression $P_\mu = -i\partial_\mu$ are the generators of translation and σ^μ are the Pauli matrices.

There are representations of a Lie superalgebra that are analogous to representations of a Lie algebra. Each Lie algebra has an associated Lie group and a Lie superalgebra can sometimes be extended into representations of a Lie supergroup.

7.3.2 The Supersymmetric Standard Model

Main article: Minimal Supersymmetric Standard Model

Incorporating supersymmetry into the Standard Model requires doubling the number of particles since there is no way that any of the particles in the Standard Model can be superpartners of each other. With the addition of new particles, there are many possible new interactions. The simplest possible supersymmetric model consistent with the Standard Model is the Minimal Supersymmetric Standard Model (MSSM) which can include the necessary additional new particles that are able to be superpartners of those in the Standard Model.

One of the main motivations for SUSY comes from the quadratically divergent contributions to the Higgs mass squared. The quantum mechanical interactions of the Higgs boson causes a large renormalization of the Higgs mass and unless there is an accidental cancellation, the natural size of the Higgs mass is the highest scale possible. This problem is known as the hierarchy problem. Supersymmetry reduces the size of the quantum corrections by having automatic cancellations between fermionic and bosonic Higgs interactions. If supersymmetry is restored at the weak scale, then the Higgs mass is related to supersymmetry breaking which can be induced from small non-perturbative effects explaining the vastly different scales in the weak interactions and gravitational interactions.

In many supersymmetric Standard Models there is a heavy stable particle (such as neutralino) which could serve as a weakly interacting massive particle (WIMP) dark matter candidate. The existence of a supersymmetric dark matter candidate is closely tied to R-parity.

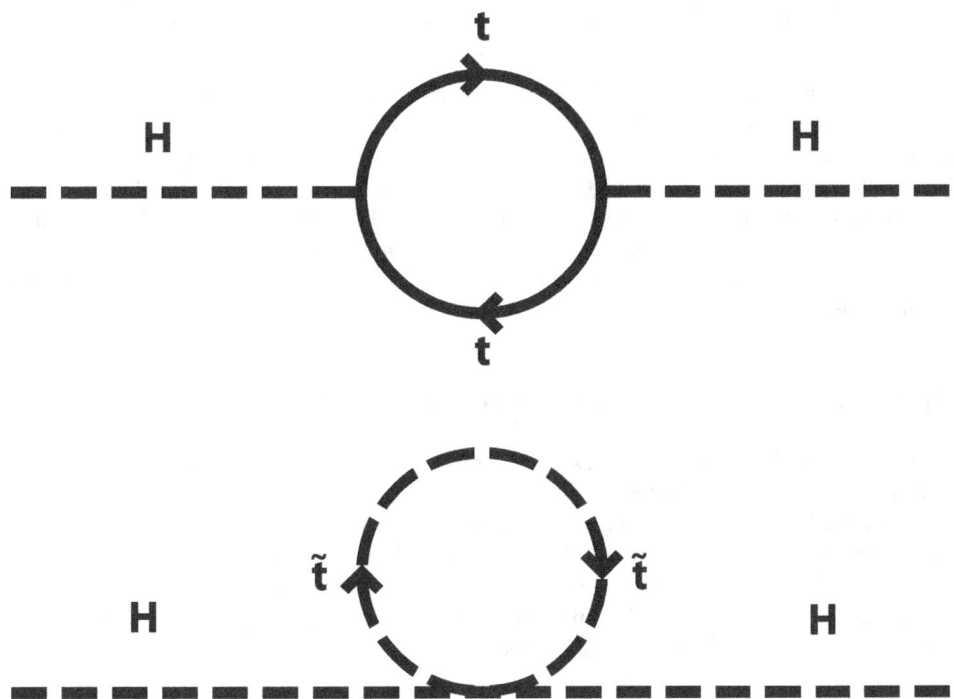

Cancellation of the Higgs boson quadratic mass renormalization between fermionic top quark loop and scalar stop squark tadpole Feynman diagrams in a supersymmetric extension of the Standard Model

The standard paradigm for incorporating supersymmetry into a realistic theory is to have the underlying dynamics of the theory be supersymmetric, but the ground state of the theory does not respect the symmetry and supersymmetry is broken spontaneously. The supersymmetry break can not be done permanently by the particles of the MSSM as they currently appear. This means that there is a new sector of the theory that is responsible for the breaking. The only constraint on this new sector is that it must break supersymmetry permanently and must give superparticles TeV scale masses. There are many models that can do this and most of their details do not matter. In order to parameterize the relevant features of supersymmetry breaking, arbitrary soft SUSY breaking terms are added to the theory which temporarily break SUSY explicitly but could never arise from a complete theory of supersymmetry breaking.

Gauge-coupling unification

Main article: Minimal Supersymmetric Standard Model § Gauge-coupling unification

One piece of evidence for supersymmetry existing is gauge coupling unification. The renormalization group evolution of the three gauge coupling constants of the Standard Model is somewhat sensitive to the present particle content of the theory. These coupling constants do not quite meet together at a common energy scale if we run the renormalization group using the Standard Model.[5] With the addition of minimal SUSY joint convergence of the coupling constants is projected at approximately 10^{16} GeV.[5]

7.3.3 Supersymmetric quantum mechanics

Main article: Supersymmetric quantum mechanics

Supersymmetric quantum mechanics adds the SUSY superalgebra to quantum mechanics as opposed to quantum field theory. Supersymmetric quantum mechanics often comes up when studying the dynamics of supersymmetric solitons, and due to the simplified nature of having fields which are only functions of time (rather than space-time), a great deal of progress has been made in this subject and it is now studied in its own right.

SUSY quantum mechanics involves pairs of Hamiltonians which share a particular mathematical relationship, which are called *partner Hamiltonians*. (The potential energy terms which occur in the Hamiltonians are then called *partner potentials*.) An introductory theorem shows that for every eigenstate of one Hamiltonian, its partner Hamiltonian has a corresponding eigenstate with the same energy. This fact can be exploited to deduce many properties of the eigenstate spectrum. It is analogous to the original description of SUSY, which referred to bosons and fermions. We can imagine a "bosonic Hamiltonian", whose eigenstates are the various bosons of our theory. The SUSY partner of this Hamiltonian would be "fermionic", and its eigenstates would be the theory's fermions. Each boson would have a fermionic partner of equal energy.

7.3.4 Supersymmetry: Applications to condensed matter physics

SUSY concepts have provided useful extensions to the WKB approximation. In addition, SUSY has been applied to disorder averaged systems both quantum and non-quantum (through statistical mechanics). The Fokker-Planck equation being an example of a non-quantum theory. The `supersymmetry' in all these systems arises from the fact that one is modelling one particle and as such the`statistics' don't matter. The use of the supersymmetry method provides a mathematical rigorous alternative to the replica trick, but only in non-interacting systems, which attempts to address the so-called `problem of the denominator' under disorder averaging. For more on the applications of supersymmetry in condensed matter physics see the book[25]

7.3.5 Supersymmetry in optics

Integrated optics was recently found[26] to provide a fertile ground on which certain ramifications of SUSY can be explored in readily-accessible laboratory settings. Making use of the analogous mathematical structure of the quantum-mechanical Schrödinger equation and the wave equation governing the evolution of light in one-dimensional settings, one may interpret the refractive index distribution of a structure as a potential landscape in which optical wave packets propagate. Along these lines, a new class of functional optical structures with possible applications in phase matching, mode conversion[27] and space-division multiplexing becomes possible. SUSY transformations have been also proposed as a way to address inverse scattering problems in optics and as a one-dimensional transformation optics [28]

7.3.6 Mathematics

SUSY is also sometimes studied mathematically for its intrinsic properties. This is because it describes complex fields satisfying a property known as holomorphy, which allows holomorphic quantities to be exactly computed. This makes supersymmetric models useful toy models of more realistic theories. A prime example of this has been the demonstration of S-duality in four-dimensional gauge theories[29] that interchanges particles and monopoles.

The proof of the Atiyah-Singer index theorem is much simplified by the use of supersymmetric quantum mechanics.

7.4 General supersymmetry

Supersymmetry appears in many different contexts in theoretical physics that are closely related. It is possible to have multiple supersymmetries and also have supersymmetric extra dimensions.

7.4.1 Extended supersymmetry

Main article: Extended supersymmetry

It is possible to have more than one kind of supersymmetry transformation. Theories with more than one supersymmetry transformation are known as extended supersymmetric theories. The more supersymmetry a theory has, the more constrained the field content and interactions are. Typically the number of copies of a supersymmetry is a power of 2, i.e. 1, 2, 4, 8. In four dimensions, a spinor has four degrees of freedom and thus the minimal number of supersymmetry generators is four in four dimensions and having eight copies of supersymmetry means that there are 32 supersymmetry generators.

The maximal number of supersymmetry generators possible is 32. Theories with more than 32 supersymmetry generators automatically have massless fields with spin greater than 2. It is not known how to make massless fields with spin greater than two interact, so the maximal number of supersymmetry generators considered is 32. This corresponds to an $N = 8$ supersymmetry theory. Theories with 32 supersymmetries automatically have a graviton.

In four dimensions there are the following theories, with the corresponding multiplets[30](CPT adds a copy, whenever they are not invariant under such symmetry)

- $N = 1$

Chiral multiplet: $(0, \frac{1}{2})$ Vector multiplet: $(\frac{1}{2}, 1)$ Gravitino multiplet: $(1, \frac{3}{2})$ Graviton multiplet: $(\frac{3}{2}, 2)$

- $N = 2$

hypermultiplet: $(-\frac{1}{2}, 0^2, \frac{1}{2})$ vector multiplet: $(0, \frac{1}{2}^2, 1)$ supergravity multiplet: $(1, \frac{3}{2}^2, 2)$

- $N = 4$

Vector multiplet: $(-1, -\frac{1}{2}^4, 0^6, \frac{1}{2}^4, 1)$ Supergravity multiplet: $(0, \frac{1}{2}^4, 1^6, \frac{3}{2}^4, 2)$

- $N = 8$

Supergravity multiplet: $(-2, -\frac{3}{2}^8, -1^{28}, -\frac{1}{2}^{56}, 0^{70}, \frac{1}{2}^{56}, 1^{28}, \frac{3}{2}^8, 2)$

7.4.2 Supersymmetry in alternate numbers of dimensions

It is possible to have supersymmetry in dimensions other than four. Because the properties of spinors change drastically between different dimensions, each dimension has its characteristic. In d dimensions, the size of spinors is roughly $2^{d/2}$ or $2^{(d-1)/2}$. Since the maximum number of supersymmetries is 32, the greatest number of dimensions in which a supersymmetric theory can exist is eleven.

7.5 Supersymmetry as a quantum group

Main article: Supersymmetry as a quantum group

Supersymmetry can be reinterpreted in the language of noncommutative geometry and quantum groups. In particular, it involves a mild form of noncommutativity, namely supercommutativity. See the main article for more details.

7.6 Supersymmetry in quantum gravity

Supersymmetry is part of a larger enterprise of theoretical physics to unify everything we know about the physical world into a single fundamental framework of physical laws, known as the quest for a Theory of Everything (TOE). A significant part of this larger enterprise is the quest for a theory of quantum gravity, which would unify the classical theory of general relativity and the Standard Model, which explains the other three basic forces in physics (electromagnetism, the strong interaction, and the weak interaction), and provides a palette of fundamental particles upon which all four forces act. Two of the most active approaches to forming a theory of quantum gravity are string theory and loop quantum gravity (LQG), although in theory, supersymmetry could be a component of other theoretical approaches as well.

For string theory to be consistent, supersymmetry appears to be required at some level (although it may be a strongly broken symmetry). In particle theory, supersymmetry is recognized as a way to stabilize the hierarchy between the unification scale and the electroweak scale (or the Higgs boson mass), and can also provide a natural dark matter candidate. String theory also requires extra spatial dimensions which have to be compactified as in Kaluza–Klein theory.

Loop quantum gravity (LQG) predicts no additional spatial dimensions, nor anything else about particle physics. These theories can be formulated in three spatial dimensions and one dimension of time, although in some LQG theories dimensionality is an emergent property of the theory, rather than a fundamental assumption of the theory. Also, LQG is a theory of quantum gravity which does not require supersymmetry. Lee Smolin, one of the originators of LQG, has proposed that a loop quantum gravity theory incorporating either supersymmetry or extra dimensions, or both, be called "loop quantum gravity II".

If experimental evidence confirms supersymmetry in the form of supersymmetric particles such as the neutralino that is often believed to be the lightest superpartner, some people believe this would be a major boost to string theory. Since supersymmetry is a required component of string theory, any discovered supersymmetry would be consistent with string theory. If the Large Hadron Collider and other major particle physics experiments fail to detect supersymmetric partners or evidence of extra dimensions, many versions of string theory which had predicted certain low mass superpartners to existing particles may need to be significantly revised. The failure of experiments to discover either supersymmetric partners or extra spatial dimensions, as of 2013, has encouraged loop quantum gravity researchers.

7.7 Falsifiability

SUSY is often criticized in that its greatest strength and weakness is that it is not falsifiable, because its breaking mechanism and the minimum mass above which it is restored are unknown. This minimum mass can be pushed upwards to arbitrarily large values, without disproving the symmetry, and a non-falsifiable theory is generally considered unscientific. However, many theoretical physicists continue to focus on supersymmetry because of its usefulness as a tool in quantum field theory, its interesting mathematical properties, and the possibility that extremely high energy physics (as in around the time of the big bang) are described by supersymmetric theories.

7.8 Current status

Supersymmetric models are constrained by a variety of experiments, including measurements of low-energy observables, for example the anomalous magnetic moment of the muon at Brookhaven; the WMAP dark matter density measurement and direct detection experiments, for example XENON−100; and by particle collider experiments, including B-physics, Higgs phenomenology and direct searches for superpartners (sparticles), at the Large Electron–Positron Collider, Tevatron and the LHC.

Historically, the tightest limits were from direct production at colliders. The first mass limits for squarks and gluinos were made at CERN by the UA1 experiment and the UA2 experiment at the Super Proton Synchrotron. LEP later set very strong limits.[31] In 2006 these limits were extended by the D0 experiment.[32][33]

From 2003, WMAP's dark matter density measurements have strongly constrained supersymmetry models, which have to be tuned to invoke a particular mechanism to sufficiently reduce the neutralino density.

Prior to the launch of the LHC, in 2009, fits of available data to CMSSM and NUHM1 indicated that squarks and gluinos were most likely to have masses in 500 to 800 GeV range, though values as high as 2.5 TeV were allowed with low probabilities. Neutralinos and sleptons were expected to be quite light, with the lightest neutralino and the lightest stau most likely to be found between 100 to 150 GeV.[34]

As of 2014, the LHC has found no evidence for supersymmetry, and, as a result, has surpassed existing experimental limits from Large Electron–Positron Collider and Tevatron and partially excluded the aforementioned expected ranges.[35][36][37][38] Based on the data sample collected by the CMS detector at the LHC through the summer of 2011, CMSSM squarks have been excluded up to the mass of 1.1 TeV and gluinos have been excluded up to 500 GeV.[39] Searches are only applicable for a finite set of tested points because simulation using the Monte Carlo method must be made so that limits for that particular model can be calculated. This complicates matters because different experiments have looked at different sets of points. Some extrapolation between points can be made within particular models but it is difficult to set general limits even for the Minimal Supersymmetric Standard Model.

In 2011 and 2012, the LHC discovered a Higgs boson with a mass of about 125 GeV, and with couplings to fermions and bosons which are consistent with the Standard Model. The MSSM predicts that the mass of the lightest Higgs boson should not be much higher than the mass of the Z boson, and, in the absence of fine tuning (with the supersymmetry breaking scale on the order of 1 TeV), should not exceed 130 GeV. Furthermore, for values of the MSSM parameter *tan* β ≤ 3, it predicts Higgs mass below 114 GeV over most of the parameter space.[40] This region of Higgs mass was excluded by LEP by 2000. The LHC result is somewhat problematic for the minimal supersymmetric model, as the value of 125 GeV is relatively large for the model and can only be achieved with large radiative loop corrections from top squarks, which many theorists consider to be "unnatural" (see naturalness and fine tuning).[41]

There are eight arguments against supersymmetry. (1) The LUX experiment for cold dark matter has not observed neutralinos. (2) The large size of the WMAP cold spot is larger than predicted by Lambda cold dark matter models. (3) The large-scale flow of galaxies is larger than predicted by Lambda CDM models. (4) The number of faint dwarf galaxies is smaller than predicted by Lambda CDM models. (5) Neither the ATLAS nor the CMS collaboration have observed gluinos and squarks. (6) The rest mass, interaction cross-section and decay rates of the Higgs boson are compatible with the standard theory, but not with earlier predictions by supersymmetric models. (7) Dirac fermions can be described by a gravitation theory which includes Cartan torsion (Einstein-Cartan theory), supersymmetry is not required. (8) The mass hierarchy problem of Grand Unified theories need not arise if Grand Unification does not exist. The proton decay predicted by Grand Unified theories has not been observed. The quantization of electric charge can be explained by theories which include Dirac magnetic monopoles, so Grand Unification is not necessary.[42]

In spite of the null searches and the heavy Higgs, a recent analysis of the constrained minimal supersymmetric Standard Model, the CMSSM, suggests that the model is still compatible with all present experimental constraints.[43] The preferred masses for squarks and gluinos is about 2 TeV. The resulting fine-tuning of the Higgs boson mass and Z-boson mass (see mu problem and little hierarchy problem), however, is considered "unnatural", and some theorists now favor extended supersymmetry models, for example, the NMSSM.

7.9 See also

- Supersymmetric gauge theory
- Wess–Zumino model
- Minimal Supersymmetric Standard Model
- Supersymmetry as a quantum group
- Quantum group
- Supercharge
- Superfield
- Supergeometry
- Supergravity
- Supergroup
- Superspace

7.10 References

[1] Sean Carroll, *Dark Matter, Dark Energy: The Dark Side of the Universe*, The Teaching Company, Guidebook Part 2 page 60, Accessed Oct. 7, 2013, "...Supersymmetry -- A hypothetical symmetry relating bosons to fermions..."

[2] Wolchover, Natalie (November 29, 2012). "Supersymmetry Fails Test, Forcing Physics to Seek New Ideas". *Scientific American*.

[3] M. Hogenboom (24 July 2013). "Ultra-rare decay confirmed in LHC". BBC. Retrieved 2013-08-18.

[4] Richard M. Weiner (2013). "Spin-statistics-quantum number connection and supersymmetry". *Physical Review D* **87** (5). arXiv:1302.0969. Bibcode:2013PhRvD..87e5003W. doi:10.1103/PhysRevD.87.055003.

[5] Gordon L. Kane, *The Dawn of Physics Beyond the Standard Model*, Scientific American, June 2003, page 60 and *The frontiers of physics*, special edition, Vol 15, #3, page 8 "Indirect evidence for supersymmetry comes from the extrapolation of interactions to high energies."

[6] H. Miyazawa (1966). "Baryon Number Changing Currents". *Prog. Theor. Phys.* **36** (6): 1266–1276. Bibcode:1966PThPh..36.1266M. doi:10.1143/PTP.36.1266.

[7] H. Miyazawa (1968). "Spinor Currents and Symmetries of Baryons and Mesons". *Phys. Rev.* **170** (5): 1586–1590. Bibcode:1968PhRv..170.1586M. doi:10.1103/PhysRev.170.1586.

[8] Michio Kaku, *Quantum Field Theory*, ISBN 0-19-509158-2, pg 663.

[9] Peter Freund, *Introduction to Supersymmetry*, ISBN 0-521-35675-X, pages 26-27, 138.

[10] Gervais, J. -L.; Sakita, B. (1971). "Field theory interpretation of supergauges in dual models". *Nuclear Physics B* **34** (2): 632. Bibcode:1971NuPhB..34..632G. doi:10.1016/0550-3213(71)90351-8.

[11] D.V. Volkov, V.P. Akulov, Pisma Zh.Eksp.Teor.Fiz. 16 (1972) 621; Phys.Lett. B46 (1973) 109; V.P. Akulov, D.V. Volkov, Teor.Mat.Fiz. 18 (1974) 39

[12] Ramond, P. (1971). "Dual Theory for Free Fermions". *Physical Review D* **3** (10): 2415. Bibcode:1971PhRvD...3.2415R. doi:10.1103/PhysRevD.3.2415.

[13] Wess, J.; Zumino, B. (1974). "Supergauge transformations in four dimensions". *Nuclear Physics B* **70**: 39. Bibcode:1974NuPhB..70...39W. doi:10.1016/0550-3213(74)90355-1.

[14] http://users.physik.fu-berlin.de/~{}kleinert/kleinert/?p=supersym suggested here

[15] Iachello, F. (1980). "Dynamical Supersymmetries in Nuclei". *Physical Review Letters* **44** (12): 772. Bibcode:1980PhRvL..44..772I. doi:10.1103/PhysRevLett.44.772.

[16] Friedan, D.; Qiu, Z.; Shenker, S. (1984). "Conformal Invariance, Unitarity, and Critical Exponents in Two Dimensions". *Physical Review Letters* **52** (18): 1575. Bibcode:1984PhRvL..52.1575F. doi:10.1103/PhysRevLett.52.1575.

[17] ATLAS SUSY search documents

[18] CMS SUSY search documents

[19] CERN latest data shows no sign of supersymmetry – yet Phys.Org, 25 July 2013

[20] Perimeter Institute and the crisis in modern physics Paul Wells, 5 Sep 2013

[21] Jonathan Feng: Supersymmetric Dark Matter *(pdf)*, University of California, Irvine, 11 May 2007

[22] Torsten Bringmann: The WIMP "Miracle" *(pdf)* University of Hamburg

[23] http://profmattstrassler.com/articles-and-posts/lhcposts/what-do-current-mid-august-2011-lhc-results-imply-about-supersymmetry/

[24] R. Haag, J. T. Lopuszanski and M. Sohnius, "All Possible Generators Of Supersymmetries Of The S Matrix", Nucl. Phys. B 88 (1975) 257

[25] *Supersymmetry in Disorder and Chaos*, Konstantin Efetov, Cambridge university press, 1997.

[26] Miri, M.-A.; Heinrich, M.; El-Ganainy, R.; Christodoulides, D. N. (2013). "Superymmetric optical structures". *Phyical Review Letters* (APS) **110** (23): 233902. arXiv:1304.6646. Bibcode:2013PhRvL.110w3902M. doi:10.1103/PhysRevLett.110.233902. Retrieved April 2014.

[27] Heinrich, M.; Miri, M.-A.; Stützer, S.; El-Ganainy, R.; Nolte, S.; Szameit, A.; Christodoulides, D. N. (2014). "Superymmetric mode converters". *Nature Communications* (NPG) **5**: 3698. arXiv:1401.5734. Bibcode:2014NatCo...5E3698H. doi:10.1038/ncomms4698 Retrieved April 2014.

[28] Miri, M.-A.; Heinrich; Christodoulides, D. N. (2014). "SUSY-inspired one-dimensional transformation optics". *Optica* (OSA) **1**: 89. arXiv:1408.0832. doi:10.1364/OPTICA.1.000089. Retrieved August 2014.

[29] Krasnitz, Michael (2002). *Correlation functions in supersymmetric gauge theories from supergravity fluctuafluctuations hHKtions*. Princeton University Department of Physics: Princeton University Department of Physics. p. 91.

[30] Polchinski,J. *String theory. Vol. 2: Superstring theory and beyond*, Appendix B

[31] LEPSUSYWG, ALEPH, DELPHI, L3 and OPAL experiments, charginos, large m0 LEPSUSYWG/01-03.1

[32] The D0-Collaboration (2009). "Search for associated production of charginos and neutralinos in the trilepton final state using 2.3 fb^{-1} of data". arXiv:0901.0646. Bibcode:2009PhLB..680...34D. doi:10.1016/j.physletb.2009.08.011.

[33] The D0 Collaboration (2006). "Search for squarks and gluinos in events with jets and missing transverse energy in $p\bar{p}$ collisions at $\sqrt{s}$ =1.96 TeV". arXiv:0712.3805. Bibcode:2008PhLB..660..449D. doi:10.1016/j.physletb.2008.01.042.

[34] O. Buchmueller et al.. "Likelihood Functions for Supersymmetric Observables in Frequentist Analyses of the CMSSM and NUHM1". arXiv:0907.5568.

[35] Implications of Initial LHC Searches for Supersymmetry

[36] Fine-tuning implications for complementary dark matter and LHC SUSY searches

[37] What LHC tells about SUSY

[38] Early SUSY searches at the LHC

[39] CMS Collaboration; Khachatryan, V.; Sirunyan, A.; Tumasyan, A.; Adam, W.; Bergauer, T.; Dragicevic, M.; Erö, J.; Fabjan, C.; Toropin, Yu; Dermenev, A; Gninenko, S; Golubev, N; Kirsanov, M; Krasnikov, N; Matveev, V; Hrubec, A; Jeitler, A; Troitsky, S; Swain, A; Liko, O; Mikulec, M; Pernicka, M.; Rahbaran, B.; Trocino, H.; Wood, R.; Zhang, J.; Anastassov, A.; Kubik, A et al. (November 2011). "Search for Supersymmetry at the LHC in Events with Jets and Missing Transverse Energy". *Physical Review Letters* **107** (22): 221804. arXiv:1109.2352. Bibcode:2011PhRvL.107v1804C. doi:10.1103/PhysRevLett.107.221804. PMID 22182023.

[40] Marcela Carena and Howard E. Haber; Haber (1970). "Higgs Boson Theory and Phenomenology". *Progress in Particle and Nuclear Physics* **50**: 63. arXiv:hep-ph/0208209v3. Bibcode:2003PrPNP..50...63C. doi:10.1016/S0146-6410(02)00177-1.

[41] Patrick Draper et al (December 2011). "Implications of a 125 GeV Higgs for the MSSM and Low-Scale SUSY Breaking". arXiv:1112.3068. Bibcode:2012PhRvD..85i5007D. doi:10.1103/PhysRevD.85.095007.

[42] R. W. Kühne: Quantum Field Theory with Electric-Magnetic Duality and Spin-Mass Duality but Without Grand Unification and Supersymmetry. African Review of Physics 6 (2011) 165-179.

[43] "Global Fits of the cMSSM and NUHM including the LHC Higgs discovery and new XENON100 constraints", C. Strege, G. Bertone, F. Feroz, M. Fornasa, R. Ruiz de Austri, R. Trotta, arXiv:1212.2636

7.11 Further reading

- Supersymmetry and Supergravity page in String Theory Wiki lists more books and reviews.

7.11.1 Theoretical introductions, free and online

- S. Martin (2011). "A Supersymmetry Primer". arXiv:hep-ph/9709356.
- Joseph D. Lykken (1996). "Introduction to Supersymmetry". arXiv:hep-th/9612114.
- Manuel Drees (1996). "An Introduction to Supersymmetry". arXiv:hep-ph/9611409.
- Adel Bilal (2001). "Introduction to Supersymmetry". arXiv:hep-th/0101055.
- An Introduction to Global Supersymmetry by Philip Arygres, 2001

7.11.2 Monographs

- Weak Scale Supersymmetry by Howard Baer and Xerxes Tata, 2006.
- Cooper, F., A. Khare and U. Sukhatme. "Supersymmetry in Quantum Mechanics." Phys. Rep. 251 (1995) 267-85 (arXiv:hep-th/9405029).
- Junker, G. *Supersymmetric Methods in Quantum and Statistical Physics*, Springer-Verlag (1996).
- Gordon L. Kane.*Supersymmetry: Unveiling the Ultimate Laws of Nature* Basic Books, New York (2001). ISBN 0-7382-0489-7.

- Gordon L. Kane and Shifman, M., eds. *The Supersymmetric World: The Beginnings of the Theory*, World Scientific, Singapore (2000). ISBN 981-02-4522-X.

- Weinberg, Steven, *The Quantum Theory of Fields, Volume 3: Supersymmetry*, Cambridge University Press, Cambridge, (1999). ISBN 0-521-66000-9.

- Wess, Julius, and Jonathan Bagger, *Supersymmetry and Supergravity*, Princeton University Press, Princeton, (1992). ISBN 0-691-02530-4.

- Duplij, Steven; Siegel, Warren; Bagger, Jonathan (eds.) (2005). *Concise Encyclopedia of Supersymmetry*, Springer, Berlin/New York, (Second printing) ISBN 978-1-4020-1338-6

7.11.3 On experiments

- Bennett GW, *et al.*; Muon (g–2) Collaboration; Bousquet; Brown; Bunce; Carey; Cushman; Danby; Debevec; Deile; Deng; Dhawan; Druzhinin; Duong; Farley; Fedotovich; Gray; Grigoriev; Grosse-Perdekamp; Grossmann; Hare; Hertzog; Huang; Hughes; Iwasaki; Jungmann; Kawall; Khazin; Krienen; Kronkvist et al. (2004). "Measurement of the negative muon anomalous magnetic moment to 0.7 ppm". *Physical Review Letters* **92** (16): 161802. arXiv:hep-ex/0401008. Bibcode:2004PhRvL..92p1802B. doi:10.1103/PhysRevLett.92.161802. PMID 15169217.

- Brookhaven National Laboratory (Jan. 8, 2004). *New g−2 measurement deviates further from Standard Model*. Press Release.

- Fermi National Accelerator Laboratory (Sept 25, 2006). *Fermilab's CDF scientists have discovered the quick-change behavior of the B-sub-s meson*. Press Release.

7.12 External links

- What do current LHC results (mid-August 2011) imply about supersymmetry? Matt Strassler
- ATLAS Experiment Supersymmetry search documents
- CMS Experiment Supersymmetry search documents
- "Particle wobble shakes up supersymmetry", *Cosmos* magazine, September 2006
- LHC results put supersymmetry theory 'on the spot' BBC news 27/8/2011
- SUSY running out of hiding places BBC news 12/11/2012
- Supersymmetry in optics? "Skulls in the Stars" blog 22/08/2013

Chapter 8

Quantum gravity

Quantum gravity (**QG**) is a field of theoretical physics that seeks to describe the force of gravity according to the principles of quantum mechanics.

The current understanding of gravity is based on Albert Einstein's general theory of relativity, which is formulated within the framework of classical physics. On the other hand, the nongravitational forces are described within the framework of quantum mechanics, a radically different formalism for describing physical phenomena based on probability.[1] The necessity of a quantum mechanical description of gravity follows from the fact that one cannot consistently couple a classical system to a quantum one.[2]

Although a quantum theory of gravity is needed in order to reconcile general relativity with the principles of quantum mechanics, difficulties arise when one attempts to apply the usual prescriptions of quantum field theory to the force of gravity.[3] From a technical point of view, the problem is that the theory one gets in this way is not renormalizable and therefore cannot be used to make meaningful physical predictions. As a result, theorists have taken up more radical approaches to the problem of quantum gravity, the most popular approaches being string theory and loop quantum gravity.[4]

Strictly speaking, the aim of quantum gravity is only to describe the quantum behavior of the gravitational field and should not be confused with the objective of unifying all fundamental interactions into a single mathematical framework. Although some quantum gravity theories such as string theory try to unify gravity with the other fundamental forces, others such as loop quantum gravity make no such attempt; instead, they make an effort to quantize the gravitational field while it is kept separate from the other forces. A theory of quantum gravity which is also a grand unification of all known interactions, is sometimes referred to as a theory of everything (TOE).

One of the difficulties of quantum gravity is that quantum gravitational effects are only expected to become apparent near the Planck scale, a scale far smaller in distance (equivalently, far larger in energy) than what is currently accessible at high energy particle accelerators. As a result, quantum gravity is a mainly theoretical enterprise, although there are speculations about how quantum gravity effects might be observed in existing experiments.[5]

8.1 Overview

Much of the difficulty in meshing these theories at all energy scales comes from the different assumptions that these theories make on how the universe works. Quantum field theory depends on particle fields embedded in the flat space-time of special relativity. General relativity models gravity as a curvature within space-time that changes as a gravitational mass moves. Historically, the most obvious way of combining the two (such as treating gravity as simply another particle field) ran quickly into what is known as the renormalization problem. In the old-fashioned understanding of renormalization, gravity particles would attract each other and adding together all of the interactions results in many infinite values which cannot easily be cancelled out mathematically to yield sensible, finite results. This is in contrast with quantum electrodynamics where, while the series still do not converge, the interactions sometimes evaluate to infinite results, but those are few enough in number to be removable via renormalization.

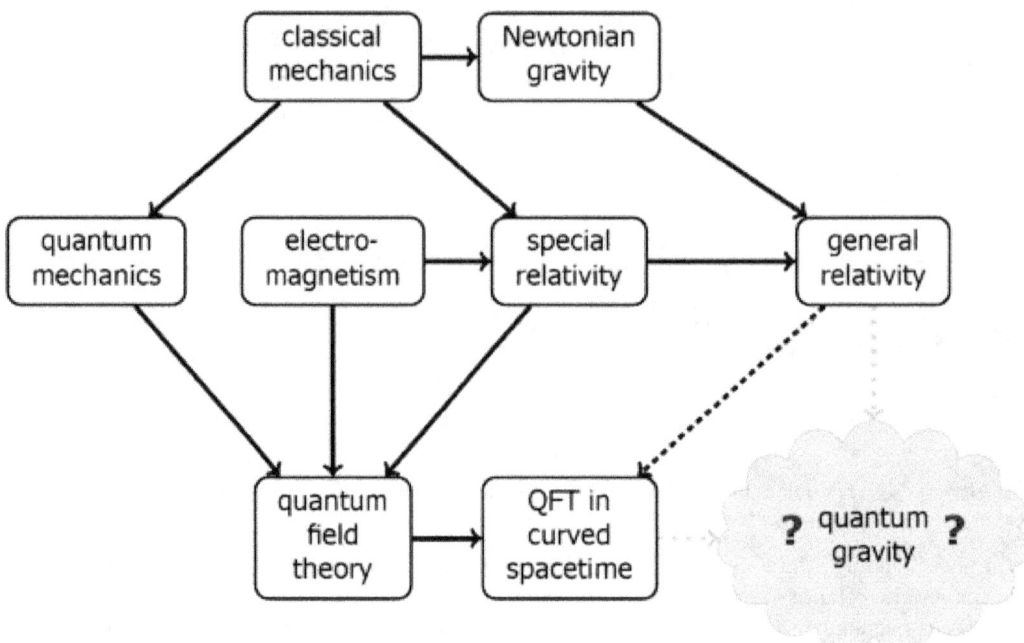

Diagram showing where quantum gravity sits in the hierarchy of physics theories

8.1.1 Effective field theories

Quantum gravity can be treated as an effective field theory. Effective quantum field theories come with some high-energy cutoff, beyond which we do not expect that the theory provides a good description of nature. The "infinities" then become large but finite quantities depending on this finite cutoff scale, and correspond to processes that involve very high energies near the fundamental cutoff. These quantities can then be absorbed into an infinite collection of coupling constants, and at energies well below the fundamental cutoff of the theory, to any desired precision; only a finite number of these coupling constants need to be measured in order to make legitimate quantum-mechanical predictions. This same logic works just as well for the highly successful theory of low-energy pions as for quantum gravity. Indeed, the first quantum-mechanical corrections to graviton-scattering and Newton's law of gravitation have been explicitly computed[6] (although they are so astronomically small that we may never be able to measure them). In fact, gravity is in many ways a much better quantum field theory than the Standard Model, since it appears to be valid all the way up to its cutoff at the Planck scale.

While confirming that quantum mechanics and gravity are indeed consistent at reasonable energies, it is clear that near or above the fundamental cutoff of our effective quantum theory of gravity (the cutoff is generally assumed to be of the order of the Planck scale), a new model of nature will be needed. Specifically, the problem of combining quantum mechanics and gravity becomes an issue only at very high energies, and may well require a totally new kind of model.

8.1.2 Quantum gravity theory for the highest energy scales

The general approach to deriving a quantum gravity theory that is valid at even the highest energy scales is to assume that such a theory will be simple and elegant and, accordingly, to study symmetries and other clues offered by current theories that might suggest ways to combine them into a comprehensive, unified theory. One problem with this approach is that it is unknown whether quantum gravity will actually conform to a simple and elegant theory, as it should resolve the dual conundrums of special relativity with regard to the uniformity of acceleration and gravity, and general relativity with regard to spacetime curvature.

Such a theory is required in order to understand problems involving the combination of very high energy and very small dimensions of space, such as the behavior of black holes, and the origin of the universe.

8.2 Quantum mechanics and general relativity

8.2.1 The graviton

Main article: Graviton

At present, one of the deepest problems in theoretical physics is harmonizing the theory of general relativity, which describes gravitation, and applications to large-scale structures (stars, planets, galaxies), with quantum mechanics, which describes the other three fundamental forces acting on the atomic scale. This problem must be put in the proper context, however. In particular, contrary to the popular claim that quantum mechanics and general relativity are fundamentally incompatible, one can demonstrate that the structure of general relativity essentially follows inevitably from the quantum mechanics of interacting theoretical spin-2 massless particles (called gravitons).[7][8][9][10][11]

While there is no concrete proof of the existence of gravitons, quantized theories of matter may necessitate their existence. Supporting this theory is the observation that all fundamental forces except gravity have one or more known messenger particles, leading researchers to believe that at least one most likely does exist; they have dubbed these hypothetical particles *gravitons*. The predicted find would result in the classification of the graviton as a "force particle" similar to the photon of the electromagnetic field. Many of the accepted notions of a unified theory of physics since the 1970s assume, and to some degree depend upon, the existence of the graviton. These include string theory, superstring theory, M-theory, and loop quantum gravity. Detection of gravitons is thus vital to the validation of various lines of research to unify quantum mechanics and relativity theory.

8.2.2 The dilaton

Main article: Dilaton

The dilaton made its first appearance in Kaluza–Klein theory, a five-dimensional theory that combined gravitation and electromagnetism. Generally, it appears in string theory. More recently, it has appeared in the lower-dimensional many-bodied gravity problem[12] based on the field theoretic approach of Roman Jackiw. The impetus arose from the fact that complete analytical solutions for the metric of a covariant N-body system have proven elusive in General Relativity. To simplify the problem, the number of dimensions was lowered to $(1+1)$ namely one spatial dimension and one temporal dimension. This model problem, known as $R=T$ theory[13] (as opposed to the general $G=T$ theory) was amenable to exact solutions in terms of a generalization of the Lambert W function. It was also found that the field equation governing the dilaton (derived from differential geometry) was the Schrödinger equation and consequently amenable to quantization.[14]

Thus, one had a theory which combined gravity, quantization and even the electromagnetic interaction, promising ingredients of a fundamental physical theory. It is worth noting that the outcome revealed a previously unknown and already existing *natural link* between general relativity and quantum mechanics. However, this theory needs to be generalized in $(2+1)$ or $(3+1)$ dimensions although, in principle, the field equations are amenable to such generalization as shown with the inclusion of a one-graviton process[15] and yielding the correct Newtonian limit in d dimensions if a dilaton is included. However, it is not yet clear what the fully generalized field equation governing the dilaton in (3+1) dimensions should be. This is further complicated by the fact that gravitons can propagate in $(3+1)$ dimensions and consequently that would imply gravitons and dilatons exist in the real world. Moreover, detection of the dilaton is expected to be even more elusive than the graviton. However, since this approach allows for the combination of gravitational, electromagnetic and quantum effects, their coupling could potentially lead to a means of vindicating the theory, through cosmology and perhaps even *experimentally*.

8.2.3 Nonrenormalizability of gravity

Further information: Renormalization

General relativity, like electromagnetism, is a classical field theory. One might expect that, as with electromagnetism, there should be a corresponding quantum field theory.

However, gravity is perturbatively nonrenormalizable.[16] [17] For a quantum field theory to be well-defined according to this understanding of the subject, it must be asymptotically free or asymptotically safe. The theory must be characterized by a choice of *finitely many* parameters, which could, in principle, be set by experiment. For example, in quantum electrodynamics, these parameters are the charge and mass of the electron, as measured at a particular energy scale.

On the other hand, in quantizing gravity, there are *infinitely many independent parameters* (counterterm coefficients) needed to define the theory. For a given choice of those parameters, one could make sense of the theory, but since we can never do infinitely many experiments to fix the values of every parameter, we do not have a meaningful physical theory:

- At low energies, the logic of the renormalization group tells us that, despite the unknown choices of these infinitely many parameters, quantum gravity will reduce to the usual Einstein theory of general relativity.

- On the other hand, if we could probe very high energies where quantum effects take over, then *every one* of the infinitely many unknown parameters would begin to matter, and we could make no predictions at all.

As explained below, there is a way around this problem by treating QG as an effective field theory.

Any meaningful theory of quantum gravity that makes sense and is predictive at all energy scales must have some deep principle that reduces the infinitely many unknown parameters to a finite number that can then be measured.

- One possibility is that normal perturbation theory is not a reliable guide to the renormalizability of the theory, and that there really *is* a UV fixed point for gravity. Since this is a question of non-perturbative quantum field theory, it is difficult to find a reliable answer, but some people still pursue this option.

- Another possibility is that there are new symmetry principles that constrain the parameters and reduce them to a finite set. This is the route taken by string theory, where all of the excitations of the string essentially manifest themselves as new symmetries.

8.2.4 QG as an effective field theory

Main article: Effective field theory

In an effective field theory, all but the first few of the infinite set of parameters in a non-renormalizable theory are suppressed by huge energy scales and hence can be neglected when computing low-energy effects. Thus, at least in the low-energy regime, the model is indeed a predictive quantum field theory.[6] (A very similar situation occurs for the very similar effective field theory of low-energy pions.) Furthermore, many theorists agree that even the Standard Model should really be regarded as an effective field theory as well, with "nonrenormalizable" interactions suppressed by large energy scales and whose effects have consequently not been observed experimentally.

Recent work[6] has shown that by treating general relativity as an effective field theory, one can actually make legitimate predictions for quantum gravity, at least for low-energy phenomena. An example is the well-known calculation of the tiny first-order quantum-mechanical correction to the classical Newtonian gravitational potential between two masses.

8.2.5 Spacetime background dependence

Main article: Background independence

A fundamental lesson of general relativity is that there is no fixed spacetime background, as found in Newtonian mechanics and special relativity; the spacetime geometry is dynamic. While easy to grasp in principle, this is the hardest idea to understand about general relativity, and its consequences are profound and not fully explored, even at the classical level. To a certain extent, general relativity can be seen to be a relational theory,[18] in which the only physically relevant information is the relationship between different events in space-time.

On the other hand, quantum mechanics has depended since its inception on a fixed background (non-dynamic) structure. In the case of quantum mechanics, it is time that is given and not dynamic, just as in Newtonian classical mechanics. In relativistic quantum field theory, just as in classical field theory, Minkowski spacetime is the fixed background of the theory.

String theory

String theory can be seen as a generalization of quantum field theory where instead of point particles, string-like objects propagate in a fixed spacetime background, although the interactions among closed strings give rise to spacetime in a dynamical way. Although string theory had its origins in the study of quark confinement and not of quantum gravity, it was soon discovered that the string spectrum contains the graviton, and that "condensation" of certain vibration modes of strings is equivalent to a modification of the original background. In this sense, string perturbation theory exhibits exactly the features one would expect of a perturbation theory that may exhibit a strong dependence on asymptotics (as seen, for example, in the AdS/CFT correspondence) which is a weak form of background dependence.

Background independent theories

Loop quantum gravity is the fruit of an effort to formulate a background-independent quantum theory.

Topological quantum field theory provided an example of background-independent quantum theory, but with no local degrees of freedom, and only finitely many degrees of freedom globally. This is inadequate to describe gravity in 3+1 dimensions which has local degrees of freedom according to general relativity. In 2+1 dimensions, however, gravity is a topological field theory, and it has been successfully quantized in several different ways, including spin networks.

8.2.6 Semi-classical quantum gravity

Quantum field theory on curved (non-Minkowskian) backgrounds, while not a full quantum theory of gravity, has shown many promising early results. In an analogous way to the development of quantum electrodynamics in the early part of the 20th century (when physicists considered quantum mechanics in classical electromagnetic fields), the consideration of quantum field theory on a curved background has led to predictions such as black hole radiation.

Phenomena such as the Unruh effect, in which particles exist in certain accelerating frames but not in stationary ones, do not pose any difficulty when considered on a curved background (the Unruh effect occurs even in flat Minkowskian backgrounds). The vacuum state is the state with the least energy (and may or may not contain particles). See Quantum field theory in curved spacetime for a more complete discussion.

8.2.7 Points of tension

There are other points of tension between quantum mechanics and general relativity.

- First, classical general relativity breaks down at singularities, and quantum mechanics becomes inconsistent with general relativity in the neighborhood of singularities (however, no one is certain that classical general relativity applies near singularities in the first place).

- Second, it is not clear how to determine the gravitational field of a particle, since under the Heisenberg uncertainty principle of quantum mechanics its location and velocity cannot be known with certainty. The resolution of these points may come from a better understanding of general relativity.[19]

- Third, there is the Problem of time in quantum gravity. Time has a different meaning in quantum mechanics and general relativity and hence there are subtle issues to resolve when trying to formulate a theory which combines the two.[20]

8.3 Candidate theories

There are a number of proposed quantum gravity theories.[21] Currently, there is still no complete and consistent quantum theory of gravity, and the candidate models still need to overcome major formal and conceptual problems. They also face the common problem that, as yet, there is no way to put quantum gravity predictions to experimental tests, although there is hope for this to change as future data from cosmological observations and particle physics experiments becomes available.[22][23]

8.3.1 String theory

Main article: String theory

One suggested starting point is ordinary quantum field theories which, after all, are successful in describing the other three basic fundamental forces in the context of the standard model of elementary particle physics. However, while this leads to an acceptable effective (quantum) field theory of gravity at low energies,[24] gravity turns out to be much more problematic at higher energies. Where, for ordinary field theories such as quantum electrodynamics, a technique known as renormalization is an integral part of deriving predictions which take into account higher-energy contributions,[25] gravity turns out to be nonrenormalizable: at high energies, applying the recipes of ordinary quantum field theory yields models that are devoid of all predictive power.[26]

One attempt to overcome these limitations is to replace ordinary quantum field theory, which is based on the classical concept of a point particle, with a quantum theory of one-dimensional extended objects: string theory.[27] At the energies reached in current experiments, these strings are indistinguishable from point-like particles, but, crucially, different modes of oscillation of one and the same type of fundamental string appear as particles with different (electric and other) charges. In this way, string theory promises to be a unified description of all particles and interactions.[28] The theory is successful in that one mode will always correspond to a graviton, the messenger particle of gravity; however, the price to pay are unusual features such as six extra dimensions of space in addition to the usual three for space and one for time.[29]

In what is called the second superstring revolution, it was conjectured that both string theory and a unification of general relativity and supersymmetry known as supergravity[30] form part of a hypothesized eleven-dimensional model known as M-theory, which would constitute a uniquely defined and consistent theory of quantum gravity.[31][32] As presently understood, however, string theory admits a very large number (10^{500} by some estimates) of consistent vacua, comprising the so-called "string landscape". Sorting through this large family of solutions remains a major challenge.

8.3.2 Loop quantum gravity

Main article: Loop quantum gravity

Loop quantum gravity is based first of all on the idea to take seriously the insight of general relativity that spacetime is a dynamical field and therefore is a quantum object. The second idea is that the quantum discreteness that determines the particle-like behavior of other field theories (for instance, the photons of the electromagnetic field) affects also the structure of space.

The main result of loop quantum gravity is the derivation of a granular structure of space at the Planck length. This is derived as follows. In the case of electromagnetism, the quantum operator representing the energy of each frequency of the field has discrete spectrum. Therefore the energy of each frequency is quantized, and the quanta are the photons. In the case of gravity, the operators representing the area and the volume of each surface or space region have discrete spectrum. Therefore area and volume of any portion of space are quantized, and the quanta are elementary quanta of space. It follows that spacetime has an elementary quantum granular structure at the Planck scale, which cuts-off the ultraviolet infinities of quantum field theory.

The quantum state of spacetime is described in the theory by means of a mathematical structure called spin networks. Spin networks were initially introduced by Roger Penrose in abstract form, and later shown by Carlo Rovelli and Lee Smolin to derive naturally from a non perturbative quantization of general relativity. Spin networks do not represent quantum states of a field in spacetime: they represent directly quantum states of spacetime.

The theory is based on the reformulation of general relativity known as Ashtekar variables, which represent geometric gravity using mathematical analogues of electric and magnetic fields.[33][34] In the quantum theory space is represented by a network structure called a spin network, evolving over time in discrete steps.[35][36][37][38]

The dynamics of the theory is today constructed in several versions. One version starts with the canonical quantization of general relativity. The analogue of the Schrödinger equation is a Wheeler–DeWitt equation, which can be defined in the theory.[39] In the covariant, or spinfoam formulation of the theory, the quantum dynamics is obtained via a sum over discrete versions of spacetime, called spinfoams. These represent histories of spin networks.

8.3.3 Other approaches

There are a number of other approaches to quantum gravity. The approaches differ depending on which features of general relativity and quantum theory are accepted unchanged, and which features are modified.[40][41] Examples include:

- Acoustic metric and other analog models of gravity
- Asymptotic safety in quantum gravity
- Euclidean quantum gravity
- Causal Dynamical Triangulation[42]
- Causal fermion systems,[43][44][45][46][47] giving quantum mechanics, general relativity and quantum field theory as limiting cases.
- Causal sets[48]
- Covariant Feynman path integral approach
- Group field theory[49]
- Wheeler-DeWitt equation
- Geometrodynamics
- Hořava–Lifshitz gravity
- MacDowell–Mansouri action
- Noncommutative geometry.
- Path-integral based models of quantum cosmology[50]
- Regge calculus
- String-nets giving rise to gapless helicity ±2 excitations with no other gapless excitations[51]
- Superfluid vacuum theory a.k.a. theory of BEC vacuum
- Supergravity
- Twistor theory[52]
- Canonical quantum gravity
- E8 Theory

8.4 Weinberg–Witten theorem

In quantum field theory, the Weinberg–Witten theorem places some constraints on theories of composite gravity/emergent gravity. However, recent developments attempt to show that if locality is only approximate and the holographic principle is correct, the Weinberg–Witten theorem would not be valid.

8.5 Experimental tests

As was emphasized above, quantum gravitational effects are extremely weak and therefore difficult to test. For this reason, the possibility of experimentally testing quantum gravity had not received much attention prior to the late 1990s. However, in the past decade, physicists have realized that evidence for quantum gravitational effects can guide the development of the theory. Since the theoretical development has been slow, Phenomenological Quantum Gravity which studies the possibility of experimental tests, has obtained increased attention.[53][54]

The most widely pursued possibilities for quantum gravity phenomenology include violations of Lorentz invariance, imprints of quantum gravitational effects in the Cosmic Microwave Background (in particular its polarization), and decoherence induced by fluctuations in the space-time foam.

The BICEP2 experiment detected primordial B-mode polarization caused by gravitational waves in the early universe. The waves were born as quantum fluctuations in gravity itself. Cosmologist Ken Olum (Tufts University) stated: "I think this is the only observational evidence that we have that actually shows that gravity is quantized....It's probably the only evidence of this that we will ever have."[55]

8.6 See also

8.7 References

[1] Griffiths, David J. (2004). *Introduction to Quantum Mechanics*. Pearson Prentice Hall. OCLC 803860989.

[2] Wald, Robert M. (1984). *General Relativity*. University of Chicago Press. p. 382. OCLC 471881415.

[3] Zee, Anthony (2010). *Quantum Field Theory in a Nutshell* (2nd ed.). Princeton University Press. p. 172. OCLC 659549695.

[4] Penrose, Roger (2007). *The road to reality : a complete guide to the laws of the universe*. Vintage. p. 1017. OCLC 716437154.

[5] Quantum effects in the early universe might have an observable effect on the structure of the present universe, for example, or gravity might play a role in the unification of the other forces. Cf. the text by Wald cited above.

[6] Donoghue (1995). "Introduction to the Effective Field Theory Description of Gravity". arXiv:9512024 [gr-qc]. (verify against ISBN 9789810229085)

[7] Kraichnan, R. H. (1955). "Special-Relativistic Derivation of Generally Covariant Gravitation Theory". *Physical Review* **98** (4): 1118–1122. Bibcode:1955PhRv...98.1118K. doi:10.1103/PhysRev.98.1118.

[8] Gupta, S. N. (1954). "Gravitation and Electromagnetism". *Physical Review* **96** (6): 1683–1685. Bibcode:1954PhRv...96.1683G. doi:10.1103/PhysRev.96.1683.

[9] Gupta, S. N. (1957). "Einstein's and Other Theories of Gravitation". *Reviews of Modern Physics* **29** (3): 334–336. Bibcode:1957RvMP...29..334G. doi:10.1103/RevModPhys.29.334.

[10] Gupta, S. N. (1962). "Quantum Theory of Gravitation". *Recent Developments in General Relativity*. Pergamon Press. pp. 251–258.

[11] Deser, S. (1970). "Self-Interaction and Gauge Invariance". *General Relativity and Gravitation* **1**: 9–18. arXiv:gr-qc/0411023. Bibcode:1970GReGr...1....9D. doi:10.1007/BF00759198.

[12] Ohta, Tadayuki; Mann, Robert (1996). "Canonical reduction of two-dimensional gravity for particle dynamics". *Classical and Quantum Gravity* **13** (9): 2585–2602. arXiv:gr-qc/9605004. Bibcode:1996CQGra..13.2585O. doi:10.1088/0264-9381/13/9/022.

[13] Sikkema, A E; Mann, R B (1991). "Gravitation and cosmology in (1+1) dimensions". *Classical and Quantum Gravity* **8**: 219–235. Bibcode:1991CQGra...8..219S. doi:10.1088/0264-9381/8/1/022.

[14] Farrugia; Mann; Scott (2007). "N-body Gravity and the Schroedinger Equation". *Classical and Quantum Gravity* **24** (18): 4647–4659. arXiv:gr-qc/0611144. Bibcode:2007CQGra..24.4647F. doi:10.1088/0264-9381/24/18/006.

[15] Mann, R B; Ohta, T (1997). "Exact solution for the metric and the motion of two bodies in (1+1)-dimensional gravity". *Physical Review D* **55** (8): 4723–4747. arXiv:gr-qc/9611008. Bibcode:1997PhRvD..55.4723M. doi:10.1103/PhysRevD.55.4723.

[16] Feynman, R. P.; Morinigo, F. B.; Wagner, W. G.; Hatfield, B. (1995). *Feynman lectures on gravitation*. Addison-Wesley. ISBN 0-201-62734-5.

[17] Hamber, H. W. (2009). *Quantum Gravitation - The Feynman Path Integral Approach*. Springer Publishing. ISBN 978-3-540-85292-6.

[18] Smolin, Lee (2001). *Three Roads to Quantum Gravity*. Basic Books. pp. 20–25. ISBN 0-465-07835-4. Pages 220–226 are annotated references and guide for further reading.

[19] Hunter Monroe (2005). "Singularity-Free Collapse through Local Inflation". arXiv:0506506 [astro-ph].

[20] Edward Anderson (2010). "The Problem of Time in Quantum Gravity". arXiv:1009.2157 [gr-qc]. (also published as chapter 4 of ISBN 9781611229578)

[21] A timeline and overview can be found in Rovelli, Carlo (2000). "Notes for a brief history of quantum gravity". arXiv:0006061 [gr-qc]. (verify against ISBN 9789812777386)

[22] Ashtekar, Abhay (2007). "Loop Quantum Gravity: Four Recent Advances and a Dozen Frequently Asked Questions". *11th Marcel Grossmann Meeting on Recent Developments in Theoretical and Experimental General Relativity*. p. 126. arXiv:0705.2222. Bibcode:2008mgm..conf..126A. doi:10.1142/9789812834300_0008.

[23] Schwarz, John H. (2007). "String Theory: Progress and Problems". *Progress of Theoretical Physics Supplement* **170**: 214–226. arXiv:hep-th/0702219. Bibcode:2007PThPS.170..214S. doi:10.1143/PTPS.170.214.

[24] Donoghue, John F.(editor), (1995). "Introduction to the Effective Field Theory Description of Gravity". In Cornet, Fernando. *Effective Theories: Proceedings of the Advanced School, Almunecar, Spain, 26 June–1 July 1995*. Singapore: World Scientific. arXiv:gr-qc/9512024. ISBN 981-02-2908-9.

[25] Weinberg, Steven (1996). "17–18". *The Quantum Theory of Fields II: Modern Applications*. Cambridge University Press. ISBN 0-521-55002-5.

[26] Goroff, Marc H.; Sagnotti, Augusto; Sagnotti, Augusto (1985). "Quantum gravity at two loops". *Physics Letters B* **160**: 81–86. Bibcode:1985PhLB..160...81G. doi:10.1016/0370-2693(85)91470-4.

[27] An accessible introduction at the undergraduate level can be found in Zwiebach, Barton (2004). *A First Course in String Theory*. Cambridge University Press. ISBN 0-521-83143-1., and more complete overviews in Polchinski, Joseph (1998). *String Theory Vol. I: An Introduction to the Bosonic String*. Cambridge University Press. ISBN 0-521-63303-6. and Polchinski, Joseph (1998b). *String Theory Vol. II: Superstring Theory and Beyond*. Cambridge University Press. ISBN 0-521-63304-4.

[28] Ibanez, L. E. (2000). "The second string (phenomenology) revolution". *Classical & Quantum Gravity* **17** (5): 1117–1128. arXiv:hep-ph/9911499. Bibcode:2000CQGra..17.1117I. doi:10.1088/0264-9381/17/5/321.

[29] For the graviton as part of the string spectrum, e.g. Green, Schwarz & Witten 1987, sec. 2.3 and 5.3; for the extra dimensions, ibid sec. 4.2.

[30] Weinberg, Steven (2000). "31". *The Quantum Theory of Fields II: Modern Applications*. Cambridge University Press. ISBN 0-521-55002-5.

[31] Townsend, Paul K. (1996). "Four Lectures on M-Theory". *1996 Summer School in High Energy Physics and Cosmology*. ICTP Series in Theoretical Physics. p. 385. arXiv:hep-th/9612121. Bibcode:1997hepcbconf..385T.

[32] Duff, Michael (1996). "M-Theory (the Theory Formerly Known as Strings)". *International Journal of Modern Physics A* **11** (32): 5623–5642. arXiv:hep-th/9608117. Bibcode:1996IJMPA..11.5623D. doi:10.1142/S0217751X96002583.

[33] Ashtekar, Abhay (1986). "New variables for classical and quantum gravity". *Physical Review Letters* **57** (18): 2244–2247. Bibcode:1986PhRvL..57.2244A. doi:10.1103/PhysRevLett.57.2244. PMID 10033673.

[34] Ashtekar, Abhay (1987). "New Hamiltonian formulation of general relativity". *Physical Review D* **36** (6): 1587–1602. Bibcode:1987PhRvD..36.1587A. doi:10.1103/PhysRevD.36.1587.

[35] Thiemann, Thomas (2006). "Loop Quantum Gravity: An Inside View". *Approaches to Fundamental Physics*. Lecture Notes in Physics **721**: 185. arXiv:hep-th/0608210. Bibcode:2007LNP...721..185T. doi:10.1007/978-3-540-71117-9_10. ISBN 978-3-540-71115-5.

[36] Rovelli, Carlo (1998). "Loop Quantum Gravity". *Living Reviews in Relativity* **1**. Retrieved 2008-03-13.

[37] Ashtekar, Abhay; Lewandowski, Jerzy (2004). "Background Independent Quantum Gravity: A Status Report". *Classical & Quantum Gravity* **21** (15): R53–R152. arXiv:gr-qc/0404018. Bibcode:2004CQGra..21R..53A. doi:10.1088/0264-9381/21/15/R01.

[38] Thiemann, Thomas (2003). "Lectures on Loop Quantum Gravity". *Lecture Notes in Physics.* Lecture Notes in Physics **631**: 41–135. arXiv:gr-qc/0210094. Bibcode:2003LNP...631...41T. doi:10.1007/978-3-540-45230-0_3. ISBN 978-3-540-40810-9.

[39] Rovelli, Carlo (2004). *Quantum Gravity.* Cambridge University Press. ISBN 0521715962.

[40] Isham, Christopher J. (1994). "Prima facie questions in quantum gravity". In Ehlers, Jürgen; Friedrich, Helmut. *Canonical Gravity: From Classical to Quantum.* Springer. arXiv:gr-qc/9310031. ISBN 3-540-58339-4.

[41] Sorkin, Rafael D. (1997). "Forks in the Road, on the Way to Quantum Gravity". *International Journal of Theoretical Physics* **36** (12): 2759–2781. arXiv:gr-qc/9706002. Bibcode:1997IJTP...36.2759S. doi:10.1007/BF02435709.

[42] Loll, Renate (1998). "Discrete Approaches to Quantum Gravity in Four Dimensions". *Living Reviews in Relativity* **1**: 13. arXiv:gr-qc/9805049. Bibcode:1998LRR.....1...13L. doi:10.12942/lrr-1998-13. Retrieved 2008-03-09.

[43] F. Finster, The Principle of the Fermionic Projector, hep-th/0001048, hep-th/0202059, hep-th/0210121, AMS/IP Studies in Advanced Mathematics, vol. **35**, American Mathematical Society, Providence, RI, 2006.

[44] F. Finster, A formulation of quantum field theory realizing a sea of interacting Dirac particles, arXiv:0911.2102 [hep-th], Lett. Math. Phys. **97** (2011), no. 2, 165–183.

[45] F. Finster, An action principle for an interacting fermion system and its analysis in the continuum limit, arXiv:0908.1542 [math-ph] (2009).

[46] F. Finster, The continuum limit of a fermion system involving neutrinos: Weak and gravitational interactions, arXiv:1211.3351 [math-ph] (2012).

[47] F. Finster, Perturbative quantum field theory in the framework of the fermionic projector, arXiv:1310.4121 [math-ph], J. Math. Phys. **55** (2014), no. 4, 042301.

[48] Sorkin, Rafael D. (2005). "Causal Sets: Discrete Gravity". In Gomberoff, Andres; Marolf, Donald. *Lectures on Quantum Gravity.* Springer. arXiv:gr-qc/0309009. ISBN 0-387-23995-2.

[49] See Daniele Oriti and references therein.

[50] Hawking, Stephen W. (1987). "Quantum cosmology". In Hawking, Stephen W.; Israel, Werner. *300 Years of Gravitation.* Cambridge University Press. pp. 631–651. ISBN 0-521-37976-8.

[51] Wen 2006

[52] See ch. 33 in Penrose 2004 and references therein.

[53] Hossenfelder, Sabine (2011). "Experimental Search for Quantum Gravity". In V. R. Frignanni. *Classical and Quantum Gravity: Theory, Analysis and Applications.* Chapter 5: Nova Publishers. ISBN 978-1-61122-957-8.

[54] Hossenfelder, Sabine (2010-10-17). "Experimental Search for Quantum Gravity". *"Classical and Quantum Gravity: Theory, Analysis and Applications," Chapter , Edited by V. R. Frignanni, Nova Publishers ()* **5** (2011). arXiv:1010.3420.

[55] Camille Carlisle. "First Direct Evidence of Big Bang Inflation". SkyandTelescope.com. Retrieved March 18, 2014.

8.8 Further reading

- Ahluwalia, D. V. (2002). "Interface of Gravitational and Quantum Realms". *Modern Physics Letters A* **17** (15–17): 1135. arXiv:gr-qc/0205121. Bibcode:2002MPLA...17.1135A. doi:10.1142/S021773230200765X.

- Ashtekar, Abhay (2005). "The winding road to quantum gravity". *Current Science* **89**: 2064–2074.

- Carlip, Steven (2001). "Quantum Gravity: a Progress Report". *Reports on Progress in Physics* **64** (8): 885–942. arXiv:gr-qc/0108040. Bibcode:2001RPPh...64..885C. doi:10.1088/0034-4885/64/8/301.

- Herbert W. Hamber (2009). *Quantum Gravitation.* Springer Publishing. doi:10.1007/978-3-540-85293-3. ISBN 978-3-540-85292-6.

- Kiefer, Claus (2007). *Quantum Gravity.* Oxford University Press. ISBN 0-19-921252-X.

- Kiefer, Claus (2005). "Quantum Gravity: General Introduction and Recent Developments". *Annalen der Physik* **15**: 129–148. arXiv:gr-qc/0508120. Bibcode:2006AnP...518..129K. doi:10.1002/andp.200510175.

- Lämmerzahl, Claus, ed. (2003). *Quantum Gravity: From Theory to Experimental Search*. Lecture Notes in Physics. Springer. ISBN 3-540-40810-X.

- Rovelli, Carlo (2004). *Quantum Gravity*. Cambridge University Press. ISBN 0-521-83733-2.

- Trifonov, Vladimir (2008). "GR-friendly description of quantum systems". *International Journal of Theoretical Physics* **47** (2): 492–510. arXiv:math-ph/0702095. Bibcode:2008IJTP...47..492T. doi:10.1007/s10773-007-9474-3.

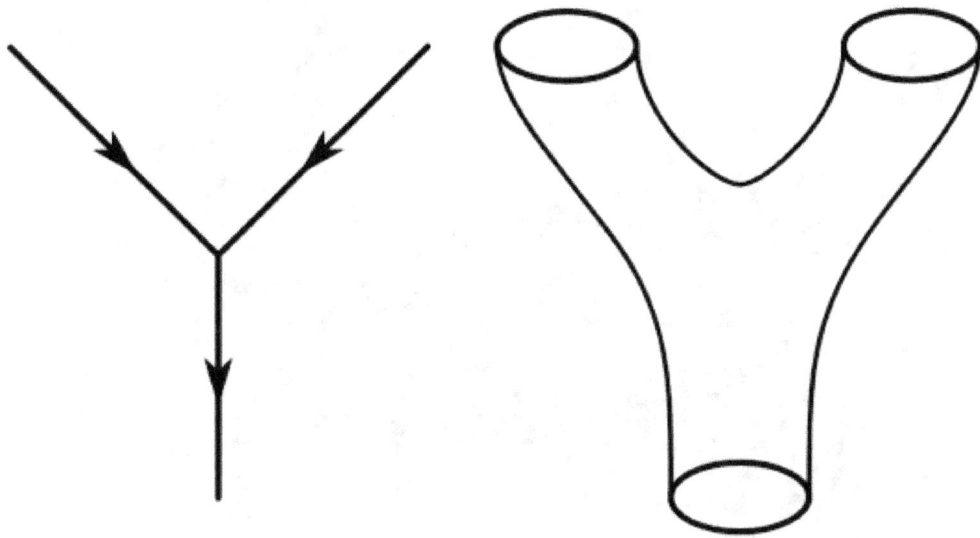

Interaction in the subatomic world: world lines of point-like particles in the Standard Model or a world sheet swept up by closed strings in string theory

Projection of a Calabi–Yau manifold, one of the ways of compactifying the extra dimensions posited by string theory

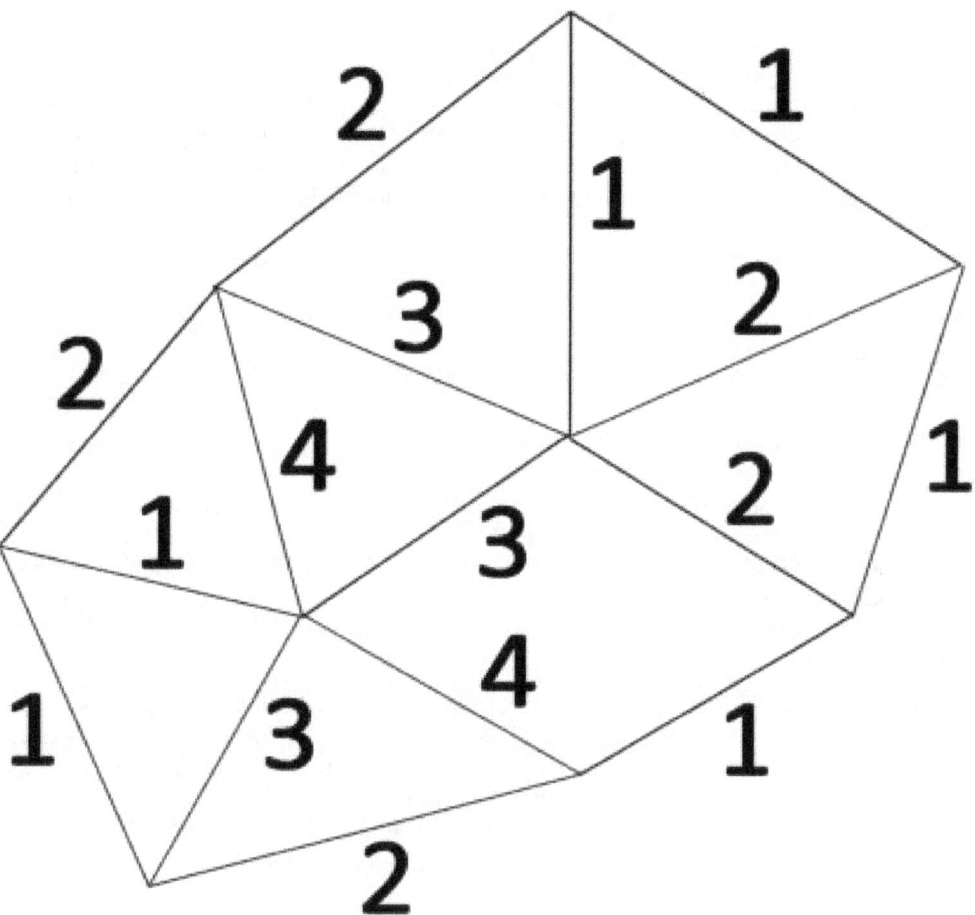

Simple spin network of the type used in loop quantum gravity

Chapter 9

Simple group

In mathematics, a **simple group** is a nontrivial group whose only normal subgroups are the trivial group and the group itself. A group that is not simple can be broken into two smaller groups, a normal subgroup and the quotient group, and the process can be repeated. If the group is finite, then eventually one arrives at uniquely determined simple groups by the Jordan–Hölder theorem. The complete classification of finite simple groups, completed in 2008, is a major milestone in the history of mathematics.

9.1 Examples

9.1.1 Finite simple groups

The cyclic group $G = \mathbf{Z}/3\mathbf{Z}$ of congruence classes modulo 3 (see modular arithmetic) is simple. If H is a subgroup of this group, its order (the number of elements) must be a divisor of the order of G which is 3. Since 3 is prime, its only divisors are 1 and 3, so either H is G, or H is the trivial group. On the other hand, the group $G = \mathbf{Z}/12\mathbf{Z}$ is not simple. The set H of congruence classes of 0, 4, and 8 modulo 12 is a subgroup of order 3, and it is a normal subgroup since any subgroup of an abelian group is normal. Similarly, the additive group $\mathbf{Z}$ of integers is not simple; the set of even integers is a non-trivial proper normal subgroup.[1]

One may use the same kind of reasoning for any abelian group, to deduce that the only simple abelian groups are the cyclic groups of prime order. The classification of nonabelian simple groups is far less trivial. The smallest nonabelian simple group is the alternating group A_5 of order 60, and every simple group of order 60 is isomorphic to A_5.[2] The second smallest nonabelian simple group is the projective special linear group PSL(2,7) of order 168, and it is possible to prove that every simple group of order 168 is isomorphic to PSL(2,7).[3][4]

9.1.2 Infinite simple groups

The infinite alternating group, i.e. the group of even permutations of the integers, A_∞ is simple. This group can be defined as the increasing union of the finite simple groups A_n with respect to standard embeddings $A_n \to A_{n+1}$. Another family of examples of infinite simple groups is given by $\text{PSL}_n(F)$, where F is a field and $n \geq 3$.

It is much more difficult to construct *finitely generated* infinite simple groups. The first example is due to Graham Higman and is a quotient of the Higman group.[5] Other examples include the infinite Thompson groups T and V. Finitely presented torsion-free infinite simple groups were constructed by Burger-Mozes.[6]

9.2 Classification

There is as yet no known classification for general simple groups.

9.2.1 Finite simple groups

Main article: list of finite simple groups
For more details on this topic, see Classification of finite simple groups.

The finite simple groups are important because in a certain sense they are the "basic building blocks" of all finite groups, somewhat similar to the way prime numbers are the basic building blocks of the integers. This is expressed by the Jordan–Hölder theorem which states that any two composition series of a given group have the same length and the same factors, up to permutation and isomorphism. In a huge collaborative effort, the classification of finite simple groups was declared accomplished in 1983 by Daniel Gorenstein, though some problems surfaced (specifically in the classification of quasithin groups, which were plugged in 2004).

Briefly, finite simple groups are classified as lying in one of 18 families, or being one of 26 exceptions:

- Z_p – cyclic group of prime order

- A_n – alternating group for $n \geq 5$

    The alternating groups may be considered as groups of Lie type over the field with one element, which unites this family with the next, and thus all families of non-abelian finite simple groups may be considered to be of Lie type.

- One of 16 families of groups of Lie type

    The Tits group is generally considered of this form, though strictly speaking it is not of Lie type, but rather index 2 in a group of Lie type.

- One of 26 exceptions, the sporadic groups, of which 20 are subgroups or subquotients of the monster group and are referred to as the "Happy Family", while the remaining 6 are referred to as pariahs.

9.3 Structure of finite simple groups

The famous theorem of Feit and Thompson states that every group of odd order is solvable. Therefore every finite simple group has even order unless it is cyclic of prime order.

The Schreier conjecture asserts that the group of outer automorphisms of every finite simple group is solvable. This can be proved using the classification theorem.

9.4 History for finite simple groups

There are two threads in the history of finite simple groups – the discovery and construction of specific simple groups and families, which took place from the work of Galois in the 1820s to the construction of the Monster in 1981; and proof that this list was complete, which began in the 19th century, most significantly took place 1955 through 1983 (when victory was initially declared), but was only generally agreed to be finished in 2004. As of 2010, work on improving the proofs and understanding continues; see (Silvestri 1979) for 19th century history of simple groups.

9.4.1 Construction

Simple groups have been studied at least since early Galois theory, where Évariste Galois realized that the fact that the alternating groups on five or more points was simple (and hence not solvable), which he proved in 1831, was the reason that one could not solve the quintic in radicals. Galois also constructed the projective special linear group of a plane over a prime finite field, PSL(2,p), and remarked that they were simple for p not 2 or 3. This is contained in his last letter to Chevalier,[7] and are the next example of finite simple groups.[8]

The next discoveries were by Camille Jordan in 1870.[9] Jordan had found 4 families of simple matrix groups over finite fields of prime order, which are now known as the classical groups.

At about the same time, it was shown that a family of five groups, called the Mathieu groups and first described by Émile Léonard Mathieu in 1861 and 1873, were also simple. Since these five groups were constructed by methods which did not yield infinitely many possibilities, they were called "sporadic" by William Burnside in his 1897 textbook.

Later Jordan's results on classical groups were generalized to arbitrary finite fields by Leonard Dickson, following the classification of complex simple Lie algebras by Wilhelm Killing. Dickson also constructed exception groups of type G_2 and E_6 as well, but not of types F_4, E_7, or E_8 (Wilson 2009, p. 2). In the 1950s the work on groups of Lie type was continued, with Claude Chevalley giving a uniform construction of the classical groups and the groups of exceptional type in a 1955 paper. This omitted certain known groups (the projective unitary groups), which were obtained by "twisting" the Chevalley construction. The remaining groups of Lie type were produced by Steinberg, Tits, and Herzig (who produced $^3D_4(q)$ and $^2E_6(q)$) and by Suzuki and Ree (the Suzuki–Ree groups).

These groups (the groups of Lie type, together with the cyclic groups, alternating groups, and the five exceptional Mathieu groups) were believed to be a complete list, but after a lull of almost a century since the work of Mathieu, in 1964 the first Janko group was discovered, and the remaining 20 sporadic groups were discovered or conjectured in 1965–1975, culminating in 1981, when Robert Griess announced that he had constructed Bernd Fischer's "Monster group". The Monster is the largest sporadic simple group having order of 808,017,424,794,512,875,886,459,904,961,710,757,005,7 The Monster has a faithful 196,883-dimensional representation in the 196,884-dimensional Griess algebra, meaning that each element of the Monster can be expressed as a 196,883 by 196,883 matrix.

9.4.2 Classification

The full classification is generally accepted as starting with the Feit–Thompson theorem of 1962/63, largely lasting until 1983, but only being finished in 2004.

Soon after the construction of the Monster in 1981, a proof, totaling more than 10,000 pages, was supplied that group theorists had successfully listed all finite simple groups, with victory declared in 1983 by Daniel Gorenstein. This was premature – some gaps were later discovered, notably in the classification of quasithin groups, which were eventually replaced in 2004 by a 1,300 page classification of quasithin groups, which is now generally accepted as complete.

9.5 Tests for nonsimplicity

Sylows' test: Let n be a positive integer that is not prime, and let p be a prime divisor of n. If 1 is the only divisor of n that is equal to 1 modulo p, then there does not exist a simple group of order n.

Proof: If n is a prime-power, then a group of order n has a nontrivial center[10] and, therefore, is not simple. If n is not a prime power, then every Sylow subgroup is proper, and, by Sylow's Third Theorem, we know that the number of Sylow p-subgroups of a group of order n is equal to 1 modulo p and divides n. Since 1 is the only such number, the Sylow p-subgroup is unique, and therefore it is normal. Since it is a proper, non-identity subgroup, the group is not simple.

Burnside: A non-Abelian finite simple group has order divisible by at least three distinct primes. This follows from Burnside's p-q theorem.

9.6 See also

- Almost simple group
- Characteristically simple group
- Quasisimple group
- Semisimple group
- List of finite simple groups

9.7 References

9.7.1 Notes

[1] Knapp (2006), p. 170

[2] Rotman (1995), p. 226

[3] Rotman (1995), p. 281

[4] Smith & Tabachnikova (2000), p. 144

[5] Higman, Graham (1951), *A finitely generated infinite simple group*, Journal of the London Mathematical Society. Second Series **26** (1): 61–64, doi:10.1112/jlms/s1-26.1.59, ISSN 0024-6107, MR 0038348

[6] M. Burger and S. Mozes. " Lattices in product of trees." *Publ. Math. IHES* **92** (2000), pp.151–194.

[7] Galois, Évariste (1846), *Lettre de Galois à M. Auguste Chevalier*, Journal de Mathématiques Pures et Appliquées **XI**: 408–415, retrieved 2009-02-04, PSL(2,p) and simplicity discussed on p. 411; exceptional action on 5, 7, or 11 points discussed on pp. 411–412; GL(v,p) discussed on p. 410

[8] Wilson, Robert (October 31, 2006), "Chapter 1: Introduction", *The finite simple groups*

[9] Jordan, Camille (1870), *Traité des substitutions et des équations algébriques*

[10] See the proof in p-group, for instance.

9.7.2 Textbooks

- Wilson, Robert A. (2009), *The finite simple groups*, Graduate Texts in Mathematics 251 **251**, Berlin, New York: Springer-Verlag, doi:10.1007/978-1-84800-988-2, ISBN 978-1-84800-987-5, Zbl 05622792, 2007 preprint.

- Burnside, William (1897), *Theory of groups of finite order*, Cambridge University Press

- Knapp, Anthony W. (2006), *Basic algebra*, Springer, ISBN 978-0-8176-3248-9

- Rotman, Joseph J. (1995), *An introduction to the theory of groups*, Graduate texts in mathematics **148**, Springer, ISBN 978-0-387-94285-8

- Smith, Geoff; Tabachnikova, Olga (2000), *Topics in group theory*, Springer undergraduate mathematics series (2 ed.), Springer, ISBN 978-1-85233-235-8

9.7.3 Papers

- Silvestri, R. (September 1979), *Simple groups of finite order in the nineteenth century*, Archive for History of Exact Sciences **20** (3–4): 313–356, doi:10.1007/BF00327738

9.8 External links

- The alternating group A_n is simple at PlanetMath.org.

Chapter 10

Georgi–Glashow model

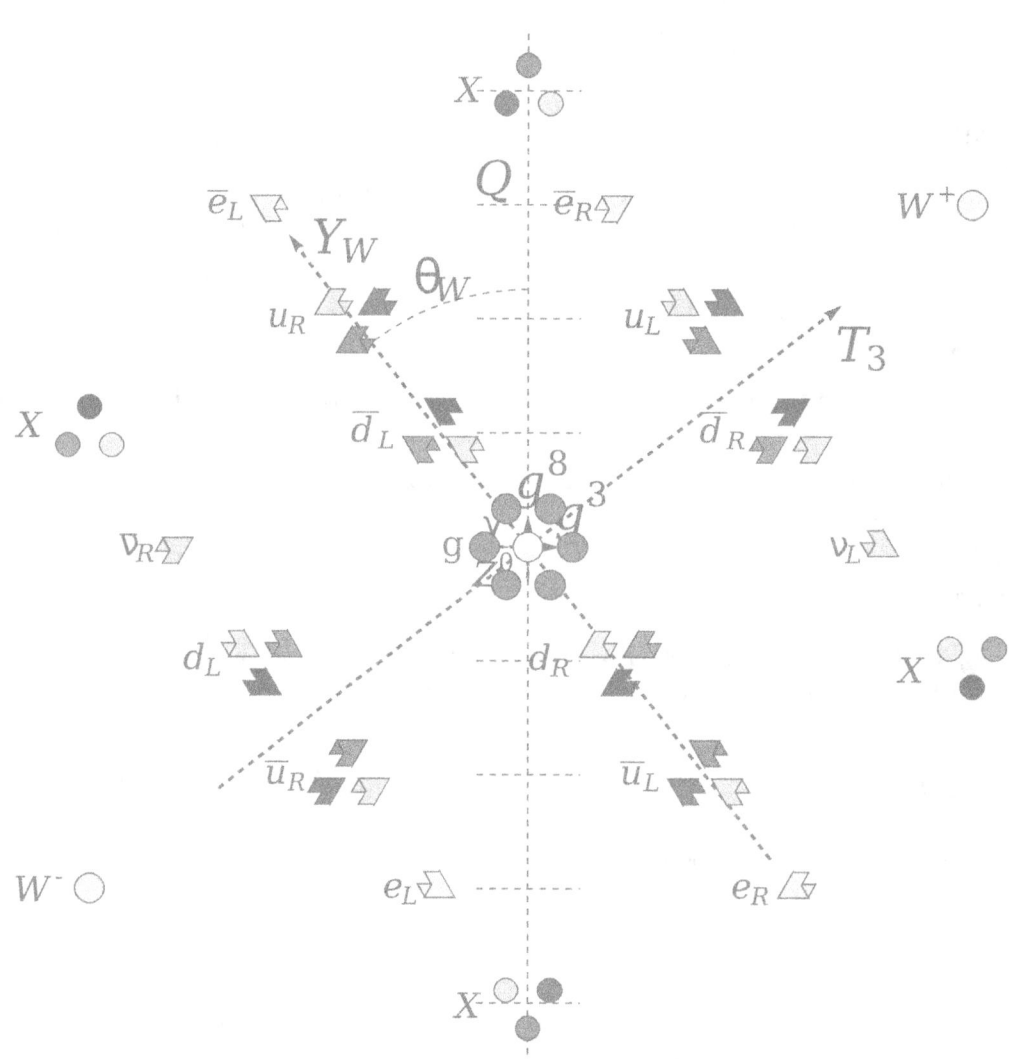

The pattern of weak isospins, weak hypercharges, and strong charges for particles in the Georgi-Glashow model, rotated by the predicted weak mixing angle, showing electric charge roughly along the vertical. In addition to Standard Model particles, the theory includes twelve colored X bosons, responsible for proton decay.

In particle physics, the **Georgi–Glashow model** is a particular grand unification theory (GUT) proposed by Howard

Georgi and Sheldon Glashow in 1974. In this model the standard model gauge groups SU(3) × SU(2) × U(1) are combined into a single simple gauge group -- SU(5). The unified group SU(5) is then thought to be spontaneously broken to the standard model subgroup at some high energy scale called the grand unification scale.

Since the Georgi–Glashow model combines leptons and quarks into single irreducible representations, there exist interactions which do not conserve baryon number, although they still conserve B-L. This yields a mechanism for proton decay, and the rate of proton decay can be predicted from the dynamics of the model. However, proton decay has not yet been observed experimentally, and the resulting lower limit on the lifetime of the proton contradicts the predictions of this model. However, the elegance of the model has led particle physicists to use it as the foundation for more complex models which yield longer proton lifetimes.

(For a more elementary introduction to how the representation theory of Lie algebras are related to particle physics, see the article Particle physics and representation theory.)

This model suffers from the doublet-triplet splitting problem.

10.1 Breaking SU(5)

SU(5) breaking occurs when a scalar field, analogous to the Higgs field, and transforming in the adjoint of SU(5) acquires a vacuum expectation value proportional to the weak hypercharge generator

$$\frac{Y}{2} = \text{diag}\left(-1/3, -1/3, -1/3, 1/2, 1/2\right)$$

When this occurs SU(5) is spontaneously broken to the subgroup of SU(5) commuting with the group generated by Y. This unbroken subgroup is just the standard model group: $[SU(3) \times SU(2) \times U(1)_Y]/\mathbb{Z}_6$.

Under the unbroken subgroup the adjoint **24** transforms as

$$24 \to (8,1)_0 \oplus (1,3)_0 \oplus (1,1)_0 \oplus (3,2)_{-\frac{5}{6}} \oplus (\bar{3},2)_{\frac{5}{6}}$$

giving the gauge bosons of the standard model. See restricted representation.

The standard model quarks and leptons fit neatly into representations of SU(5). Specifically, the left-handed fermions combine into 3 generations of $\bar{5} \oplus 10 \oplus 1$. Under the unbroken subgroup these transform as

$$\bar{5} \to (\bar{3},1)_{\frac{1}{3}} \oplus (1,2)_{-\frac{1}{2}} \text{ (d}^c \text{ and l)}$$
$$10 \to (3,2)_{\frac{1}{6}} \oplus (\bar{3},1)_{-\frac{2}{3}} \oplus (1,1)_1 \text{ (q, u}^c \text{ and e}^c\text{)}$$
$$1 \to (1,1)_0 \text{ (}\nu^c\text{)}$$

giving precisely the left-handed fermionic content of the standard model, where for every generation d^c, u^c, e^c and ν^c stand for anti-down-type quark, anti-up-type quark, anti-down-type lepton and anti-up-type lepton, respectively, and q and l stand for quark and lepton. Note that fermions transforming as a **1** under SU(5) are now thought to be necessary because of the evidence for neutrino oscillations. Actually though, it is possible for there to be only left-handed neutrinos without any right-handed neutrinos if we could somehow introduce a tiny Majorana coupling for the left-handed neutrinos.

Since the homotopy group

$$\pi_2\left(\frac{SU(5)}{[SU(3) \times SU(2) \times U(1)_Y]/\mathbb{Z}_6}\right) = \mathbb{Z}$$

this model predicts 't Hooft–Polyakov monopoles.

These monopoles have quantized Y magnetic charges. Since the electromagnetic charge Q is a linear combination of some SU(2) generator with Y/2, these monopoles also have quantized magnetic charges, where by magnetic here, we mean electromagnetic magnetic charges.

10.2 Minimal supersymmetric SU(5)

10.2.1 Spacetime

The N=1 superspace extension of 3+1 Minkowski spacetime.

10.2.2 Spatial symmetry

N=1 SUSY over 3+1 Minkowski spacetime without R-symmetry.

10.2.3 Gauge symmetry group

SU(5)

10.2.4 Global internal symmetry

$\mathbb{Z}_2$ (matter parity)

10.2.5 Matter parity

To prevent unwanted couplings in the supersymmetric version of the model, we assign a $\mathbb{Z}_2$ matter parity to the chiral superfields with the matter fields having odd parity and the Higgs having even parity. This is unnecessary in the nonsupersymmetric version, but then, we can't protect the electroweak Higgs from quadratic radiative mass corrections. See hierarchy problem. In the nonsupersymmetric version the action is invariant under a similar $\mathbb{Z}_2$ symmetry because the matter fields are all fermionic and thus must appear in the action in pairs, while the Higgs fields are bosonic.

10.2.6 Vector superfields

Those associated with the SU(5) gauge symmetry

10.2.7 Chiral superfields

As complex representations:

10.2.8 Superpotential

A generic invariant renormalizable superpotential is a (complex) $SU(5) \times \mathbb{Z}_2$ invariant cubic polynomial in the superfields. It is a linear combination of the following terms:

$$\begin{array}{ll} \Phi^2 & \Phi^A_B \Phi^B_A \\ \Phi^3 & \Phi^A_B \Phi^B_C \Phi^C_A \\ H_d H_u & H_{dA} H_u^A \\ H_d \Phi H_u & H_{dA} \Phi^A_B H_u^B \\ H_u 10_i\, 10_j & \epsilon_{ABCDE} H_u^A 10_i^{BC} 10_j^{DE} \\ H_d \bar{5}_i\, 10_j & H_{dA} \bar{5}_{Bi} 10_j^{AB} \\ H_u \bar{5}_i N_j^c & H_u^A \bar{5}_{Ai} N_j^c \\ N_i^c N_j^c & N_i^c N_j^c \end{array}$$

The first column is an Abbreviation of the second column (neglecting proper normalization factors), where capital indices are SU(5) indices, and i and j are the generation indices.

The last two rows presupposes the multiplicity of N^c is not zero (i.e. that a sterile neutrino exists). The coupling $H_u\, 10_i\, 10_j$ has coefficients which are symmetric in i and j. The coupling $N^c_i N^c_j$ has coefficients which are symmetric in

10.2. MINIMAL SUPERSYMMETRIC SU(5)

i and j. Note that the number of sterile neutrino generations need not be three, unless the SU(5) is embedded in a higher unification scheme such as SO(10).

10.2.9 Vacua

The vacua correspond to the mutual zeros of the F and D terms. Let's first look at the case where the VEVs of all the chiral fields are zero except for Φ.

The Φ sector

$$W = Tr[a\Phi^2 + b\Phi^3]$$

The F zeros corresponds to finding the stationary points of W subject to the traceless constraint $Tr[\Phi] = 0$. So, $2a\Phi + 3b\Phi^2 = \lambda \mathbf{1}$ where λ is a Lagrange multiplier.

Up to an SU(5) (unitary) transformation,

$$\Phi = \begin{cases} \text{diag}(0,0,0,0,0) \\ \text{diag}(\frac{2a}{9b}, \frac{2a}{9b}, \frac{2a}{9b}, \frac{2a}{9b}, -\frac{8a}{9b}) \\ \text{diag}(\frac{4a}{3b}, \frac{4a}{3b}, \frac{4a}{3b}, -\frac{2a}{b}, -\frac{2a}{b}) \end{cases}$$

The three cases are called case I, II and III and they break the gauge symmetry into SU(5), $[SU(4) \times U(1)]/\mathbb{Z}_4$ and $[SU(3) \times SU(2) \times U(1)]/\mathbb{Z}_6$ respectively (the stabilizer of the VEV).

In other words, there at least three different superselection sections, which is typical for supersymmetric theories.

Only case III makes any phenomenological sense and so, we will focus on this case from now onwards.

It can be verified that this solution together with zero VEVs for all the other chiral multiplets is a zero of the F-terms and D-terms. The matter parity remains unbroken (right up to the TeV scale).

Decomposition

The gauge algebra **24** decomposes as $\begin{pmatrix} (8,1)_0 \\ (1,3)_0 \\ (1,1)_0 \\ (3,2)_{-\frac{5}{6}} \\ (\bar{3},2)_{\frac{5}{6}} \end{pmatrix}$. This **24** is a real representation, so the last two terms need explanation. Both $(3,2)_{-\frac{5}{6}}$ and $(\bar{3},2)_{\frac{5}{6}}$ are complex representations. However, the direct sum of both representation decomposes into two irreducible real representations and we only take half of the direct sum, i.e. one of the two real irreducible copies. The first three components are left unbroken. The adjoint Higgs also has a similar decomposition, except that it is complex. The Higgs mechanism causes one real HALF of the $(3,2)_{-\frac{5}{6}}$ and $(\bar{3},2)_{\frac{5}{6}}$ of the adjoint Higgs to be absorbed. The other real half acquires a mass coming from the D-terms. And the other three components of the adjoint Higgs, $(8,1)_0$, $(1,3)_0$ and $(1,1)_0$ acquire GUT scale masses coming from self pairings of the superpotential, $a\Phi^2 + b<\Phi>\Phi^2$.

The sterile neutrinos, if any exists, would also acquire a GUT scale Majorana mass coming from the superpotential coupling ν^{c2}.

Because of matter parity, the matter representations $\bar{5}$ and **10** remain chiral.

It's the Higgs fields 5H and $\bar{5}_H$ which are interesting.

The two relevant superpotential terms here are $5_H \bar{5}_H$ and $<24> 5_H \bar{5}_H$. Unless there happens to be some fine tuning, we would expect both the triplet terms and the doublet terms to pair up, leaving us with no light electroweak doublets. This is in complete disagreement with phenomenology. See doublet-triplet splitting problem for more details.

Fermion masses

See Georgi-Jarlskog mass relation.

10.3 Lee Smolin's view of SU(5)

In his book *The Trouble with Physics*, Smolin states:

> After some twenty-five years, we are still waiting. No protons have decayed. We have been waiting long enough to know that SU(5) grand unification is wrong. It's a beautiful idea, but one that nature seems not to have adopted. Page 64.
>
> Indeed, it would be hard to underestimate the implications of this negative result. SU(5) is the most elegant way imaginable of unifying quarks with leptons, and it leads to a codification of the properties of the standard model in simple terms. Even after twenty-five years, I still find it stunning that SU(5) doesn't work. Page 65.
>
> Smolin, Lee (2007). *The Trouble with Physics*.

10.4 Popular culture

When the filmmaker Sandy Bates (played by Woody Allen) in the 1980 Woody Allen film *Stardust Memories* launches a depressive soliloquy with the quote, "Did anybody read on the front page of *The Times* that matter is decaying?", this was almost certainly a reference to the Georgi–Glashow model, given the film's period, the importance of the Georgi–Glashow model at the time and the many contemporary layperson articles in circulation about some of the model's most striking consequences, particularly its mechanism for proton decay. An actual *New York Times* article[1] appeared two years later, fulfilling Allen's blackly humorous foreshadowing of a world whose news was so baleful that the mainstream media were systematically reporting its material demise.

10.5 References

[1] Physics sometimes takes G.U.T.s, New York Times, September 19, 1982

- Howard Georgi and Sheldon Glashow, *Unity of All Elementary-Particle Forces*, Physical Review Letters, **32** (1974) 438.

- J.C. Baez, J. Huerta (2009). "The Algebra of Grand Unified Theories". arXiv:0904.1556 [hep-th].

Chapter 11

Simple Lie group

In group theory, a **simple Lie group** is a connected non-abelian Lie group G which does not have nontrivial connected normal subgroups.

A **simple Lie algebra** is a non-abelian Lie algebra whose only ideals are 0 and itself. A direct sum of simple Lie algebras is called a semisimple Lie algebra.

An equivalent definition of a simple Lie group follows from the Lie correspondence: a connected Lie group is simple if its Lie algebra is simple. An important technical point is that a simple Lie group may contain *discrete* normal subgroups, hence being a simple Lie group is different from being simple as an abstract group.

Simple Lie groups include many classical Lie groups, which provide a group-theoretic underpinning for spherical geometry, projective geometry and related geometries in the sense of Felix Klein's Erlangen programme. It emerged in the course of classification of simple Lie groups that there exist also several exceptional possibilities not corresponding to any familiar geometry. These *exceptional groups* account for many special examples and configurations in other branches of mathematics, as well as contemporary theoretical physics.

While the notion of a simple Lie group is satisfying from the axiomatic perspective, in applications of Lie theory, such as the theory of Riemannian symmetric spaces, somewhat more general notions of semisimple and reductive Lie groups proved to be even more useful. In particular, every connected compact Lie group is reductive, and the study of representations of general reductive groups is a major branch of representation theory.

11.1 Comments on the definition

Unfortunately there is no single standard definition of a simple Lie group. The definition given above is sometimes varied in the following ways:

- Connectedness: Usually simple Lie groups are connected by definition. This excludes discrete simple groups (these are zero-dimensional Lie groups that are simple as abstract groups) as well as disconnected orthogonal groups.

- Center: Usually simple Lie groups are allowed to have a discrete center; for example, SL(2, **R**) has a center of order 2, but is still counted as a simple Lie group. If the center is non-trivial (and not the whole group) then the simple Lie group is not simple as an abstract group. Some authors require that the center of a simple Lie group be finite (or trivial); the universal cover of SL(2, **R**) is an example of a simple Lie group with infinite center.

- **R**: Usually the group **R** of real numbers under addition (and its quotient **R/Z**) are not counted as simple Lie groups, even though they are connected and have a Lie algebra with no proper non-zero ideals. Occasionally authors define simple Lie groups in such a way that **R** is simple, though this sometimes seems to be an accident caused by overlooking this case.

- Matrix groups: Some authors restrict themselves to Lie groups that can be represented as groups of finite matrices. The metaplectic group is an example of a simple Lie group that cannot be represented in this way.

- Complex Lie algebras: The definition of a simple Lie algebra is not stable under the *extension of scalars*. The complexification of a complex simple Lie algebra, such as **sl**(*n*, **C**) is semisimple, but not simple.

The most common definition is the one above: simple Lie groups have to be connected, they are allowed to have non-trivial centers (possibly infinite), they need not be representable by finite matrices, and they must be non-abelian.

11.2 Method of classification

Main article: list of simple Lie groups

Such groups are classified using the prior classification of the complex simple Lie algebras: for which see the page on root systems. It is shown that a simple Lie group has a simple Lie algebra that will occur on the list given there, once it is complexified (that is, made into a complex vector space rather than a real one). This reduces the classification to two further matters.

11.3 Real forms

The groups SO(p,q,**R**) and SO($p+q$,**R**), for example, give rise to different real Lie algebras, but having the same Dynkin diagram. In general there may be different *real forms* of the same complex Lie algebra.

11.4 Relationship of simple Lie algebras to groups

Secondly the Lie algebra only determines uniquely the simply connected (universal) cover G^* of the component containing the identity of a Lie group G. It may well happen that G^* isn't actually a simple group, for example having a non-trivial center. We have therefore to worry about the global topology, by computing the fundamental group of G (an abelian group: a Lie group is an H-space). This was done by Élie Cartan.

For an example, take the special orthogonal groups in even dimension. With the non-identity matrix $-I$ in the center, these aren't actually simple groups; and having a twofold spin cover, they aren't simply-connected either. They lie 'between' G^* and G, in the notation above.

11.5 Classification by Dynkin diagram

Main article: root system

According to Dynkin's classification, we have as possibilities these only, where n is the number of nodes:

A_n ○—○—○------○—○ F_4 ○—○⇒○—○ G_2 ○⇛○

B_n ○—○------○—○⇒○

C_n ○—○------○—○⇐○ E_6 ○—○—○—○—○

D_n ○—○------⟨ E_7 ○—○—○—○—○—○

                                                        E_8 ○—○—○—○—○—○—○

11.6 Infinite series

11.6.1 A series

$A_1, A_2, \ldots$

A_r corresponds to the special unitary group, $SU(r + 1)$.

11.6.2 B series

$B_2, B_3, \ldots$

B_r corresponds to the special orthogonal group, $SO(2r + 1)$.

11.6.3 C series

$C_3, C_4, \ldots$

C_r corresponds to the symplectic group, $Sp(2r)$.

11.6.4 D series

$D_4, D_5, \ldots$

D_r corresponds to the special orthogonal group, $SO(2r)$. Note that $SO(4)$ is not a simple group, though. The Dynkin diagram has two nodes that are not connected. There is a surjective homomorphism from $SO(3)^* \times SO(3)^*$ to $SO(4)$ given by quaternion multiplication; see quaternions and spatial rotation. Therefore the simple groups here start with D_3, which as a diagram straightens out to A_3. With D_4 there is an 'exotic' symmetry of the diagram, corresponding to so-called triality.

11.7 Exceptional cases

For the so-called exceptional cases see G_2, F_4, E_6, E_7, and E_8. These cases are deemed 'exceptional' because they do not fall into infinite series of groups of increasing dimension. From the point of view of each group taken separately, there is nothing so unusual about them. These exceptional groups were discovered around 1890 in the classification of the simple Lie algebras, over the complex numbers (Wilhelm Killing, re-done by Élie Cartan). For some time it was a research issue to find concrete ways in which they arise, for example as a symmetry group of a differential system.

See also $E_7\frac{1}{2}$.

11.8 Simply laced groups

A **simply laced group** is a Lie group whose Dynkin diagram only contain simple links, and therefore all the nonzero roots of the corresponding Lie algebra have the same length. The A, D and E series groups are all simply laced, but no group of type B, C, F, or G is simply laced.

11.9 See also

- Cartan matrix
- Coxeter matrix
- Weyl group
- Coxeter group
- Kac–Moody algebra

- Catastrophe theory

11.10 References

- Jacobson, Nathan (1971-06-01). *Exceptional Lie Algebras* (1 ed.). CRC Press. ISBN 0-8247-1326-5.

Chapter 12

Group representation

Not to be confused with Presentation of a group.

In the mathematical field of representation theory, **group representations** describe abstract groups in terms of linear transformations of vector spaces; in particular, they can be used to represent group elements as matrices so that the group operation can be represented by matrix multiplication. Representations of groups are important because they allow many group-theoretic problems to be reduced to problems in linear algebra, which is well understood. They are also important in physics because, for example, they describe how the symmetry group of a physical system affects the solutions of equations describing that system.

The term *representation of a group* is also used in a more general sense to mean any "description" of a group as a group of transformations of some mathematical object. More formally, a "representation" means a homomorphism from the group to the automorphism group of an object. If the object is a vector space we have a *linear representation*. Some people use *realization* for the general notion and reserve the term *representation* for the special case of linear representations. The bulk of this article describes linear representation theory; see the last section for generalizations.

12.1 Branches of group representation theory

The representation theory of groups divides into subtheories depending on the kind of group being represented. The various theories are quite different in detail, though some basic definitions and concepts are similar. The most important divisions are:

- *Finite groups* — Group representations are a very important tool in the study of finite groups. They also arise in the applications of finite group theory to crystallography and to geometry. If the field of scalars of the vector space has characteristic p, and if p divides the order of the group, then this is called *modular representation theory*; this special case has very different properties. See Representation theory of finite groups.

- *Compact groups or locally compact groups* — Many of the results of finite group representation theory are proved by averaging over the group. These proofs can be carried over to infinite groups by replacement of the average with an integral, provided that an acceptable notion of integral can be defined. This can be done for locally compact groups, using Haar measure. The resulting theory is a central part of harmonic analysis. The Pontryagin duality describes the theory for commutative groups, as a generalised Fourier transform. See also: Peter–Weyl theorem.

- *Lie groups* — Many important Lie groups are compact, so the results of compact representation theory apply to them. Other techniques specific to Lie groups are used as well. Most of the groups important in physics and chemistry are Lie groups, and their representation theory is crucial to the application of group theory in those fields. See Representations of Lie groups and Representations of Lie algebras.

- *Linear algebraic groups* (or more generally *affine group schemes*) — These are the analogues of Lie groups, but over more general fields than just **R** or **C**. Although linear algebraic groups have a classification that is very similar to that of Lie groups, and give rise to the same families of Lie algebras, their representations are rather different (and much less well understood). The analytic techniques used for studying Lie groups must be replaced by techniques from algebraic geometry, where the relatively weak Zariski topology causes many technical complications.

- *Non-compact topological groups* — The class of non-compact groups is too broad to construct any general representation theory, but specific special cases have been studied, sometimes using ad hoc techniques. The *semisimple Lie groups* have a deep theory, building on the compact case. The complementary *solvable* Lie groups cannot in the same way be classified. The general theory for Lie groups deals with semidirect products of the two types, by means of general results called *Mackey theory*, which is a generalization of Wigner's classification methods.

Representation theory also depends heavily on the type of vector space on which the group acts. One distinguishes between finite-dimensional representations and infinite-dimensional ones. In the infinite-dimensional case, additional structures are important (e.g. whether or not the space is a Hilbert space, Banach space, etc.).

One must also consider the type of field over which the vector space is defined. The most important case is the field of complex numbers. The other important cases are the field of real numbers, finite fields, and fields of p-adic numbers. In general, algebraically closed fields are easier to handle than non-algebraically closed ones. The characteristic of the field is also significant; many theorems for finite groups depend on the characteristic of the field not dividing the order of the group.

12.2 Definitions

A **representation** of a group G on a vector space V over a field K is a group homomorphism from G to GL(V), the general linear group on V. That is, a representation is a map

$$\rho : G \to \mathrm{GL}(V)$$

such that

$$\rho(g_1 g_2) = \rho(g_1)\rho(g_2), \quad \text{all for } g_1, g_2 \in G.$$

Here V is called the **representation space** and the dimension of V is called the **dimension** of the representation. It is common practice to refer to V itself as the representation when the homomorphism is clear from the context.

In the case where V is of finite dimension n it is common to choose a basis for V and identify GL(V) with GL(n, K), the group of n-by-n invertible matrices on the field K.

- If G is a topological group and V is a topological vector space, a **continuous representation** of G on V is a representation ϱ such that the application $\Phi : G \times V \to V$ defined by $\Phi(g, v) = \varrho(g)(v)$ is continuous.

- The **kernel** of a representation ϱ of a group G is defined as the normal subgroup of G whose image under ϱ is the identity transformation:

  $$\ker \rho = \{g \in G \mid \rho(g) = \mathrm{id}\}.$$

  A faithful representation is one in which the homomorphism $G \to \mathrm{GL}(V)$ is injective; in other words, one whose kernel is the trivial subgroup $\{e\}$ consisting only of the group's identity element.

- Given two K vector spaces V and W, two representations $\varrho : G \to \mathrm{GL}(V)$ and $\pi : G \to \mathrm{GL}(W)$ are said to be **equivalent** or **isomorphic** if there exists a vector space isomorphism $\alpha : V \to W$ so that for all g in G,

  $$\alpha \circ \rho(g) \circ \alpha^{-1} = \pi(g).$$

12.3 Examples

Consider the complex number $u = e^{2\pi i/3}$ which has the property $u^3 = 1$. The cyclic group $C_3 = \{1, u, u^2\}$ has a representation ρ on $\mathbf{C}^2$ given by:

$$\rho(1) = \begin{bmatrix} 1 & 0 \\ 0 & 1 \end{bmatrix} \quad \rho(u) = \begin{bmatrix} 1 & 0 \\ 0 & u \end{bmatrix} \quad \rho(u^2) = \begin{bmatrix} 1 & 0 \\ 0 & u^2 \end{bmatrix}.$$

This representation is faithful because ρ is a one-to-one map.

An isomorphic representation for C_3 is

$$\rho(1) = \begin{bmatrix} 1 & 0 \\ 0 & 1 \end{bmatrix} \quad \rho(u) = \begin{bmatrix} u & 0 \\ 0 & 1 \end{bmatrix} \quad \rho(u^2) = \begin{bmatrix} u^2 & 0 \\ 0 & 1 \end{bmatrix}.$$

The group C_3 may also be faithfully represented on $\mathbf{R}^2$ by

$$\rho(1) = \begin{bmatrix} 1 & 0 \\ 0 & 1 \end{bmatrix} \quad \rho(u) = \begin{bmatrix} a & -b \\ b & a \end{bmatrix} \quad \rho(u^2) = \begin{bmatrix} a & b \\ -b & a \end{bmatrix}$$

where

$$a = \text{Re}(u) = -\tfrac{1}{2}, \qquad b = \text{Im}(u) = \tfrac{\sqrt{3}}{2}.$$

12.4 Reducibility

Main article: Irreducible representation

A subspace W of V that is invariant under the group action is called a *subrepresentation*. If V has exactly two subrepresentations, namely the zero-dimensional subspace and V itself, then the representation is said to be *irreducible*; if it has a proper subrepresentation of nonzero dimension, the representation is said to be *reducible*. The representation of dimension zero is considered to be neither reducible nor irreducible, just like the number 1 is considered to be neither composite nor prime.

Under the assumption that the characteristic of the field K does not divide the size of the group, representations of finite groups can be decomposed into a direct sum of irreducible subrepresentations (see Maschke's theorem). This holds in particular for any representation of a finite group over the complex numbers, since the characteristic of the complex numbers is zero, which never divides the size of a group.

In the example above, the first two representations given are both decomposable into two 1-dimensional subrepresentations (given by span$\{(1,0)\}$ and span$\{(0,1)\}$), while the third representation is irreducible.

12.5 Generalizations

12.5.1 Set-theoretical representations

A *set-theoretic representation* (also known as a group action or *permutation representation*) of a group G on a set X is given by a function $\rho : G \to X^X$, the set of functions from X to X, such that for all g_1, g_2 in G and all x in X:

$\rho(1)[x] = x$

$\rho(g_1 g_2)[x] = \rho(g_1)[\rho(g_2)[x]].$

This condition and the axioms for a group imply that $\rho(g)$ is a bijection (or permutation) for all g in G. Thus we may equivalently define a permutation representation to be a group homomorphism from G to the symmetric group SX of X.

For more information on this topic see the article on group action.

12.5.2 Representations in other categories

Every group G can be viewed as a category with a single object; morphisms in this category are just the elements of G. Given an arbitrary category C, a *representation* of G in C is a functor from G to C. Such a functor selects an object X in C and a group homomorphism from G to Aut(X), the automorphism group of X.

In the case where C is **Vect**K, the category of vector spaces over a field K, this definition is equivalent to a linear representation. Likewise, a set-theoretic representation is just a representation of G in the category of sets.

When C is **Ab**, the category of abelian groups, the objects obtained are called G-modules.

For another example consider the category of topological spaces, **Top**. Representations in **Top** are homomorphisms from G to the homeomorphism group of a topological space X.

Two types of representations closely related to linear representations are:

- projective representations: in the category of projective spaces. These can be described as "linear representations up to scalar transformations".

- affine representations: in the category of affine spaces. For example, the Euclidean group acts affinely upon Euclidean space.

12.6 See also

- Character theory
- List of harmonic analysis topics
- List of representation theory topics
- Representation theory of finite groups

12.7 References

- Fulton, William; Harris, Joe (1991), *Representation theory. A first course*, Graduate Texts in Mathematics, Readings in Mathematics **129**, New York: Springer-Verlag, ISBN 978-0-387-97495-8, MR 1153249, ISBN 978-0-387-97527-6. Introduction to representation theory with emphasis on Lie groups.

- Yurii I. Lyubich. *Introduction to the Theory of Banach Representations of Groups*. Translated from the 1985 Russian-language edition (Kharkov, Ukraine). Birkhäuser Verlag. 1988.

Chapter 13
SO(10) (physics)

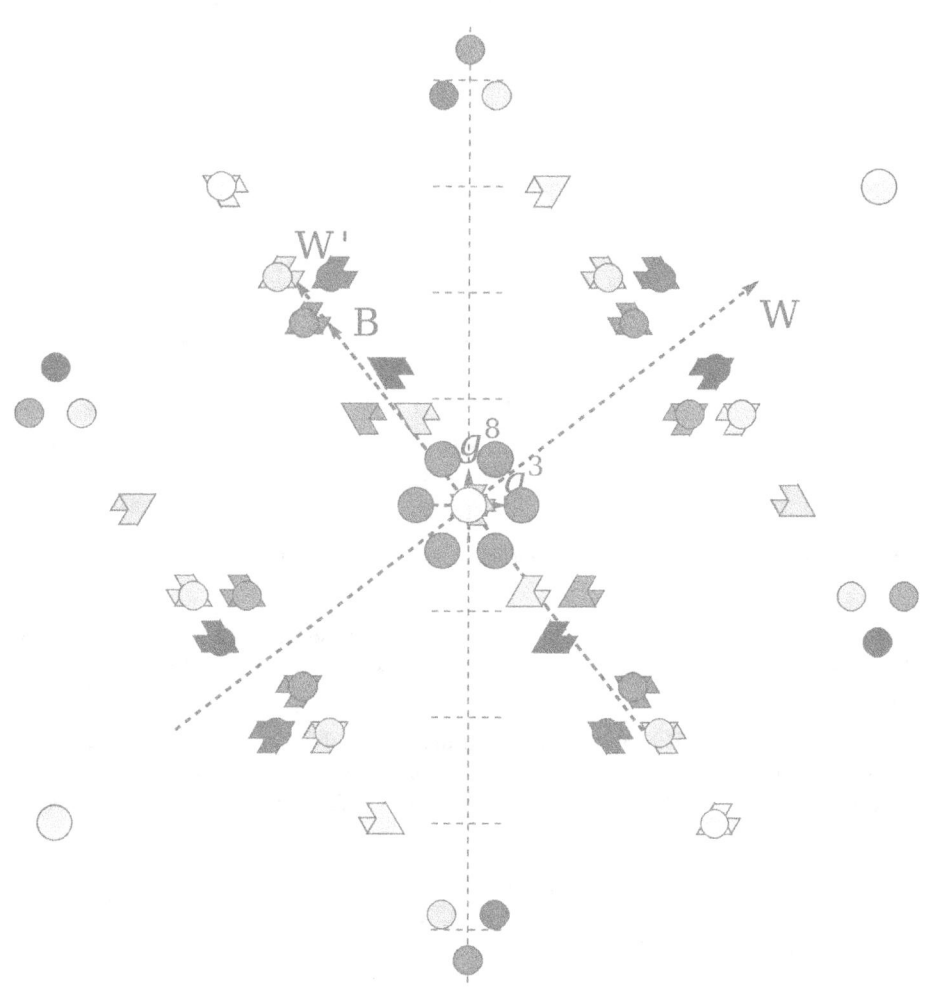

The pattern of weak isospin, W, weaker isospin, W', strong g3 and g8, and baryon minus lepton, B, charges for particles in the SO(10) model, rotated to show the embedding of the Georgi–Glashow model and Standard Model, with electric charge roughly along the vertical. In addition to Standard Model particles, the theory includes thirty colored X bosons, responsible for proton decay, and two W' bosons.

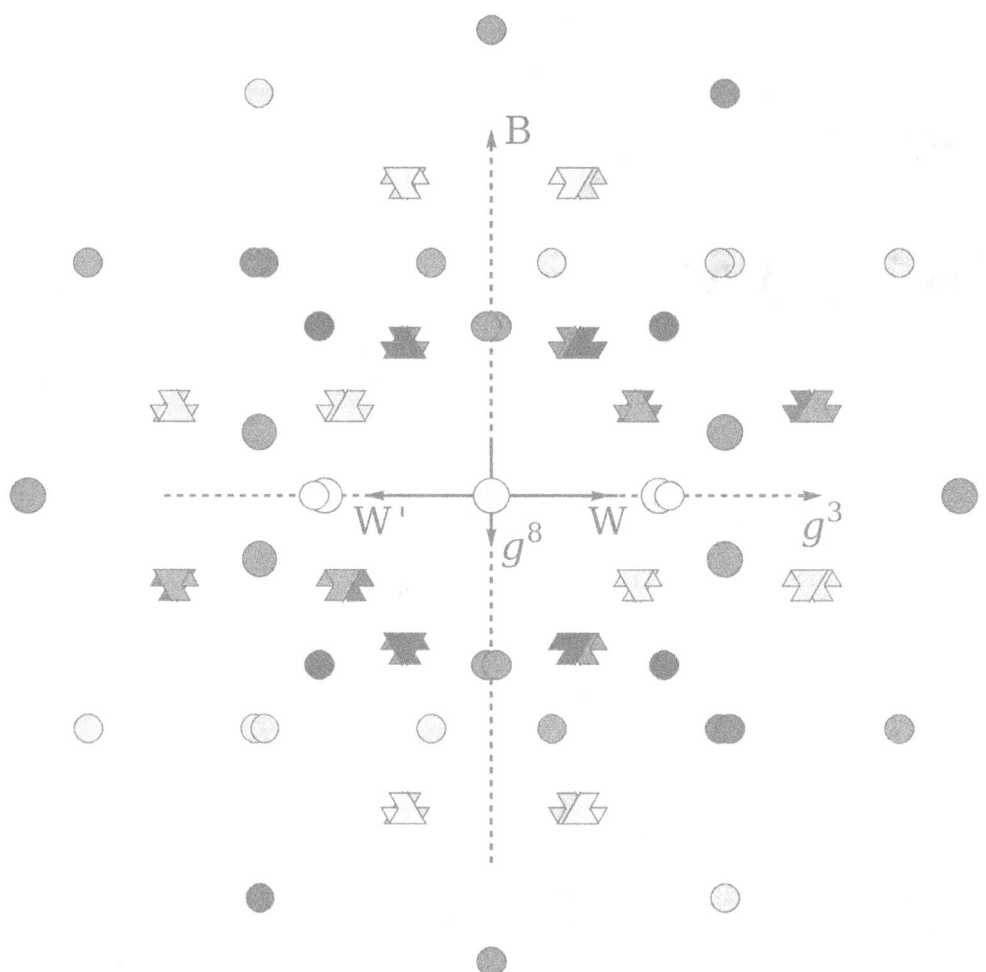

The pattern of charges for particles in the SO(10) model, rotated to show the embedding in E6.

In particle physics, one of the grand unified theories (GUT) is based on the **SO(10)** Lie group. (The Lie group involved is not really the special orthogonal group SO(10), but rather its double cover Spin(10); but calling it SO(10) is the standard convention.)

Before SU(5), Harald Fritzsch and Peter Minkowski and independently Howard Georgi found that all the matter contents are incorporated into a single representation, spinorial 16 of SO(10). (Historical note: the *before* in the previous sentence is misleading: Georgi found the SO(10) theory a few hours before finding SU(5) at the end of 1973.[1])

13.1 Important subgroups

It has the branching rules to [SU(5)×U(1)$_\chi$]/**Z**$_5$.

$45 \to 24_0 \oplus 10_{-4} \oplus \overline{10}_4 \oplus 1_0$

$16 \to 10_1 \oplus \bar{5}_{-3} \oplus 1_5$.

$10 \to 5_{-2} \oplus \bar{5}_2$.

If the hypercharge is contained within SU(5), this is the conventional Georgi–Glashow model, with the 16 as the matter fields, the 10 as the electroweak Higgs field and the 24 within the 45 as the GUT Higgs field. The superpotential may then include renormalizable terms of the form $Tr(45 \cdot 45)$; $Tr(45 \cdot 45 \cdot 45)$; $10 \cdot 45 \cdot 10$, $10 \cdot 16^* \cdot 16$ and $16^* \cdot 16$. The first three are responsible to the gauge symmetry breaking at low energies and give the Higgs mass, and the latter two give the matter particles masses and their Yukawa couplings to the Higgs.

There is another possible branching, under which the hypercharge is a linear combination of an SU(5) generator and χ. This is known as flipped SU(5).

Another important subgroup is either [SU(4) × SU(2)L × SU(2)R]/$\mathbf{Z}_2$ or $\mathbf{Z}_2 \rtimes$ [SU(4) × SU(2)L × SU(2)R]/$\mathbf{Z}_2$ depending upon whether or not the left-right symmetry is broken, yielding the Pati–Salam model, whose branching rule is

$$16 \to (4, 2, 1) \oplus (\bar{4}, 1, 2).$$

13.2 Spontaneous symmetry breaking

The symmetry breaking of SO(10) is usually done with a combination of ((a 45H OR a 54H) AND ((a 16H AND a $\overline{16}_H$) OR (a 126H AND a $\overline{126}_H$))).

Let's say we choose a 54H. When this Higgs field acquires a GUT scale VEV, we have a symmetry breaking to $\mathbf{Z}_2 \rtimes$ [SU(4) × SU(2)L × SU(2)R]/$\mathbf{Z}_2$, i.e. the Pati–Salam model with a $\mathbf{Z}_2$ left-right symmetry.

If we have a 45H instead, this Higgs field can acquire any VEV in a two dimensional subspace without breaking the standard model. Depending on the direction of this linear combination, we can break the symmetry to SU(5)×U(1), the Georgi–Glashow model with a U(1) (diag(1,1,1,1,1,−1,−1,−1,−1,−1)), flipped SU(5) (diag(1,1,1,−1,−1,−1,−1,−1,1,1)), SU(4)×SU(2)×U(1) (diag(0,0,0,1,1,0,0,0,−1,−1)), the minimal left-right model (diag(1,1,1,0,0,−1,−1,−1,0,0)) or SU(3)×SU(2)×U(1)×U(1) for any other nonzero VEV.

The choice diag(1,1,1,0,0,−1,−1,−1,0,0) is called the Dimopoulos-Wilczek mechanism aka the missing VEV mechanism and it is proportional to B−L.

The choice of a 16H and a $\overline{16}_H$ breaks the gauge group down to the Georgi–Glashow SU(5). The same comment applies to the choice of a 126H and a $\overline{126}_H$.

It is the combination of BOTH a 45/54 and a 16/ $\overline{16}$ or 126/ $\overline{126}$ which breaks SO(10) down to the Standard Model.

13.3 The electroweak Higgs and the doublet-triplet splitting problem

The electroweak Higgs doublets come from an SO(10) 10H. Unfortunately, this same 10 also contains triplets. The masses of the doublets have to be stabilized at the electroweak scale, which is many orders of magnitude smaller than the GUT scale whereas the triplets have to be really heavy in order to prevent triplet-mediated proton decays. See doublet-triplet splitting problem.

Among the solutions for it is the Dimopoulos-Wilczek mechanism, or the choice of diag(0,0,0,1,1,0,0,0,−1,−1) of <45>. Unfortunately, this is not stable once the 16/ $\overline{16}$ or 126/ $\overline{126}$ sector interacts with the 45 sector.[2]

13.4 Matter

The matter representations come in three copies (generations) of the 16 representation. The Yukawa coupling is 10H 16_f 16_f. This includes a right-handed neutrino. We can either include three copies of singlet representations φ and a Yukawa coupling $< \overline{16}_H > 16_f \phi$ (see double seesaw mechanism) or add the Yukawa interaction $< \overline{126}_H > 16_f 16_f$ or add the nonrenormalizable coupling $< \overline{16}_H >< \overline{16}_H > 16_f 16_f$. See seesaw mechanism.

13.5 Proton decay

- These graphics refer to the X bosons and Higgs bosons.
- Dimension 6 proton decay mediated by the X boson in SU(5) GUT
- Dimension 6 proton decay mediated by the X boson in flipped SU(5) GUT

Note that SO(10) contains both the Georgi–Glashow SU(5) and flipped SU(5).

13.6 See also

- Flipped SO(10)

13.7 Notes

[1] This story is told in various places; see for example, Yukawa-Tomonaga 100th Birthday Celebration; Fritzsch and Minkowski analyzed SO(10) in 1974.

[2]
- J.C. Baez, J. Huerta (2009). "The Algebra of Grand Unified Theories". arXiv:0904.1556 [hep-th].

13.7. NOTES

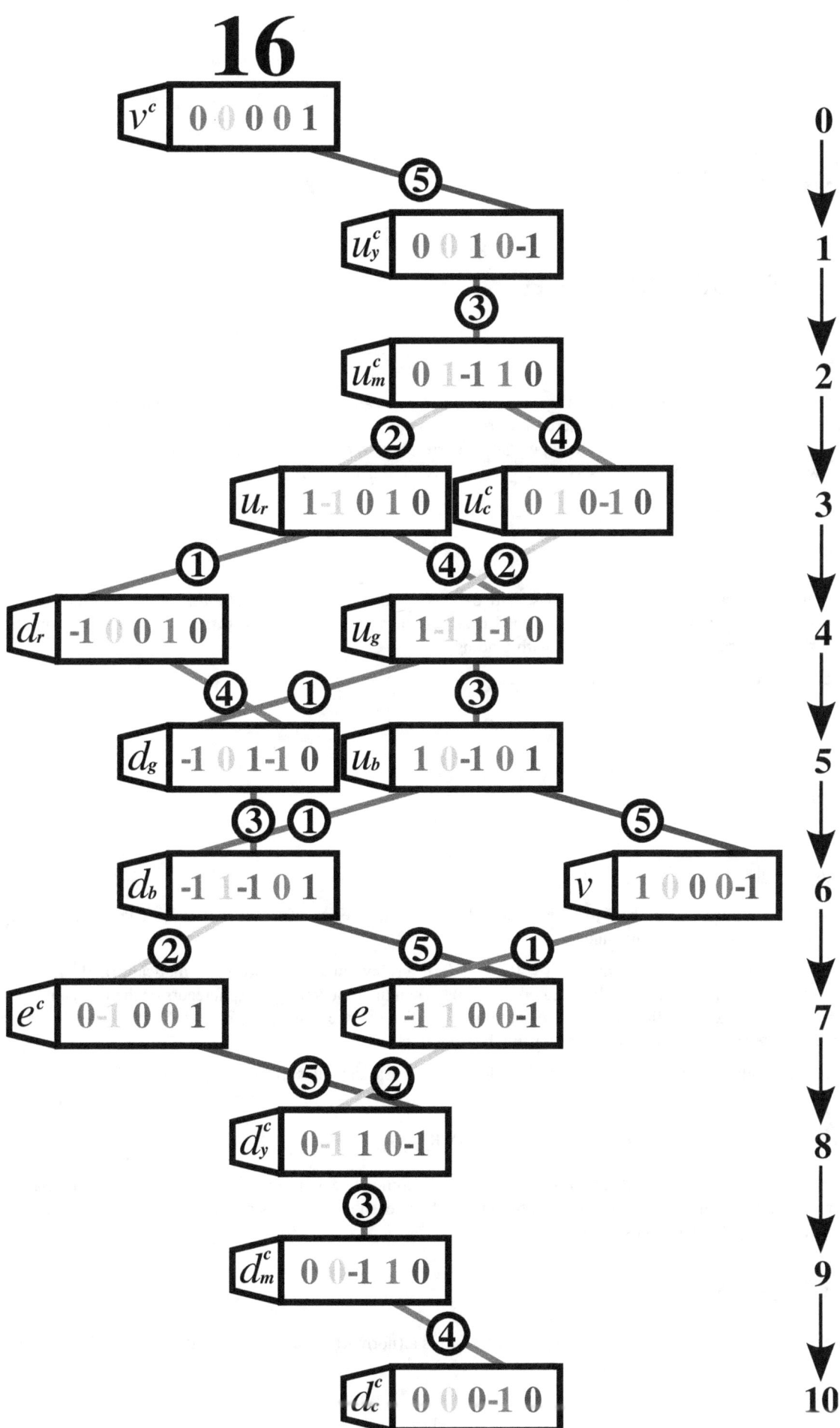

Chapter 14

Lie superalgebra

In mathematics, a **Lie superalgebra** is a generalisation of a Lie algebra to include a $\mathbf{Z}_2$-grading. Lie superalgebras are important in theoretical physics where they are used to describe the mathematics of supersymmetry. In most of these theories, the *even* elements of the superalgebra correspond to bosons and *odd* elements to fermions (but this is not always true; for example, the BRST supersymmetry is the other way around).

14.1 Definition

Formally, a **Lie superalgebra** is a (nonassociative) $\mathbf{Z}_2$-graded algebra, or *superalgebra*, over a commutative ring (typically $\mathbf{R}$ or $\mathbf{C}$) whose product $[\cdot,\cdot]$, called the **Lie superbracket** or **supercommutator**, satisfies the two conditions (analogs of the usual Lie algebra axioms, with grading):

Super skew-symmetry:

$$[x,y] = -(-1)^{|x||y|}[y,x].$$

The super Jacobi identity:

$$[x,[y,z]] = [[x,y],z] + (-1)^{|x||y|}[y,[x,z]]$$

where x, y, and z are pure in the $\mathbf{Z}_2$-grading. Here, |x| denotes the degree of x (either 0 or 1). The degree of [x,y] is the sum of degree of x and y modulo 2.

One also sometimes adds the axioms $[x,x] = 0$ for |x|=0 (if 2 is invertible this follows automatically) and $[[x,x],x] = 0$ for |x|=1 (if 3 is invertible this follows automatically). When the ground ring is the integers or the Lie superalgebra is a free module, these conditions are equivalent to the condition that the Poincaré–Birkhoff–Witt theorem holds (and, in general, they are necessary conditions for the theorem to hold).

Just as for Lie algebras, the universal enveloping algebra of the Lie superalgebra can be given a Hopf algebra structure.

14.1.1 Distinction from graded Lie algebra

A graded Lie algebra (say, graded by $\mathbf{Z}$ or $\mathbf{N}$) that is anticommutative and Jacobi in the graded sense also has a Z_2 grading (which is called "rolling up" the algebra into odd and even parts), but is not referred to as "super". See note at graded Lie algebra for discussion.

14.1.2 Even and odd parts

Note that the even subalgebra of a Lie superalgebra forms a (normal) Lie algebra as all the signs disappear, and the superbracket becomes a normal Lie bracket.

One way of thinking about a Lie superalgebra is to consider its even and odd parts, L_0 and L_1 separately. Then, L_0 is a Lie algebra, L_1 is a linear representation of L_0, and there exists a symmetric L_0-equivariant linear map $\{\cdot,\cdot\} : L_1 \otimes L_1 \to L_0$ such that for all x,y and z in L_1,

$$\{x,y\}[z] + \{y,z\}[x] + \{z,x\}[y] = 0.$$

14.1.3 Involution

A *** Lie superalgebra** is a complex Lie superalgebra equipped with an involutive antilinear map from itself to itself which respects the $\mathbf{Z}_2$ grading and satisfies $[x,y]^* = [y^*,x^*]$ for all x and y in the Lie superalgebra. (Some authors prefer the convention $[x,y]^* = (-1)^{|x||y|}[y^*,x^*]$; changing * to −* switches between the two conventions.) Its universal enveloping algebra would be an ordinary *-algebra.

14.2 Examples

Given any associative superalgebra A one can define the supercommutator on homogeneous elements by

$$[x,y] = xy - (-1)^{|x||y|}yx$$

and then extending by linearity to all elements. The algebra A together with the supercommutator then becomes a Lie superalgebra.

The Whitehead product on homotopy groups gives many examples of Lie superalgebras over the integers.

14.3 Classification

The simple complex finite-dimensional Lie superalgebras were classified by Victor Kac.

The basic classical compact Lie superalgebras (that are not Lie algebras) are:

SU(m/n) These are the superunitary Lie algebras which have invariants:

$$z.\bar{z} + iw.\bar{w}$$

This gives two orthosymplectic (see below) invariants if we take the m z variables and n w variables to be non-commuative and we take the real and imaginary parts. Therefore we have

$$SU(m/n) = OSp(2m/2n) \cap OSp(2n/2m)$$

SU(n/n)/U(1) A special case of the superunitary Lie algebras where we remove one U(1) generator to make the algebra simple.

OSp(m/2n) These are the Orthosymplectic groups. They have invariants given by:

$$x.x + y.z - z.y$$

for m commutative variables (x) and n pairs of anti-commutative variables (y,z). They are important symmetries in supergravity theories.

D(2/1; α) This is a set of superalgebras parameterised by the variable α. It has dimension 17 and is a sub-algebra of OSp(9|8). The even part of the group is O(3)xO(3)xO(3). So the invariants are:

$$A_\mu A_\mu + B_\mu B_\mu + C_\mu C_\mu + \psi^{\alpha\beta\gamma}\psi^{\alpha'\beta'\gamma'}\varepsilon_{\alpha\alpha'}\varepsilon_{\beta\beta'}\varepsilon_{\gamma\gamma'}$$

$$A_{\{1}A_2A_{3\}} + B_{\{1}B_2B_{3\}} + C_{\{1}C_2C_{3\}} + A_\mu\Gamma_\mu^{\alpha\alpha'}\psi\psi + B_\mu\Gamma_\mu^{\beta\beta'}\psi\psi + C_\mu\Gamma_\mu^{\gamma\gamma'}\psi\psi$$

for particular constants γ.

F(4) This exceptional Lie superalgebra has dimension 40 and is a sub-algebra of OSp(24|16). The even part of the group is O(3)xSO(7) so three invariants are:

$$B_{\mu\nu} + B_{\nu\mu} = 0$$

$$A_\mu A_\mu + B_{\mu\nu}B_{\mu\nu} + \psi_{\{1}^\alpha\psi_{2\}}^\alpha$$

$$A_{\{1}A_2A_{3\}} + B_{\{\mu\nu}B_{\nu\tau}B_{\tau\mu\}} + B_{\mu\nu}\sigma_{\mu\nu}^{\alpha\beta}\psi_k^\alpha\psi_k^\beta + A_\mu\Gamma_\mu^{\alpha\beta}\psi_\alpha^k\psi_\beta^k + (sym.)$$

This group is related to the octonions by considering the 16 component spinors as two component octonion spinors and the gamma matrices acting on the upper indices as unit octonions. We then have $f^{\mu\nu\tau}\sigma_{\nu\tau} \equiv \gamma_\mu$ where f is the structure constants of octonion multiplication.

G(3) This exceptional Lie superalgebra has dimension 31 and is a sub-algebra of OSp(17|14). The even part of the group is O(3)xG2. The invariants are similar to the above (it being a subalgebra of the F(4)?) so the first invariant is:

$$A_\mu A_\mu + C_\alpha^\mu C_\alpha^\mu + \psi_{\{1}^\mu\psi_{2\}}^\nu$$

There are also two so-called **strange** series called **p(n)** and **q(n)**.

14.4 Classification of infinite-dimensional simple linearly compact Lie superalgebras

The classification consists of the 10 series **W(m, n)**, **S(m, n)** ((m, n) $\neq$ (1, 1)), **H(2m, n)**, **K(2m+1, n)**, **HO(m,m)** (m $\geq$ 2), **SHO(m,m)** (m $\geq$ 3), **KO(m,m + 1)**, **SKO(m,m + 1; β)** (m $\geq$ 2), **SHO~(2m,2m)**, **SKO~(2m+1,2m + 3)** and the 5 exceptional algebras:

E(1,6), E(5,10), E(4,4), E(3,6), E(3,8)

The last two are particularly interesting (according to Kac) because they have the standard model gauge group **SU(3)xSU(2)xU(1)** as their zero level algebra. Infinite-dimensional (affine) Lie superalgebras are important symmetries in superstring theory.

14.5 Category-theoretic definition

In category theory, a **Lie superalgebra** can be defined as a nonassociative superalgebra whose product satisfies

- $[\cdot,\cdot] \circ (id + \tau_{A,A}) = 0$
- $[\cdot,\cdot] \circ ([\cdot,\cdot] \otimes id) \circ (id + \sigma + \sigma^2) = 0$

where σ is the cyclic permutation braiding $(id \otimes \tau_{A,A}) \circ (\tau_{A,A} \otimes id)$. In diagrammatic form:

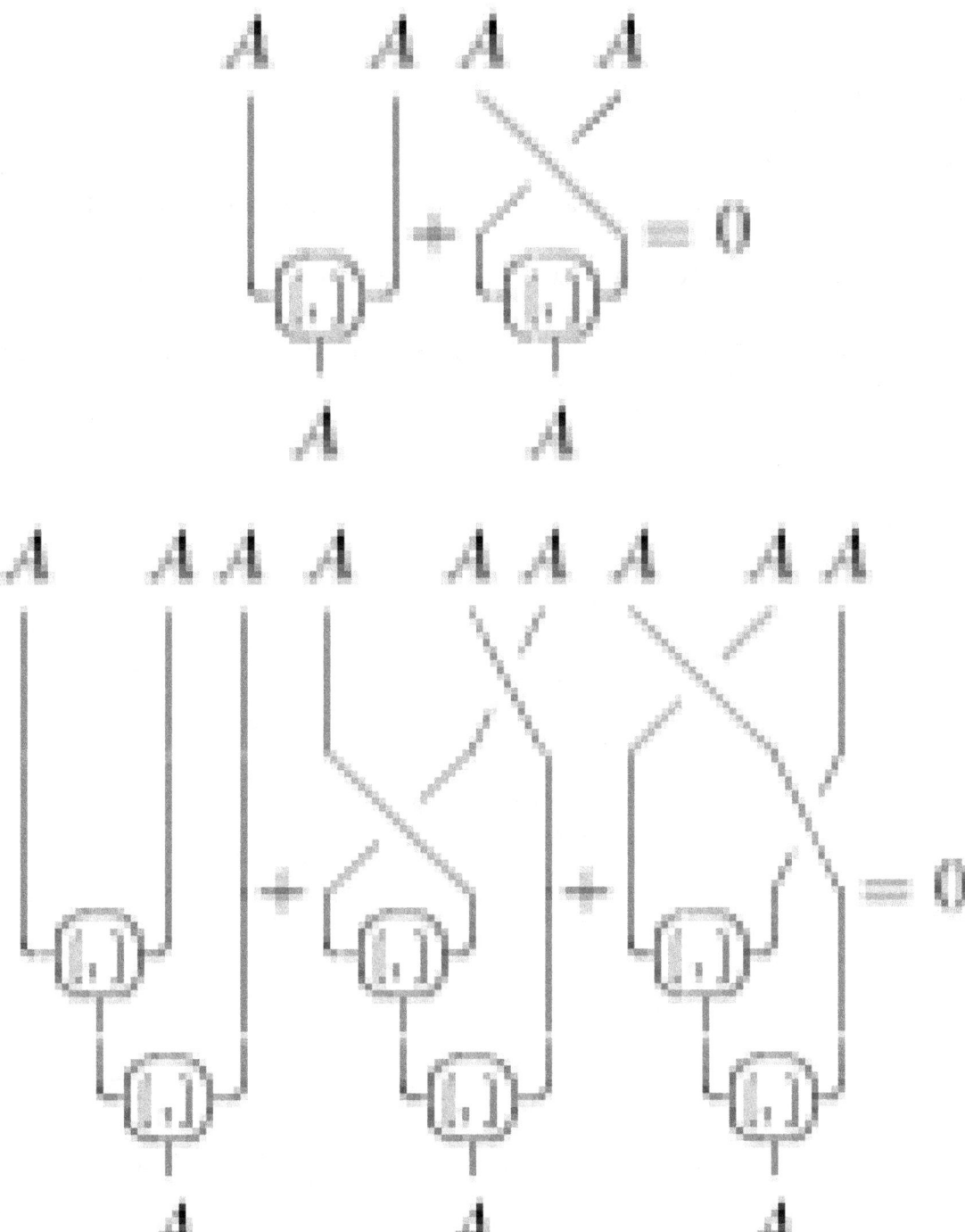

14.6 See also

- Anyonic Lie algebra
- Grassmann algebra
- Representation of a Lie superalgebra
- Superspace
- Supergroup

- Universal enveloping algebra

14.7 References

- Kac, V. G. *Lie superalgebras.* Advances in Math. 26 (1977), no. 1, 8-–96.

- Manin, Yuri I. *Gauge field theory and complex geometry.* Grundlehren der Mathematischen Wissenschaften, 289. Springer-Verlag, Berlin, 1997. ISBN 3-540-61378-1

- Pavel Grozman, Dimitry Leites and Irina Shchepochkina. "LIE SUPERALGEBRAS OF STRING THEORIES"

- Lie Superalgebras and Enveloping Algebras Ian M. Musson, Graduate Studies in Mathematics 2012; 488 pp; hardcover Volume: 131 ISBN 978-0-8218-6867-6

14.8 External links

- Irving Kaplansky + Lie Superalgebras

Chapter 15

Yang–Mills theory

Yang–Mills theory is a gauge theory based on the SU(N) group, or more generally any compact, semi-simple Lie group. Yang–Mills theory seeks to describe the behavior of elementary particles using these non-Abelian Lie groups and is at the core of the unification of the electromagnetic force and weak (i.e. U(1) × SU(2)) as well as Quantum Chromodynamics, the theory of the strong force (based on SU(3)). Thus it forms the basis of our understanding of particle physics, the Standard Model.

15.1 History and theoretical description

In a private correspondence, Wolfgang Pauli formulated in 1953 a six-dimensional theory of Einstein's field equations of general relativity, extending the five-dimensional theory of Kaluza, Klein, Fock and others to a higher-dimensional internal space.[1] However, there is no evidence that Pauli developed the Lagrangian of a gauge field or the quantization of it. Because Pauli found that his theory "leads to some rather unphysical shadow particles", he refrained from publishing his results formally.[1] Although Pauli did not publish his six-dimensional theory, he gave two talks about it in Zürich.[2] Recent research shows that an extended Kaluza–Klein theory is in general not equivalent to Yang–Mills theory, as the former contains additional terms.[3]

In early 1954, Chen Ning Yang and Robert Mills [4] extended the concept of gauge theory for abelian groups, e.g. quantum electrodynamics, to nonabelian groups to provide an explanation for strong interactions. The idea by Yang–Mills was criticized by Pauli,[5] as the quanta of the Yang–Mills field must be massless in order to maintain gauge invariance. The idea was set aside until 1960, when the concept of particles acquiring mass through symmetry breaking in massless theories was put forward, initially by Jeffrey Goldstone, Yoichiro Nambu, and Giovanni Jona-Lasinio.

This prompted a significant restart of Yang–Mills theory studies that proved successful in the formulation of both electroweak unification and quantum chromodynamics (QCD). The electroweak interaction is described by SU(2) × U(1) group while QCD is an SU(3) Yang–Mills theory. The electroweak theory is obtained by combining SU(2) with U(1), where quantum electrodynamics (QED) is described by a U(1) group, and is replaced in the unified electroweak theory by a U(1) group representing a weak hypercharge rather than electric charge. The massless bosons from the SU(2) × U(1) theory mix after spontaneous symmetry breaking to produce the 3 massive weak bosons, and the photon field. The Standard Model combines the strong interaction, with the unified electroweak interaction (unifying the weak and electromagnetic interaction) through the symmetry group SU(2) × U(1) × SU(3). In the current epoch the strong interaction is not unified with the electroweak interaction, but from the observed running of the coupling constants it is believed they all converge to a single value at very high energies.

Phenomenology at lower energies in quantum chromodynamics is not completely understood due to the difficulties of managing such a theory with a strong coupling. This may be the reason why confinement has not been theoretically proven, though it is a consistent experimental observation. Proof that QCD confines at low energy is a mathematical problem of great relevance, and an award has been proposed by the Clay Mathematics Institute for whoever is also able to show that the Yang–Mills theory has a mass gap and its existence.

15.2 Mathematical overview

Yang–Mills theories are a special example of gauge theory with a non-abelian symmetry group given by the Lagrangian

$$\mathcal{L}_{\text{gf}} = -\frac{1}{2}\text{Tr}(F^2) = -\frac{1}{4}F^{a\mu\nu}F^a_{\mu\nu}$$

with the generators of the Lie algebra corresponding to the F-quantities (the curvature or field-strength form) satisfying

$$\text{Tr}(T^a T^b) = \frac{1}{2}\delta^{ab}, \quad [T^a, T^b] = if^{abc}T^c$$

and the covariant derivative defined as

$$D_\mu = I\partial_\mu - igT^a A^a_\mu$$

where I is the identity for the group generators, A^a_μ is the vector potential, and g is the coupling constant. In four dimensions, the coupling constant g is a pure number and for a SU(N) group one has $a, b, c = 1 \ldots N^2 - 1$.

The relation

$$F^a_{\mu\nu} = \partial_\mu A^a_\nu - \partial_\nu A^a_\mu + gf^{abc}A^b_\mu A^c_\nu$$

can be derived by the commutator

$$[D_\mu, D_\nu] = -igT^a F^a_{\mu\nu}.$$

The field has the property of being self-interacting and equations of motion that one obtains are said to be semi-linear, as nonlinearities are both with and without derivatives. This means that one can manage this theory only by perturbation theory, with small nonlinearities.

Note that the transition between "upper" ("contravariant") and "lower" ("covariant") vector or tensor components is trivial for a indices (e.g. $f^{abc} = f_{abc}$), whereas for μ and ν it is nontrivial, corresponding e.g. to the usual Lorentz signature, $\eta_{\mu\nu} = \text{diag}(+---)$.

From the given Lagrangian one can derive the equations of motion given by

$$\partial^\mu F^a_{\mu\nu} + gf^{abc}A^{\mu b}F^c_{\mu\nu} = 0.$$

Putting $F_{\mu\nu} = T^a F^a_{\mu\nu}$, these can be rewritten as

$$(D^\mu F_{\mu\nu})^a = 0.$$

A Bianchi identity holds

$$(D_\mu F_{\nu\kappa})^a + (D_\kappa F_{\mu\nu})^a + (D_\nu F_{\kappa\mu})^a = 0$$

which is equivalent to the Jacobi identity

$$[D_\mu, [D_\nu, D_\kappa]] + [D_\kappa, [D_\mu, D_\nu]] + [D_\nu, [D_\kappa, D_\mu]] = 0$$

since $[D_\mu, F^a_{\nu\kappa}] = D_\mu F^a_{\nu\kappa}$. Define the dual strength tensor $\tilde{F}^{\mu\nu} = \frac{1}{2}\varepsilon^{\mu\nu\rho\sigma} F_{\rho\sigma}$, then the Bianchi identity can be rewritten as

$$D_\mu \tilde{F}^{\mu\nu} = 0.$$

A source J^a_μ enters into the equations of motion as

$$\partial^\mu F^a_{\mu\nu} + g f^{abc} A^{b\mu} F^c_{\mu\nu} = -J^a_\nu.$$

Note that the currents must properly change under gauge group transformations.

We give here some comments about the physical dimensions of the coupling. We note that, in D dimensions, the field scales as $[A] = [L^{\frac{2-D}{2}}]$ and so the coupling must scale as $[g^2] = [L^{D-4}]$. This implies that Yang–Mills theory is not renormalizable for dimensions greater than four. Further, we note that, for $D = 4$, the coupling is dimensionless and both the field and the square of the coupling have the same dimensions of the field and the coupling of a massless quartic scalar field theory. So, these theories share the scale invariance at the classical level.

15.3 Quantization of Yang–Mills theory

A method of quantizing the Yang–Mills theory is by functional methods, i.e. path integrals. One introduces a generating functional for n-point functions as

$$Z[j] = \int [dA] \exp\left[-\frac{i}{2}\int d^4x \, \text{Tr}(F^{\mu\nu} F_{\mu\nu}) + i\int d^4x \, j^a_\mu(x) A^{a\mu}(x)\right],$$

but this integral has no meaning as it is because the potential vector can be arbitrarily chosen due to the gauge freedom. This problem was already known for quantum electrodynamics but here becomes more severe due to non-abelian properties of the gauge group. A way out has been given by Ludvig Faddeev and Victor Popov with the introduction of a **ghost field** (see Faddeev–Popov ghost) that has the property of being unphysical since, although it agrees with Fermi–Dirac statistics, it is a complex scalar field, which violates the spin-statistics theorem. So, we can write the generating functional as

$$Z[j, \bar{\varepsilon}, \varepsilon] = \int [dA][d\bar{c}][dc] \exp\{iS_F[\partial A, A] + iS_{gf}[\partial A] + iS_g[\partial c, \partial \bar{c}, c, \bar{c}, A]\}$$
$$\exp\left\{i\int d^4x \, j^a_\mu(x) A^{a\mu}(x) + i\int d^4x[\bar{c}^a(x)\varepsilon^a(x) + \bar{\varepsilon}^a(x)c^a(x)]\right\}$$

being

$$S_F = -\frac{1}{2}\text{Tr}(F^{\mu\nu} F_{\mu\nu})$$

for the field,

$$S_{gf} = -\frac{1}{2\xi}(\partial \cdot A)^2$$

for the gauge fixing and

$$S_g = -(\bar{c}^a \partial_\mu \partial^\mu c^a + g\bar{c}^a f^{abc} \partial_\mu A^{b\mu} c^c)$$

for the ghost. This is the expression commonly used to derive Feynman's rules (see Feynman diagram). Here we have c^a for the ghost field while α fixes the gauge's choice for the quantization. Feynman's rules obtained from this functional are the following gluon propagator: $\quad D^{ab}_{\mu\nu}(p) = \dfrac{-i\delta^{ab}}{p^2+i0}\left[\eta_{\mu\nu} - \dfrac{(1-\xi)p_\mu p_\nu}{p^2+i0}\right]$

3-gluon vertex: $\quad \Gamma^{abc}_{\mu\nu\lambda}(p,q,r) = -gf^{abc}[(p-q)_\lambda \eta_{\mu\nu} + (q-r)_\mu \eta_{\nu\lambda} + (r-p)_\nu \eta_{\mu\lambda}]$

4-gluon vertex: $\quad \Gamma^{abcd}_{\mu\nu\lambda\sigma} = -ig^2 f^{abe}f^{cde}(\eta_{\mu\lambda}\eta_{\nu\sigma} - \eta_{\mu\sigma}\eta_{\nu\lambda})$
$\qquad -ig^2 f^{ace}f^{bde}(\eta_{\mu\nu}\eta_{\sigma\lambda} - \eta_{\mu\sigma}\eta_{\nu\lambda})$
$\qquad -ig^2 f^{ade}f^{bce}(\eta_{\mu\nu}\eta_{\sigma\lambda} - \eta_{\mu\lambda}\eta_{\nu\sigma})$

ghost propagator: $\quad C'^{ab}(p) = \dfrac{i\delta^{ab}}{p^2+i0}$

$\bar{c}cg$ − vertex: $\quad \Gamma^{abc}(p) = gf^{abc}p_\mu$

These rules for Feynman diagrams can be obtained when the generating functional given above is rewritten as

$$Z[j,\bar{\varepsilon},\varepsilon] = \exp\left(-ig \int d^4x \frac{\delta}{i\delta\bar{\varepsilon}^a(x)} f^{abc} \partial_\mu \frac{i\delta}{\delta j^b_\mu(x)} \frac{i\delta}{\delta\varepsilon^c(x)}\right)$$
$$\times \exp\left(-ig \int d^4x f^{abc} \partial_\mu \frac{i\delta}{\delta j^a_\nu(x)} \frac{i\delta}{\delta j^b_\mu(x)} \frac{i\delta}{\delta j^{c\nu}(x)}\right)$$
$$\times \exp\left(-i\frac{g^2}{4} \int d^4x f^{abc} f^{ars} \frac{i\delta}{\delta j^b_\mu(x)} \frac{i\delta}{\delta j^c_\nu(x)} \frac{i\delta}{\delta j^{r\mu}(x)} \frac{i\delta}{\delta j^{s\nu}(x)}\right)$$
$$\times Z_0[j,\bar{\varepsilon},\varepsilon]$$

with

$$Z_0[j,\bar{\varepsilon},\varepsilon] = \exp\left(-\int d^4x d^4y \bar{\varepsilon}^a(x) C^{ab}(x-y)\varepsilon^b(y)\right) \exp\left(\tfrac{1}{2} \int d^4x d^4y j^a_\mu(x) D^{ab\mu\nu}(x-y) j^b_\nu(y)\right)$$

being the generating functional of the free theory. Expanding in g and computing the functional derivatives, we are able to obtain all the n-point functions with perturbation theory. Using LSZ reduction formula we get from the n-

point functions the corresponding process amplitudes, cross sections and decay rates. The theory is renormalizable and corrections are finite at any order of perturbation theory.

For quantum electrodynamics the ghost field decouples because the gauge group is abelian. This can be seen from the coupling between the gauge field and the ghost field that is $\bar{c}^a f^{abc} \partial_\mu A^{b\mu} c^c$. For the abelian case, all the structure constants f^{abc} are zero and so there is no coupling. In the non-abelian case, the ghost field appears as a useful way to rewrite the quantum field theory without physical consequences on the observables of the theory such as cross sections or decay rates.

One of the most important results obtained for Yang–Mills theory is asymptotic freedom. This result can be obtained by assuming that the coupling constant g is small (so small nonlinearities), as for high energies, and applying perturbation theory. The relevance of this result is due to the fact that a Yang–Mills theory that describes strong interaction and asymptotic freedom permits proper treatment of experimental results coming from deep inelastic scattering.

To obtain the behavior of the Yang–Mills theory at high energies, and so to prove asymptotic freedom, one applies perturbation theory assuming a small coupling. This is verified a posteriori in the ultraviolet limit. In the opposite limit, the infrared limit, the situation is the opposite, as the coupling is too large for perturbation theory to be reliable. Most of the difficulties that research meets is just managing the theory at low energies. That is the interesting case, being inherent to the description of hadronic matter and, more generally, to all the observed bound states of gluons and quarks and their confinement (see hadrons). The most used method to study the theory in this limit is to try to solve it on computers (see lattice gauge theory). In this case, large computational resources are needed to be sure the correct limit of infinite volume (smaller lattice spacing) is obtained. This is the limit the results must be compared with. Smaller spacing and larger coupling are not independent of each other, and larger computational resources are needed for each. As of today, the situation appears somewhat satisfactory for the hadronic spectrum and the computation of the gluon and ghost propagators, but the glueball and hybrids spectra are yet a questioned matter in view of the experimental observation of such exotic states. Indeed, the σ resonance[6][7] is not seen in any of such lattice computations and contrasting interpretations have been put forward. This is a hotly debated issue.

15.4 Propagators

In order to understand the behavior of the theory at large and small momenta, a key quantity is the propagator. For a Yang–Mills theory we have to consider both the gluon and the ghost propagators. At large momenta (ultraviolet limit), the question was completely settled with the discovery of the asymptotic freedom.[8][9] In this case it is seen that the theory becomes free (trivial ultraviolet fixed point for renormalization group) and both the gluon and ghost propagators are those of a free massless particle. The asymptotic states of the theory are represented by massless gluons that carry the interaction. The coupling runs to zero as we will see in the next section.

At low momenta (infrared limit) the question has been more involved to settle. The reason is that the theory becomes strongly coupled in this case and perturbation theory cannot be applied. The only reliable approach to get an answer is performing lattice computation on a computer powerful enough to afford large volumes. An answer to this question is a fundamental one as it would provide an understanding to the problem of confinement. On the other side, it should not be forgotten that propagators are gauge-dependent quantities and so, they must be managed carefully when one wants to get meaningful physical results.

On the other side, theoretical approaches were conceived to get an understanding of the theory in this case. Pioneering works were due to Vladimir Gribov and Daniel Zwanziger. Gribov uncovered the question of the gauge-fixing in a Yang–Mills theory: He showed that, even once a gauge is fixed, a freedom is left yet (Gribov ambiguity).[10] Besides, he was able to provide a functional form for the gluon propagator in the Landau gauge

$$D^{ab}_{\mu\nu}(p) = \delta^{ab} \left(\eta_{\mu\nu} - \frac{p_\mu p_\nu}{p^2} \right) \frac{p^2}{p^4 + M^4}.$$

This propagator cannot be correct in this way as it would violate causality. On the other side, it provides a linear rising potential, $V(r) \propto r$, that would give reason to quark confinement. An important aspect of this functional form is that *the gluon propagator appears to go to zero with momenta*. This will become a crucial point in the following. From these studies by Gribov, Zwanziger extended his approach.[11][12] The inescapable conclusion was that the gluon propagator should go to zero with momenta while the ghost propagator should be enhanced with respect to the free case running to infinity.[13][14] This became known in literature as the Gribov-Zwanziger scenario. When this scenario was proposed, computational resources were insufficient to decide if it was correct or not. Rather, people pursued a different approach using the Dyson-Schwinger equations. This is a set of coupled equations for the n-point functions of the theory forming a hierarchy. This means that the equation for the n-point function will depend on the (n+1)-point function. So, to solve them one needs a proper truncation. On the other side, these equation are non-perturbative and could permit to obtain the behavior of the n-point functions in any regime. A solution to this hierarchy through truncation was proposed by Reinhard Alkofer, Andreas Hauck and Lorenz von Smekal.[15] This paper and the following publications from this group, the German group, set the agenda for the determination of the behavior of the propagators in the Landau gauge in the subsequent years. The main conclusions these authors arrived to were to confirm the Gribov-Zwanziger scenario and that the running coupling should reach a finite non-null fixed point when momenta runs to zero. This paper represents the birth of the so-called *scaling solution* as the propagators are seen to obey scaling laws with given exponents. A proposal in the eighties by John Cornwall was in contrast with this scenario rather showing that the gluons get massive when momenta goes to zero and the propagator should be finite and non-null there[16] but went ignored at that time because the theoretical evidence appeared overwhelming for the Gribov-Zwanziger scenario. Attempts to solve the Dyson-Schwinger equations numerically seemed to provide a different scenario[17][18] but this could have been due to the way truncation and approximations were applied.

The significant improvement in the computational resources made possible to unveil the proper behavior of the propagators in the Landau gauge. These results where firstly announced in Regensburg at the Lattice 2007 Conference. The results were somewhat unexpected and an example is given in the following figure for the gluon propagator [19]

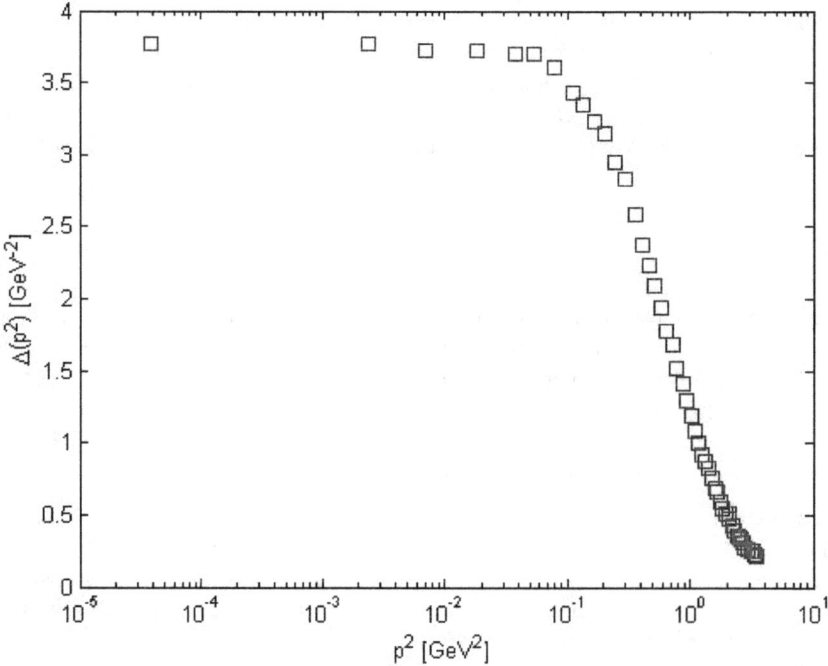

that was obtained for the SU(2) case with a lattice of 128^4 points reaching momenta in the very deep infrared. This result from a huge lattice shows that the gluon propagator never goes to zero with momenta but rather reaches a plateau with a finite value at zero momenta. This went called the *decoupling solution* in literature. Similarly, the ghost propagator is seen to behave as that of a free particle. The ghost field just decouples from the gauge field and becomes free in the deep infrared. Other groups at the same conference confirmed similar results.[20][21]

The decoupling scenario is consistent with a Yukawa-like propagator in the very deep infrared

$$D^{ab}_{\mu\nu}(p) \stackrel{p\to 0}{=} \delta^{ab}\left(\eta_{\mu\nu} - \frac{p_\mu p_\nu}{p^2}\right)\frac{Z}{p^2 - M^2 + i0},$$

with Z a constant. The gluon field develops a mass gap parametrized by M in the above formula, while the BRST symmetry appears to be dynamically broken. These results hold in dimensions greater than 2 while for two dimensions the scaling solution holds.[22] Today, this scenario is generally accepted as the correct one for Yang-Mills theories

15.5 Beta function and running coupling

One of the key properties of a quantum field theory is the behavior over all the energy range of the running coupling. Such a behavior can be obtained from a theory once its beta function is known. Our ability to extract results from a quantum field theory relies on perturbation theory. Once the beta function is known, the behavior at all energy scales of the running coupling is obtained through the equation

$$\mu^2 \frac{d\alpha_s}{d\mu^2} = \beta(\alpha_s).$$

being $\alpha_s = g^2/4\pi$. Yang–Mills theory has the property of being asymptotically free in the large energy limit (ultraviolet limit). This means that, in this limit, the beta function has a minus sign driving the behavior of the running coupling toward even smaller values as the energy increases. Perturbation theory permits to evaluate beta function in this limit producing the following result for SU(N)

$$\beta(\alpha_s) = -\frac{11N}{12\pi}\alpha_s^2 - \frac{17N^2}{24\pi^2}\alpha_s^3 + O\left(\alpha_s^4\right).$$

In the opposite limit of low energies (infrared limit), the beta function is not known. It is note the exact one for a supersymmetric Yang–Mills theory. This has been obtained by Novikov, Shifman, Vainshtein and Zakharov[23] and can be written as

$$\beta(\alpha_s) = -\frac{\alpha_s^2}{4\pi}\frac{3N}{1-\frac{N\alpha_s}{2\pi}}.$$

With this starting point, Thomas Ryttov and Francesco Sannino were able to postulate a non-supersymmetric version of it writing down[24]

$$\beta(\alpha_s) = -\alpha_s^2\frac{11N}{12\pi}\frac{1}{1-\frac{17N}{11}\frac{\alpha_s}{2\pi}}.$$

As can be seen from the beta function of the supersymmetric theory, the limit of a large coupling (infrared limit) implies

$$\beta(\alpha_s) \approx \frac{3}{2}\alpha_s.$$

and so the running coupling in the deep infrared limit goes to zero making this theory trivial. This implies that the coupling reaches a maximum at some value of the energy turning again to zero as the energy is lowered. Then, if Ryttov and Sannino hypothesis is correct, the same should be true for ordinary Yang–Mills theory. This would be in agreement with recent lattice computations.[25]

15.6 Open problems

Yang–Mills theories met with general acceptance in the physics community after Gerard 't Hooft, in 1972, worked out their renormalization, relying on a formulation of the problem worked out by his advisor Martinus Veltman. (Their work[26] was recognized by the 1999 Nobel prize in physics.) Renormalizability is obtained even if the gauge bosons described by this theory are massive, as in the electroweak theory, provided the mass is only an "acquired" one, generated by the Higgs mechanism.

Concerning the mathematics, it should be noted that presently, i.e. in 2014, the Yang–Mills theory is a very active field of research, yielding e.g. invariants of differentiable structures on four-dimensional manifolds via work of Simon Donaldson. Furthermore, the field of Yang–Mills theories was included in the Clay Mathematics Institute's list of "Millennium Prize Problems". Here the prize-problem consists, especially, in a proof of the conjecture that the lowest excitations of a pure Yang–Mills theory (i.e. without matter fields) have a finite mass-gap with regard to the vacuum state. Another open problem, connected with this conjecture, is a proof of the confinement property in the presence of additional Fermion particles.

In physics the survey of Yang–Mills theories does not usually start from perturbation analysis or analytical methods, but more recently from systematic application of numerical methods to lattice gauge theories.

15.7 See also

- Yang–Mills existence and mass gap
- Aharonov–Bohm effect
- Coulomb gauge
- Electroweak theory
- Field theoretical formulation of the standard model
- Gauge covariant derivative
- Kaluza–Klein theory
- Lorenz gauge
- $N = 4$ supersymmetric Yang–Mills theory
- Quantum chromodynamics
- Quantum gauge theory
- Symmetry in physics
- Weyl gauge
- Yang–Mills–Higgs equations
- Propagator
- Lattice gauge theory

15.8 References

[1] Straumann, N (2000). "On Pauli's invention of non-abelian Kaluza-Klein Theory in 1953". arXiv:gr-qc/0012054 [gr-qc].

[2] See Abraham Pais' account of this period as well as L. Susskind's "Superstrings, Physics World on the first non-abelian gauge theory" where Susskind wrote that Yang–Mills was "rediscovered" only because Pauli had chosen not to publish.

[3] Reifler, N (2007). "Conditions for exact equivalence of Kaluza-Klein and Yang–Mills theories". arXiv:gr-qc/0707.3790 [gr-qc].

[4] Yang, C. N.; Mills, R. (1954). "Conservation of Isotopic Spin and Isotopic Gauge Invariance". *Physical Review* **96** (1): 191–195. Bibcode:1954PhRv...96..191Y. doi:10.1103/PhysRev.96.191.

[5] An Anecdote by C. N. Yang

[6] Caprini, I.; Colangelo, G.; Leutwyler, H. (2006). "Mass and width of the lowest resonance in QCD". *Physical Review Letters* **96** (13): 132001. arXiv:hep-ph/0512364. Bibcode:2006PhRvL..96m2001C. doi:10.1103/PhysRevLett.96.132001.

[7] Yndurain, F. J.; Garcia-Martin, R.; Pelaez, J. R. (2007). "Experimental status of the $\pi\pi$ isoscalar S wave at low energy: $f_0(600)$ pole and scattering length". *Physical Review D* **76** (7): 074034. arXiv:hep-ph/0701025. Bibcode:2007PhRvD..76g4034G. doi:10.1103/PhysRevD.76.074034.

[8] D.J. Gross, F. Wilczek (1973). "Ultraviolet behavior of non-abelian gauge theories". *Physical Review Letters* **30** (26): 1343–1346. Bibcode:1973PhRvL..30.1343G. doi:10.1103/PhysRevLett.30.1343.

[9] H.D. Politzer (1973). "Reliable perturbative results for strong interactions". *Physical Review Letters* **30** (26): 1346–1349. Bibcode:1973PhRvL..30.1346P. doi:10.1103/PhysRevLett.30.1346.

[10] V.N. Gribov (1978). "Quantization of non-Abelian gauge theories". *Nuclear Physics B* **139** (1-2): 1–19. Bibcode:1978NuPhB.139....1G. doi:10.1016/0550-3213(78)90175-X.

[11] Daniel Zwanziger (1981). "Covariant quantization of gauge fields without Gribov ambiguity". *Nuclear Physics B* **192** (1): 259–269. Bibcode:1981NuPhB.192..259Z. doi:10.1016/0550-3213(81)90202-9.

[12] Daniel Zwanziger (1982). "Non-perturbative modification of the Faddeev-Popov formula and banishment of the naive vacuum". *Nuclear Physics B* **209** (2): 336–348. Bibcode:1982NuPhB.209..336Z. doi:10.1016/0550-3213(82)90260-7.

[13] Daniel Zwanziger (1989). "Local and renormalizable action from the gribov horizon". *Nuclear Physics B* **323** (3): 513–544. Bibcode:1989NuPhB.323..513Z. doi:10.1016/0550-3213(89)90122-3.

[14] Daniel Zwanziger (1993). "Renormalizability of the critical limit of lattice gauge theory by BRS invariance". *Nuclear Physics B* **399** (2): 477–513. Bibcode:1993NuPhB.399..477Z. doi:10.1016/0550-3213(93)90506-K.

[15] Reinhard Alkofer, Andreas Hauck, Lorenz von Smekal (1997). "Infrared Behavior of Gluon and Ghost Propagators in Landau Gauge QCD". *Physical Review Letters* **79** (19): 3591–3594. arXiv:hep-ph/9705242. Bibcode:1997PhRvL..79.3591V. doi:10.1103/PhysRevLett.79.3591.

[16] John Cornwall (1982). "Dynamical mass generation in continuum quantum chromodynamics". *Physical Review D* **26** (6): 1453–1478. Bibcode:1982PhRvD..26.1453C. doi:10.1103/PhysRevD.26.1453.

[17] A.C. Aguilar, A.A. Natale (2004). "A dynamical gluon mass solution in a coupled system of the Schwinger-Dyson equations". *Journal of High Energy Physics* (8): 057. arXiv:hep-ph/0408254. Bibcode:2004JHEP...08..057A. doi:10.1088/1126-6708/2004/08/057.

[18] Philippe Boucaud, Thorsten Brüntjen, Jean Pierre Leroy, Alain Le Yaouanc, Alexey Lokhov, Jacques Micheli, Olivier Pène, Jose Rodriguez-Quintero (2006). "Is the QCD ghost dressing function finite at zero momentum ?". *Journal of High Energy Physics* (6): 001. arXiv:hep-ph/0604056. Bibcode:2006JHEP...06..001B. doi:10.1088/1126-6708/2006/06/001.

[19] Cucchieri, Attilio; Mendes, Tereza (2007). "What's up with IR gluon and ghost propagators in Landau gauge? A puzzling answer from huge lattices" (PDF). Lattice 2007. Trieste: Proceedings of Science. p. 297. arXiv:0710.0412. Retrieved 2013-11-18.

[20] Bogolubsky, I.L.; Ilgenfritz, E. M.; Müller-Preussker, M.; Sternbeck, A. (2007). "Landau-gauge gluon and ghost propagators in 4D SU(3) gluodynamics on large lattice volumes". Lattice 2007. Trieste: Proceedings of Science. p. 290. arXiv:0710.1968. Retrieved 2013-11-18.

[21] Oliveira, O.; Silva, P.J.; Ilgenfritz, E. M.; Sternbeck, A. (2007). "The gluon propagator from large asymmetric lattices". Lattice 2007. Trieste: Proceedings of Science. p. 323. arXiv:0710.1424. Retrieved 2013-11-18.

[22] Markus Huber, Axel Maas, Lorenz von Smekal (2012). "Two- and three-point functions in two-dimensional Landau-gauge Yang-Mills theory: continuum results". *Journal of High Energy Physics* **2012**: 035. arXiv:1207.0222. Bibcode:2012JHEP...11..035H. doi:10.1007/JHEP11(2012)035.

[23] Novikov, V. A.; Shifman, M. A.; A. I. Vainshtein, A. I.; Zakharov, V. I. (1983). "Exact Gell-Mann-Low Function Of Supersymmetric Yang–Mills Theories From Instanton Calculus". *Nuclear Physics B* **229** (2): 381–393. Bibcode:1983NuPhB.229..381N. doi:10.1016/0550-3213(83)90338-3.

[24] Ryttov, T.; Sannino, F. (2008). "Supersymmetry Inspired QCD Beta Function". *Physical Review D* **78** (6): 065001. arXiv:0711.3745. Bibcode:2008PhRvD..78f5001R. doi:10.1103/PhysRevD.78.065001.

[25] Bogolubsky, I. L.; Ilgenfritz, E.-M.; A. I. Müller-Preussker, M.; Sternbeck, A. (2009). "Lattice gluodynamics computation of Landau-gauge Green's functions in the deep infrared". *Physics Letters B* **676** (1-3): 69–73. arXiv:0901.0736. Bibcode:2009PhLB..676...69B. doi:10.1016/j.physletb.2009.04.076.

[26] 't Hooft, G.; Veltman, M. (1972). "Regularization and renormalization of gauge fields". *Nuclear Physics B* **44**: 189. Bibcode:1972NuPhB..44..189T. doi:10.1016/0550-3213(72)90279-9.

15.9 Further reading

Books

- Frampton, P. (2008). *Gauge Field Theories* (3rd ed.). Wiley-VCH. ISBN 978-3-527-40835-1.
- Cheng, T.-P.; Li, L.-F. (1983). *Gauge Theory of Elementary Particle Physics*. Oxford University Press. ISBN 0-19-851961-3.
- 't Hooft, Gerardus (2005). *50 Years of Yang–Mills theory*. World Scientific. ISBN 981-238-934-2.

Articles

- Svetlichny, George (1999). "Preparation for Gauge Theory". arXiv:math-ph/9902027 [math-ph].
- Gross, D. (1992). "Gauge theory - Past, Present and Future". Retrieved 2009-04-23.

15.10 External links

- Hazewinkel, Michiel, ed. (2001), "Yang-Mills field", *Encyclopedia of Mathematics*, Springer, ISBN 978-1-55608-010-4
- Yang–Mills theory on DispersiveWiki
- The Clay Mathematics Institute
- The Millennium Prize Problems

Chapter 16

Minimal Supersymmetric Standard Model

The **Minimal Supersymmetric Standard Model** (**MSSM**) is a minimal extension to the Standard Model that realizes supersymmetry, although non-minimal extensions do exist. Supersymmetry pairs bosons with fermions; therefore every Standard Model particle has a partner that has yet to be discovered. If the superparticles are found, it may be analogous to discovering dark matter [1] and depending on the details of what might be found, it could provide evidence for grand unification and might even, in principle, provide hints as to whether string theory describes nature. The failure of the Large Hadron Collider to find evidence for supersymmetry has led some physicists to suggest that the theory should be abandoned.[2]

The MSSM was originally proposed in 1981 to stabilize the weak scale, solving the hierarchy problem.[3] The Higgs boson mass of the Standard Model is unstable to quantum corrections and the theory predicts that weak scale should be much weaker than what is observed to be. In the MSSM, the Higgs boson has a fermionic superpartner, the Higgsino, that has the same mass as it would if supersymmetry were an exact symmetry. Because fermion masses are radiatively stable, the Higgs mass inherits this stability. However, in MSSM there is a need for more than one Higgs field, as described below.

The only unambiguous way to claim discovery of supersymmetry is to produce superparticles in the laboratory. Because superparticles are expected to be 100 to 1000 times heavier than the proton, it requires a huge amount of energy to make these particles that can only be achieved at particle accelerators. The Tevatron was actively looking for evidence of the production of supersymmetric particles before it was shut down on 30 September 2011. Most physicists believe that supersymmetry must be discovered at the LHC if it is responsible for stabilizing the weak scale. There are five classes of particle that superpartners of the Standard Model fall into: squarks, gluinos, charginos, neutralinos, and sleptons. These superparticles have their interactions and subsequent decays described by the MSSM and each has characteristic signatures.

The MSSM imposes R-parity to explain the stability of the proton. It adds supersymmetry breaking by introducing explicit soft supersymmetry breaking operators into the Lagrangian that is communicated to it by some unknown (and unspecified) dynamics. This means that there are 120 new parameters in the MSSM. Most of these parameters lead to unacceptable phenomenology such as large flavor changing neutral currents or large electric dipole moments for the neutron and electron. To avoid these problems, the MSSM takes all of the soft supersymmetry breaking to be diagonal in flavor space and for all of the new CP violating phases to vanish.

16.1 Theoretical motivations

There are three principal motivations for the MSSM over other theoretical extensions of the Standard Model, namely:

- Naturalness
- Gauge coupling unification
- Dark Matter

These motivations come out without much effort and they are the primary reasons why the MSSM is the leading candidate for a new theory to be discovered at collider experiments such as the Tevatron or the LHC.

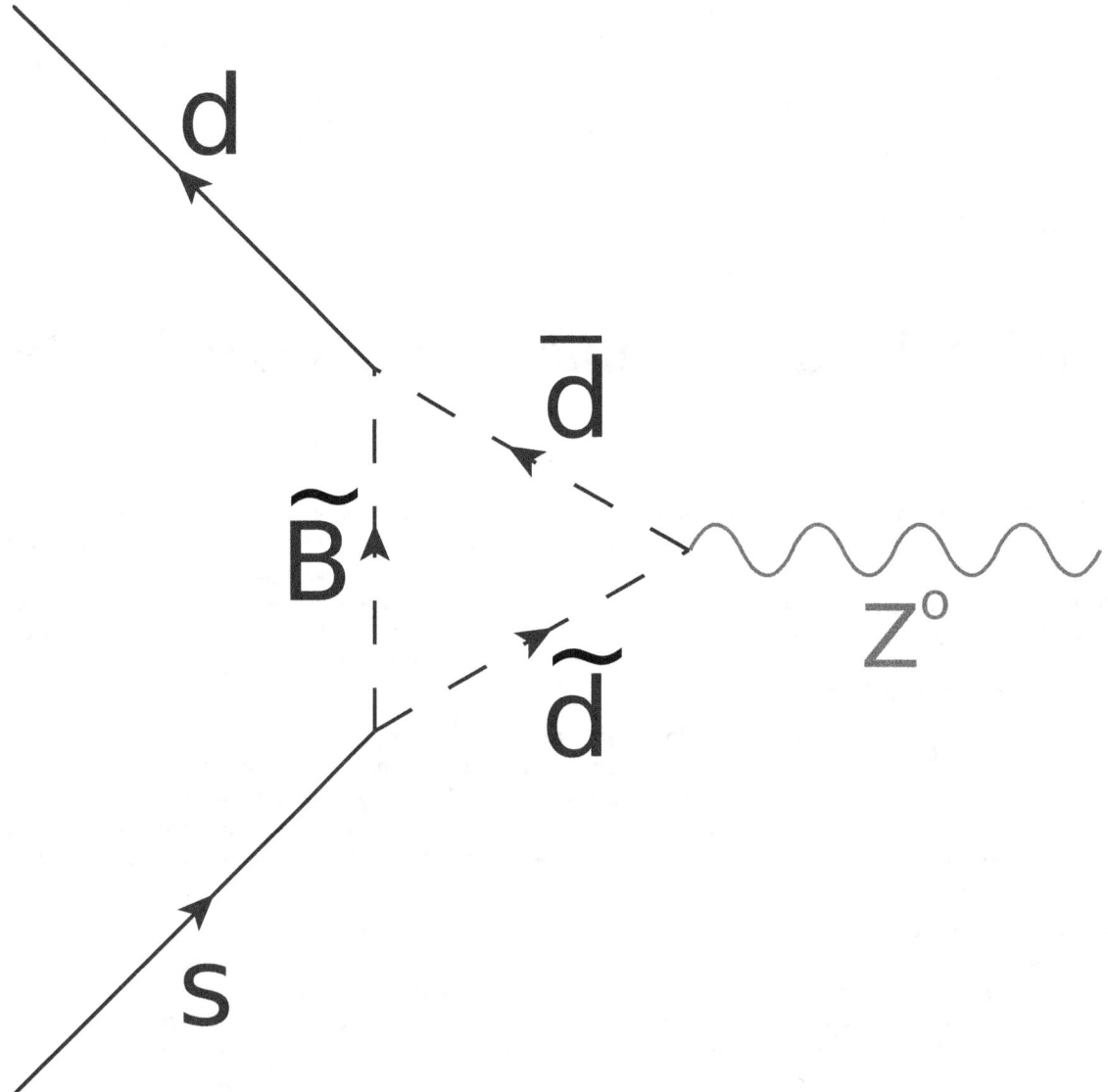

An example of a flavor changing neutral current process in MSSM. A strange quark emits a bino, turning into a sdown-type quark, which then emits a Z boson and reabsorbs the bino, turning into a down quark. If the MSSM squark masses are flavor violating, such a process can occur.

16.1.1 Naturalness

The original motivation for proposing the MSSM was to stabilize the Higgs mass to radiative corrections that are quadratically divergent in the Standard Model (hierarchy problem). In supersymmetric models, scalars are related to fermions and have the same mass. Since fermion masses are logarithmically divergent, scalar masses inherit the same radiative stability. The Higgs vacuum expectation value is related to the negative scalar mass in the Lagrangian. In order for the radiative corrections to the Higgs mass to not be dramatically larger than the actual value, the mass of the superpartners of the Standard Model should not be significantly heavier than the Higgs VEV—roughly 100 GeV. In 2012, the Higgs particle was discovered at the LHC, and its mass was found to be 125-127 GeV.

16.1.2 Gauge-coupling unification

If the superpartners of the Standard Model are near the TeV scale, then measured gauge couplings of the three gauge groups unify at high energies.[4] [5] [6] The beta-functions for the MSSM gauge couplings are given by

where α_1^{-1} is measured in SU(5) normalization—a factor of $\frac{3}{5}$ different than the Standard Model's normalization and predicted by Georgi–Glashow SU(5) .

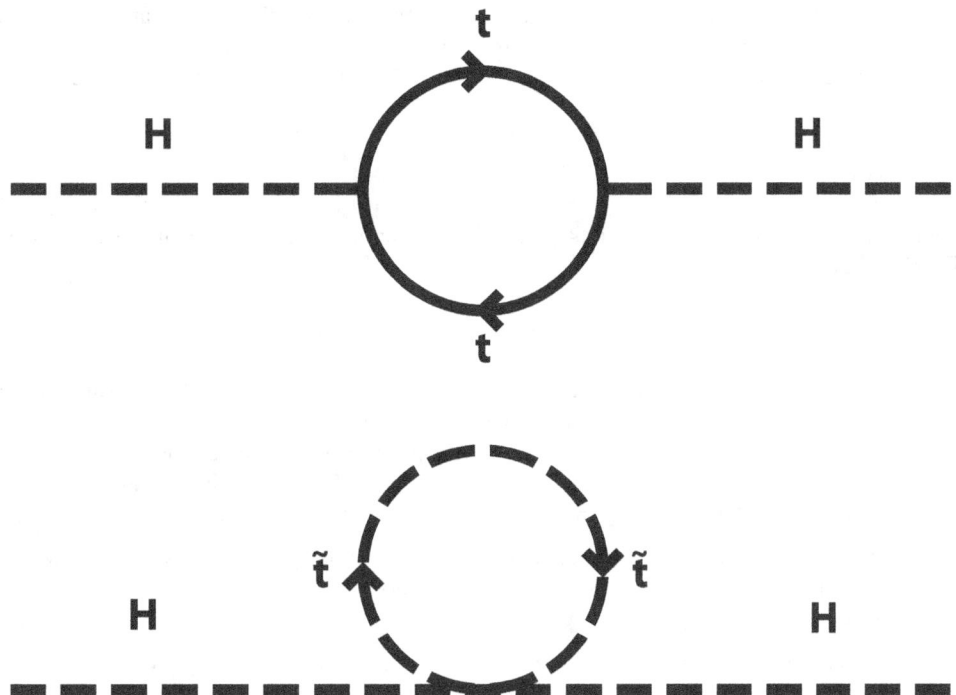

Cancellation of the Higgs boson quadratic mass renormalization between fermionic top quark loop and scalar top squark Feynman diagrams in a supersymmetric extension of the Standard Model

The condition for gauge coupling unification at one loop is whether the following expression is satisfied $\frac{\alpha_3^{-1}-\alpha_2^{-1}}{\alpha_2^{-1}-\alpha_1^{-1}} = \frac{b_{0\,3}-b_{0\,2}}{b_{0\,2}-b_{0\,1}}$.

Remarkably, this is precisely satisfied to experimental errors in the values of $\alpha^{-1}(M_{Z^0})$. There are two loop corrections and both TeV-scale and GUT-scale threshold corrections that alter this condition on gauge coupling unification, and the results of more extensive calculations reveal that gauge coupling unification occurs to an accuracy of 1%, though this is about 3 standard deviations from the theoretical expectations.

This prediction is generally considered as indirect evidence for both the MSSM and SUSY GUTs.[7] It should be noted that gauge coupling unification does not necessarily imply grand unification and there exist other mechanisms to reproduce gauge coupling unification. However, if superpartners are found in the near future, the apparent success of gauge coupling unification would suggest that a supersymmetric grand unified theory is a promising candidate for high scale physics.

16.1.3 Dark matter

If R-parity is preserved, then the lightest superparticle (LSP) of the MSSM is stable and is a Weakly interacting massive particle (WIMP) — i.e. it does not have electromagnetic or strong interactions. This makes the LSP a good dark matter candidate and falls into the category of cold dark matter (CDM) particle.

16.2 Predictions of the MSSM regarding hadron colliders

The Tevatron and LHC have active experimental programs searching for supersymmetric particles. Since both of these machines are hadron colliders — proton antiproton for the Tevatron and proton proton for the LHC — they

16.2.1 Neutralinos

There are four neutralinos that are fermions and are electrically neutral, the lightest of which is typically stable. They are typically labeled N0
1, N0
2, N0
3, N0
4 (although sometimes $\tilde{\chi}_1^0, \ldots, \tilde{\chi}_4^0$ is used instead). These four states are mixtures of the Bino and the neutral Wino (which are the neutral electroweak Gauginos), and the neutral Higgsinos. As the neutralinos are Majorana fermions, each of them is identical with its antiparticle. Because these particles only interact with the weak vector bosons, they are not directly produced at hadron colliders in copious numbers. They primarily appear as particles in cascade decays of heavier particles usually originating from colored supersymmetric particles such as squarks or gluinos.

In R-parity conserving models, the lightest neutralino is stable and all supersymmetric cascades decays end up decaying into this particle which leaves the detector unseen and its existence can only be inferred by looking for unbalanced momentum in a detector.

The heavier neutralinos typically decay through a Z0 to a lighter neutralino or through a W± to chargino. Thus a typical decay is

The mass splittings between the different Neutralinos will dictate which patterns of decays are allowed.

16.2.2 Charginos

There are two Charginos that are fermions and are electrically charged. They are typically labeled C $\tilde{\chi}$ ±
1 and C $\tilde{\chi}$ ±
2 (although sometimes $\tilde{\chi}_1^\pm$ and $\tilde{\chi}_2^\pm$ is used instead). The heavier chargino can decay through Z0 to the lighter chargino. Both can decay through a W± to neutralino.

16.2.3 Squarks

The squarks are the scalar superpartners of the quarks and there is one version for each Standard Model quark. Due to phenomenological constraints from flavor changing neutral currents, typically the lighter two generations of squarks have to be nearly the same in mass and therefore are not given distinct names. The superpartners of the top and bottom quark can be split from the lighter squarks and are called *stop* and *sbottom*.

On the other way, there may be a remarkable left-right mixing of the stops $\tilde{t}$ and of the sbottoms $\tilde{b}$ because of the high masses of the partner quarks top and bottom: [8]

- $\tilde{t}_1 = e^{+i\phi} \cos(\theta) \tilde{t_L} + \sin(\theta) \tilde{t_R}$

- $\tilde{t}_2 = e^{-i\phi} \cos(\theta) \tilde{t_R} - \sin(\theta) \tilde{t_L}$

Same holds for bottom $\tilde{b}$ with its own parameters ϕ and θ.

Squarks can be produced through strong interactions and therefore are easily produced at hadron colliders. They decay to quarks and neutralinos or charginos which further decay. In R-parity conserving scenarios, squarks are pair produced and therefore a typical signal is

- $\tilde{q}\tilde{\bar{q}} \to q\tilde{N}_1^0 \bar{q}\tilde{N}_1^0 \to 2$ jets + missing energy

- $\tilde{q}\tilde{\bar{q}} \to q\tilde{N}_2^0 \bar{q}\tilde{N}_1^0 \to q\tilde{N}_1^0 \ell\bar{\ell} \bar{q}\tilde{N}_1^0 \to 2$ jets + 2 leptons + missing energy

16.2.4 Gluinos

Gluinos are Majorana fermionic partners of the gluon which means that they are their own antiparticles. They interact strongly and therefore can be produced significantly at the LHC. They can only decay to a quark and a squark and thus a typical gluino signal is

- $\tilde{g}\tilde{g} \to (q\tilde{\bar{q}})(\bar{q}\tilde{q}) \to (q\bar{q}\tilde{N}_1^0)(\bar{q}q\tilde{N}_1^0) \to 4$ jets + Missing energy

Because gluinos are Majorana, gluinos can decay to either a quark+anti-squark or an anti-quark+squark with equal probability. Therefore pairs of gluinos can decay to

- $\tilde{g}\tilde{g} \to (\bar{q}\tilde{q})(\bar{q}\tilde{q}) \to (q\bar{q}\tilde{C}_1^+)(q\bar{q}\tilde{C}_1^+) \to (q\bar{q}W^+)(q\bar{q}W^+) \to 4$ jets+ $\ell^+\ell^+$ + Missing energy

This is a distinctive signature because it has same-sign di-leptons and has very little background in the Standard Model.

16.2.5 Sleptons

Sleptons are the scalar partners of the leptons of the Standard Model. They are not strongly interacting and therefore are not produced very often at hadron colliders unless they are very light.

Because of the high mass of the tau lepton there will be left-right mixing of the stau similar to that of stop and sbottom (see above).

Sfermions will typically be found in decays of a charginos and neutralinos if they are light enough to be a decay product

- $\tilde{C}^+ \to \tilde{\ell}^+ \nu$
- $\tilde{N}^0 \to \tilde{\ell}^+ \ell^-$

16.3 MSSM fields

Fermions have bosonic superpartners (called sfermions), and bosons have fermionic superpartners (called bosinos). For most of the Standard Model particles, doubling is very straightforward. However, for the Higgs boson, it is more complicated.

A single Higgsino (the fermionic superpartner of the Higgs boson) would lead to a gauge anomaly and would cause the theory to be inconsistent. However if two Higgsinos are added, there is no gauge anomaly. The simplest theory is one with two Higgsinos and therefore two scalar Higgs doublets. Another reason for having two scalar Higgs doublets rather than one is in order to have Yukawa couplings between the Higgs and both down-type quarks and up-type quarks; these are the terms responsible for the quarks' masses. In the Standard Model the down-type quarks couple to the Higgs field (which has Y=−1/2) and the up-type quarks to its complex conjugate (which has Y=+1/2). However in a supersymmetric theory this is not allowed, so two types of Higgs fields are needed.

16.3.1 MSSM superfields

In supersymmetric theories, every field and its superpartner can be written together as a superfield. The superfield formulation of supersymmetry is very convenient to write down manifestly supersymmetric theories (i.e. one does not have to tediously check that the theory is supersymmetric term by term in the Lagrangian). The MSSM contains vector superfields associated with the Standard Model gauge groups which contain the vector bosons and associated gauginos. It also contains chiral superfields for the Standard Model fermions and Higgs bosons (and their respective superpartners).

16.3.2 MSSM Higgs Mass

The MSSM Higgs Mass is a prediction of the Minimal Supersymmetric Standard Model. The mass of the lightest Higgs boson is set by the Higgs *quartic coupling*. Quartic couplings are not soft supersymmetry-breaking parameters since they lead to a quadratic divergence of the Higgs mass. Furthermore, there are no supersymmetric parameters to make the Higgs mass a free parameter in the MSSM (though not in non-minimal extensions). This means that Higgs mass is a prediction of the MSSM. The LEP II and the IV experiments placed a lower limit on the Higgs mass of 114.4 GeV. This lower limit is significantly above where the MSSM would typically predict it to be, and while it does not rule out the MSSM, the discovery of the Higgs with a mass of 125 GeV makes proponents of the MSSM nervous.[9][10]

Formulas

The only susy-preserving operator that creates a quartic coupling for the Higgs in the MSSM arise for the D-terms of the SU(2) and U(1) gauge sector and the magnitude of the quartic coupling is set by the size of the gauge couplings.

This leads to the prediction that the Standard Model-like Higgs mass (the scalar that couples approximately to the vev) is limited to be less than the Z mass

$$m_{h^0}^2 \leq m_{Z^0}^2 \cos^2 2\beta \ .$$

Since supersymmetry is broken, there are radiative corrections to the quartic coupling that can increase the Higgs mass. These dominantly arise from the 'top sector'

$$m_{h^0}^2 \leq m_{Z^0}^2 \cos^2 2\beta + \frac{3}{\pi^2} \frac{m_t^4 \sin^4 \beta}{v^2} \log \frac{m_{\tilde{t}}}{m_t}$$

where m_t is the top mass and $m_{\tilde{t}}$ is the mass of the top squark. This result can be interpreted as the RG running of the Higgs quartic coupling from the scale of supersymmetry to the top mass—however since the top squark mass should be relatively close to the top mass, this is usually a fairly modest contribution and increases the Higgs mass to roughly the LEP II bound of 114 GeV before the top squark becomes too heavy.

Finally there is a contribution from the top squark A-terms

$$\mathcal{L} = y_t \, m_{\tilde{t}} \, a \, h_u \tilde{q}_3 \tilde{u}_3^c$$

where a is a dimensionless number. This contributes an additional term to the Higgs mass at loop level, but is not logarithmically enhanced

$$m_{h^0}^2 \leq m_{Z^0}^2 \cos^2 2\beta + \frac{3}{\pi^2} \frac{m_t^4 \sin^4 \beta}{v^2} \left(\log \frac{m_{\tilde{t}}}{m_t} + a^2(1 - a^2/12) \right)$$

by pushing $a \to \sqrt{6}$ (known as 'maximal mixing') it is possible to push the Higgs mass to 125 GeV without decoupling the top squark or adding new dynamics to the MSSM.

As the Higgs was found at around 125 GeV (along with no other superparticles) at the LHC, this strongly hints at new dynamics beyond the MSSM, such as the 'Next to Minimal Supersymmetric Standard Model' (NMSSM); and suggests some correlation to the little hierarchy problem.

16.4 The MSSM Lagrangian

The Lagrangian for the MSSM contains several pieces.

- The first is the Kähler potential for the matter and Higgs fields which produces the kinetic terms for the fields.
- The second piece is the gauge field superpotential that produces the kinetic terms for the gauge bosons and gauginos.
- The next term is the superpotential for the matter and Higgs fields. These produce the Yukawa couplings for the Standard Model fermions and also the mass term for the Higgsinos. After imposing R-parity, the renormalizable, gauge invariant operators in the superpotential are

$$W = \mu H_u H_d + y_u H_u Q U^c + y_d H_d Q D^c + y_l H_d L E^c$$

The constant term is unphysical in global supersymmetry (as opposed to supergravity).

16.4.1 Soft Susy breaking

Main article: Soft SUSY breaking

The last piece of the MSSM Lagrangian is the soft supersymmetry breaking Lagrangian. The vast majority of the parameters of the MSSM are in the susy breaking Lagrangian. The soft susy breaking are divided into roughly three pieces.

- The first are the gaugino masses

$$\mathcal{L} \supset m_{\frac{1}{2}} \tilde{\lambda}\tilde{\lambda} + \text{h.c.}$$

Where $\tilde{\lambda}$ are the gauginos and $m_{\frac{1}{2}}$ is different for the wino, bino and gluino.

- The next are the soft masses for the scalar fields

$$\mathcal{L} \supset m_0^2 \phi^\dagger \phi$$

where ϕ are any of the scalars in the MSSM and m_0 are 3×3 hermitean matrices for the squarks and sleptons of a given set of gauge quantum numbers. The eigenvalues of these matrices are actually the masses squared, rather than the masses.

- There are the A and B terms which are given by

$$\mathcal{L} \supset B_\mu h_u h_d + A h_u \tilde{q} \tilde{u}^c + A h_d \tilde{q} \tilde{d}^c + A h_d \tilde{l} \tilde{e}^c + \text{h.c.}$$

The A terms are 3×3 complex matrices much as the scalar masses are.

- Although not often mentioned with regard to soft terms, to be consistent with observation, one must also include Gravitino and Goldstino soft masses given by

$$\mathcal{L} \supset m_{3/2} \Psi_\mu^\alpha (\sigma^{\mu\nu})_\alpha^\beta \Psi_\beta + m_{3/2} G^\alpha G_\alpha + \text{h.c.}$$

The reason these soft terms are not often mentioned are that they arise through local supersymmetry and not global supersymmetry, although they are required otherwise if the Goldstino were massless it would contradict observation. The Goldstino mode is eaten by the Gravitino to become massive, through a gauge shift, which also absorbs the would-be "mass" term of the Goldstino.

16.5 Problems with the MSSM

There are several problems with the MSSM — most of them falling into understanding the parameters.

- The mu problem: The Higgsino mass parameter μ appears as the following term in the superpotential: μH$_u$H$_d$. It should have the same order of magnitude as the electroweak scale, many orders of magnitude smaller than that of the planck scale, which is the natural cutoff scale. The soft supersymmetry breaking terms should also be of the same order of magnitude as the electroweak scale. This brings about a problem of naturalness: why are these scales so much smaller than the cutoff scale yet happen to fall so close to each other?

- Flavor universality of soft masses and A-terms: since no flavor mixing additional to that predicted by the standard model has been discovered so far, the coefficients of the additional terms in the MSSM Lagrangian must be, at least approximately, flavor invariant (i.e. the same for all flavors).

- Smallness of CP violating phases: since no CP violation additional to that predicted by the standard model has been discovered so far, the additional terms in the MSSM Lagrangian must be, at least approximately, CP invariant, so that their CP violating phases are small.

16.6 Theories of supersymmetry breaking

A large amount of theoretical effort has been spent trying to understand the mechanism for soft supersymmetry breaking that produces the desired properties in the superpartner masses and interactions. The three most extensively studied mechanisms are:

16.6.1 Gravity-mediated supersymmetry breaking

Gravity-mediated supersymmetry breaking is a method of communicating supersymmetry breaking to the supersymmetric Standard Model through gravitational interactions. It was the first method proposed to communicate supersymmetry breaking. In gravity-mediated supersymmetry-breaking models, there is a part of the theory that only interacts with the MSSM through gravitational interaction. This hidden sector of the theory breaks supersymmetry. Through the supersymmetric version of the Higgs mechanism, the gravitino, the supersymmetric version of the graviton, acquires a mass. After the gravitino has a mass, gravitational radiative corrections to soft masses are incompletely cancelled beneath the gravitino's mass.

It is currently believed that it is not generic to have a sector completely decoupled from the MSSM and there should be higher dimension operators that couple different sectors together with the higher dimension operators suppressed by the Planck scale. These operators give as large of a contribution to the soft supersymmetry breaking masses as the gravitational loops; therefore, today people usually consider gravity mediation to be gravitational sized direct interactions between the hidden sector and the MSSM.

mSUGRA stands for minimal supergravity. The construction of a realistic model of interactions within $N = 1$ supergravity framework where supersymmetry breaking is communicated through the supergravity interactions was carried out by Ali Chamseddine, Richard Arnowitt, and Pran Nath in 1982.[11] mSUGRA is one of the most widely investigated models of particle physics due to its predictive power requiring only 4 input parameters and a sign, to determine the low energy phenomenology from the scale of Grand Unification. The most widely used set of parameters is:

Gravity-Mediated Supersymmetry Breaking was assumed to be flavor universal because of the universality of gravity; however, in 1986 Hall, Kostelecky, and Raby [12] showed that Planck-scale physics that are necessary to generate the Standard-Model Yukawa couplings spoil the universality of the supersymmetry breaking.

16.6.2 Gauge-mediated supersymmetry breaking (GMSB)

Gauge-mediated supersymmetry breaking is method of communicating supersymmetry breaking to the supersymmetric Standard Model through the Standard Model's gauge interactions. Typically a hidden sector breaks supersymmetry and communicates it to massive messenger fields that are charged under the Standard Model. These messenger fields induce a gaugino mass at one loop and then this is transmitted on to the scalar superpartners at two loops. Requiring stop squarks below 2 TeV, the maximum Higg's boson mass predicted is just 121.5GeV.[13] With the Higgs being discovered at 125GeV - this model requires stops above 2 TeV.

16.6.3 Anomaly-mediated supersymmetry breaking (AMSB)

Anomaly-mediated supersymmetry breaking is a special type of gravity mediated supersymmetry breaking that results in supersymmetry breaking being communicated to the supersymmetric Standard Model through the conformal anomaly.[14][15] Requiring stop squarks below 2 TeV, the maximum Higg's boson mass predicted is just 121.0GeV.[13] With the Higgs being discovered at 125GeV - this scenario requires stops heavier than 2 TeV.

16.7 Phenomenological MSSM (pMSSM)

The unconstrained MSSM has more than 100 parameters in addition to the Standard Model parameters. This makes any phenomenological analysis (e.g. finding regions in parameter space consistent with observed data) impractical. Under the following three assumptions:

- no new source of CP-violation

- no Flavour Changing Neutral Currents
- first and second generation universality

one can reduce the number of additional parameters to the following 19 quantities of the phenomenological MSSM (pMSSM):[16] The large parameter space of pMSSM makes searches in pMSSM extremely challenging and makes pMSSM difficult to exclude.

16.8 See also

- MSSM Higgs Mass
- Desert (particle physics)

16.9 References

[1] Murayama, Hitoshi (2000). "Supersymmetry phenomenology". arXiv:hep-ph/0002232.

[2] Wolchover, Natalie (November 29, 2012). "Supersymmetry Fails Test, Forcing Physics to Seek New Ideas". *Scientific American*.

[3] S. Dimopoulos, H. Georgi; Georgi (1981). "Softly Broken Supersymmetry and SU(5)". *Nuclear Physics B* **193**: 150. Bibcode:1981NuPhB.193..150D. doi:10.1016/0550-3213(81)90522-8.

[4] S. Dimopoulos, S. Raby and F. Wilczek; Raby; Wilczek (1981). "Supersymmetry and the Scale of Unification". *Physical Review D* **24** (6): 1681–1683. Bibcode:1981PhRvD..24.1681D. doi:10.1103/PhysRevD.24.1681.

[5] L.E. Ibanez and G.G. Ross; Ross (1981). "Low-energy predictions in supersymmetric grand unified theories". *Physics Letters B* **105** (6): 439. Bibcode:1981PhLB..105..439I. doi:10.1016/0370-2693(81)91200-4.

[6] W.J. Marciano and G. Senjanovic; Senjanović (1982). "Predictions of supersymmetric grand unified theories". *Physical Review D* **25** (11): 3092. Bibcode:1982PhRvD..25.3092M. doi:10.1103/PhysRevD.25.3092.

[7] Gordon Kane, "The Dawn of Physics Beyond the Standard Model", *Scientific American*, June 2003, page 60 and *The frontiers of physics*, special edition, Vol 15, #3, page 8 "Indirect evidence for supersymmetry comes from the extrapolation of interactions to high energies."

[8] Bartl, A.; Hesselbach, S.; Hidaka, K.; Kernreiter, T.; Porod, W. (2003). "Impact of SUSY CP Phases on Stop and Sbottom Decays in the MSSM". arXiv:0306281 [hep-ph].

[9] "Interpreting the LHC Higgs search results in the MSSM". arXiv:1112.3026v3. Bibcode:2012PhLB..710..201H. doi:10.1016/j.physletb.2012.0

[10] "MSSM Higgs boson searches at the evatron and the LHC: Impact of different benchmark scenarios". arXiv:hep-ph/0511023. Bibcode:2006EPJC...45..797C. doi:10.1140/epjc/s2005-02470-y.

[11] A. Chamseddine, R. Arnowitt, P. Nath; Arnowitt; Nath (1982). "Locally Supersymmetric Grand Unification". *Physical Review Letters* **49** (14): 970–974. Bibcode:1982PhRvL..49..970C. doi:10.1103/PhysRevLett.49.970.

[12] L.J. Hall, V.A. Kostelecky, S. Raby; Kostelecky; Raby (1986). "New Flavor Violations in Supergravity Models". *Nuclear Physics B* **267** (2): 415. Bibcode:1986NuPhB.267..415H. doi:10.1016/0550-3213(86)90397-4.

[13] Arbey, A.; Battaglia, M.; Djouadi, A.; Mahmoudi, F.; Quevillon, J. (2011). "Implications of a 125 GeV Higgs for supersymmetric models". *Physics Letters B*. 3 **708** (2012): 162–169. arXiv:1112.3028. Bibcode:2012PhLB..708..162A. doi:10.1016/j.physletb.2012.01.053.

[14] L. Randall, R. Sundrum; Sundrum (1999). "Out of this world supersymmetry breaking". *Nuclear Physics B* **557**: 79–118. arXiv:hep-th/9810155. Bibcode:1999NuPhB.557...79R. doi:10.1016/S0550-3213(99)00359-4.

[15] G. Giudice, M. Luty, H. Murayama, R. Rattazzi; Rattazzi; Luty; Murayama (1998). "Gaugino mass without singlets". *Journal of High Energy Physics* **9812** (12): 027. arXiv:hep-ph/9810442. Bibcode:1998JHEP...12..027G. doi:10.1088/1126-6708/1998/12/027.

[16] Djouadi, A.; Rosier-Lees, S.; Bezouh, M.; Bizouard, M. A.; Boehm, C.; Borzumati, F.; Briot, C.; Carr, J. et al. (1999). "The Minimal Supersymmetric Standard Model: Group Summary Report". arXiv:hep-ph/9901246.

16.10 External links

- MSSM on arxiv.org
- Stephen P. Martin (1997). "A Supersymmetry Primer". arXiv:hep-ph/9709356.
- Particle Data Group review of MSSM and search for MSSM predicted particles
- Ian J R Aitchison (2005). "Supersymmetry and the MSSM: An Elementary Introduction". arXiv:hep-ph/0505105.

Chapter 17

Grand unification energy

The **grand unification energy** Λ_{GUT}, or the **GUT scale**, is the energy level above which, it is believed, the electromagnetic force, weak force, and strong force become equal in strength and unify to one force governed by a simple Lie group. Specific Grand unified theories (GUTs) can predict the grand unification energy but, usually, with large uncertainties due to model dependent details such as the choice of the gauge group, the Higgs sector, the matter content or further free parameters. Furthermore, at the moment it seems fair to state that there is no agreed *minimal GUT*.

The unification of the electroweak forces and the strong force with the gravitational force in a so-called "Theory of Everything" requires an even higher energy level which is generally assumed to be close to the Planck Scale. In theory, at such short distances, gravity becomes comparable in strength to the other three forces of nature known to date. This statement is modified if there exist additional dimensions of space at intermediate scales. In this case, the strength of gravitational interactions increases faster at smaller distances, and the energy scale at which all known forces of nature unify, can be considerably lower. This effect is exploited in models of large extra dimensions.

The exact value of the grand unification energy (if grand unification is indeed realized in nature) depends on the precise physics present at shorter distance scales not yet explored by experiments. If one assumes the Desert and supersymmetry, it is at around 10^{16} GeV.

The most powerful collider to date, the LHC, is designed to reach a center of mass energy of 1.4×10^4 GeV in proton-proton collisions. The scale 10^{16} GeV is only a few orders of magnitude below the Planck scale, and thus not within reach of man-made earth bound colliders. [1]

17.1 See also

- Desert (particle physics)
- Timeline of the Big Bang

17.2 References

[1] Ross, G. (1984). *Grand Unified Theories*. Westview Press. ISBN 978-0-8053-6968-7.

Chapter 18

Hierarchy problem

In theoretical physics, the **hierarchy problem** is the large discrepancy between aspects of the weak force and gravity.[1] Physicists are unable to explain, for example, why the weak force is 10^{32} times stronger than gravity.

18.1 Technical definition

A hierarchy problem occurs when the fundamental parameters, such as coupling constants or masses, of some Lagrangian are vastly different than the parameters measured by experiment. This can happen because measured parameters are related to the fundamental parameters by a prescription known as renormalization. Typically the renormalization parameters are closely related to the fundamental parameters, but in some cases, it appears that there has been a delicate cancellation between the fundamental quantity and the quantum corrections to it. Hierarchy problems are related to fine-tuning problems and problems of naturalness.

Studying the renormalization in hierarchy problems is difficult, because such quantum corrections are usually power-law divergent, which means that the shortest-distance physics are most important. Because we do not know the precise details of the shortest-distance theory of physics, we cannot even address how this delicate cancellation between two large terms occurs. Therefore, researchers postulate new physical phenomena that resolve hierarchy problems without fine tuning.

18.2 The Higgs mass

In particle physics, the most important **hierarchy problem** is the question that asks why the weak force is 10^{32} times stronger than gravity. Both of these forces involve constants of nature, Fermi's constant for the weak force and Newton's constant for gravity. Furthermore if the Standard Model is used to calculate the quantum corrections to Fermi's constant, it appears that Fermi's constant is surprisingly large and is expected to be closer to Newton's constant, unless there is a delicate cancellation between the bare value of Fermi's constant and the quantum corrections to it.

More technically, the question is why the Higgs boson is so much lighter than the Planck mass (or the grand unification energy, or a heavy neutrino mass scale): one would expect that the large quantum contributions to the square of the Higgs boson mass would inevitably make the mass huge, comparable to the scale at which new physics appears, unless there is an incredible fine-tuning cancellation between the quadratic radiative corrections and the bare mass.

It should be remarked that the problem cannot even be formulated in the strict context of the Standard Model, for the Higgs mass cannot be calculated. In a sense, the problem amounts to the worry that a future theory of fundamental particles, in which the Higgs boson mass will be calculable, should not have excessive fine-tunings.

One proposed solution, popular amongst many physicists, is that one may solve the hierarchy problem via supersymmetry. Supersymmetry can explain how a tiny Higgs mass can be protected from quantum corrections. Supersymmetry removes the power-law divergences of the radiative corrections to the Higgs mass and solves the hierarchy problem as long as the supersymmetric particles are light enough to satisfy the Barbieri–Giudice criterion.[2] This still leaves open the mu problem, however. Currently the tenets of supersymmetry are being tested at the LHC, although no evidence has been found so far for supersymmetry.

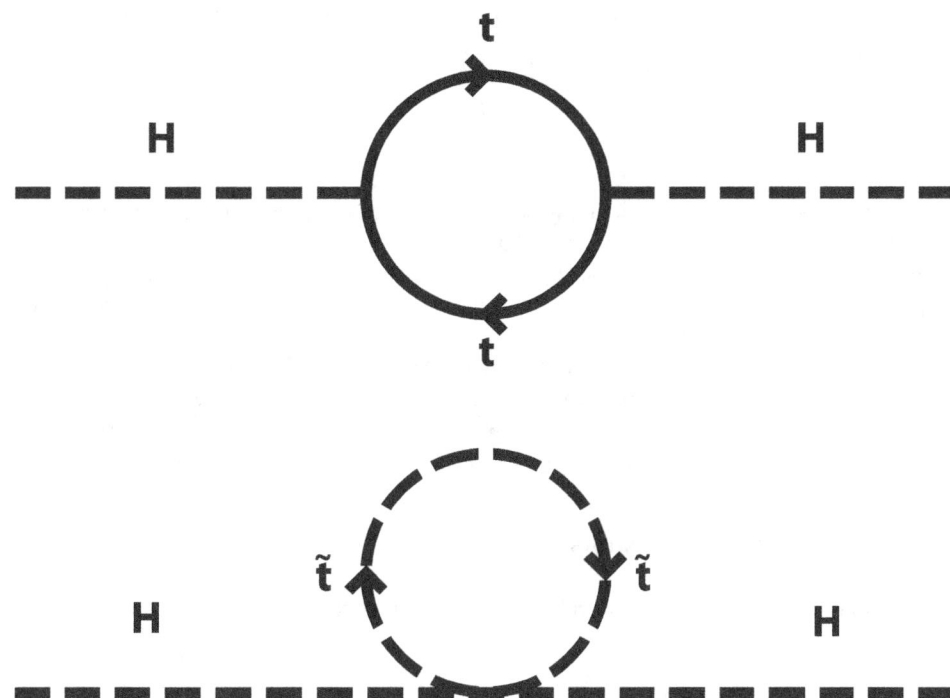

Cancellation of the Higgs boson quadratic mass renormalization between fermionic top quark loop and scalar stop squark tadpole Feynman diagrams in a supersymmetric extension of the Standard Model

18.3 Theoretical solutions

18.3.1 Supersymmetric solution

Each particle that couples to the Higgs field has a Yukawa coupling λ_f. The coupling with the Higgs field for fermions gives an interaction term $\mathcal{L}_{\text{Yukawa}} = -\lambda_f \bar{\psi} H \psi$, with ψ being the Dirac Field and H the Higgs Field. Also, the mass of a fermion is proportional to its Yukawa coupling, meaning that the Higgs boson will couple most to the most massive particle. This means that the most significant corrections to the Higgs mass will originate from the heaviest particles, most prominently the top quark. By applying the Feynman rules, one gets the quantum corrections to the Higgs mass squared from a fermion to be:

$$\Delta m_H^2 = -\frac{|\lambda_f|^2}{8\pi^2}[\Lambda_{\text{UV}}^2 + ...].$$

The Λ_{UV} is called the ultraviolet cutoff and is the scale up to which the Standard Model is valid. If we take this scale to be the Planck scale, then we have the quadratically diverging Lagrangian. However, suppose there existed two complex scalars (taken to be spin 0) such that:

$\lambda_S = |\lambda_f|^2$ (the couplings to the Higgs are exactly the same).

Then by the Feynman rules, the correction (from both scalars) is:

$$\Delta m_H^2 = 2 \times \frac{\lambda_S}{16\pi^2}[\Lambda_{\text{UV}}^2 + ...].$$

(Note that the contribution here is positive. This is because of the spin-statistics theorem, which means that fermions will have a negative contribution and bosons a positive contribution. This fact is exploited.)

This gives a total contribution to the Higgs mass to be zero if we include both the fermionic and bosonic particles. Supersymmetry is an extension of this that creates 'superpartners' for all Standard Model particles.

This section adapted from Stephen P. Martin's "A Supersymmetry Primer" on arXiv.[3]

18.3.2 Conformal solution

Without supersymmetry, a solution to the hierarchy problem has been proposed using just the Standard Model. The idea can be traced back to the fact that the term in the Higgs field that produces the uncontrolled quadratic correction upon renormalization is the quadratic one. If the Higgs field had no mass term, then no hierarchy problem arises. But by missing a quadratic term in the Higgs field, one must find a way to recover the breaking of electroweak symmetry through a non-null vacuum expectation value. This can be obtained using the Weinberg-Coleman mechanism with terms in the Higgs potential arising from quantum corrections. Mass obtained in this way is far too small with respect to what is seen in accelerator facilities and so a conformal Standard Model needs more than one Higgs particle. This proposal has been put forward in 2006 by Krzysztof Meissner and Hermann Nicolai[4] and is currently under scrutiny. But if no further excitation is observed beyond the one seen so far at LHC, this model would have to be abandoned.

18.3.3 Solution via extra dimensions

If we live in a 3+1 dimensional world, then we calculate the Gravitational Force via Gauss' law for gravity:

$$\mathbf{g}(\mathbf{r}) = -Gm\frac{\mathbf{e_r}}{r^2}$$

which is simply Newton's law of gravitation. Note that Newton's constant G can be rewritten in terms of the Planck mass.

$$\frac{1}{M_{\text{Pl}}^2}$$

If we extend this idea to δ extra dimensions, then we get:

$$\mathbf{g}(\mathbf{r}) = -m\frac{\mathbf{e_r}}{M_{\text{Pl}_{3+1+\delta}}^{2+\delta} r^{2+\delta}}$$

where $M_{\text{Pl}_{3+1+\delta}}$ is the 3+1+ δ dimensional Planck mass. However, we are assuming that these extra dimensions are the same size as the normal 3+1 dimensions. Let us say that the extra dimensions are of size $n \lll$ than normal dimensions. If we let $r \ll n$, then we get (2). However, if we let $r \gg n$, then we get our usual Newton's law. However, when $r \gg n$, the flux in the extra dimensions becomes a constant, because there is no extra room for gravitational flux to flow through. Thus the flux will be proportional to n^δ because this is the flux in the extra dimensions. The formula is:

$$\mathbf{g}(\mathbf{r}) = -m\frac{\mathbf{e_r}}{M_{\text{Pl}_{3+1+\delta}}^{2+\delta} r^2 n^\delta}$$

$$-m\frac{\mathbf{e_r}}{M_{\text{Pl}}^2 r^2} = -m\frac{\mathbf{e_r}}{M_{\text{Pl}_{3+1+\delta}}^{2+\delta} r^2 n^\delta}$$

which gives:

$$\frac{1}{M_{\text{Pl}}^2 r^2} = \frac{1}{M_{\text{Pl}_{3+1+\delta}}^{2+\delta} r^2 n^\delta} \Rightarrow$$

$$M_{\text{Pl}}^2 = M_{\text{Pl}_{3+1+\delta}}^{2+\delta} n^\delta.$$

Thus the fundamental Planck mass (the extra dimensional one) could actually be small, meaning that gravity is actually strong, but this must be compensated by the number of the extra dimensions and their size. Physically, this means that gravity is weak because there is a loss of flux to the extra dimensions.

This section adapted from "Quantum Field Theory in a Nutshell" by A. Zee.[5]

Braneworld models

Main article: Brane cosmology

In 1998 Nima Arkani-Hamed, Savas Dimopoulos, and Gia Dvali proposed the **ADD model**, also known as the model with large extra dimensions, an alternative scenario to explain the weakness of gravity relative to the other forces.[6][7] This theory requires that the fields of the Standard Model are confined to a four-dimensional membrane, while gravity propagates in several additional spatial dimensions that are large compared to the Planck scale.[8]

In 1998/99 Merab Gogberashvili published on the arXiv (and subsequently in peer-reviewed journals) a number of articles where he showed that if the Universe is considered as a thin shell (a mathematical synonym for "brane") expanding in 5-dimensional space then it is possible to obtain one scale for particle theory corresponding to the 5-dimensional cosmological constant and Universe thickness, and thus to solve the hierarchy problem.[9][10][11] It was also shown that four-dimensionality of the Universe is the result of stability requirement since the extra component of the Einstein field equations giving the localized solution for matter fields coincides with the one of the conditions of stability.

Subsequently, there were proposed the closely related Randall–Sundrum scenarios which offered their solution to the hierarchy problem.

Empirical tests

Until now, no experimental or observational evidence of extra dimensions has been officially reported. An analysis of results from the Large Hadron Collider in December 2010 severely constrains theories with large extra dimensions.[12]

18.4 The cosmological constant

In physical cosmology, current observations in favor of an accelerating universe imply the existence of a tiny, but nonzero cosmological constant. This is a hierarchy problem very similar to that of the Higgs boson mass problem, since the cosmological constant is also very sensitive to quantum corrections. It is complicated, however, by the necessary involvement of general relativity in the problem and may be a clue that we do not understand gravity on long distance scales (such as the size of the universe today). While quintessence has been proposed as an explanation of the acceleration of the Universe, it does not actually address the cosmological constant hierarchy problem in the technical sense of addressing the large quantum corrections. Supersymmetry does not address the cosmological constant problem, since supersymmetry cancels the $M^4 P_{lanck}$ contribution, but not the $M^2 P_{lanck}$ one (quadratically diverging).

18.5 See also

- Little hierarchy problem
- Quantum triviality

18.6 References

[1] http://profmattstrassler.com/articles-and-posts/particle-physics-basics/the-hierarchy-problem/

[2] R. Barbieri, G. F. Giudice (1988). "Upper Bounds on Supersymmetric Particle Masses". *Nucl. Phys.* B **306**: 63. Bibcode:1988NuPhB.306...63B. doi:10.1016/0550-3213(88)90171-X.

[3] Stephen P. Martin, A Supersymmetry Primer

[4] K. Meissner, H. Nicolai (2006). "Conformal Symmetry and the Standard Model". *Physics Letters* **B648**: 312–317. arXiv:hep-th/0612165. Bibcode:2007PhLB..648..312M. doi:10.1016/j.physletb.2007.03.023.

[5] Zee, A. (2003). "Quantum field theory in a nutshell". Princeton University Press. Bibcode:2003qftn.book.....Z.

[6] N. Arkani-Hamed, S. Dimopoulos, G. Dvali (1998). "The Hierarchy problem and new dimensions at a millimeter". *Physics Letters* **B429**: 263–272. arXiv:hep-ph/9803315. Bibcode:1998PhLB..429..263A. doi:10.1016/S0370-2693(98)00466-3.

[7] N. Arkani-Hamed, S. Dimopoulos, G. Dvali (1999). "Phenomenology, astrophysics and cosmology of theories with submillimeter dimensions and TeV scale quantum gravity". *Physical Review* **D59**: 086004. arXiv:hep-ph/9807344. Bibcode:1999PhRvD..59h6004A. doi:10.1103/PhysRevD.59.086004.

[8] For a pedagogical introduction, see M. Shifman (2009). "Large Extra Dimensions: Becoming acquainted with an alternative paradigm". Crossing the boundaries: Gauge dynamics at strong coupling. Singapore: World Scientific. arXiv:0907.3074.

[9] M. Gogberashvili, *Hierarchy problem in the shell universe model*, Arxiv:hep-ph/9812296.

[10] M. Gogberashvili, *Our world as an expanding shell*, Arxiv:hep-ph/9812365.

[11] M. Gogberashvili, *Four dimensionality in noncompact Kaluza-Klein model*, Arxiv:hep-ph/9904383.

[12] CMS Collaboration, "Search for Microscopic Black Hole Signatures at the Large Hadron Collider," http://arxiv.org/abs/1012.3375

Chapter 19

Left–right symmetry

Left–right symmetry is a general principle in physics which holds that valid physical laws must not produce a different result for a motion that is left-handed than motion that is right-handed. The most common application is expressed as equal treatment of clockwise and counter-clockwise rotations from a fixed frame of reference.

The general principle is often referred to by the name **chiral symmetry**. The rule is absolutely valid in the classical mechanics of Newton and Einstein, but results from quantum mechanical experiments show a difference in the behavior of left-chiral versus right-chiral subatomic particles.

19.1 Particle Physics

In theoretical physics, the electroweak model breaks parity maximally. All its fermions are chiral Weyl fermions, which means that the charged weak gauge bosons only couple to left-handed quarks and leptons. (Note that the neutral electroweak Z boson already couples to left *and* right-handed fermions.) Some theorists found this objectionable, and so proposed a GUT extension of the weak force which has new, high energy W' and Z' bosons which couple with right handed quarks and leptons.

$$\frac{[SU(2)_W \times U(1)_Y]}{\mathbb{Z}_2}$$

to

$$\frac{SU(2)_L \times SU(2)_R \times U(1)_{B-L}}{\mathbb{Z}_2}.$$

Here, SU(2)L (pronounced SU(2) left) is none other than SU(2)W and B–L is the baryon number minus the lepton number. An advantage of this model over the Standard Model is that the electric charge formula in this model is given by

$$Q = I_{3L} + I_{3R} + \frac{B - L}{2}$$

where $I_{3L,R}$ are the weak isospin values of the fields in the theory.

There is also the chromodynamic SU(3)C. The idea was to restore parity by introducing a **left-right symmetry**. This is a group extension of $\mathbb{Z}_2$ (the left-right symmetry) by

$$\frac{SU(3)_C \times SU(2)_L \times SU(2)_R \times U(1)_{B-L}}{\mathbb{Z}_6}$$

to the semidirect product

$$\frac{SU(3)_C \times SU(2)_L \times SU(2)_R \times U(1)_{B-L}}{\mathbb{Z}_6} \rtimes \mathbb{Z}_2.$$

This has two connected components where $\mathbb{Z}_2$ acts as an automorphism, which is the composition of an involutive outer automorphism of SU(3)C with the interchange of the left and right copies of SU(2) with the reversal of U(1)B-L. It was shown by Rabindra N. Mohapatra and Goran Senjanovic in 1975 that left-right symmetry can be spontaneously broken to give a chiral low energy theory, which is the Standard Model of Glashow, Weinberg and Salam and it also connects the small observed neutrino masses to the breaking of left-right symmetry via the seesaw mechanism.

In this setting the chiral quarks

$$(3,2,1)_{1/3}$$

and

$$(\bar{3},1,2)_{-\frac{1}{3}}$$

are unified into an irrep

$$(3,2,1)_{\frac{1}{3}} \oplus (\bar{3},1,2)_{-\frac{1}{3}}.$$

The leptons are also unified into an irrep

$$(1,2,1)_{-1} \oplus (1,1,2)_1.$$

The Higgs bosons needed to implement the breaking of left-right symmetry down to the Standard Model are

$$(1,3,1)_2 \oplus (1,1,3)_2.$$

This then predicts three sterile neutrinos, which is perfectly consistent with current neutrino oscillation data. Within the seesaw mechanism, the sterile neutrinos become superheavy without affecting physics at low energies.

Because the left-right symmetry is spontaneously broken, left-right models predict domain walls.

This left-right symmetry idea first appeared in the Pati–Salam model (1974), Mohapatra–Pati models (1975) and later in trinification (1984).

Chapter 20

Trinification

In physics, the **trinification model** is a GUT theory.
It states that the gauge group is either

$$SU(3)_C \times SU(3)_L \times SU(3)_R$$

or

$$[SU(3)_C \times SU(3)_L \times SU(3)_R]/\mathbb{Z}_3$$

and that the fermions form three families, each consisting of the representations : $(3, \bar{3}, 1)$, $(\bar{3}, 1, 3)$ and : $(1, 3, \bar{3})$.

This includes the right-handed neutrino, which can account for observed neutrino masses (but has not yet been proved to exist). See neutrino oscillations.

There is also a $(1, 3, \bar{3})$ and maybe also a $(1, \bar{3}, 3)$ scalar field called the Higgs field which acquires a VEV. This results in a spontaneous symmetry breaking from

$$SU(3)_L \times SU(3)_R \text{ to } [SU(2) \times U(1)]/\mathbb{Z}_2$$

and also,

$(3, \bar{3}, 1) \to (3, 2)_{\frac{1}{6}} \oplus (3, 1)_{-\frac{1}{3}}$

$(\bar{3}, 1, 3) \to 2\,(\bar{3}, 1)_{\frac{1}{3}} \oplus (\bar{3}, 1)_{-\frac{2}{3}}$

$(1, 3, \bar{3}) \to 2\,(1, 2)_{-\frac{1}{2}} \oplus (1, 2)_{\frac{1}{2}} \oplus 2\,(1, 1)_0 \oplus (1, 1)_1$

$(8, 1, 1) \to (8, 1)_0$

$(1, 8, 1) \to (1, 3)_0 \oplus (1, 2)_{\frac{1}{2}} \oplus (1, 2)_{-\frac{1}{2}} \oplus (1, 1)_0$

$(1, 1, 8) \to 4\,(1, 1)_0 \oplus 2\,(1, 1)_1 \oplus 2\,(1, 1)_{-1}$

See restricted representation.

Note that there are **two** Majorana neutrinos per generation (which is consistent with neutrino oscillations). Also, a copy of

$(3, 1)_{-\frac{1}{3}}$ and $(\bar{3}, 1)_{\frac{1}{3}}$

as well as

$(1, 2)_{\frac{1}{2}}$ and $(1, 2)_{-\frac{1}{2}}$

per generation decouple at the GUT breaking scale due to the couplings

$(1, 3, \bar{3})_H (3, \bar{3}, 1)(\bar{3}, 1, 3)$

and

$(1, 3, \bar{3})_H (1, 3, \bar{3})(1, 3, \bar{3})$.

Note that calling the representations things like $(3, \bar{3}, 1)$ and $(8,1,1)$ is purely a physicist's convention, not a mathematician's convention, where representations are either labelled by Young tableaux or Dynkin diagrams with numbers on their vertices, but still, it is standard among GUT theorists.

Since the homotopy group

$$\pi_2 \left(\frac{SU(3) \times SU(3)}{[SU(2) \times U(1)]/\mathbb{Z}_2} \right) = \mathbb{Z} ,$$

this model predicts monopoles. See 't Hooft–Polyakov monopole.

This model is suggested by Sheldon Lee Glashow, Howard Georgi and Alvaro de Rujula, in 1984. This is one of the maximal subalgebra of E6, whose matter representation 27 has exactly the same representation as above.

20.1 References

Chapter 21

SU(6) (physics)

SU(6) is a grand unified theory which includes the Georgi–Glashow SU(5) gauge group.
The fermionic matter content of this model comes in three generations (copies) of

$$15 \oplus \bar{6} \oplus \bar{6}$$

as SU(6) representations.

This gauge group is broken down to the Georgi–Glashow model by a pair of $6H/\bar{6}_H$ Higgs fields.

$$SU(6) \xrightarrow{6_H/\bar{6}_H} SU(5) \times U(1)$$

The fermions decompose as

$$15 \to 10 \oplus 5$$

$$\bar{6} \to \bar{5} \oplus 1$$

The Yukawa coupling $\bar{6}\bar{6}_H 15$ causes 5 and a $\bar{5}$ SU(5)-rep of fermions to pair up and acquire GUT scale masses for each generation.

An adjoint Higgs **35H** breaks the model down further to the standard model.

Most important, SU(6) is also used as a classification group for baryons and mesons. In fact it takes in account both SU(3) flavour and SU(2) spin. Then, quarks u,d,s, with spin up and down belong to the fundamental representation **6**. Mesons belong to the **35** representation and baryons to the **56** representation obtained from the product quark-antiquark and quark-quark-quark, respectively.

Recently, a new type of grand unified theories based on SU(6) group has been proposed to realize a unification of strong and electroweak interactions.[1] The work provides a complete form of 35 generators in SU(6) group.

21.1 References

[1] A. Hartanto and L.T. Handoko, Grand Unified Theory based on the SU(6) symmetry, *Physical Review D71 (2005) 095013*

Chapter 22

E6 (mathematics)

In mathematics, **E_6** is the name of some closely related Lie groups, linear algebraic groups or their Lie algebras $\mathfrak{e}_6$, all of which have dimension 78; the same notation E_6 is used for the corresponding root lattice, which has rank 6. The designation E_6 comes from the Cartan–Killing classification of the complex simple Lie algebras, which fall into four infinite series labeled An, Bn, Cn, Dn, and five exceptional cases labeled E_6, E_7, E_8, F_4, and G_2. The E_6 algebra is thus one of the five exceptional cases.

The fundamental group of the complex form, compact real form, or any algebraic version of E_6 is the cyclic group **Z/3Z**, and its outer automorphism group is the cyclic group **Z/2Z**. Its fundamental representation is 27-dimensional (complex), and a basis is given by the 27 lines on a cubic surface. The dual representation, which is inequivalent, is also 27-dimensional.

In particle physics, E_6 plays a role in some grand unified theories.

22.1 Real and complex forms

There is a unique complex Lie algebra of type E_6, corresponding to a complex group of complex dimension 78. The complex adjoint Lie group E_6 of complex dimension 78 can be considered as a simple real Lie group of real dimension 156. This has fundamental group **Z/3Z**, has maximal compact subgroup the compact form (see below) of E_6, and has an outer automorphism group non-cyclic of order 4 generated by complex conjugation and by the outer automorphism which already exists as a complex automorphism.

As well as the complex Lie group of type E_6, there are five real forms of the Lie algebra, and correspondingly five real forms of the group with trivial center (all of which have an algebraic double cover, and three of which have further non-algebraic covers, giving further real forms), all of real dimension 78, as follows:

- The compact form (which is usually the one meant if no other information is given), which has fundamental group **Z/3Z** and outer automorphism group **Z/2Z**.

- The split form, EI (or $E_{6(6)}$), which has maximal compact subgroup $Sp(4)/(\pm 1)$, fundamental group of order 2 and outer automorphism group of order 2.

- The quasi-split form EII (or $E_{6(2)}$), which has maximal compact subgroup $SU(2) \times SU(6)/(\text{center})$, fundamental group cyclic of order 6 and outer automorphism group of order 2.

- EIII (or $E_{6(-14)}$), which has maximal compact subgroup $SO(2) \times Spin(10)/(\text{center})$, fundamental group **Z** and trivial outer automorphism group.

- EIV (or $E_{6(-26)}$), which has maximal compact subgroup F_4, trivial fundamental group cyclic and outer automorphism group of order 2.

The EIV form of E_6 is the group of collineations (line-preserving transformations) of the octonionic projective plane OP^2.[1] It is also the group of determinant-preserving linear transformations of the exceptional Jordan algebra. The exceptional Jordan algebra is 27-dimensional, which explains why the compact real form of E_6 has a 27-dimensional complex representation. The compact real form of E_6 is the isometry group of a 32-dimensional Riemannian manifold

22.2 E$_6$ as an algebraic group

By means of a Chevalley basis for the Lie algebra, one can define E$_6$ as a linear algebraic group over the integers and, consequently, over any commutative ring and in particular over any field: this defines the so-called split (sometimes also known as "untwisted") adjoint form of E$_6$. Over an algebraically closed field, this and its triple cover are the only forms; however, over other fields, there are often many other forms, or "twists" of E$_6$, which are classified in the general framework of Galois cohomology (over a perfect field k) by the set $H^1(k, \text{Aut}(E_6))$ which, because the Dynkin diagram of E$_6$ (see below) has automorphism group $\mathbf{Z}/2\mathbf{Z}$, maps to $H^1(k, \mathbf{Z}/2\mathbf{Z}) = \text{Hom}(\text{Gal}(k), \mathbf{Z}/2\mathbf{Z})$ with kernel $H^1(k, E_{6,\text{ad}})$.[2]

Over the field of real numbers, the real component of the identity of these algebraically twisted forms of E$_6$ coincide with the three real Lie groups mentioned above, but with a subtlety concerning the fundamental group: all adjoint forms of E$_6$ have fundamental group $\mathbf{Z}/3\mathbf{Z}$ in the sense of algebraic geometry, with Galois action as on the third roots of unity; this means that they admit exactly one triple cover (which may be trivial on the real points); the further non-compact real Lie group forms of E$_6$ are therefore not algebraic and admit no faithful finite-dimensional representations. The compact real form of E$_6$ as well as the noncompact forms EI=E$_{6(6)}$ and EIV=E$_{6(-26)}$ are said to be *inner* or of type ^{1}E$_6$ meaning that their class lies in $H^1(k, E_{6,\text{ad}})$ or that complex conjugation induces the trivial automorphism on the Dynkin diagram, whereas the other two real forms are said to be *outer* or of type ^{2}E$_6$.

Over finite fields, the Lang–Steinberg theorem implies that $H^1(k, E_6) = 0$, meaning that E$_6$ has exactly one twisted form, known as ^{2}E$_6$: see below.

22.3 Algebra

22.3.1 Dynkin diagram

The Dynkin diagram for E$_6$ is given by ○—○—○—○—○ , which may also be drawn as ○—○—○—○—○—○ or ○—○—○—○—○—○ .

22.3.2 Roots of E$_6$

Although they span a six-dimensional space, it is much more symmetrical to consider them as vectors in a six-dimensional subspace of a nine-dimensional space.

$(1,-1,0;0,0,0;0,0,0)$, $(-1,1,0;0,0,0;0,0,0)$,
$(-1,0,1;0,0,0;0,0,0)$, $(1,0,-1;0,0,0;0,0,0)$,
$(0,1,-1;0,0,0;0,0,0)$, $(0,-1,1;0,0,0;0,0,0)$,
$(0,0,0;1,-1,0;0,0,0)$, $(0,0,0;-1,1,0;0,0,0)$,
$(0,0,0;-1,0,1;0,0,0)$, $(0,0,0;1,0,-1;0,0,0)$,
$(0,0,0;0,1,-1;0,0,0)$, $(0,0,0;0,-1,1;0,0,0)$,
$(0,0,0;0,0,0;1,-1,0)$, $(0,0,0;0,0,0;-1,1,0)$,
$(0,0,0;0,0,0;-1,0,1)$, $(0,0,0;0,0,0;1,0,-1)$,
$(0,0,0;0,0,0;0,1,-1)$, $(0,0,0;0,0,0;0,-1,1)$,

All 27 combinations of $(\mathbf{3};\mathbf{3};\mathbf{3})$ where $\mathbf{3}$ is one of $\left(\frac{2}{3},-\frac{1}{3},-\frac{1}{3}\right)$, $\left(-\frac{1}{3},\frac{2}{3},-\frac{1}{3}\right)$, $\left(-\frac{1}{3},-\frac{1}{3},\frac{2}{3}\right)$

All 27 combinations of $(\bar{\mathbf{3}};\bar{\mathbf{3}};\bar{\mathbf{3}})$ where $\bar{\mathbf{3}}$ is one of $\left(-\frac{2}{3},\frac{1}{3},\frac{1}{3}\right)$, $\left(\frac{1}{3},-\frac{2}{3},\frac{1}{3}\right)$, $\left(\frac{1}{3},\frac{1}{3},-\frac{2}{3}\right)$

Simple roots

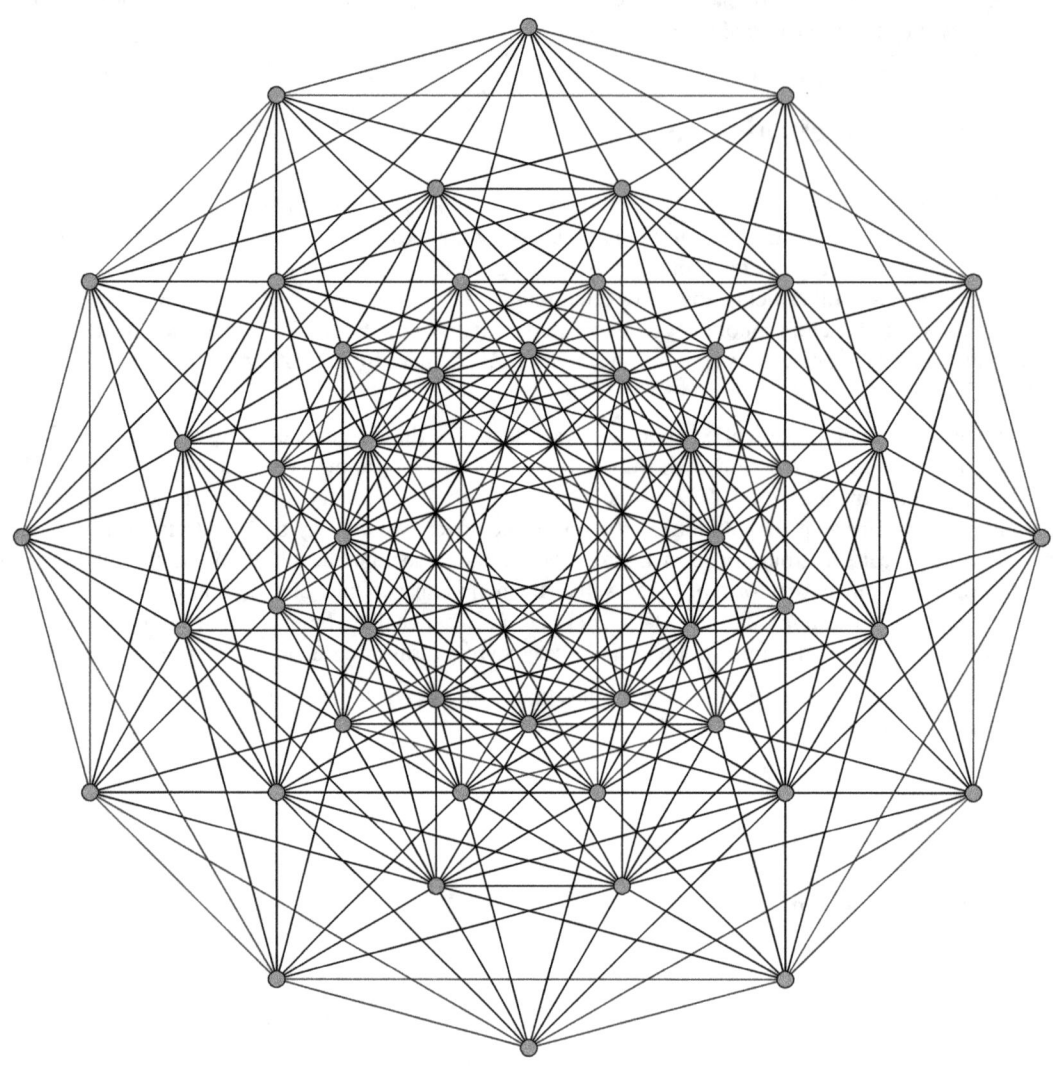

The 72 vertices of the 1_{22} polytope represent the root vectors of the E_6, as shown in this Coxeter plane projection. Coxeter-Dynkin diagram:

$(0,0,0;0,0,0;0,1,-1)$

$(0,0,0;0,0,0;1,-1,0)$

$(0,0,0;0,1,-1;0,0,0)$

$(0,0,0;1,-1,0;0,0,0)$

$(0,1,-1;0,0,0;0,0,0)$

$\left(\frac{1}{3}, -\frac{2}{3}, \frac{1}{3}; -\frac{2}{3}, \frac{1}{3}, \frac{1}{3}; -\frac{2}{3}, \frac{1}{3}, \frac{1}{3}\right)$

An alternative description

An alternative (6-dimensional) description of the root system, which is useful in considering $E_6 \times SU(3)$ as a subgroup of E_8, is the following:

22.3. ALGEBRA

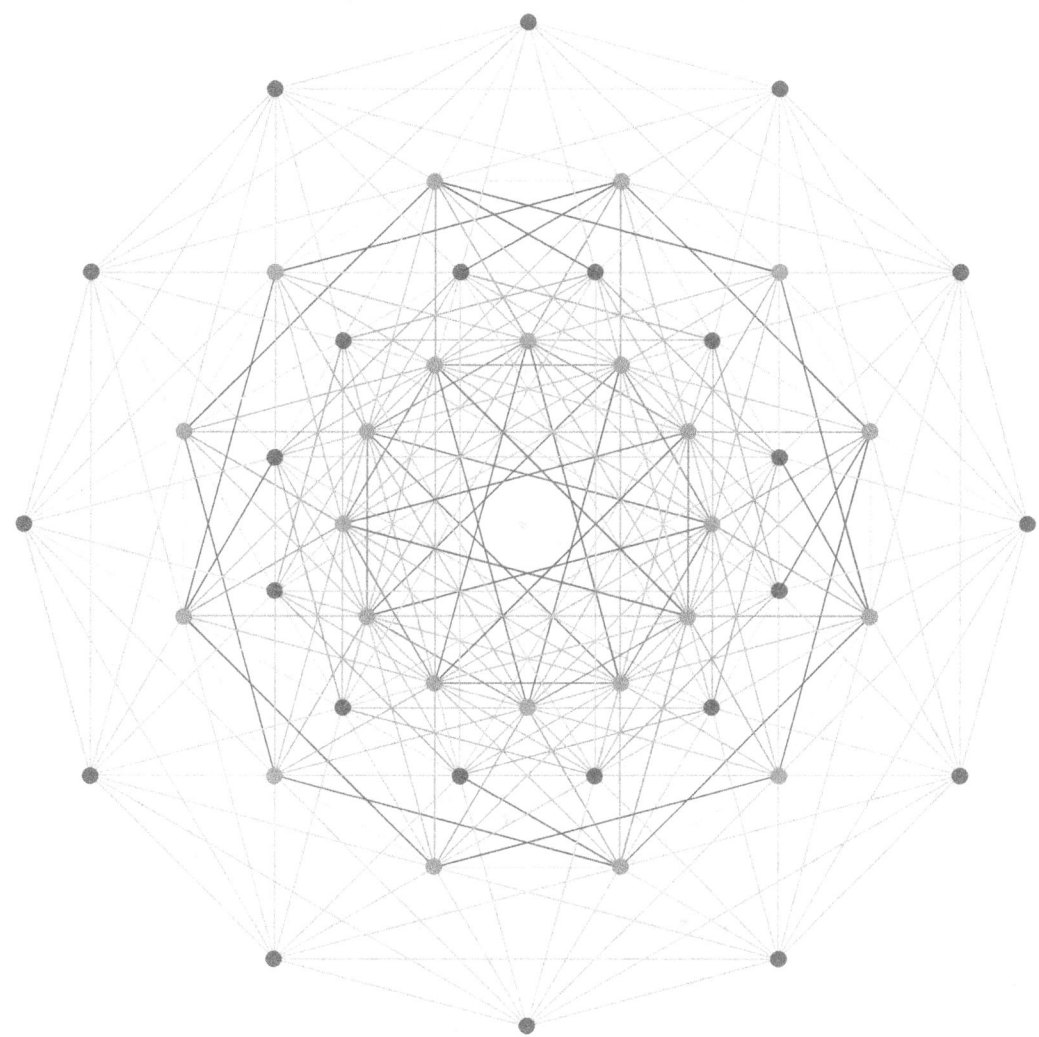

Graph of E6 as a subgroup of E8 projected into the Coxeter plane

All $4 \times \binom{5}{2}$ permutations of

$$(\pm 1, \pm 1, 0, 0, 0, 0)$$

and all of the following roots with an odd number of plus signs

$$\left(\pm\frac{1}{2}, \pm\frac{1}{2}, \pm\frac{1}{2}, \pm\frac{1}{2}, \pm\frac{1}{2}, \pm\frac{\sqrt{3}}{2}\right).$$

Thus the 78 generators consist of the following subalgebras:

> A 45-dimensional SO(10) subalgebra, including the above $4 \times \binom{5}{2}$ generators plus the five Cartan generators corresponding to the first five entries.
>
> Two 16-dimensional subalgebras that transform as a Weyl spinor of spin(10) and its complex conjugate. These have a non-zero last entry.
>
> 1 generator which is their chirality generator, and is the sixth Cartan generator.

Hasse diagram of E6 root poset with edge labels identifying added simple root position

One choice of simple roots for E_6 is given by the rows of the following matrix, indexed in the order ①—②—③ ④—⑤ with ⑥ above ③:

$$\begin{bmatrix} 1 & -1 & 0 & 0 & 0 & 0 \\ 0 & 1 & -1 & 0 & 0 & 0 \\ 0 & 0 & 1 & -1 & 0 & 0 \\ 0 & 0 & 0 & 1 & 1 & 0 \\ -\frac{1}{2} & -\frac{1}{2} & -\frac{1}{2} & -\frac{1}{2} & -\frac{1}{2} & \frac{\sqrt{3}}{2} \\ 0 & 0 & 0 & 1 & -1 & 0 \end{bmatrix}$$

22.3.3 Weyl group

The Weyl group of E_6 is of order 51840: it is the automorphism group of the unique simple group of order 25920 (which can be described as any of: $PSU_4(2)$, $PS\Omega_6^-(2)$, $PSp_4(3)$ or $PS\Omega_5(3)$).[3]

22.3.4 Cartan matrix

$$\begin{bmatrix} 2 & -1 & 0 & 0 & 0 & 0 \\ -1 & 2 & -1 & 0 & 0 & 0 \\ 0 & -1 & 2 & -1 & 0 & -1 \\ 0 & 0 & -1 & 2 & -1 & 0 \\ 0 & 0 & 0 & -1 & 2 & 0 \\ 0 & 0 & -1 & 0 & 0 & 2 \end{bmatrix}$$

22.4 Important subalgebras and representations

The Lie algebra E_6 has an F_4 subalgebra, which is the fixed subalgebra of an outer automorphism, and an SU(3) × SU(3) × SU(3) subalgebra. Other maximal subalgebras which have an importance in physics (see below) and can be read off the Dynkin diagram, are the algebras of SO(10) × U(1) and SU(6) × SU(2).

In addition to the 78-dimensional adjoint representation, there are two dual 27-dimensional "vector" representations.

The characters of finite dimensional representations of the real and complex Lie algebras and Lie groups are all given by the Weyl character formula. The dimensions of the smallest irreducible representations are (sequence A121737 in OEIS):

1, 27 (twice), 78, 351 (four times), 650, 1728 (twice), 2430, 2925, 3003 (twice), 5824 (twice), 7371 (twice), 7722 (twice), 17550 (twice), 19305 (four times), 34398 (twice), 34749, 43758, 46332 (twice), 51975 (twice), 54054 (twice), 61425 (twice), 70070, 78975 (twice), 85293, 100386 (twice), 105600, 112320 (twice), 146432 (twice), 252252 (twice), 314496 (twice), 359424 (four times), 371800 (twice), 386100 (twice), 393822 (twice), 412776 (twice), 442442 (twice)...

The underlined terms in the sequence above are the dimensions of those irreducible representations possessed by the adjoint form of E_6 (equivalently, those whose weights belong to the root lattice of E_6), whereas the full sequence gives the dimensions of the irreducible representations of the simply connected form of E_6.

The symmetry of the Dynkin diagram of E_6 explains why many dimensions occur twice, the corresponding representations being related by the non-trivial outer automorphism; however, there are sometimes even more representations than this, such as four of dimension 351, two of which are fundamental and two of which are not.

The fundamental representations have dimensions 27, 351, 2925, 351, 27 and 78 (corresponding to the seven nodes in the Dynkin diagram in the order chosen for the Cartan matrix above, i.e., the nodes are read in the five-node chain first, with the last node being connected to the middle one).

22.5 E6 polytope

The **E_6 polytope** is the convex hull of the roots of E_6. It therefore exists in 6 dimensions; its symmetry group contains the Coxeter group for E_6 as an index 2 subgroup.

22.6 Chevalley and Steinberg groups of type E_6 and $^2E_6$

Main article: $^2E_6$

The groups of type E_6 over arbitrary fields (in particular finite fields) were introduced by Dickson (1901, 1908).

The points over a finite field with q elements of the (split) algebraic group E_6 (see above), whether of the adjoint (centerless) or simply connected form (its algebraic universal cover), give a finite Chevalley group. This is closely connected to the group written $E_6(q)$, however there is ambiguity in this notation, which can stand for several things:

- the finite group consisting of the points over $\mathbf{F}q$ of the simply connected form of E_6 (for clarity, this can be written $E_{6,\text{sc}}(q)$ or more rarely $\tilde{E}_6(q)$ and is known as the "universal" Chevalley group of type E_6 over $\mathbf{F}q$),

- (rarely) the finite group consisting of the points over $\mathbf{F}q$ of the adjoint form of E_6 (for clarity, this can be written $E_{6,\text{ad}}(q)$, and is known as the "adjoint" Chevalley group of type E_6 over $\mathbf{F}q$), or

- the finite group which is the image of the natural map from the former to the latter: this is what will be denoted by $E_6(q)$ in the following, as is most common in texts dealing with finite groups.

From the finite group perspective, the relation between these three groups, which is quite analogous to that between $SL(n,q)$, $PGL(n,q)$ and $PSL(n,q)$, can be summarized as follows: $E_6(q)$ is simple for any q, $E_{6,sc}(q)$ is its Schur cover, and $E_{6,ad}(q)$ lies in its automorphism group; furthermore, when $q-1$ is not divisible by 3, all three coincide, and otherwise (when q is congruent to 1 mod 3), the Schur multiplier of $E_6(q)$ is 3 and $E_6(q)$ is of index 3 in $E_{6,ad}(q)$, which explains why $E_{6,sc}(q)$ and $E_{6,ad}(q)$ are often written as $3 \cdot E_6(q)$ and $E_6(q) \cdot 3$. From the algebraic group perspective, it is less common for $E_6(q)$ to refer to the finite simple group, because the latter is not in a natural way the set of points of an algebraic group over $\mathbf{F}q$ unlike $E_{6,sc}(q)$ and $E_{6,ad}(q)$.

Beyond this "split" (or "untwisted") form of E_6, there is also one other form of E_6 over the finite field $\mathbf{F}q$, known as $^2E_6$, which is obtained by twisting by the non-trivial automorphism of the Dynkin diagram of E_6. Concretely, $^2E_6(q)$, which is known as a Steinberg group, can be seen as the subgroup of $E_6(q^2)$ fixed by the composition of the non-trivial diagram automorphism and the non-trivial field automorphism of $\mathbf{F}q^2$. Twisting does not change the fact that the algebraic fundamental group of $^2E_{6,ad}$ is $\mathbf{Z}/3\mathbf{Z}$, but it does change those q for which the covering of $^2E_{6,ad}$ by $^2E_{6,sc}$ is non-trivial on the $\mathbf{F}q$-points. Precisely: $^2E_{6,sc}(q)$ is a covering of $^2E_6(q)$, and $^2E_{6,ad}(q)$ lies in its automorphism group; when $q+1$ is not divisible by 3, all three coincide, and otherwise (when q is congruent to 2 mod 3), the degree of $^2E_{6,sc}(q)$ over $^2E_6(q)$ is 3 and $^2E_6(q)$ is of index 3 in $^2E_{6,ad}(q)$, which explains why $^2E_{6,sc}(q)$ and $^2E_{6,ad}(q)$ are often written as $3 \cdot {}^2E_6(q)$ and $^2E_6(q) \cdot 3$.

Two notational issues should be raised concerning the groups $^2E_6(q)$. One is that this is sometimes written $^2E_6(q^2)$, a notation which has the advantage of transposing more easily to the Suzuki and Ree groups, but the disadvantage of deviating from the notation for the $\mathbf{F}q$-points of an algebraic group. Another is that whereas $^2E_{6,sc}(q)$ and $^2E_{6,ad}(q)$ are the $\mathbf{F}q$-points of an algebraic group, the group in question also depends on q (e.g., the points over $\mathbf{F}q^2$ of the same group are the untwisted $E_{6,sc}(q^2)$ and $E_{6,ad}(q^2)$).

The groups $E_6(q)$ and $^2E_6(q)$ are simple for any q,[4][5] and constitute two of the infinite families in the classification of finite simple groups. Their order is given by the following formula (sequence A008872 in OEIS):

$$|E_6(q)| = \frac{1}{\gcd(3, q-1)} q^{36}(q^{12}-1)(q^9-1)(q^8-1)(q^6-1)(q^5-1)(q^2-1)$$

$$|{}^2E_6(q)| = \frac{1}{\gcd(3, q+1)} q^{36}(q^{12}-1)(q^9+1)(q^8-1)(q^6-1)(q^5+1)(q^2-1)$$

(sequence A008916 in OEIS). The order of $E_{6,sc}(q)$ or $E_{6,ad}(q)$ (both are equal) can be obtained by removing the dividing factor $\gcd(3,q-1)$ from the first formula (sequence A008871 in OEIS), and the order of $^2E_{6,sc}(q)$ or $^2E_{6,ad}(q)$ (both are equal) can be obtained by removing the dividing factor $\gcd(3,q+1)$ from the second (sequence A008915 in OEIS).

The Schur multiplier of $E_6(q)$ is always $\gcd(3,q-1)$ (i.e., $E_{6,sc}(q)$ is its Schur cover). The Schur multiplier of $^2E_6(q)$ is $\gcd(3,q+1)$ (i.e., $^2E_{6,sc}(q)$ is its Schur cover) outside of the exceptional case $q=2$ where it is $2^2 \cdot 3$ (i.e., there is an additional 2^2-fold cover). The outer automorphism group of $E_6(q)$ is the product of the diagonal automorphism group $\mathbf{Z}/\gcd(3,q-1)\mathbf{Z}$ (given by the action of $E_{6,ad}(q)$), the group $\mathbf{Z}/2\mathbf{Z}$ of diagram automorphisms, and the group of field automorphisms (i.e., cyclic of order f if $q=p^f$ where p is prime). The outer automorphism group of $^2E_6(q)$ is the product of the diagonal automorphism group $\mathbf{Z}/\gcd(3,q+1)\mathbf{Z}$ (given by the action of $^2E_{6,ad}(q)$) and the group of field automorphisms (i.e., cyclic of order f if $q=p^f$ where p is prime).

22.7 Importance in physics

N=8 supergravity in five dimensions, which is a dimensional reduction from 11 dimensional supergravity, admits an E_6 bosonic global symmetry and an Sp(8) bosonic local symmetry. The fermions are in representations of Sp(8), the gauge fields are in a representation of E_6, and the scalars are in a representation of both (Gravitons are singlets with respect to both). Physical states are in representations of the coset E_6/Sp(8).

In grand unification theories, E_6 appears as a possible gauge group which, after its breaking, gives rise to the SU(3) × SU(2) × U(1) gauge group of the standard model (also see Importance in physics of E8). One way of achieving this is through breaking to SO(10) × U(1). The adjoint 78 representation breaks, as explained above, into an adjoint 45, spinor 16 and $\overline{16}$ as well as a singlet of the SO(10) subalgebra. Including the U(1) charge we have

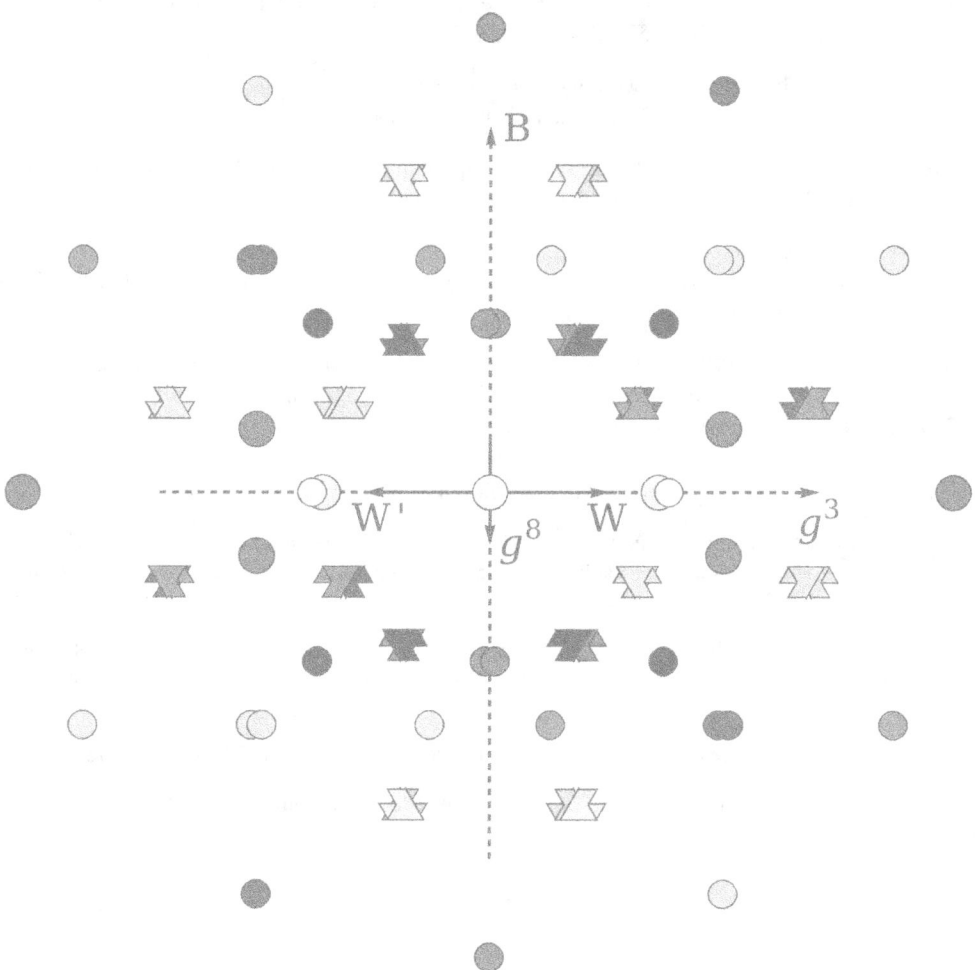

The pattern of weak isospin, W, weaker isospin, W', strong g3 and g8, and baryon minus lepton, B, charges for particles in the SO(10) Grand Unified Theory, rotated to show the embedding in E_6.

$78 \to 45_0 \oplus 16_{-3} \oplus \overline{16}_3 + 1_0$.

Where the subscript denotes the U(1) charge.

22.8 See also

- En (Lie algebra)

- ADE classification

- Freudenthal magic square

22.9 References

- Adams, J. Frank (1996), *Lectures on exceptional Lie groups*, Chicago Lectures in Mathematics, University of Chicago Press, ISBN 978-0-226-00526-3, MR 1428422

- Baez, John (2002). "The Octonions, Section 4.4: E_6". *Bull. Amer. Math. Soc.* **39** (2): 145–205. doi:10.1090/S0273-0979-01-00934-X. ISSN 0273-0979.. Online HTML version at .

- Cremmer, E.; J. Scherk and J. H. Schwarz (1979). "Spontaneously Broken N=8 Supergravity". *Phys. Lett. B* **84** (1): 83–86. doi:10.1016/0370-2693(79)90654-3. . Online scanned version at .

- Dickson, Leonard Eugene (1901), *A class of groups in an arbitrary realm connected with the configuration of the 27 lines on a cubic surface*, The quarterly journal of pure and applied mathematics **33**: 145–173, Reprinted in volume 5 of his collected works

- Dickson, Leonard Eugene (1908), *A class of groups in an arbitrary realm connected with the configuration of the 27 lines on a cubic surface (second paper)*, The quarterly journal of pure and applied mathematics **39**: 205–209, Reprinted in volume VI of his collected works

[1] Rosenfeld, Boris (1997), *Geometry of Lie Groups* (theorem 7.4 on page 335, and following paragraph).

[2] Платонов, Владимир П.; Рапинчук, Андрей С. (1991). *Алгебраические группы и теория чисел*. Наука. ISBN 5-02-014191-7. (English translation: Platonov, Vladimir P.; Rapinchuk, Andrei S. (1994). *Algebraic groups and number theory*. Academic Press. ISBN 0-12-558180-7.), §2.2.4

[3] Conway, John Horton; Curtis, Robert Turner; Norton, Simon Phillips; Parker, Richard A; Wilson, Robert Arnott (1985). *Atlas of Finite Groups: Maximal Subgroups and Ordinary Characters for Simple Groups*. Oxford University Press. p. 26. ISBN 0-19-853199-0.

[4] Carter, Roger W. (1989). *Simple Groups of Lie Type*. Wiley Classics Library. John Wiley & Sons. ISBN 0-471-50683-4.

[5] Wilson, Robert A. (2009). *The Finite Simple Groups*. Graduate Texts in Mathematics **251**. Springer-Verlag. ISBN 1-84800-987-9.

Chapter 23

331 model

The **331 model** in particle physics offers an explanation of why there must be three families of quarks and leptons. One curious feature of the Standard Model is that the gauge anomalies cancel exactly for each of the three known quark-lepton families independently. The Standard Model thus offers no explanation of why there are three families, or indeed why there is more than one family.

The idea behind the 331 model is to extend the standard model such that all three families are required for anomaly cancellation. More specifically, in this model the three families transform differently under an extended gauge group. The perfect cancellation of the anomalies within each family is ruined, but the anomalies of the extended gauge group cancel when all three families are present. The cancellation will persist for 6, 9, ... families, so having only the three families observed in nature is the simplest possible choice of matter content.

Such a construction necessarily requires the addition of further gauge bosons and chiral fermions, which then provide testable predictions of the model in the form of elementary particles. These particles could be found experimentally at masses above the electroweak scale, which is around 100 GeV. The minimal 331 model predicts singly and doubly charged spin-one bosons, bileptons, which could show up in electron-electron scattering when it is studied at TeV energy scales and may also be produced in multi-TeV proton–proton scattering at the Large Hadron Collider.

The 331 model is an extension of the electroweak gauge symmetry from $SU(2)_W \times U(1)_Y$ to $SU(3)_L \times U(1)_X$ with $SU(2)_W \subset SU(3)_W$ and the hypercharge $Y = \beta T_8 + IX$ and the electric charge $Q = Y/2 + T_3/2$ where T_3 and T_8 are the Gell-Mann matrices of SU(3)L and β and I are parameters of the model. The name 331 comes from the full gauge symmetry group $SU(3)_C \times SU(3)_L \times U(1)_X$.

23.1 References

- Frampton, P.H (1992). "Chiral Dilepton Model and the Flavor Question". *Physical Review Letters* **69** (20): 2889–2891. Bibcode:1992PhRvL..69.2889F. doi:10.1103/PhysRevLett.69.2889. PMID 10046667.

- Pisano, F.; Pleitez, V. (1992). "An SU(3) x U(1) model for electroweak interactions". *Physical Review D* **46**: 410–417. arXiv:hep-ph/9206242. Bibcode:1992PhRvD..46..410P. doi:10.1103/PhysRevD.46.410.

- Foot, R.; Hernandez, O.F.; Pisano, F.; Pleitez, V. (1992). "Lepton masses in an SU(3)L x U(1)N gauge model". *Physical Review D* **47** (9): 4158–4161. arXiv:hep-ph/9207264. Bibcode:1993PhRvD..47.4158F. doi:10.1103/PhysRevD.47.4158.

23.2 See also

- Standard Model
- Standard model (basic details)
- Beyond the Standard Model

Chapter 24

Chiral color

In particle physics phenomenology, **chiral color** is a speculative model[1] which extends quantum chromodynamics (QCD), the generally accepted theory for the strong interactions of quarks. QCD is a gauge field theory based on a gauge group known as color SU(3)C with an octet of colored gluons known as **gluons** acting as the force carriers between a triplet of colored quarks.

In Chiral Color, QCD is extended to a gauge group which is SU(3)L × SU(3)R and leads to a second octet of force carriers. SU(3)C is identified with a diagonal subgroup of these two factors. The gluons correspond to the unbroken gauge bosons and the color octet axigluons -- which couple strongly to the quarks—are massive. Hence the name is Chiral Color. Although Chiral Color has presently no experimental support—which is not really any worse than most models currently investigated—it has the "aesthetic" advantage of rendering the Standard Model more similar in its treatment of the two short range forces, strong and weak interactions.

Unlike gluons, the axigluons are predicted to be massive. Extensive searches for axigluons at CERN[2] and Fermilab[3][4] have successfully placed a lower bound on the axigluon mass of about 1 TeV. Axigluons may be discovered when collisions are studied with higher energy at the Large Hadron Collider.

24.1 References

[1] Paul H. Frampton and Sheldon L. Glashow (1987). "Chiral Color: An Alternative to the Standard Model". *Physics Letters B* **190**: 157. Bibcode:1987PhLB..190..157F. doi:10.1016/0370-2693(87)90859-8.

[2] C. Albajar et al. (UA1 collaboration) (1988). "Two-jet mass distributions at the CERN proton-antiproton collider". *Physics Letters B* **209**: 127. Bibcode:1988PhLB..209..127A. doi:10.1016/0370-2693(88)91843-6.

[3] F. Abe et al. (CDF Collaboration) (1990). "Two-jet invariant mass distribution at $s^{1/2}$ = 1.8 TeV". *Physical Review D* **41** (5): 1722. Bibcode:1990PhRvD..41.1722A. doi:10.1103/PhysRevD.41.1722.

[4] F. Abe et al. (CDF Collaboration) (1997). "Search for new particles decaying to dijets at CDF". *Physical Review D* **55** (9): 5263. arXiv:hep-ex/9702004. Bibcode:1997PhRvD..55.5263A. doi:10.1103/PhysRevD.55.R5263.

Chapter 25

Flipped SU(5)

The **Flipped SU(5) model** is a Grand Unified Theory (GUT) theory first contemplated by Stephen Barr in 1982,[1] and by Dimitri Nanopoulos and others in 1984.[2][3] Ignatios Antoniadis, John Ellis, John Hagelin, and Nanopoulos developed the supersymmetric flipped SU(5), derived from the deeper-level superstring.[4]

Current efforts to explain the theoretical underpinnings for observed neutrino masses are being developed in the context of supersymmetric flipped SU(5).[5]

Flipped SU(5) is not a fully unified model, because the U(1)Y factor of the SM gauge group is within the U(1) factor of the GUT group. The addition of states below M_X in this model, while solving certain threshold correction issues in string theory, makes the model merely descriptive, rather than predictive.[6]

25.1 The Model

The Flipped SU(5) model states that the gauge group is:

$(SU(5) \times U(1)_\chi)/Z_5$

Fermions form three families, each consisting of the representations

$\mathbf{5}_{-3}$ for the lepton doublet, L, and the up quarks u^c;

$\mathbf{10}_1$ for the quark doublet, Q, the down quark, d^c and the right-handed neutrino, N;

$\mathbf{1}_5$ for the charged leptons, e^c.

This assignment includes three right-handed neutrinos, which have never been observed, but are often postulated to explain the lightness of the observed neutrinos and neutrino oscillations. There is also a $\mathbf{10}_1$ and/or $\mathbf{10}_{-1}$ called the Higgs fields which acquire a VEV, yielding the spontaneous symmetry breaking

$(SU(5) \times U(1)_\chi)/Z_5 \to (SU(3) \times SU(2) \times U(1)_Y)/Z_6$

The SU(5) representations transform under this subgroup as the reducible representation as follows:

$\bar{\mathbf{5}}_{-3} \to (\bar{3}, 1)_{-\frac{2}{3}} \oplus (1, 2)_{-\frac{1}{2}}$ (u^c and l)

$\mathbf{10}_1 \to (3, 2)_{\frac{1}{6}} \oplus (\bar{3}, 1)_{\frac{1}{3}} \oplus (1, 1)_0$ (q, d^c and v^c)

$\mathbf{1}_5 \to (1, 1)_1$ (e^c)

$\mathbf{24}_0 \to (8, 1)_0 \oplus (1, 3)_0 \oplus (1, 1)_0 \oplus (3, 2)_{\frac{1}{6}} \oplus (\bar{3}, 2)_{-\frac{1}{6}}$.

25.2 Comparison with the standard SU(5)

The name "flipped" SU(5) arose in comparison to the "standard" SU(5) Georgi–Glashow model, in which u^c and d^c quark are respectively assigned to the **10** and **5** representation. In comparison with the standard SU(5), the flipped SU(5) can accomplish the spontaneous symmetry breaking using Higgs fields of dimension 10, while the standard SU(5) requires both a 5- and 45-dimensional Higgs.

The sign convention for $U(1)_\chi$ varies from article/book to article.

The hypercharge Y/2 is a linear combination (sum) of the following:

$$\begin{pmatrix} \frac{1}{15} & 0 & 0 & 0 & 0 \\ 0 & \frac{1}{15} & 0 & 0 & 0 \\ 0 & 0 & \frac{1}{15} & 0 & 0 \\ 0 & 0 & 0 & -\frac{1}{10} & 0 \\ 0 & 0 & 0 & 0 & -\frac{1}{10} \end{pmatrix} \in \mathrm{SU}(5), \qquad \chi/5.$$

There are also the additional fields $\mathbf{5}_{-2}$ and $\mathbf{5}_2$ containing the electroweak Higgs doublets.

Calling the representations for example, $\mathbf{5}_{-3}$ and $\mathbf{24}_0$ is purely a physicist's convention, not a mathematician's convention, where representations are either labelled by Young tableaux or Dynkin diagrams with numbers on their vertices, and is a standard used by GUT theorists.

Since the homotopy group

$$\pi_2\left(\frac{[SU(5) \times U(1)_\chi]/\mathbf{Z}_5}{[SU(3) \times SU(2) \times U(1)_Y]/\mathbf{Z}_6}\right) = 0$$

this model does not predict monopoles. See Hooft–Polyakov monopole.

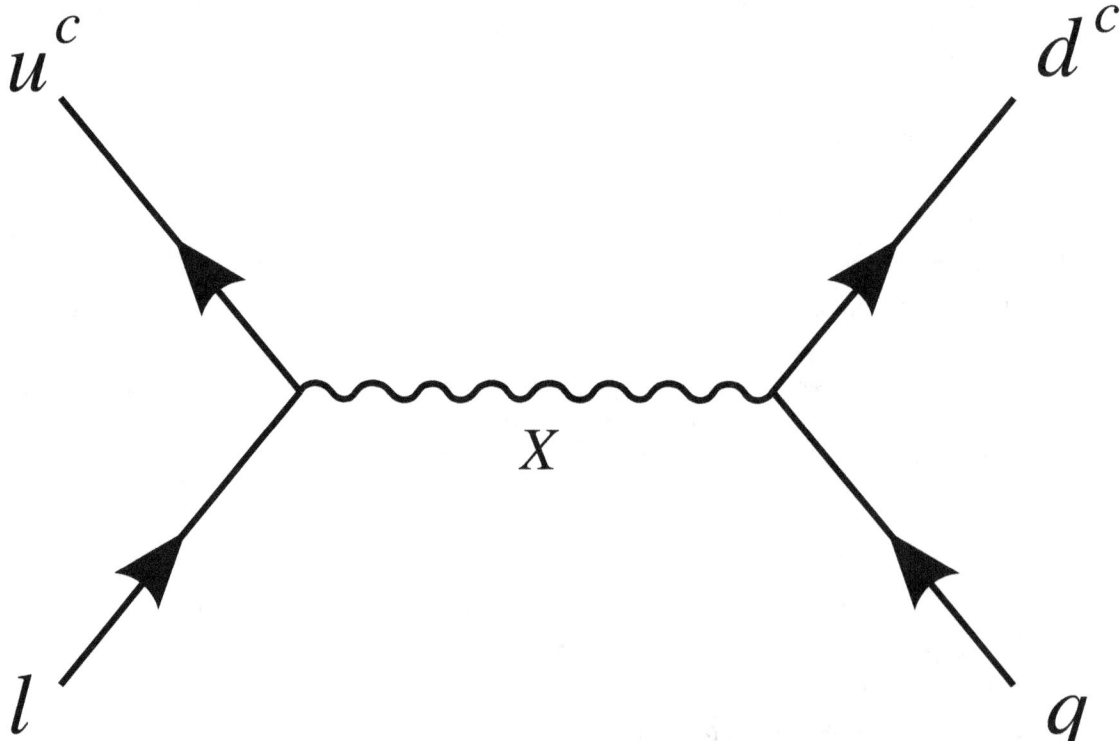

Dimension 6 proton decay mediated by the X boson $(3, 2)_{\frac{1}{6}}$ in flipped SU(5) GUT

25.3 spacetime

The $N = 1$ superspace extension of $3 + 1$ Minkowski spacetime

25.4 spatial symmetry

$N = 1$ SUSY over $3 + 1$ Minkowski spacetime with R-symmetry

25.5 gauge symmetry group

$(SU(5) \times U(1)_\chi)/\mathbf{Z}_5$

25.6 global internal symmetry

$\mathbf{Z}_2$ (matter parity) not related to U(1)R in any way for this particular model

25.7 vector superfields

Those associated with the $SU(5) \times U(1)\chi$ gauge symmetry

25.8 chiral superfields

As complex representations:

25.9 Superpotential

A generic invariant renormalizable superpotential is a (complex) $SU(5) \times U(1)\chi \times \mathbf{Z}_2$ invariant cubic polynomial in the superfields which has an R-charge of 2. It is a linear combination of the following terms:

$$
\begin{array}{ll}
S & S \\
S 10_H \overline{10}_H & S 10_H^{\alpha\beta} \overline{10}_{H\alpha\beta} \\
10_H 10_H H_d & \epsilon_{\alpha\beta\gamma\delta\epsilon} 10_H^{\alpha\beta} 10_H^{\gamma\delta} H_d^\epsilon \\
\overline{10}_H \overline{10}_H H_u & \epsilon^{\alpha\beta\gamma\delta\epsilon} \overline{10}_{H\alpha\beta} \overline{10}_{H\gamma\delta} H_{u\epsilon} \\
H_d 10 10 & \epsilon_{\alpha\beta\gamma\delta\epsilon} H_d^\alpha 10_i^{\beta\gamma} 10_j^{\delta\epsilon} \\
H_d \bar{5} 1 & H_d^\alpha \bar{5}_{i\alpha} 1_j \\
H_u 10 \bar{5} & H_{u\alpha} 10_i^{\alpha\beta} \bar{5}_{j\beta} \\
\overline{10}_H 10 \phi & \overline{10}_{H\alpha\beta} 10_i^{\alpha\beta} \phi_j
\end{array}
$$

The second column expands each term in index notation (neglecting the proper normalization coefficient). i and j are the generation indices. The coupling H_d **10***i* **10***j* has coefficients which are symmetric in i and j.

In those models without the optional φ sterile neutrinos, we add the nonrenormalizable couplings instead.

$$
\begin{array}{ll}
(\overline{10}_H 10)(\overline{10}_H 10) & \overline{10}_{H\alpha\beta} 10_i^{\alpha\beta} \overline{10}_{H\gamma\delta} 10_j^{\gamma\delta} \\
\overline{10}_H 10 \overline{10}_H 10 & \overline{10}_{H\alpha\beta} 10_i^{\beta\gamma} \overline{10}_{H\gamma\delta} 10_j^{\delta\alpha}
\end{array}
$$

These couplings do break the R-symmetry.

25.10 References

[1] Barr SM. Phys.Lett. B 112 (1982) 219

[2] Derendinger JP, Kim JE, and Nanopoulos DV. Phys.Lett. B 139 (1984) 170

[3] Stenger, Victor J., *Quantum Gods: Creation, Chaos and the Search for Cosmic Consciouness*, Prometheus Books (2009) ISBN 978-1-59102-713-3 p 61.

[4] Freedman DH. "The new theory of everything", *Discover*, 1991, pp 54–61.

[5] Rizos J and Tamvakis K. Hierarchical Neutrino Masses and Mixing in Flipped-SU(5). arXiv.org > hep-ph > arXiv:0912.3997v1 (Dec. 20, 2009)

[6] Barcow, Timothy et al, *Electroweak symmetry breaking and new physics at the TeV scale* World Scientific, 1996 ISBN 978-981-02-2631-2 p 194

Chapter 26

Pati–Salam model

In physics, the **Pati–Salam model** is a Grand Unification Theory (GUT) proposed in 1974 by nobel laureate Abdus Salam and Jogesh Pati. The unification is based on there being four quark color charges, dubbed red, green, blue and violet (or lilac), instead of the conventional three, with the new "violet" quark being identified with the leptons. The model also has Left–right symmetry and predicts the existence of a high energy right handed weak interaction with heavy W' and Z' bosons.

Originally the fourth color was labelled "lilac" to alliterate with "lepton". Pati–Salam is a mainstream theory and a viable alternative to the Georgi–Glashow SU(5) unification. It can be embedded within an SO(10) unification model (as can SU(5)).

26.1 Core theory

The Pati–Salam model states that the gauge group is either SU(4) × SU(2)L × SU(2)R or (SU(4) × SU(2)L × SU(2)R)/$\mathbf{Z}_2$ and the fermions form three families, each consisting of the representations (**4**, **2**, **1**) and (**4**, **1**, **2**). This needs some explanation. The center of SU(4) × SU(2)L × SU(2)R is $\mathbf{Z}_4 \times \mathbf{Z}_2L \times \mathbf{Z}_2R$. The $\mathbf{Z}_2$ in the quotient refers to the two element subgroup generated by the element of the center corresponding to the two element of $\mathbf{Z}_4$ and the 1 elements of $\mathbf{Z}_2L$ and $\mathbf{Z}_2R$. This includes the right-handed neutrino, which is now likely believed to exist. See neutrino oscillations. There is also a (**4**, **1**, **2**) and/or a (**4**, **1**, **2**) scalar field called the Higgs field which acquires a VEV. This results in a spontaneous symmetry breaking from SU(4) × SU(2)L × SU(2)R to SU(3) × SU(2) × U(1)Y/$\mathbf{Z}_3$ or from (SU(4) × SU(2)L × SU(2)R)/$\mathbf{Z}_2$ to (SU(3) × SU(2) × U(1)Y)/$\mathbf{Z}_6$ and also,

(**4**, **2**, **1**) → (**3**, **2**)$_{1/6}$ ⊕ (**1**, **2**)$_{-1/2}$ (q & l)

(**4**, **1**, **2**) → (**3**, **1**)$_{1/3}$ ⊕ (**3**, **1**)$_{-2/3}$ ⊕ (**1**, **1**)$_1$ ⊕ (**1**, **1**)$_0$ (d^c, u^c, e^c & ν^c)

(**6**, **1**, **1**) → (**3**, **1**)$_{-1/3}$ ⊕ (**3**, **1**)$_{1/3}$

(**1**, **3**, **1**) → (**1**, **3**)$_0$

(**1**, **1**, **3**) → (**1**, **1**)$_1$ ⊕ (**1**, **1**)$_0$ ⊕ (**1**, **1**)$_{-1}$

See restricted representation. Of course, calling the representations things like (**4**, **1**, **2**) and (**6**, **1**, **1**) is purely a physicist's convention, not a mathematician's convention, where representations are either labelled by Young tableaux or Dynkin diagrams with numbers on their vertices, but still, it is standard among GUT theorists.

The weak hypercharge, Y, is the sum of the two matrices:

$$\begin{pmatrix} \frac{1}{3} & 0 & 0 & 0 \\ 0 & \frac{1}{3} & 0 & 0 \\ 0 & 0 & \frac{1}{3} & 0 \\ 0 & 0 & 0 & -1 \end{pmatrix} \in \mathrm{SU}(4), \qquad \begin{pmatrix} 1 & 0 \\ 0 & -1 \end{pmatrix} \in \mathrm{SU}(2)_R$$

Actually, it is possible to extend the Pati–Salam group so that it has two connected components. The relevant group is now the semidirect product $([SU(4) \times SU(2)_L \times SU(2)_R]/\mathbf{Z}_2) \rtimes \mathbf{Z}_2$. The last $\mathbf{Z}_2$ also needs explaining. It corresponds to an automorphism of the (unextended) Pati–Salam group which is the composition of an involutive outer automorphism of SU(4) which isn't an inner automorphism with interchanging the left and right copies of SU(2). This explains the name left and right and is one of the main motivations for originally studying this model. This extra "left-right symmetry" restores the concept of parity which had been shown not to hold at low energy scales for the weak interaction. In this extended model, **(4, 2, 1)** ⊕ **(4, 1, 2)** is an irrep and so is **(4, 1, 2)** ⊕ **(4, 2, 1)**. This is the simplest extension of the minimal left-right model unifying QCD with B–L.

Since the homotopy group

$$\pi_2\left(\frac{SU(4) \times SU(2)}{[SU(3) \times U(1)]/\mathbf{Z}_3}\right) = \mathbf{Z},$$

this model predicts monopoles. See 't Hooft–Polyakov monopole.

This model was invented by Jogesh Pati and Abdus Salam.

This model doesn't predict gauge mediated proton decay (unless it is embedded within an even larger GUT group).

26.2 Differences from the SU(5) unification

As mentioned above, both the Pati–Salam and Georgi–Glashow SU(5) unification models can be embedded in a SO(10) unification. The difference between the two models then lies in the way that the SO(10) symmetry is broken, generating different particles that may or may not be important at low scales and accessible by current experiments. If we look at the individual models, the most important difference is in the origin of the weak hypercharge. In the SU(5) model by itself there is no left-right symmetry (although there could be one in a larger unification in which the model is embedded), and the weak hypercharge is treated separately from the color charge. In the Pati–Salam model, part of the weak hypercharge (often called U(1)B-L) starts being unified with the color charge in the SU(4)C group, while the other part of the weak hypercharge is in the SU(2)R. When those two groups break then the two parts together eventually unify into the usual weak hypercharge U(1)Y.

26.3 Minimal supersymmetric Pati–Salam

26.3.1 Spacetime

The $N = 1$ superspace extension of 3 + 1 Minkowski spacetime

26.3.2 Spatial symmetry

N=1 SUSY over 3 + 1 Minkowski spacetime with R-symmetry

26.3.3 Gauge symmetry group

(SU(4) × SU(2)L × SU(2)R)/$\mathbf{Z}_2$

26.3.4 Global internal symmetry

U(1)A

26.3.5 Vector superfields

Those associated with the SU(4) × SU(2)L × SU(2)R gauge symmetry

26.3.6 Chiral superfields

As complex representations:

26.3.7 Superpotential

A generic invariant renormalizable superpotential is a (complex) SU(4) × SU(2)L × SU(2)R and U(1)R invariant cubic polynomial in the superfields. It is a linear combination of the following terms:

$$S$$
$$S(4,1,2)_H(\bar{4},1,2)_H$$
$$S(1,2,2)_H(1,2,2)_H$$
$$(6,1,1)_H(4,1,2)_H(4,1,2)_H$$
$$(6,1,1)_H(\bar{4},1,2)_H(\bar{4},1,2)_H$$
$$(1,2,2)_H(4,2,1)_i(\bar{4},1,2)_j$$
$$(4,1,2)_H(\bar{4},1,2)_i\phi_j$$

i and j are the generation indices.

26.3.8 Left-right extension

We can extend this model to include left-right symmetry. For that, we need the additional chiral multiplets (**4, 2, 1**)H and (**4, 2, 1**)H.

26.4 Sources

- Graham G. Ross, *Grand Unified Theories*, Benjamin/Cummings, 1985, ISBN 0-8053-6968-6
- Anthony Zee, *Quantum Field Theory in a Nutshell*, Princeton U. Press, Princeton, 2003, ISBN 0-691-01019-6

26.5 References

- J. Pati and A. Salam, Phys. Rev. D10 (1974), 275. Lepton number as the fourth "color"
- J.C. Baez, J. Huerta (2009). "The Algebra of Grand Unified Theories". arXiv:0904.1556 [hep-th].

26.6 External links

- Pati-Salam model on Scholarpedia
- Proton decay, annihilation or fusion? by Wu, Dan-Di; Li, Tie-Zhong, *Zeitschrift für Physik C*, Volume 27, Issue 2, pp. 321–323 preview Fusion of all three quarks is the only decay mechanism mediated by the Higgs particle, not the gauge bosons, in the Pati–Salam model
- The Algebra of Grand Unified Theories John Huerta. Slide show: contains an overview of Pati–Salam
- the Pati-Salam model Motivation for the Pati–Salam model

Chapter 27

Flipped SO(10)

Flipped SO(10) is a grand unified theory which is to standard **SO(10)** as flipped **SU(5)** is to **SU(5)**.

In conventional **SO(10)** models, the fermions lie in three spinorial 16 representations, one for each generation, which decomposes under $[SU(5) \times U(1)_\chi]/Z_5$ as

$$16 \to 10_1 \oplus \bar{5}_{-3} \oplus 1_5$$

This can either be the Georgi–Glashow **SU(5)** or flipped **SU(5)**.

In flipped **SO(10)** models, however, the gauge group is not just **SO(10)** but **SO(10)F** × **U(1)B** or $[SO(10)F \times U(1)B]/Z_4$. The fermion fields are now three copies of

$$16_1 \oplus 10_{-2} \oplus 1_4$$

These contain the Standard Model fermions as well as additional vector fermions with GUT scale masses. If we suppose $[SU(5) \times U(1)_\chi]/Z_5$ is a subgroup of **SO(10)F**, then we have the intermediate scale symmetry breaking $[SO(10)F \times U(1)B]/Z_5 \to [SU(5) \times U(1)_\chi]/Z_5$ where

$$\chi = -\frac{A}{4} + \frac{5B}{4}$$

In that case,

$$16_1 \to 10_1 \oplus \bar{5}_2 \oplus 1_0$$
$$10_{-2} \to 5_{-2} \oplus \bar{5}_{-3}$$
$$1_4 \to 1_5$$

note that the Standard Model fermion fields (including the right handed neutrinos) come from all three $[SO(10)F \times U(1)B]/Z_5$ representations. In particular, they happen to be the 10_1 of 16_1, the $\bar{5}_{-3}$ of 10_{-2} and the 1_5 of 1_4 (my apologies for mixing up **SO(10)** × **U(1)** notation with **SU(5)** × **U(1)** notation, but it would be really cumbersome if we have to spell out which group any given notation happens to refer to. It is left up to the reader to determine the group from the context. This is a standard practice in the GUT model building literature anyway).

The other remaining fermions are vectorlike. To see this, note that with a 16_1H and a $\overline{16}_{-1H}$ Higgs field, we can have VEVs which breaks the GUT group down to $[SU(5) \times U(1)_\chi]/Z_4$. The Yukawa coupling $16_1H\, 16_1\, 10_{-2}$ will pair up the 5_{-2} and $\bar{5}_2$ fermions. And we can always introduce a sterile neutrino φ which is invariant under $[SO(10) \times U(1)B]/Z_4$ and add the Yukawa coupling

$$<\overline{16}_{-1H}> 16_1 \varphi$$

OR we can add the nonrenormalizable term

$$<\overline{16}_{-1H}><\overline{16}_{-1H}> 16_1 16_1$$

Either way, the 1_0 component of the fermion 16_1 gets taken care of so that it is no longer chiral.

It has been left unspecified so far whether $[\mathbf{SU}(5) \times \mathbf{U}(1)_\chi]/\mathbf{Z}_5$ is the Georgi–Glashow $\mathbf{SU}(5)$ or the flipped $\mathbf{SU}(5)$. This is because both alternatives lead to reasonable GUT models.

One reason for studying flipped $\mathbf{SO}(10)$ is because it can be derived from an $\mathbf{E}_6$ GUT model.

27.1 References

Chapter 28

Little Higgs

In particle physics, **little Higgs** models are based on the idea that the Higgs boson is a pseudo-Goldstone boson arising from some global symmetry breaking at a TeV energy scale. The main goal of little Higgs models is to use the spontaneous breaking of such approximate global symmetries to stabilize the mass of the Higgs boson(s) responsible for electroweak symmetry breaking.

Although the idea was first suggested in the 1970s,[1][2][3] a viable model was only constructed by Nima Arkani-Hamed, Andy Cohen, and Howard Georgi in the spring of 2001.[4] The idea was explored further by Nima Arkani-Hamed, Andy Cohen, Thomas Gregoire, and Jay Wacker in the spring of 2002.[5] Also in 2002, several other papers appeared that refined the ideas of little Higgs theories, notably the Littlest Higgs by Nima Arkani-Hamed, Andy Cohen, Emmanuel Katz, and Ann Nelson.[6]

Little Higgs theories were an outgrowth of dimensional deconstruction. In these theories, the gauge group has the form of a direct product of several copies of the same factor, for example **SU**(2) × **SU**(2). Each **SU**(2) factor may be visualised as the **SU**(2) group living at a particular point along an additional dimension of space. Consequently, many virtues of extra-dimensional theories may be reproduced even though the little Higgs theory is 3+1-dimensional. The little Higgs models are able to predict a naturally-light Higgs particle.

The main idea behind the little Higgs models is that the one-loop contribution to the tachyonic Higgs boson mass coming from the top quark cancels (the other one-loop contributions are small enough that they don't really matter; the top Yukawa coupling is huge (because related to its mass) and all the other Yukawa couplings and gauge couplings are small). The reason is that, simplifying, a loop is proportional to the coupling constant (following the example above) of one of the **SU**(2) groups. Because of the symmetries of the theory the contributions cancel until you have a two-loop contribution involving both groups. This protects the Higgs boson mass for about one order of magnitude, which is good enough to evade many of the precision electroweak constraints.

In 2005 Martin Schmaltz and David Tucker-Smith posted on arXiv.org a pedagogical review on Little Higgs models.[7]

28.1 References

[1] Steven Weinberg, "Approximate Symmetries and Pseudo-Goldstone Bosons", *Physical Review Letters* **29**:1698 (1972).

[2] Howard Georgi and A. Pais, "Calculability and naturalness in gauge theories", *Physical Review* **D10**:539 (1974).

[3] Howard Georgi and A. Pais, "Vacuum symmetry and the pseudo-Goldstone phenomenon", *Physical Review* **D12**:508 (1975).

[4] Nima Arkani-Hamed, Andrew G. Cohen, Howard Georgi, "Electroweak symmetry breaking from dimensional deconstruction", *Physics Letters* **B513**:232 (2001).

[5] Nima Arkani-Hamed, Andrew G. Cohen, Thomas Gregoire, Jay G. Wacker, "Phenomenology of electroweak symmetry breaking from theory space", *JHEP* **0208**:020 (2002).

[6] N. Arkani-Hamed, A. G. Cohen, E. Katz, A. E. Nelson, "The Littlest Higgs", *JHEP* **0207**:034 (2002).

[7] Martin Schmaltz and David Tucker-Smith, "Little Higgs Review", Ann.Rev.Nucl.Part.Sci.55:229-270,2005.

Chapter 29

Preon

For the protein diseases, see Prion. For the Freon trade name, see Chlorofluorocarbon.

In particle physics, **preons** are "point-like" particles, conceived to be subcomponents of quarks and leptons.[1] The word was coined by Jogesh Pati and Abdus Salam in 1974. Interest in preon models peaked in the 1980s but has slowed as the Standard Model of particle physics continues to describe the physics mostly successfully, and no direct experimental evidence for lepton and quark compositeness has been found.

Note that in the hadronic sector there are some intriguing open questions and some effects considered anomalies within the Standard Model. For example, four very important open questions are the proton spin puzzle, the EMC effect, the distributions of electric charges inside the nucleons as found by Hofstadter in 1956, and the ad hoc CKM matrix elements.

29.1 Background

Before the Standard Model (SM) was developed in the 1970s (the key elements of the Standard Model known as quarks were proposed by Murray Gell-Mann and George Zweig in 1964), physicists observed hundreds of different kinds of particles in particle accelerators. These were organized into relationships on their physical properties in a largely ad-hoc system of hierarchies, not entirely unlike the way taxonomy grouped animals based on their physical features. Not surprisingly, the huge number of particles was referred to as the "particle zoo".

The Standard Model, which is now the prevailing model of particle physics, dramatically simplified this picture by showing that most of the observed particles were mesons, which are combinations of two quarks, or baryons which are combinations of three quarks, plus a handful of other particles. The particles being seen in the ever-more-powerful accelerators were, according to the theory, typically nothing more than combinations of these quarks.

Within the Standard Model, there are several different types of particles. One of these, the quarks, has six different types, of which there are three varieties in each (dubbed "colors", red, green, and blue, giving rise to quantum chromodynamics). Additionally, there are six different types of what are known as leptons. Of these six leptons, there are three charged particles: the electron, muon, and tau. The neutrinos comprise the other three leptons, and for each neutrino there is a corresponding member from the other set of three leptons. In the Standard Model, there are also bosons, including the photons; W^+, W^-, and Z bosons; gluons and the Higgs boson; and an open space left for the graviton. Almost all of these particles come in "left-handed" and "right-handed" versions (see *chirality*). The quarks, leptons and W boson all have antiparticles with opposite electric charge.

The Standard Model also has a number of problems which have not been entirely solved. In particular, no successful theory of gravitation based on a particle theory has yet been proposed. Although the Model assumes the existence of a graviton, all attempts to produce a consistent theory based on them have failed. Additionally, mass remains a mystery in the Standard Model. Additionally Kalman [2] notes that according to the concept of atomism, the fundamental building blocks of nature are invisible and indivisible bits of matter that are ungenerated and indestructible. Quarks are not indestructible, some can decay into other quarks. Thus on fundamental grounds- quarks must be composed of fundamental quantities-preons. Although the mass of each successive particle follows certain patterns, predictions of the rest mass of most particles cannot be made precisely, except for the masses of almost all baryons which have been recently described very well by the model of de Souza.[3] The Higgs boson explains why particles show inertial mass (but does not explain rest mass).

The Standard Model also has problems predicting the large scale structure of the universe. For instance, the SM generally predicts equal amounts of matter and antimatter in the universe, something that is observably not the case. A number of attempts have been made to "fix" this through a variety of mechanisms, but to date none have won widespread support. Likewise, basic adaptations of the Model suggest the presence of proton decay, which has not yet been observed.

Preon theory is motivated by a desire to replicate the achievements of the periodic table, and the later Standard Model which tamed the "particle zoo", by finding more fundamental answers to the huge number of arbitrary constants present in the Standard Model. It is one of several models to have been put forward in an attempt to provide a more fundamental explanation of the results in experimental and theoretical particle physics. The preon model has attracted comparatively little interest to date among the particle physics community.

29.2 Motivations

Preon research is motivated by the desire to explain already known facts (retrodiction), which include

- To reduce the large number of particles, many that differ only in charge, to a smaller number of more fundamental particles. For example, the electron and positron are identical except for charge, and preon research is motivated by explaining that electrons and positrons are composed of similar preons with the relevant difference accounting for charge. The hope is to reproduce the reductionist strategy that has worked for the periodic table of elements.

- To explain the three generations of fermions.

- To calculate parameters that are currently unexplained by the Standard Model, such as particle masses, electric charges, and color charges, and reduce the number of experimental input parameters required by the Standard Model.

- To provide reasons for the very large differences in energy-masses observed in supposedly fundamental particles, from the electron neutrino to the top quark.

- To provide alternative explanations for the electro-weak symmetry breaking without invoking a Higgs field, which in turn possibly needs a supersymmetry to correct the theoretical problems involved with the Higgs field. Supersymmetry itself has theoretical problems.

- To account for neutrino oscillation and mass.

- The desire to make new nontrivial predictions, for example, to provide possible cold dark matter candidates.

- To explain why there exists only the observed variety of particle species and not something else and to reproduce only these observed particles (since the prediction of non-observed particles is one of the major theoretical problems, as, for example, with supersymmetry).

29.3 History

A number of physicists have attempted to develop a theory of "pre-quarks" (from which the name *preon* derives) in an effort to justify theoretically the many parts of the Standard Model that are known only through experimental data.

Other names which have been used for these proposed fundamental particles (or particles intermediate between the most fundamental particles and those observed in the Standard Model) include *prequarks*, *subquarks*, *maons*,[4] *alphons*, *quinks*, *rishons*, *tweedles*, *helons*, *haplons*, *Y-particles*,[5] and *primons*.[6] Preon is the leading name in the physics community.

Efforts to develop a substructure date at least as far back as 1974 with a paper by Pati and Salam in *Physical Review*.[7] Other attempts include a 1977 paper by Terazawa, Chikashige and Akama,[8] similar, but independent, 1979 papers by Ne'eman,[9] Harari,[10] and Shupe,[11] a 1981 paper by Fritzsch and Mandelbaum,[12] and a 1992 book by D'Souza and Kalman.[1] None of these has gained wide acceptance in the physics world. However, in a recent work[13] de Souza has shown that his model describes well all weak decays of hadrons according to selection rules dictated by a quantum number derived from his compositeness model. In his model leptons are elementary particles and each quark is composed of two *primons*, and thus, all quarks are described by four *primons*. Therefore, there is no need for the Standard Model Higgs boson and each quark mass is derived from the interaction between each pair of *primons* by means of three Higgs-like bosons. In his 1989 Nobel Prize acceptance lecture, Hans Dehmelt described a most fundamental elementary particle, with definable properties, which he called the *cosmon*, as the likely end result of a long but finite chain of increasingly more elementary particles.[14]

Each of the preon models postulates a set of fewer fundamental particles than those of the Standard Model, together with the rules governing how those fundamental particles operate. Based on these rules, the preon models try to explain the Standard Model, often predicting small discrepancies with this model and generating new particles and certain phenomena, which do not belong to the Standard Model. The Rishon model illustrates some of the typical efforts in the field.

Many of the preon models theorize that the apparent imbalance of matter and antimatter in the universe is in fact illusory, with large quantities of preon level antimatter confined within more complex structures.

Many preon models either do not account for the Higgs boson or rule it out, and propose that electro-weak symmetry is broken not by a scalar Higgs field but by composite preons.[15] For example, Fredriksson preon theory does not need the Higgs boson, and explains the electro-weak breaking as the rearrangement of preons, rather than a Higgs-mediated field. In fact, Fredriksson preon model and de Souza model predict that the Standard Model Higgs boson does not exist.

When the term "preon" was coined, it was primarily to explain the two families of spin-1/2 fermions: leptons and quarks. More-recent preon models also account for spin-1 bosons, and are still called "preons".

29.4 Rishon model

Main article: Rishon model

The *rishon model* (RM) is the earliest effort to develop a preon model to explain the phenomenon appearing in the Standard Model (SM) of particle physics. It was first developed by Haim Harari and Michael A. Shupe (independently of each other), and later expanded by Harari and his then-student Nathan Seiberg.

The model has two kinds of fundamental particles called **rishons** (which means "primary" in Hebrew). They are **T** ("Third" since it has an electric charge of ⅓ e, or Tohu which means "unformed" in Hebrew Genesis) and **V** ("Vanishes", since it is electrically neutral, or Vohu which means "void" in Hebrew Genesis). All leptons and all flavours of quarks are three-rishon ordered triplets. These groups of three rishons have spin-½.

29.5 Criticisms

29.5.1 The mass paradox

One preon model started as an internal paper at the Collider Detector at Fermilab (CDF) around 1994. The paper was written after an unexpected and inexplicable excess of jets with energies above 200 GeV were detected in the 1992–1993 running period. However, scattering experiments have shown that quarks and leptons are "pointlike" down to distance scales of less than 10^{-18} m (or 1/1000 of a proton diameter). The momentum uncertainty of a preon (of whatever mass) confined to a box of this size is about 200 GeV/c, 50,000 times larger than the rest mass of an up-quark and 400,000 times larger than the rest mass of an electron.

Heisenberg's uncertainty principle states that $\Delta x \Delta p \geq \hbar/2$ and thus anything confined to a box smaller than Δx would have a momentum uncertainty proportionally greater. Thus, the preon model proposed particles smaller than the elementary particles they make up, since the momentum uncertainty Δp should be greater than the particles themselves. And so the preon model represents a mass paradox: How could quarks or electrons be made of smaller particles that would have many orders of magnitude greater mass-energies arising from their enormous momenta? This paradox is resolved by postulating a large binding force between preons cancelling their mass-energies.

29.5.2 Constraints

Any candidate preon theory must address particle chirality and the 't Hooft Chiral anomaly constraints, and would ideally have simpler theoretical structure than the Standard Model itself.

29.6 Conflicts with observed physics

Preon models propose additional unobserved forces or dynamics to account for the observed properties of elementary particles, which may have implications in conflict with observation.

For example, now that the LHC's observation of a Higgs boson is confirmed, the observation contradicts the predictions of many preon models that did not include it.

Preon theories require that quarks and electrons should have a finite size. It is possible that the Large Hadron Collider will observe this when raised to higher energies.

29.7 Popular culture

- In the 1948 reprint/edit of his 1930 novel *Skylark Three*, E. E. Smith postulated a series of 'subelectrons of the first and second type' with the latter being fundamental particles that were associated with the gravitation force. While this may not have been an element of the original novel (the scientific basis of some of the other novels in the series was revised extensively due to the additional eighteen years of scientific development), even the edited publication may be the first, or one of the first, mentions of the possibility that electrons are not fundamental particles.

- In the novelized version of the 1982 motion picture *Star Trek II: The Wrath of Khan*, written by Vonda McIntyre, two of Dr. Carol Marcus' Genesis project team, Vance Madison and Delwyn March, have studied sub-elementary particles they've named "boojums" and "snarks", in a field they jokingly call "kindergarten physics" because it is lower than "elementary" (analogy to school levels).

- James P. Hogan's novel *Voyage from Yesteryear* discussed preons (called *tweedles*), the physics of which became central to the plot. Hogan's "tweedle" physics was patently derived from the Rishon model.

29.8 See also

- Technicolor (physics)
- Preon star
- Preon-degenerate matter

29.9 Notes

[1] D'Souza, I.A.; Kalman, C.S. (1992). *Preons: Models of Leptons, Quarks and Gauge Bosons as Composite Objects*. World Scientific. ISBN 978-981-02-1019-9.

[2] Kalman, C. S. (2005). *Nuclear Physics B (Proc. Suppl.)* **142**: 235–237.

[3] de Souza, M.E. (2010). "Calculation of almost all energy levels of baryons". *Papers in Physics* **3**: 030003–1. doi:10.4279/PIP.030003.

[4] Overbye, D. (5 December 2006). "China Pursues Major Role in Particle Physics". *The New York Times*. Retrieved 2011-09-12.

[5] Yershov, V.N. (2005). "Equilibrium Configurations of Tripolar Charges". *Few-Body Systems* **37** (1–2): 79–106. arXiv:physics/0609185. Bibcode:2005FBS....37...79Y. doi:10.1007/s00601-004-0070-2.

[6] de Souza, M.E. (2005). "The Ultimate Division of Matter". *Scientia Plena* **1** (4): 83.

[7] Pati, J.C.; Salam, A. (1974). "Lepton number as the fourth "color"". *Physical Review D* **10**: 275–289. Bibcode:1974PhRvD..10..275P. doi:10.1103/PhysRevD.10.275.

    with erratum published as *Physical Review D* **11** (3): 703. 1975. Bibcode:1975PhRvD..11..703P. doi:10.1103/PhysRevD.11.703.2.

[8] Terazawa, H.; Chikashige, Y.; Akama, K. (1977). "Unified model of the Nambu-Jona-Lasinio type for all elementary particles". *Physical Review D* **15** (2): 480–487. Bibcode:1977PhRvD..15..480T. doi:10.1103/PhysRevD.15.480.

[9] Ne'eman, Y. (1979). "Irreducible gauge theory of a consolidated Weinberg-Salam model". *Physics Letters B* **81** (2): 190–194. Bibcode:1979PhLB...81..190N. doi:10.1016/0370-2693(79)90521-5.

[10] Harari, H. (1979). "A schematic model of quarks and leptons". *Physics Letters B* **86**: 83–6. Bibcode:1979PhLB...86...83H. doi:10.1016/0370-2693(79)90626-9.

[11] Shupe, M.A. (1979). "A composite model of leptons and quarks". *Physics Letters B* **86**: 87–92. Bibcode:1979PhLB...86...87S. doi:10.1016/0370-2693(79)90627-0.

[12] Fritzsch, H.; Mandelbaum, G. (1981). "Weak interactions as manifestations of the substructure of leptons and quarks". *Physics Letters B* **102** (5): 319. Bibcode:1981PhLB..102..319F. doi:10.1016/0370-2693(81)90626-2.

[13] de Souza, M.E. (2008). "Weak decays of hadrons reveal compositeness of quarks". *Scientia Plena* **4** (6): 064801–1.

[14] Dehmelt, H.G. (1989). "Experiments with an Isolated Subatomic Particle at Rest". *Nobel Lecture*. The Nobel Foundation. See also references therein.

[15] Dugne, J.-J.; Fredriksson, S.; Hansson, J.; Predazzi, E. (1997). "Higgs pain? Take a preon!". arXiv:hep-ph/9709227 [hep-ph].

29.10 Further reading

- Ball, P. (2007). "Splitting the quark". *Nature*. doi:10.1038/news.2007.292.

- Have We Hit Bottom Yet?- an article about preons and minuteness

Chapter 30

M-theory

Not to be confused with membrane theory of shells.
For a more accessible and less technical introduction to this topic, see Introduction to M-theory.

M-theory is a theory in physics that unifies all consistent versions of superstring theory. The existence of such a theory was first conjectured by Edward Witten at the string theory conference at the University of Southern California in the summer of 1995. Witten's announcement initiated a flurry of research activity known as the second superstring revolution.

Prior to Witten's announcement, string theorists had identified five different versions of superstring theory. Although these theories appeared at first to be very different, work by a number of different physicists including Ashoke Sen, Chris Hull, Paul Townsend, and Michael Duff showed that the theories were related in intricate and nontrivial ways. In particular, physicists found that the apparently distinct theories were identified by mathematical transformations called S-duality and T-duality. Witten's conjecture was based in part on the existence of these dualities and in part on the relationship of the string theories to a gravitational theory called eleven-dimensional supergravity.

Although a complete formulation of M-theory is not known, the theory should describe two- and five-dimensional objects called branes and should be approximated by eleven-dimensional supergravity at low energies. Modern attempts to formulate M-theory are typically based on matrix theory or the AdS/CFT correspondence. According to Witten, the M in M-theory can stand for "magic", "mystery", or "matrix" according to taste, and the true meaning of the title should be decided when a more fundamental formulation of the theory is known.

Investigations of the mathematical structure of M-theory have spawned a number of important theoretical results in physics and mathematics. More speculatively, M-theory may provide a framework for developing a unified theory of all of the fundamental forces of nature. Attempts to connect M-theory to experiment typically focus on compactifying its extra dimensions to construct approximate models of our four-dimensional world.

30.1 Background

30.1.1 Quantum gravity and strings

Main articles: Quantum gravity and String theory
 One of the deepest problems in modern physics is the problem of quantum gravity. Our current understanding of gravity is based on Albert Einstein's general theory of relativity, which is formulated within the framework of classical physics. However, nongravitational forces are described within the framework of quantum mechanics, a radically different formalism for describing physical phenomena based on probability.[1] A quantum theory of gravity is needed in order to reconcile general relativity with the principles of quantum mechanics,[2] but difficulties arise when one attempts to apply the usual prescriptions of quantum theory to the force of gravity.[3]

String theory is a theoretical framework that attempts to reconcile gravity and quantum mechanics. In string theory, the point-like particles of particle physics are replaced by one-dimensional objects called strings. These strings look like small segments or loops of ordinary string. String theory describes how strings propagate through space and interact with each other. On distance scales larger than the string scale, a string will look just like an ordinary particle,

The fundamental objects of string theory are open and closed strings.

with its mass, charge, and other properties determined by the vibrational state of the string. One of the vibrational states of a string gives rise to the graviton, a quantum mechanical particle that mediates gravitational interactions.[4]

There are several versions of string theory known as type I, type IIA, type IIB, and the two flavors of heterotic string theory (SO(32) and $E_8 \times E_8$). The different theories allow different types of strings, and the particles that arise at low energies exhibit different symmetries. For example, the type I theory includes both open strings (which are segments with endpoints) and closed strings (which form closed loops), while the type II theories include only closed strings. Each of these five string theories arises as a special limiting case of M-theory. This theory, like its string theory predecessors, is an example of a quantum theory of gravity. It describes a force just like the familiar gravitational force subject to the rules of quantum mechanics.

30.1.2 Number of dimensions

Main article: Compactification (physics)

In everyday life, there are three familiar dimensions of space (up/down, left/right, and forward/backward), and there is one dimension of time (later/earlier). Thus, in the language of modern physics, one says that spacetime is four-dimensional.[5]

Despite the obvious relevance of four-dimensional spacetime for describing the physical world, there are several reasons why physicists often consider theories in other dimensions. In some cases, by modeling spacetime in a different number of dimensions, a theory becomes more mathematically tractable, and one can perform calculations and gain general insights more easily.[6] There are also situations where theories in two or three spacetime dimensions are useful for describing phenomena in condensed matter physics. Finally, there exist scenarios in which there could actually be more than four dimensions of spacetime which have nonetheless managed to escape detection.[7]

One notable feature of string theory and M-theory is that these theories require extra dimensions of spacetime for their mathematical consistency. In string theory, spacetime is ten-dimensional, while in M-theory it is eleven-dimensional. In order to describe real physical phenomena using these theories, one must therefore imagine scenarios in which these extra dimensions would not be observed in experiments.[8]

Compactification is one way of modifying the number of dimensions in a physical theory. In compactification, some of the extra dimensions are assumed to "close up" on themselves to form circles.[9] In the limit where these curled up dimensions become very small, one obtains a theory in which spacetime has effectively a lower number of dimensions. A standard analogy for this is to consider a multidimensional object such as a garden hose. If the hose is viewed from

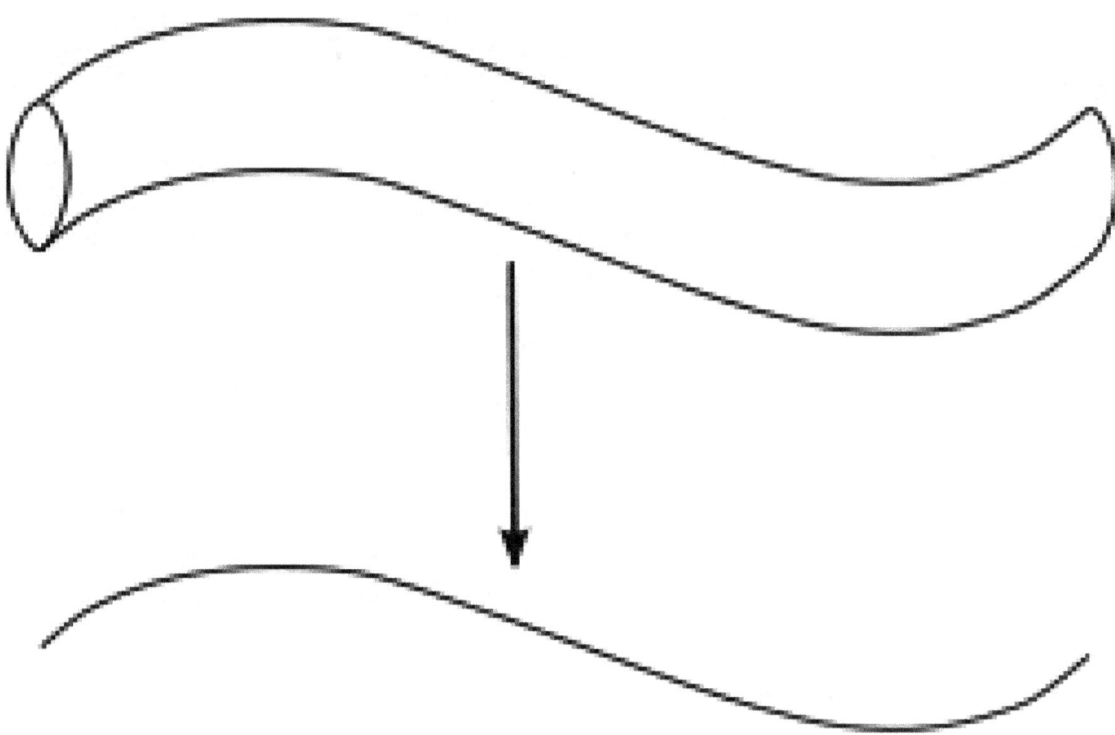

An example of compactification: At large distances, a two dimensional surface with one circular dimension looks one-dimensional.

a sufficient distance, it appears to have only one dimension, its length. However, as one approaches the hose, one discovers that it contains a second dimension, its circumference. Thus, an ant crawling on the surface of the hose would move in two dimensions.[10]

30.1.3 Dualities

Main articles: S-duality and T-duality

Theories that arise as different limits of M-theory turn out to be related in highly nontrivial ways. One of the relationships that can exist between these different physical theories is called S-duality. This is a relationship which says that a collection of strongly interacting particles in one theory can, in some cases, be viewed as a collection of weakly interacting particles in a completely different theory. For example, type I string theory turns out to be equivalent by S-duality to the SO(32) heterotic string theory. Similarly, type IIB string theory is related to itself in a nontrivial way by S-duality.[11]

Another relationship between different string theories is T-duality. Here one considers strings propagating around a circular extra dimension. T-duality states that a string propagating around a circle of radius R is equivalent to a string propagating around a circle of radius $1/R$ in the sense that all observable quantities in one description are identified with quantities in the dual description. For example, a string has momentum as it propagates around a circle, and it can also wind around the circle one or more times. The number of times the string winds around a circle is called the winding number. If a string has momentum p and winding number n in one description, it will have momentum n and winding number p in the dual description. For example, one can show that type IIA string theory is equivalent to type IIB string theory via T-duality and also that the two versions of heterotic string theory are related by T-duality.[12]

In general, the term *duality* refers to a situation where two seemingly different physical systems turn out to be equivalent in a nontrivial way. If two theories are related by a duality, it means that one theory can be transformed in some way so that it ends up looking just like the other theory. The two theories are then said to be *dual* to one another under the transformation. Put differently, the two theories are mathematically different descriptions of the same phenomena.

30.1. BACKGROUND

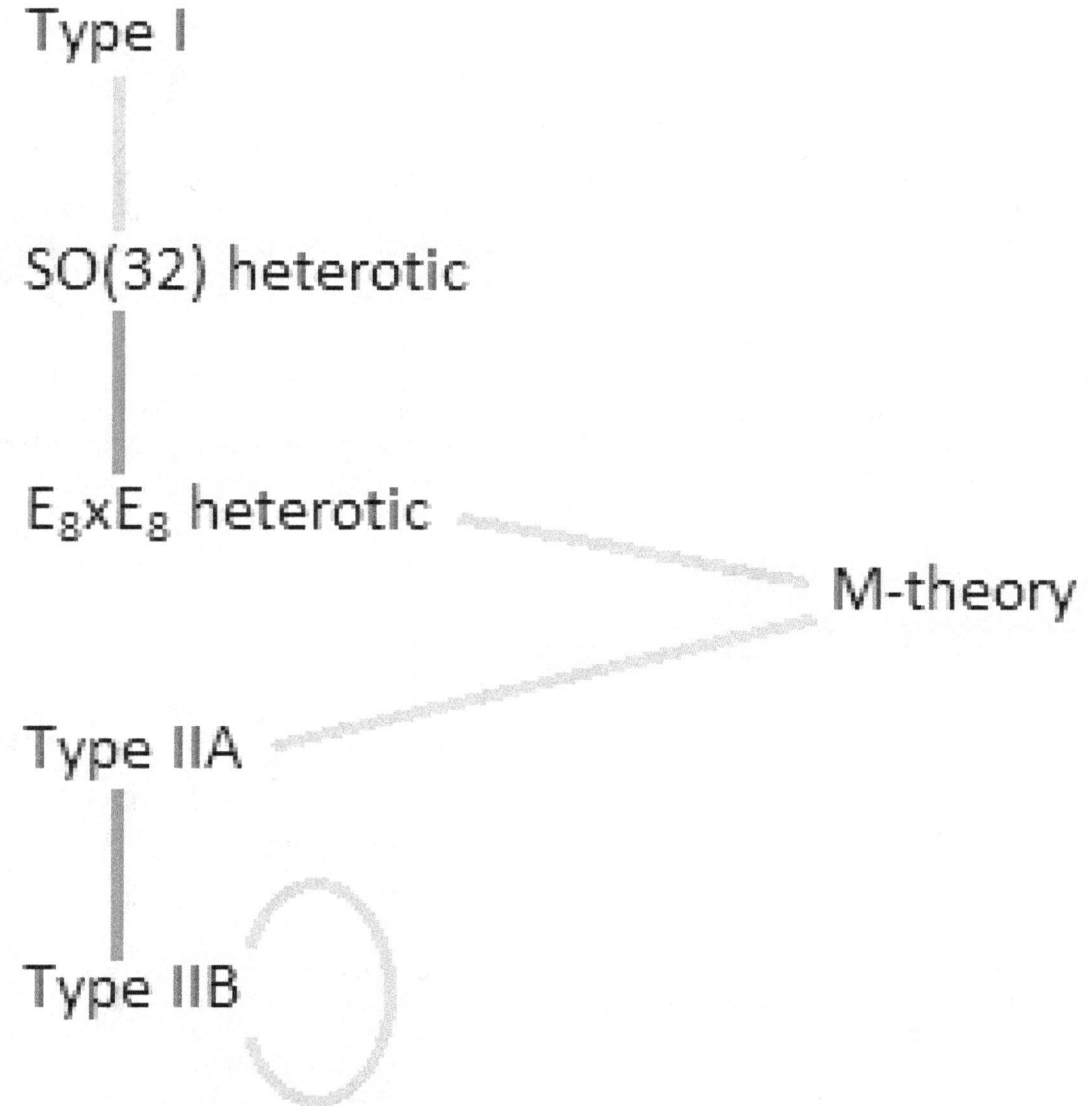

A diagram of string theory dualities. Yellow lines indicate S-duality. Blue lines indicate T-duality.

30.1.4 Supersymmetry

Main article: Supersymmetry

Another important theoretical idea that plays a role in M-theory is supersymmetry. This is a mathematical relation that exists in certain physical theories between a class of particles called bosons and a class of particles called fermions. Roughly speaking, bosons are the constituents of radiation, while fermions are the constituents of matter. In theories with supersymmetry, each boson has a counterpart which is a fermion, and vice versa. When supersymmetry is imposed as a local symmetry, one automatically obtains a quantum mechanical theory that includes gravity. Such a theory is called a supergravity theory.[13]

A theory of strings that incorporates the idea of supersymmetry is called a superstring theory. There are several different versions of superstring theory which are all subsumed within the M-theory framework. At low energies, the superstring theories are approximated by supergravity in ten spacetime dimensions. Similarly, M-theory is approximated at low energies by supergravity in eleven dimensions.

30.1.5 Branes

Main article: Brane

In string theory and related theories such as supergravity theories, a brane is a physical object that generalizes the notion of a point particle to higher dimensions. For example, a point particle can be viewed as a brane of dimension zero, while a string can be viewed as a brane of dimension one. It is also possible to consider higher-dimensional branes. In dimension p, these are called p-branes. Branes are dynamical objects which can propagate through spacetime according to the rules of quantum mechanics. They have mass and can have other attributes such as charge. A p-brane sweeps out a $(p+1)$-dimensional volume in spacetime called its *worldvolume*. Physicists often study fields analogous to the electromagnetic field which live on the worldvolume of a brane. The word brane comes from the word "membrane" which refers to a two-dimensional brane.[14]

In string theory, the fundamental objects that give rise to elementary particles are the one-dimensional strings. Although the physical phenomena described by M-theory are still poorly understood, physicists know that the theory describes two- and five-dimensional branes. Much of the current research in M-theory attempts to better understand the properties of these branes.

30.2 History and development

30.2.1 Early work on supergravity

Main article: Supergravity
General relativity does not place any limits on the possible dimensions of spacetime. Although the theory is typically formulated in four dimensions, one can write down the same equations for the gravitational field in any number of dimensions. Supergravity is more restrictive because it places an upper limit on the number of dimensions.[15] In 1978, work of Werner Nahm showed that the maximum spacetime dimension in which one can formulate a consistent supersymmetric theory is eleven.[16] In the same year, Eugene Cremmer, Bernard Julia, and Joel Scherk of the École Normale Supérieure showed that supergravity not only permits up to eleven dimensions but is in fact most elegant in this maximal number of dimensions.[17][18]

Initially, many physicists hoped that by compactifying eleven-dimensional supergravity, it might be possible to construct realistic models of our four-dimensional world. The hope was that such models would provide a unified description of the four fundamental forces of nature: electromagnetism, the strong and weak nuclear forces, and gravity. Interest in eleven-dimensional supergravity soon waned, however, as various flaws in this scheme were discovered. One of the problems was that the laws of physics appear to distinguish between left and right, a phenomenon known as chirality. As emphasized by Edward Witten and others, this chirality property cannot be readily derived by compactifying from eleven dimensions.[19]

In the first superstring revolution in 1984, many physicists turned to string theory as a unified theory of particle physics and quantum gravity. Unlike supergravity theory, string theory was able to accommodate the chirality of the standard model, and it provided a theory of gravity consistent with quantum effects.[20] Another feature of string theory that many physicists were drawn to in the 1980s and 1990s was its high degree of uniqueness. In ordinary particle theories, one can consider any collection of elementary particles whose classical behavior is described by an arbitrary Lagrangian. In string theory, the possibilities are much more constrained, and there are only a few consistent formulations of the theory. Indeed, by the 1990s, physicists had identified five consistent supersymmetric versions of the theory.

30.2.2 Relationships between string theories

Although there was only a handful of consistent superstring theories, it remained a mystery why there was not just one consistent formulation. However, as physicists began to examine string theory more closely, they began to realize that these theories are related in intricate and nontrivial ways.

In the late 1970s, Claus Montonen and David Olive,[21] had conjectured a special property of a quantum field theory called N = 4 supersymmetric Yang–Mills theory. This theory describes particles similar to the quarks and gluons that make up atomic nuclei. The strength with which the particles of this theory interact is measured by a number called the coupling constant. The result of Montonen and Olive, now known as Montonen–Olive duality, states that N=4

Edward Witten

supersymmetric Yang–Mills theory with coupling constant g is equivalent to the same theory with coupling constant $1/g$. In other words, a system of strongly interacting particles (large coupling constant) has an equivalent description as a system of weakly interacting particles (small coupling constant) and vice versa.[22]

In 1990, several theorists generalized Montonen–Olive duality to a relationship called S-duality which connects different string theories. For example, type IIB string theory with a large coupling constant is equivalent via S-duality to the same theory with small coupling constant. Theorists also found that different string theories may be related by a totally different kind of duality known as T-duality. This duality implies that strings propagating on completely different spacetime geometries may be physically equivalent.[23]

30.2.3 Membranes and fivebranes

Michael Duff

String theory extends ordinary quantum field theory by promoting zero-dimensional point particles to one-dimensional objects called strings. In the late 1980s, it was natural for theorists to attempt to formulate other extensions of quantum field theory in which particles are replaced by two-dimensional supermembranes or by higher-dimensional objects called branes. Such objects had been considered as early as 1962 by Paul Dirac, and they were reconsidered by a small but enthusiastic group of physicists in the 1980s.[24]

30.2. HISTORY AND DEVELOPMENT

Supersymmetry severely restricts the possible number of dimensions of a brane. In 1987, Eric Bergshoeff, Ergin Sezgin, and Paul Townsend showed that eleven-dimensional supergravity includes two-dimensional branes.[25] Intuitively, these objects look like sheets or membranes propagating through the eleven-dimensional spacetime. Shortly after this discovery, Michael Duff, Paul Howe, Takeo Inami, Kellogg Stelle considered a particular compactification of eleven-dimensional supergravity with one of the dimensions curled up into a circle.[26] In this setting, one can imagine the membrane wrapping around the circular dimension. If the radius of the circle is sufficiently small, then this membrane looks just like a string in ten-dimensional spacetime. In fact, Duff and his collaborators showed that this construction reproduces exactly the strings appearing in type IIA superstring theory.[27]

In 1990, Andrew Strominger published a similar result which suggested that strongly interacting strings in ten dimensions might have an equivalent description in terms of weakly interacting five-dimensional branes.[28] Initially, physicists were unable to prove this relationship for two important reasons. On the one hand, the Montonen–Olive duality was still unproven, and so Strominger's conjecture was even more tenuous. On the other hand, there were many technical issues related to the quantum properties of five-dimensional branes.[29] The first of these problems was solved in 1993 when Ashoke Sen established that certain physical theories require the existence of objects with both electric and magnetic charge which were predicted by the work of Montonen and Olive.[30]

In spite of this progress, the relationship between strings and five-dimensional branes remained conjectural because theorists were unable to quantize the branes. Starting in 1991, a team of researchers including Michael Duff, Ramzi Khuri, Jianxin Lu, and Ruben Minasian considered a special compactification of string theory in which four of the ten dimensions curl up. If one considers a five-dimensional brane wrapped around these extra dimensions, then the brane looks just like a one-dimensional string. In this way, the conjectured relationship between strings and branes was reduced to a relationship between strings and strings, and the latter could be tested using already established theoretical techniques.[31]

30.2.4 Second superstring revolution

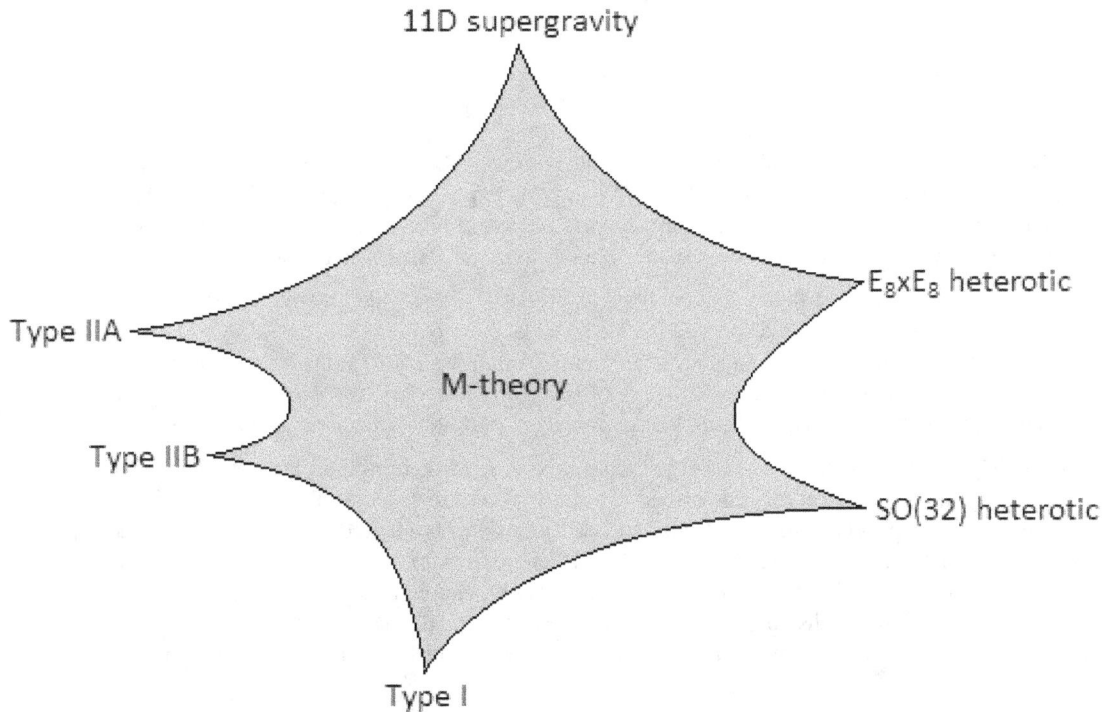

A schematic illustration of the relationship between M-theory, the five superstring theories, and eleven-dimensional supergravity. These last six theories arise as special limiting cases of M-theory.

Main article: Second superstring revolution

Speaking at the string theory conference at the University of Southern California in 1995, Edward Witten of the Institute for Advanced Study made the surprising suggestion that all five superstring theories were in fact just different limiting cases of a single theory in eleven spacetime dimensions. Witten's announcement drew together all of the previous results on S- and T-duality and the appearance of two- and five-dimensional branes in string theory.[32] In the months following Witten's announcement, hundreds of new papers appeared on the Internet confirming that the new theory involved membranes in an important way.[33] Today this flurry of work is known as the second superstring revolution.

One of the important developments following Witten's announcement was Witten's work in 1996 with string theorist Petr Hořava.[34][35] Witten and Hořava studied M-theory on a special spacetime geometry with two ten-dimensional boundary components. Their work shed light on the mathematical structure of M-theory and suggested possible ways of connecting M-theory to real world physics.[36]

30.2.5 Origin of the term

Initially, some physicists suggested that the new theory was a fundamental theory of membranes, but Witten was skeptical of the role of membranes in the theory. In a paper from 1996, Hořava and Witten wrote

> As it has been proposed that the eleven-dimensional theory is a supermembrane theory but there are some reasons to doubt that interpretation, we will non-committally call it the M-theory, leaving to the future the relation of M to membranes.[37]

In the absence of an understanding of the true meaning and structure of M-theory, Witten has suggested that the M should stand for "magic", "mystery", or "matrix" according to taste, and the true meaning of the title should be decided when a more fundamental formulation of the theory is known.[38]

30.3 AdS/CFT correspondence

30.3.1 Overview

Main article: AdS/CFT correspondence
The application of quantum mechanics to physical objects such as the electromagnetic field, which are extended in space and time, is known as quantum field theory.[39] In particle physics, quantum field theories form the basis for our understanding of elementary particles, which are modeled as excitations in the fundamental fields. Quantum field theories are also used throughout condensed matter physics to model particle-like objects called quasiparticles.[40]

One approach to formulating M-theory and studying its properties is provided by the anti-de Sitter/conformal field theory (AdS/CFT) correspondence. Proposed by Juan Maldacena in late 1997, the AdS/CFT correspondence is a theoretical result which implies that M-theory is in some cases equivalent to a quantum field theory.[41] In addition to providing insights into the mathematical structure of string and M-theory, the AdS/CFT correspondence has shed light on many aspects of quantum field theory in regimes where traditional calculational techniques are ineffective.[42]

In the AdS/CFT correspondence, one considers string theory or M-theory on an anti-de Sitter background. This means that the geometry of spacetime is described in terms of a certain vacuum solution of Einstein's equation called anti-de Sitter space.[43] In very elementary terms, anti-de Sitter space is a mathematical model of spacetime in which the notion of distance between points (the metric) is different from the notion of distance in ordinary Euclidean geometry. It is closely related to hyperbolic space, which can be viewed as a disk as illustrated on the left.[44] This image shows a tessellation of a disk by triangles and squares. One can define the distance between points of this disk in such a way that all the triangles and squares are the same size and the circular outer boundary is infinitely far from any point in the interior.[45]

Now imagine a stack of hyperbolic disks where each disk represents the state of the universe at a given time. The resulting geometric object is three-dimensional anti-de Sitter space.[46] It looks like a solid cylinder in which any cross section is a copy of the hyperbolic disk. Time runs along the vertical direction in this picture. The surface of this cylinder plays an important role in the AdS/CFT correspondence. As with the hyperbolic plane, anti-de Sitter space is curved in such a way that any point in the interior is actually infinitely far from this boundary surface.[47]

This construction describes a hypothetical universe with only two space and one time dimension, but it can be generalized to any number of dimensions. Indeed, hyperbolic space can have more than two dimensions and one can "stack up" copies of hyperbolic space to get higher-dimensional models of anti-de Sitter space.[48]

30.3. ADS/CFT CORRESPONDENCE

A tessellation of the hyperbolic plane by triangles and squares.

An important feature of anti-de Sitter space is its boundary (which looks like a cylinder in the case of three-dimensional anti-de Sitter space). One property of this boundary is that, locally around any point, it looks just like Minkowski space, the model of spacetime used in nongravitational physics.[49] One can therefore consider an auxiliary theory in which "spacetime" is given by the boundary of anti-de Sitter space. This observation is the starting point for AdS/CFT correspondence, which states that the boundary of anti-de Sitter space can be regarded as the "spacetime" for a quantum field theory. The claim is that this quantum field theory is equivalent to the gravitational theory on the bulk anti-de Sitter space in the sense that there is a "dictionary" for translating calculations in one theory into calculations in the other. Every entity in one theory has a counterpart in the other theory. For example, a single particle in the gravitational theory might correspond to some collection of particles in the boundary theory. In addition, the predictions in the two theories are quantitatively identical so that if two particles have a 40 percent chance of colliding in the gravitational theory, then the corresponding collections in the boundary theory would also have a 40 percent chance of colliding.[50]

30.3.2 6D (2,0) superconformal field theory

Main article: 6D (2,0) superconformal field theory

One particular realization of the AdS/CFT correspondence states that M-theory on the product space $AdS_7 \times S^4$ is equivalent to the so-called (2,0)-theory on the six-dimensional boundary.[51] In this example, the spacetime of

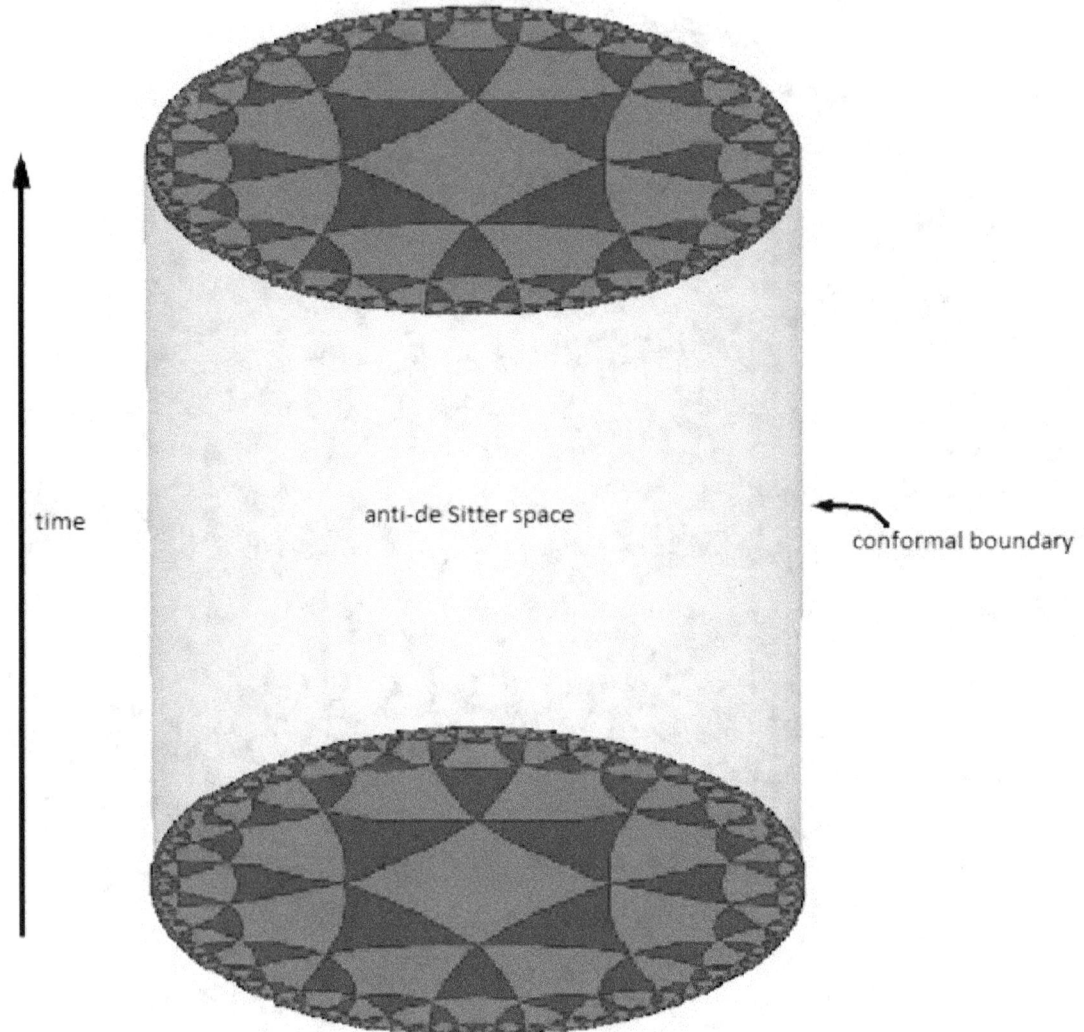

Three-dimensional anti-de Sitter space is like a stack of hyperbolic disks, each one representing the state of the universe at a given time. The resulting spacetime looks like a solid cylinder.

the gravitational theory is effectively seven-dimensional (hence the notation AdS_7), and there are four additional "compact" dimensions (encoded by the S^4 factor). In the real world, spacetime is four-dimensional, at least macroscopically, so this version of the correspondence does not provide a realistic model of gravity. Likewise, the dual theory is not a viable model of any real-world system since it describes a world with six spacetime dimensions.

Nevertheless, the (2,0)-theory has proven to be important for studying the general properties of quantum field theories.[52] Indeed, this theory subsumes a large number of mathematically interesting effective quantum field theories and points to new dualities relating these theories. For example, Luis Alday, Davide Gaiotto, and Yuji Tachikawa showed that by compactifying this theory on a surface, one obtains a four-dimensional quantum field theory, and there is a duality known as the AGT correspondence which relates the physics of this theory to certain physical concepts associated with the surface itself.[53] More recently, theorists have extended these ideas to study the theories obtained by compactifying down to three dimensions.[54][55]

In addition to its applications in quantum field theory, the (2,0)-theory has spawned a number of important results in pure mathematics. For example, the existence of the (2,0)-theory was used by Witten to give a "physical" explanation for a conjectural relationship in mathematics called the geometric Langlands correspondence. Witten's results also gave an intuitive geometric explanation of a well known fact from classical electrodynamics, namely the invariance of Maxwell's equations under the interchange of electric and magnetic fields.[56] In subsequent work, Witten showed that the (2,0)-theory could be used to understand a concept in mathematics called Khovanov homology.[57] Developed by Mikhail Khovanov around 2000, Khovanov homology provides a tool in knot theory, the branch of mathematics that studies and classifies the different shapes of knots.[58]

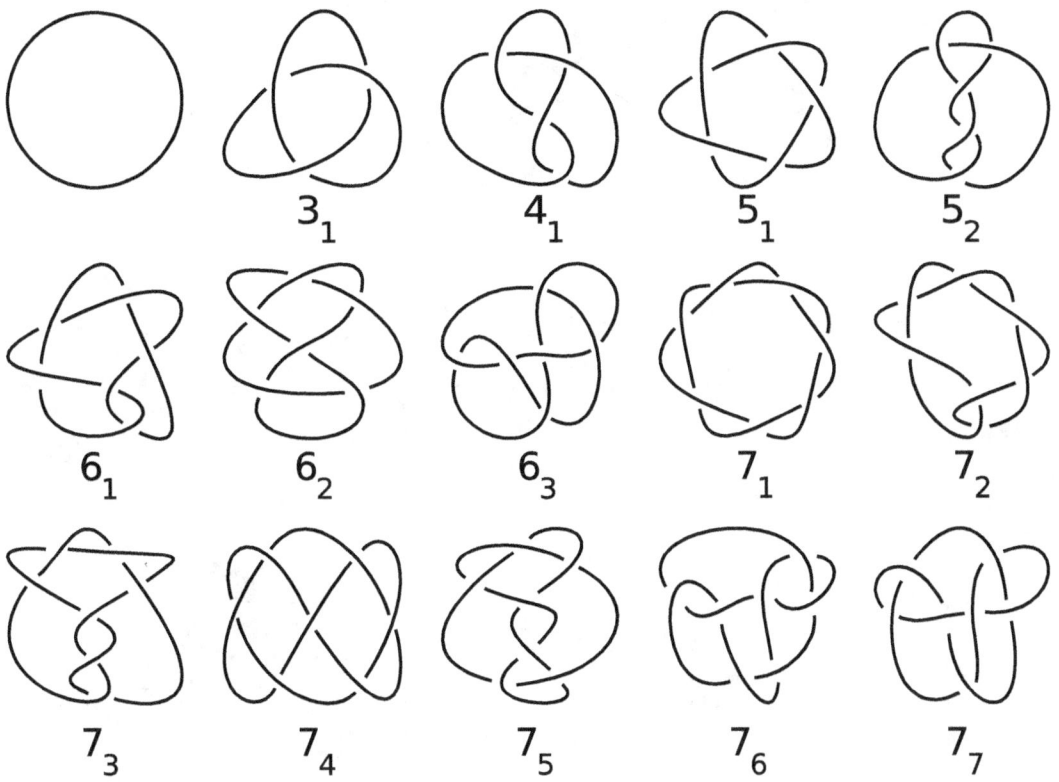

The six-dimensional (2,0)-theory has been used to understand results from the mathematical theory of knots.

30.3.3 ABJM superconformal field theory

Main article: ABJM superconformal field theory

Another realization of the AdS/CFT correspondence states that M-theory on $AdS_4 \times S^7$ is equivalent to a quantum field theory called the ABJM theory in three dimensions. In this version of the correspondence, seven of the dimensions of M-theory are curled up, leaving four non-compact dimensions. Since the spacetime of our universe is four-dimensional, this version of the correspondence provides a somewhat more realistic description of gravity.[59]

The ABJM theory appearing in this version of the correspondence is also interesting for a variety of reasons. Introduced by Aharony, Bergman, Jafferis, and Maldacena, it is closely related to another quantum field theory called Chern-Simons theory. The latter theory was popularized by Witten in the late 1980s because of its applications to knot theory.[60] In addition, the ABJM theory serves as a useful toy model for solving problems that arise in condensed matter physics.[61]

30.4 Phenomenology

30.4.1 Overview

Main article: String phenomenology
 In addition to being an idea of considerable theoretical interest, M-theory provides a framework for constructing models of real world physics that combine general relativity with the standard model of particle physics. Phenomenology is the branch of theoretical physics in which physicists construct realistic models of nature from more abstract theoretical ideas. String phenomenology is the part of string theory that attempts to construct realistic models of particle physics based on string and M-theory.

Typically, such models are based on the idea of compactification.[62] Starting with the ten- or eleven-dimensional

A cross section of a Calabi–Yau manifold

spacetime of string or M-theory, physicists postulate a shape for the extra dimensions. By choosing this shape appropriately, they can construct models roughly similar to the standard model of particle physics, together with additional undiscovered particles.[63] One popular way of deriving realistic physics from string theory is to start with the heterotic theory in ten dimensions and assume that the six extra dimensions of spacetime are shaped like a six-dimensional Calabi–Yau manifold. This is a special kind of geometric object named after mathematicians Eugenio Calabi and Shing-Tung Yau.[64] Calabi–Yau manifolds offer many ways of extracting realistic physics from string theory. Other similar methods can be used to construct realistic models of our four-dimensional world based on M-theory.[65]

Partly because of theoretical and mathematical difficulties and partly because of the extremely high energies needed to test these theories experimentally, there is so far no experimental evidence that would unambiguously point to any of these models being a correct fundamental description of nature. This has led some in the community to criticize these approaches to unification and question the value of continued research on these problems.[66]

30.4.2 Compactification on G_2 manifolds

In one approach to M-theory phenomenology, theorists assume that the seven extra dimensions of M-theory are shaped like a G_2 manifold. This is a special kind of seven-dimensional shape constructed by mathematician Dominic

Joyce of the University of Oxford.[67] These G_2 manifolds are still poorly understood mathematically, and this fact has made it difficult for physicists to fully develop this approach to phenomenology.[68]

For example, physicists and mathematicians often assume that space has a mathematical property called smoothness, but this property cannot be assumed in the case of a G_2 manifold if one wishes to recover the physics of our four-dimensional world. Another problem is that G_2 manifolds are not complex manifolds, so theorists are unable to use tools from the branch of mathematics known as complex analysis. Finally, there are many open questions about the existence, uniqueness, and other mathematical properties of G_2 manifolds, and mathematicians lack a systematic way of searching for these manifolds.[69]

30.4.3 Heterotic M-theory

Because of the difficulties with G_2 manifolds, most attempts to construct realistic theories of physics based on M-theory have taken a more indirect approach to compactifying eleven-dimensional spacetime. One approach, pioneered by Witten, Hořava, Burt Ovrut, and others, is known as heterotic M-theory. In this approach, one imagines that one of the eleven dimensions of M-theory is shaped like a circle. If this circle is very small, then the spacetime becomes effectively ten-dimensional. One then assumes that six of the ten dimensions form a Calabi–Yau manifold. If this Calabi–Yau manifold is also taken to be small, one is left with a theory in four-dimensions.[70]

30.5 Notes

[1] For a standard introduction to quantum mechanics, see Griffiths 2004.

[2] The necessity of a quantum mechanical description of gravity follows from the fact that one cannot consistently couple a classical system to a quantum one. See Wald 1984, p. 382.

[3] From a technical point of view, the problem is that the theory one gets in this way is not renormalizable and therefore cannot be used to make meaningful physical predictions. For more information, see Zee 2010, p. 72.

[4] For an accessible introduction to string theory, see Greene 2000.

[5] Wald 1984, p. 4

[6] For example, in the context of the AdS/CFT correspondence, theorists often formulate and study theories of gravity in unphysical numbers of spacetime dimensions.

[7] Zwiebach 2009, p. 9

[8] Zwiebach 2009, p. 8

[9] Yau and Nadis 2010, Ch. 6

[10] This analogy is used for example in Greene 2000, p. 186

[11] Becker, Becker, and Schwarz 2007

[12] Becker, Becker, and Schwarz 2007

[13] Duff 1998, p. 64

[14] Moore 2005

[15] Duff 1998, p. 64

[16] Nahm 1978

[17] Cremmer, Julia, and Scherk 1978

[18] Duff 1998, p. 65

[19] Duff 1998, p. 65

[20] Duff 1998, p. 65

[21] Montonen and Olive 1977

[22] Duff 1998, p. 66

[23] Duff 1998, p. 67

[24] Duff 1998, p. 65

[25] Bergshoeff, Sezgin, and Townsend 1987

[26] Duff et al. 1987

[27] Duff 1998, p. 66

[28] Strominger 1990

[29] Duff 1998, pp 66–7

[30] Sen 1993

[31] Duff 1998, p. 67

[32] Witten 1995

[33] Duff 1998, pp. 67–8

[34] Hořava and Witten 1996a

[35] Hořava and Witten 1996b

[36] Duff 1998, p. 68

[37] Hořava and Witten 1996a

[38] Duff 1996, sec. 1

[39] A standard text is Peskin and Schroeder 1995.

[40] For an introduction to the applications of quantum field theory to condensed matter physics, see Zee 2010.

[41] Maldacena 1998

[42] Klebanov and Maldacena 2009

[43] Klebanov and Maldacena 2009, p. 28

[44] Maldacena 2005, p. 60

[45] Maldacena 2005, p. 61

[46] Maldacena 2005, p. 60

[47] Maldacena 2005, p. 61

[48] Maldacena 2005, p. 60

[49] Zwiebach 2009, p. 552

[50] Maldacena 2005, pp. 61–62

[51] Maldacena 1998

[52] For a review of the (2,0)-theory, see Moore 2012.

[53] Alday, Gaiotto, and Tachikawa 2010

[54] Dimofte, Gaiotto, Gukov 2010

[55] Terashima and Masahito 2011

[56] Witten 2009

[57] Witten 2012

[58] Khovanov 2000

[59] Aharony et al. 2008

[60] Witten 1989

[61] Aharony et al. 2008

[62] Brane world scenarios provide an alternative way of recovering real world physics from string theory.

[63] Candelas et al. 1985

[64] Yau and Nadis 2010, p. ix

[65] Yau and Nadis 2010, pp. 147–150

[66] Woit 2006

[67] Yau and Nadis 2010, p. 149

[68] Yau and Nadis 2010, p. 150

[69] Yau and Nadis 2010, p. 150

[70] Yau and Nadis 2010, p. 150

30.6 References

- Aharony, Ofer; Bergman, Oren; Jafferis, Daniel Louis; Maldacena, Juan (2008). "N=6 superconformal Chern-Simons-matter theories, M2-branes and their gravity duals". *Journal of High Energy Physics* **2008** (10): 091. arXiv:0806.1218. Bibcode:2008JHEP...10..091A. doi:10.1088/1126-6708/2008/10/091.

- Alday, Luis; Gaiotto, Davide; Tachikawa, Yuji (2010). "Liouville correlation functions from four-dimensional gauge theories". *Letters in Mathematical Physics* **91** (2): 167–197. arXiv:0906.3219. Bibcode:2010LMaPh..91..167A. doi:10.1007/s11005-010-0369-5.

- Becker, Katrin; Becker, Melanie; Schwarz, John (2007). *String theory and M-theory: A modern introduction*. Cambridge University Press.

- Bergshoeff, Eric; Sezgin, Ergin; Townsend, Paul (1987). "Supermembranes and eleven-dimensional supergravity". *Physics Letters B* **189** (1): 75–78. Bibcode:1987PhLB..189...75B. doi:10.1016/0370-2693(87)91272-X.

- Candelas, Philip; Horowitz, Gary; Strominger, Andrew; Witten, Edward (1985). "Vacuum configurations for superstrings". *Nuclear Physics B* **258**: 46–74. Bibcode:1985NuPhB.258...46C. doi:10.1016/0550-3213(85)90602-9.

- Cremmer, Eugene; Julia, Bernard; Scherk, Joel (1978). "Supergravity theory in eleven dimensions". *Physics Letters B* **76** (4): 409–412. Bibcode:1978PhLB...76..409C. doi:10.1016/0370-2693(78)90894-8.

- Dimofte, Tudor; Gaiotto, Davide; Gukov, Sergei (2010). "Gauge theories labelled by three-manifolds". *Communications in Mathematical Physics* **3251** (2): 1367–419.

- Duff, Michael (1996). "M-theory (the theory formerly known as strings)". *International Journal of Modern Physics A* **11** (32): 6523–41.

- Duff, Michael (1998). "The theory formerly known as strings". *Scientific American* **278** (2): 64–9. doi:10.1038/scientificamerican 64.

- Duff, Michael; Howe, Paul; Inami, Takeo; Stelle, Kellogg (1987). "Superstrings in $D = 10$ from supermembranes in $D = 11$". *Nuclear Physics B* **191** (1): 70–74. doi:10.1016/0370-2693(87)91323-2.

- Greene, Brian (2000). *The Elegant Universe: Superstrings, Hidden Dimensions, and the Quest for the Ultimate Theory*. Random House. ISBN 978-0-9650888-0-0.

- Griffiths, David (2004). *Introduction to Quantum Mechanics*. Pearson Prentice Hall. ISBN 978-0-13-111892-8.

- Hořava, Petr; Witten, Edward (1996a). "Heterotic and Type I string dynamics from eleven dimensions". *Nuclear Physics B* **460** (3): 506–524. arXiv:hep-th/9510209. Bibcode:1996NuPhB.460..506H. doi:10.1016/0550-3213(95)00621-4.

- Hořava, Petr; Witten, Edward (1996b). "Eleven dimensional supergravity on a manifold with boundary". *Nuclear Physics B* **475** (1): 94–114. arXiv:hep-th/9603142. Bibcode:1996NuPhB.475...94H. doi:10.1016/0550-3213(96)00308-2.

- Khovanov, Mikhail (2000). "A categorification of the Jones polynomial". *Duke Mathematical Journal* **1011** (3): 359–426.

- Klebanov, Igor; Maldacena, Juan (2009). "Solving Quantum Field Theories via Curved Spacetimes" (PDF). *Physics Today* **62**: 28. Bibcode:2009PhT....62a..28K. doi:10.1063/1.3074260. Retrieved May 2013.

- Maldacena, Juan (1998). "The Large N limit of superconformal field theories and supergravity". *Advances in Theoretical and Mathematical Physics* **2**: 231–252. arXiv:hep-th/9711200. Bibcode:1998AdTMP...2..231M. doi:10.1063/1.59653.

- Maldacena, Juan (2005). "The Illusion of Gravity". *Scientific American* **293** (5): 56–63. Bibcode:2005SciAm.293e..56M. doi:10.1038/scientificamerican1105-56. PMID 16318027. Retrieved July 2013.

- Montonen, Claus; Olive, David (1977). "Magnetic monopoles as gauge particles?". *Physics Letters B* **72** (1): 117–120. Bibcode:1977PhLB...72..117M. doi:10.1016/0370-2693(77)90076-4.

- Moore, Gregory (2005). "What is ... a Brane?" (PDF). *Notices of the AMS* **52**: 214. Retrieved June 2013.

- Moore, Gregory (2012). "Applications of the six-dimensional (2,0) theories to Physical Mathematics". Retrieved 14 August 2013.

- Nahm, Walter (1978). "Supersymmetries and their representations". *Nuclear Physics B* **135** (1): 149–166. Bibcode:1978NuPhB.135..149N. doi:10.1016/0550-3213(78)90218-3.

- Peskin, Michael; Schroeder, Daniel (1995). *An Introduction to Quantum Field Theory*. Westview Press. ISBN 978-0-201-50397-5.

- Sen, Ashoke (1993). "Electric-magnetic duality in string theory". *Nuclear Physics B* **404** (1): 109–126. arXiv:hep-th/9207053. Bibcode:1993NuPhB.404..109S. doi:10.1016/0550-3213(93)90475-5.

- Strominger, Andrew (1990). "Heterotic solitons". *Nuclear Physics B* **343** (1): 167–184. Bibcode:1990NuPhB.343..167S. doi:10.1016/0550-3213(90)90599-9.

- Terashima, Yuji; Masahito, Yamazaki (2011). "$SL(2,\mathbb{R})$ Chern-Simons theory, Liouville, and gauge theory on duality walls". *Journal of High Energy Physics*. 20111 (8): 11–46.

- Wald, Robert (1984). *General Relativity*. University of Chicago Press. ISBN 978-0-226-87033-5.

- Witten, Edward (1989). "Quantum Field Theory and the Jones Polynomial". *Commun. Math. Phys.* **121** (3): 351–399. Bibcode:1989CMaPh.121..351W. doi:10.1007/BF01217730. MR 0990772.

- Witten, Edward (1995). "String theory dynamics in various dimensions". *Nuclear Physics B* **443** (1): 85–126. arXiv:hep-th/9503124. Bibcode:1995NuPhB.443...85W. doi:10.1016/0550-3213(95)00158-O.

- Witten, Edward (2009). "Geometric Langlands from six dimensions". arXiv:0905.2720 [hep-th].

- Witten, Edward (2012). "Fivebranes and knots". *Quantum Topology* **3** (1): 1–137. doi:10.4171/qt/26.

- Woit, Peter (2006). *Not Even Wrong: The Failure of String Theory and the Search for Unity in Physical Law*. New York: Basic Books. p. 105. ISBN 0-465-09275-6.

- Yau, Shing-Tung; Nadis, Steve (2010). *The Shape of Inner Space: String Theory and the Geometry of the Universe's Hidden Dimensions*. Basic Books. ISBN 978-0-465-02023-2.

- Zee, Anthony (2010). *Quantum Field Theory in a Nutshell* (2nd ed.). Princeton University Press. ISBN 978-0-691-14034-6.

- Zwiebach, Barton (2009). *A First Course in String Theory*. Cambridge University Press. ISBN 978-0-521-88032-9.

30.7 Further reading

- Greene, Brian (2000). *The Elegant Universe: Superstrings, Hidden Dimensions, and the Quest for the Ultimate Theory*. Random House. ISBN 978-0-9650888-0-0.

- Miemiec, André; Schnakenburg, Igor (2006). "Basics of M-theory". *Fortschritte der Physik* **54** (1): 5–72. arXiv:hep-th/0509137. Bibcode:2006ForPh..54....5M. doi:10.1002/prop.200510256.

- Witten, Edward (1998). "Magic, mystery, and matrix". *Notices of the AMS* **45** (9): 1124–1129.

30.8 External links

- The Elegant Universe—A three-hour miniseries with Brian Greene on the series *Nova* (original PBS broadcast dates: October 28, 8–10 p.m. and November 4, 8–9 p.m., 2003). Various images, texts, videos and animations explaining string theory and M-theory.

- Superstringtheory.com—The "Official String Theory Web Site", created by Patricia Schwarz. References on string theory and M-theory for the layperson and expert.

- Not Even Wrong—Peter Woit's blog on physics in general, and string theory in particular.

Chapter 31

Loop quantum gravity

Loop quantum gravity (**LQG**) is a theory that attempts to describe the quantum properties of gravity. It is also a theory of quantum space and quantum time, because, according to general relativity, the geometry of spacetime is a manifestation of gravity. LQG is an attempt to merge and adapt standard quantum mechanics and standard general relativity. The main output of the theory is a physical picture of space where space is granular. The granularity is a direct consequence of the quantization. It has the same nature as the granularity of the photons in the quantum theory of electromagnetism or the discrete levels of the energy of the atoms. But here, it is space itself that is discrete.

More precisely, space can be viewed as an extremely fine fabric or network "woven" of finite loops. These networks of loops are called spin networks. The evolution of a spin network over time is called a spin foam. The predicted size of this structure is the Planck length, which is approximately 10^{-35} meters. According to the theory, there is no meaning to distance at scales smaller than the Planck scale. Therefore, LQG predicts that not just matter, but also space itself has an atomic structure.

Today LQG is a vast area of research, developing in several directions, which involves about 30 research groups worldwide.[1] They all share the basic physical assumptions and the mathematical description of quantum space. The full development of the theory is being pursued in two directions: the more traditional canonical loop quantum gravity, and the newer covariant loop quantum gravity, more commonly called spin foam theory.

Research into the physical consequences of the theory is proceeding in several directions. Among these, the most well-developed is the application of LQG to cosmology, called loop quantum cosmology (LQC). LQC applies LQG ideas to the study of the early universe and the physics of the Big Bang. Its most spectacular consequence is that the evolution of the universe can be continued beyond the Big Bang. The Big Bang appears thus to be replaced by a sort of cosmic Big Bounce.

31.1 History

Main article: History of loop quantum gravity

In 1986, Abhay Ashtekar reformulated Einstein's general relativity in a language closer to that of the rest of fundamental physics. Shortly after, Ted Jacobson and Lee Smolin realized that the formal equation of quantum gravity, called the Wheeler–DeWitt equation, admitted solutions labelled by loops, when rewritten in the new Ashtekar variables, and Carlo Rovelli and Lee Smolin defined a nonperturbative and background-independent quantum theory of gravity in terms of these loop solutions. Jorge Pullin and Jerzy Lewandowski understood that the intersections of the loops are essential for the consistency of the theory, and the theory should be formulated in terms of intersecting loops, or graphs.

In 1994, Rovelli and Smolin showed that the quantum operators of the theory associated to area and volume have a discrete spectrum. That is, **geometry is quantized**. This result defines an explicit basis of states of quantum geometry, which turned out to be labelled by Roger Penrose's spin networks, which are graphs labelled by spins.

The canonical version of the dynamics was put on firm ground by Thomas Thiemann, who defined an anomaly-free Hamiltonian operator, showing the existence of a mathematically consistent background-independent theory. The covariant or spinfoam version of the dynamics developed during several decades, and crystallized in 2008, from the joint work of research groups in France, Canada, UK, Poland, and Germany, lead to the definition of a family of transition amplitudes, which in the classical limit can be shown to be related to a family of truncations of general relativity.[2] The finiteness of these amplitudes was proven in 2011.[3][4] It requires the existence of a positive cosmological constant, and this is consistent with observed acceleration in the expansion of the Universe.

31.2 General covariance and background independence

Main articles: General covariance, background-independent and diffeomorphism

In theoretical physics, general covariance is the invariance of the form of physical laws under arbitrary differentiable coordinate transformations. The essential idea is that coordinates are only artifices used in describing nature, and hence should play no role in the formulation of fundamental physical laws. A more significant requirement is the principle of General Relativity that states that the laws of physics take the same form in all reference systems. This is a generalization of the principle of special relativity which states that the laws of physics take the same form in all inertial frames.

In mathematics, a diffeomorphism is an isomorphism in the category of smooth manifolds. It is an invertible function that maps one differentiable manifold to another, such that both the function and its inverse are smooth. These are the defining symmetry transformations of General Relativity since the theory is formulated only in terms of a differentiable manifold.

In General Relativity, General covariance is intimately related to "diffeomorphism invariance". This symmetry is one of the defining features of the theory. However, it is a common misunderstanding that "diffeomorphism invariance" refers to the invariance of the physical predictions of a theory under arbitrary coordinate transformations; this is untrue and in fact every physical theory is invariant under coordinate transformations this way. Diffeomorphisms, as mathematicians define them, correspond to something much more radical; intuitively a way they can be envisaged is as simultaneously dragging all the physical fields (including the gravitational field) over the bare differentiable manifold while staying in the same coordinate system; diffeomorphisms are the true symmetry transformations of General Relativity, and come about from the assertion that the formulation of the theory is based on a bare differentiable manifold, but not on any prior geometry - the theory is background-independent (this is a profound shift, as all physical theories before general relativity had as part of their formulation a prior geometry). What is preserved under such transformations are the coincidences between the values the gravitational field take at such and such a "place" and the values the matter fields take there, from these relationships one can form a notion of matter being located with respect to the gravitational field, or vice versa. This is what Einstein discovered, physical entities are located with respect to one another only and not with respect to the spacetime manifold - as Carlo Rovelli puts it: "No more fields on spacetime: just fields on fields.".[5] This is the true meaning of the saying "The stage disappears and becomes one of the actors"; space-time as a "container" over which physics takes place has no objective physical meaning and instead the gravitational interaction is represented as just one of the fields forming the world. This is known as the relationalist interpretation of space-time. The realization by Einstein that General Relativity should be interpreted this way is the origin of his remark "Beyond my wildest expectations".

In LQG this aspect of General Relativity is taken seriously and this symmetry is preserved by requiring that the physical states remain invariant under the generators of diffeomorphisms. The interpretation of this condition is well understood for purely spatial diffeomorphisms. However, the understanding of diffeomorphisms involving time (the Hamiltonian constraint) is more subtle because it is related to dynamics and the so-called "problem of time" in general relativity.[6] A generally accepted calculational framework to account for this constraint has yet to be found.[7][8] A plausible candidate for the quantum hamiltonian constraint is the operator introduced by Thiemann.[9]

LQG is formally background independent. The equations of LQG are not embedded in, or presuppose, space and time (except for its invariant topology). Instead, they are expected to give rise to space and time at distances which are large compared to the Planck length. The issue of background independence in LQG still has some unresolved subtleties. For example, some derivations require a fixed choice of the topology, while any consistent quantum theory of gravity should include topology change as a dynamical process.

31.3 Constraints and their Poisson Bracket Algebra

Main articles: Poisson bracket and Hamiltonian constraint

31.3.1 The constraints of classical canonical general relativity

Main article: Lie derivative

In the Hamiltonian formulation of ordinary classical mechanics the Poisson bracket is an important concept. A "canonical coordinate system" consists of canonical position and momentum variables that satisfy canonical Poisson-bracket relations,

$\{q_i, p_j\} = \delta_{ij}$

where the Poisson bracket is given by

$$\{f, g\} = \sum_{i=1}^{N} \left(\frac{\partial f}{\partial q_i} \frac{\partial g}{\partial p_i} - \frac{\partial f}{\partial p_i} \frac{\partial g}{\partial q_i} \right).$$

for arbitrary phase space functions $f(q_i, p_j)$ and $g(q_i, p_j)$. With the use of Poisson brackets, the Hamilton's equations can be rewritten as,

$\dot{q}_i = \{q_i, H\}$,

$\dot{p}_i = \{p_i, H\}$.

These equations describe a "flow" or orbit in phase space generated by the Hamiltonian H. Given any phase space function $F(q, p)$, we have

$\frac{d}{dt} F(q_i, p_i) = \{F, H\}$.

Let us consider constrained systems, of which General relativity is an example. In a similar way the Poisson bracket between a constraint and the phase space variables generates a flow along an orbit in (the unconstrained) phase space generated by the constraint. There are three types of constraints in Ashtekar's reformulation of classical general relativity:

$SU(2)$ Gauss gauge constraints

The Gauss constraints

$G_j(x) = 0$.

This represents an infinite number of constraints one for each value of x. These come about from re-expressing General relativity as an SU(2) Yang–Mills type gauge theory (Yang–Mills is a generalization of Maxwell's theory where the gauge field transforms as a vector under Gauss transformations, that is, the Gauge field is of the form $A_a^i(x)$ where i is an internal index. See Ashtekar variables). These infinite number of Gauss gauge constraints can be smeared with test fields with internal indices, $\lambda^j(x)$,

$G(\lambda) = \int d^3 x G_j(x) \lambda^j(x)$.

which we demand vanish for any such function. These smeared constraints defined with respect to a suitable space of smearing functions give an equivalent description to the original constraints.

In fact Ashtekar's formulation may be thought of as ordinary SU(2) Yang–Mills theory together with the following special constraints, resulting from diffeomorphism invariance, and a Hamiltonian that vanishes. The dynamics of such a theory are thus very different from that of ordinary Yang–Mills theory.

Spatial diffeomorphisms constraints

The spatial diffeomorphism constraints

$C_a(x) = 0$

can be smeared by the so-called shift functions $\vec{N}(x)$ to give an equivalent set of smeared spatial diffeomorphism constraints,

$C(\vec{N}) = \int d^3x C_a(x) N^a(x)$.

These generate spatial diffeomorphisms along orbits defined by the shift function $N^a(x)$.

Hamiltonian constraints

The Hamiltonian

$H(x) = 0$

can be smeared by the so-called lapse functions $N(x)$ to give an equivalent set of smeared Hamiltonian constraints,

$H(N) = \int d^3x H(x) N(x)$.

These generate time diffeomorphisms along orbits defined by the lapse function $N(x)$.

In Ashtekar formulation the gauge field $A_a^i(x)$ is the configuration variable (the configuration variable being analogous to q in ordinary mechanics) and its conjugate momentum is the (densitized) triad (electrical field) $\tilde{E}_i^a(x)$. The constraints are certain functions of these phase space variables.

We consider the action of the constraints on arbitrary phase space functions. An important notion here is the Lie derivative, $\mathcal{L}_V$, which is basically a derivative operation that infinitesimally "shifts" functions along some orbit with tangent vector V .

31.3.2 The Poisson bracket algebra

Of particular importance is the Poisson bracket algebra formed between the (smeared) constraints themselves as it completely determines the theory. In terms of the above the smeared constraints the constraint algebra amongst the Gauss' law reads,

$\{G(\lambda), G(\mu)\} = G([\lambda, \mu])$

where $[\lambda, \mu]^k = \lambda_i \mu_j \epsilon^{ijk}$. And so we see that the Poisson bracket of two Gauss' law is equivalent to a single Gauss' law evaluated on the commutator of the smearings. The Poisson bracket amongst spatial diffeomorphisms constraints reads

$\{C(\vec{N}), C(\vec{M})\} = C(\mathcal{L}_{\vec{N}} \vec{M})$

and we see that its effect is to "shift the smearing". The reason for this is that the smearing functions are not functions of the canonical variables and so the spatial diffeomorphism does not generate diffeomorphims on them. They do however generate diffeomorphisms on everything else. This is equivalent to leaving everything else fixed while shifting the smearing .The action of the spatial diffeomorphism on the Gauss law is

$\{C(\vec{N}), G(\lambda)\} = G(\mathcal{L}_{\vec{N}} \lambda)$,

again, it shifts the test field λ . The Gauss law has vanishing Poisson bracket with the Hamiltonian constraint. The spatial diffeomorphism constraint with a Hamiltonian gives a Hamiltonian with its smearing shifted,

$\{C(\vec{N}), H(M)\} = H(\mathcal{L}_{\vec{N}} M)$.

Finally, the poisson bracket of two Hamiltonians is a spatial diffeomorphism,

$\{H(N), H(M)\} = C(K)$

where K is some phase space function. That is, it is a sum over infinitesimal spatial diffeomorphisms constraints where the coefficients of proportionality are not constants but have non-trivial phase space dependence.

A (Poisson bracket) Lie algebra, with constraints C_I , is of the form

$\{C_I, C_J\} = f_{IJ}^K C_K$

where f_{IJ}^K are constants (the so-called structure constants). The above Poisson bracket algebra for General relativity does not form a true Lie algebra as we have structure functions rather than structure constants for the Poisson bracket between two Hamiltonians. This leads to difficulties.

31.3.3 Dirac observables

The constraints define a constraint surface in the original phase space. The gauge motions of the constraints apply to all phase space but have the feature that they leave the constraint surface where it is, and thus the orbit of a point in the hypersurface under gauge transformations will be an orbit entirely within it. Dirac observables are defined as phase space functions, O, that Poisson commute with all the constraints when the constraint equations are imposed,

$$\{G_j, O\}_{G_j=C_a=H=0} = \{C_a, O\}_{G_j=C_a=H=0} = \{H, O\}_{G_j=C_a=H=0} = 0 ,$$

that is, they are quantities defined on the constraint surface that are invariant under the gauge transformations of the theory.

Then, solving only the constraint $G_j = 0$ and determining the Dirac observables with respect to it leads us back to the ADM phase space with constraints H, C_a. The dynamics of general relativity is generated by the constraints, it can be shown that six Einstein equations describing time evolution (really a gauge transformation) can be obtained by calculating the Poisson brackets of the three-metric and its conjugate momentum with a linear combination of the spatial diffeomorphism and Hamiltonian constraint. The vanishing of the constraints, giving the physical phase space, are the four other Einstein equations.[10]

31.4 Quantization of the constraints - the equations of Quantum General Relativity

31.4.1 Pre-history and Ashtekar new variables

Main articles: Frame fields in general relativity, Ashtekar variables and Self-dual Palatini action

Many of the technical problems in canonical quantum gravity revolve around the constraints. Canonical general relativity was originally formulated in terms of metric variables, but there seemed to be insurmountable mathematical difficulties in promoting the constraints to quantum operators because of their highly non-linear dependence on the canonical variables. The equations were much simplified with the introduction of Ashtekars new variables. Ashtekar variables describe canonical general relativity in terms of a new pair canonical variables closer to that of gauge theories. The first step consists of using densitized triads $\tilde{E}_i^a$ (a triad E_i^a is simply three orthogonal vector fields labeled by $i = 1, 2, 3$ and the densitized triad is defined by $\tilde{E}_i^a = \sqrt{\det(q)} E_i^a$) to encode information about the spatial metric,

$$\det(q) q^{ab} = \tilde{E}_i^a \tilde{E}_j^b \delta^{ij} .$$

(where δ^{ij} is the flat space metric, and the above equation expresses that q^{ab}, when written in terms of the basis E_i^a, is locally flat). (Formulating general relativity with triads instead of metrics was not new.) The densitized triads are not unique, and in fact one can perform a local in space rotation with respect to the internal indices i. The canonically conjugate variable is related to the extrinsic curvature by $K_a^i = K_{ab} \tilde{E}^{ai} / \sqrt{\det(q)}$. But problems similar to using the metric formulation arise when one tries to quantize the theory. Ashtekar's new insight was to introduce a new configuration variable,

$$A_a^i = \Gamma_a^i - i K_a^i$$

that behaves as a complex SU(2) connection where Γ_a^i is related to the so-called spin connection via $\Gamma_a^i = \Gamma_{ajk} \epsilon^{jki}$. Here A_a^i is called the chiral spin connection. It defines a covariant derivative $\mathcal{D}_a$. It turns out that $\tilde{E}_i^a$ is the conjugate momentum of A_a^i, and together these form Ashtekar's new variables.

The expressions for the constraints in Ashtekar variables; the Gauss's law, the spatial diffeomorphism constraint and the (densitized) Hamiltonian constraint then read:

$$G^i = \mathcal{D}_a \tilde{E}_i^a = 0$$
$$C_a = \tilde{E}_i^b F_{ab}^i - A_a^i (\mathcal{D}_b \tilde{E}_i^b) = V_a - A_a^i G^i = 0 ,$$
$$\tilde{H} = \epsilon_{ijk} \tilde{E}_i^a \tilde{E}_j^b F_{ab}^i = 0$$

respectively, where F_{ab}^i is the field strength tensor of the connection A_a^i and where V_a is referred to as the vector constraint. The above-mentioned local in space rotational invariance is the original of the SU(2) gauge invariance here expressed by the Gauss law. Note that these constraints are polynomial in the fundamental variables, unlike as with the constraints in the metric formulation. This dramatic simplification seemed to open up the way to quantizing the constraints. (See the article Self-dual Palatini action for a derivation of Ashtekar's formulism).

With Ashtekar's new variables, given the configuration variable A_a^i, it is natural to consider wavefunctions $\Psi(A_a^i)$. This is the connection representation. It is analogous to ordinary quantum mechanics with configuration variable q and wavefunctions $\psi(q)$. The configuration variable gets promoted to a quantum operator via:

$\hat{A}_a^i \Psi(A) = A_a^i \Psi(A)$,

(analogous to $\hat{q}\psi(q) = q\psi(q)$) and the triads are (functional) derivatives,

$\hat{\tilde{E}}_i^a \Psi(A) = -i \frac{\delta \Psi(A)}{\delta A_a^i}$.

(analogous to $\hat{p}\psi(q) = -i\hbar d\psi(q)/dq$). In passing over to the quantum theory the constraints become operators on a kinematic Hilbert space (the unconstrained SU(2) Yang–Mills Hilbert space). Note that different ordering of the A's and $\tilde{E}$'s when replacing the $\tilde{E}$'s with derivatives give rise to different operators - the choice made is called the factor ordering and should be chosen via physical reasoning. Formally they read

$\hat{G}_j |\psi\rangle = 0$

$\hat{C}_a |\psi\rangle = 0$

$\hat{\tilde{H}} |\psi\rangle = 0$.

There are still problems in properly defining all these equations and solving them. For example the Hamiltonian constraint Ashtekar worked with was the densitized version instead of the original Hamiltonian, that is, he worked with $\tilde{H} = \sqrt{\det(q)} H$. There were serious difficulties in promoting this quantity to a quantum operator. Moreover, although Ashtekar variables had the virtue of simplifying the Hamiltonian, they are complex. When one quantizes the theory, it is difficult to ensure that one recovers real general relativity as opposed to complex general relativity.

31.4.2 Quantum constraints as the equations of quantum general relativity

We now move on to demonstrate an important aspect of the quantum constraints. We consider Gauss' law only. First we state the classical result that the Poisson bracket of the smeared Gauss' law $G(\lambda) = \int d^3x \lambda^j (D_a E^a)^j$ with the connections is

$\{G(\lambda), A_a^i\} = \partial_a \lambda^i + g\epsilon^{ijk} A_a^j \lambda^k = (D_a \lambda)^i$.

The quantum Gauss' law reads

$\hat{G}_j \Psi(A) = -i D_a \frac{\delta \Psi[A]}{\delta A_a^j} = 0$.

If one smears the quantum Gauss' law and study its action on the quantum state one finds that the action of the constraint on the quantum state is equivalent to shifting the argument of Ψ by an infinitesimal (in the sense of the parameter λ small) gauge transformation,

$\left[1 + \int d^3x \lambda^j(x) \hat{G}_j\right] \Psi(A) = \Psi[A + D\lambda] = \Psi[A]$,

and the last identity comes from the fact that the constraint annihilates the state. So the constraint, as a quantum operator, is imposing the same symmetry that its vanishing imposed classically: it is telling us that the functions $\Psi[A]$ have to be gauge invariant functions of the connection. The same idea is true for the other constraints.

Therefore the two step process in the classical theory of solving the constraints $C_I = 0$ (equivalent to solving the admissibility conditions for the initial data) and looking for the gauge orbits (solving the `evolution' equations) is replaced by a one step process in the quantum theory, namely looking for solutions Ψ of the quantum equations $\hat{C}_I \Psi = 0$. This is because it obviously solves the constraint at the quantum level and it simultaneously looks for states that are gauge invariant because $\hat{C}_I$ is the quantum generator of gauge transformations (gauge invariant functions are constant along the gauge orbits and thus characterize them).[11] Recall that, at the classical level, solving the admissibility conditions and evolution equations was equivalent to solving all of Einstein's field equations, this underlines the central role of the quantum constraint equations in canonical quantum gravity.

31.4.3 Introduction of the loop representation

Main articles: Holonomy, Wilson loop and Knot invariant

It was in particular the inability to have good control over the space of solutions to the Gauss' law and spacial diffeomorphism constraints that led Rovelli and Smolin to consider a new representation - the The loop representation in gauge theories and quantum gravity.[12]

We need the notion of a holonomy. A holonomy is a measure of how much the initial and final values of a spinor or vector differ after parallel transport around a closed loop; it is denoted

$h_\gamma[A]$.

Knowledge of the holonomies is equivalent to knowledge of the connection, up to gauge equivalence. Holonomies can also be associated with an edge; under a Gauss Law these transform as

$(h'_e)_{\alpha\beta} = U^{-1}_{\alpha\gamma}(x)(h_e)_{\gamma\sigma}U_{\sigma\beta}(y)$.

For a closed loop $x = y$ if we take the trace of this, that is, putting $\alpha = \beta$ and summing we obtain

$(h'_e)_{\alpha\alpha} = U^{-1}_{\alpha\gamma}(x)(h_e)_{\gamma\sigma}U_{\sigma\alpha}(x) = [U_{\sigma\alpha}(x)U^{-1}_{\alpha\gamma}(x)](h_e)_{\gamma\sigma} = \delta_{\sigma\gamma}(h_e)_{\gamma\sigma} = (h_e)_{\gamma\gamma}$

or

$\text{Tr } h'_\gamma = \text{Tr } h_\gamma$. .

The trace of an holonomy around a closed loop and is written

$W_\gamma[A]$

and is called a Wilson loop. Thus Wilson loop are gauge invariant. The explicit form of the Holonomy is

$h_\gamma[A] = \mathcal{P} \exp\left\{ -\int_{\gamma_0}^{\gamma_1} ds \dot\gamma^a A^i_a(\gamma(s))T_i \right\}$

where γ is the curve along which the holonomy is evaluated, and s is a parameter along the curve, $\mathcal{P}$ denotes path ordering meaning factors for smaller values of s appear to the left, and T_i are matrices that satisfy the SU(2) algebra

$[T^i, T^j] = 2i\epsilon^{ijk}T^k$.

The Pauli matrices satisfy the above relation. It turns out that there are infinitely many more examples of sets of matrices that satisfy these relations, where each set comprises $(N+1) \times (N+1)$ matrices with $N = 1, 2, 3, \ldots$, and where non of these can be thought to `decompose' into two or more examples of lower dimension. They are called different irreducible representations of the SU(2) algebra. The most fundamental representation being the Pauli matrices. The holonomy is labelled by a half integer $N/2$ according to the irreducible representation used.

The use of Wilson loops explicitly solves the Gauss gauge constraint. To handle the spatial diffeomorphism constraint we need to go over to the loop representation. As Wilson loops form a basis we can formally expand any Gauss gauge invariant function as,

$\Psi[A] = \sum_\gamma \Psi[\gamma] W_\gamma[A]$.

This is called the loop transform. We can see the analogy with going to the momentum representation in quantum mechanics(see Position and momentum space). There one has a basis of states $\exp(ikx)$ labelled by a number k and one expands

$\psi[x] = \int dk \psi(k) \exp(ikx)$.

and works with the coefficients of the expansion $\psi(k)$.

The inverse loop transform is defined by

$\Psi[\gamma] = \int [dA] \Psi[A] W_\gamma[A]$.

This defines the loop representation. Given an operator $\hat{O}$ in the connection representation,

$\Phi[A] = \hat{O}\Psi[A] \qquad Eq\ 1$,

one should define the corresponding operator $\hat{O}'$ on $\Psi[\gamma]$ in the loop representation via,

$\Phi[\gamma] = \hat{O}'\Psi[\gamma] \qquad Eq\ 2$,

where $\Phi[\gamma]$ is defined by the usual inverse loop transform,

$\Phi[\gamma] = \int [dA] \Phi[A] W_\gamma[A] \qquad Eq\ 3$. .

A transformation formula giving the action of the operator $\hat{O}'$ on $\Psi[\gamma]$ in terms of the action of the operator $\hat{O}$ on $\Psi[A]$ is then obtained by equating the R.H.S. of $Eq\ 2$ with the R.H.S. of $Eq\ 3$ with $Eq\ 1$ substituted into $Eq\ 3$, namely

$$\hat{O}'\Psi[\gamma] = \int [dA] W_\gamma[A] \hat{O}\Psi[A],$$

or

$$\hat{O}'\Psi[\gamma] = \int [dA] (\hat{O}^\dagger W_\gamma[A])\Psi[A],$$

where by $\hat{O}^\dagger$ we mean the operator $\hat{O}$ but with the reverse factor ordering (remember from simple quantum mechanics where the product of operators is reversed under conjugation). We evaluate the action of this operator on the Wilson loop as a calculation in the connection representation and rearranging the result as a manipulation purely in terms of loops (one should remember that when considering the action on the Wilson loop one should choose the operator one wishes to transform with the opposite factor ordering to the one chosen for its action on wavefunctions $\Psi[A]$). This gives the physical meaning of the operator $\hat{O}'$. For example if $\hat{O}^\dagger$ corresponded to a spatial diffeomorphism, then this can be thought of as keeping the connection field A of $W_\gamma[A]$ where it is while performing a spatial diffeomorphism on γ instead. Therefore the meaning of $\hat{O}'$ is a spatial diffeomorphism on γ, the argument of $\Psi[\gamma]$.

In the loop representation we can then solve the spatial diffeomorphism constraint by considering functions of loops $\Psi[\gamma]$ that are invariant under spatial diffeomorphisms of the loop γ. That is, we construct what mathematicians call knot invariants. This opened up an unexpected connection between knot theory and quantum gravity.

What about the Hamiltonian constraint? Let us go back to the connection representation. Any collection of non-intersecting Wilson loops satisfy Ashtekar's quantum Hamiltonian constraint. This can be seen from the following. With a particular ordering of terms and replacing $\tilde{E}_i^a$ by a derivative, the action of the quantum Hamiltonian constraint on a Wilson loop is

$$\hat{\tilde{H}}^\dagger W_\gamma[A] = -\epsilon_{ijk} \hat{F}_{ab}^k \frac{\delta}{\delta A_a^i} \frac{\delta}{\delta A_b^j} W_\gamma[A].$$

When a derivative is taken it brings down the tangent vector, $\dot{\gamma}^a$, of the loop, γ. So we have something like

$$\hat{F}_{ab}^i \dot{\gamma}^a \dot{\gamma}^b.$$

However, as F_{ab}^i is anti-symmetric in the indices a and b this vanishes (this assumes that γ is not discontinuous anywhere and so the tangent vector is unique). Now let us go back to the loop representation.

We consider wavefunctions $\Psi[\gamma]$ that vanish if the loop has discontinuities and that are knot invariants. Such functions solve the Gauss law, the spatial diffeomorphism constraint and (formally) the Hamiltonian constraint. Thus we have identified an infinite set of exact (if only formal) solutions to all the equations of quantum general relativity![12] This generated a lot of interest in the approach and eventually led to LQG.

31.4.4 Geometric operators, the need for intersecting Wilson loops and spin network states

The easiest geometric quantity is the area. Let us choose coordinates so that the surface Σ is characterized by $x^3 = 0$. The area of small parallelogram of the surface Σ is the product of length of each side times $\sin\theta$ where θ is the angle between the sides. Say one edge is given by the vector $\vec{u}$ and the other by $\vec{v}$ then,

$$A = \|\vec{u}\| \|\vec{v}\| \sin\theta = \sqrt{\|\vec{u}\|^2 \|\vec{v}\|^2 (1 - \cos^2\theta)} = \sqrt{\|\vec{u}\|^2 \|\vec{v}\|^2 - (\vec{u}\cdot\vec{v})^2}$$

From this we get the area of the surface Σ to be given by

$$A_\Sigma = \int_\Sigma dx^1 dx^2 \sqrt{\det(q^{(2)})}$$

where $\det(q^{(2)}) = q_{11}q_{22} - q_{12}^2$ and is the determinant of the metric induced on Σ. This can be rewritten as

$$\det(q^{(2)}) = \frac{\epsilon^{3ab}\epsilon^{3cd} q_{ac} q_{bc}}{2}.$$

The standard formula for an inverse matrix is

$$q^{ab} = \frac{\epsilon^{acd}\epsilon^{bef} q_{ce} q_{df}}{3! \det(q)}$$

Note the similarity between this and the expression for $\det(q^{(2)})$. But in Ashtekar variables we have $\tilde{E}_i^a \tilde{E}^{bi} = \det(q) q^{ab}$. Therefore

$$A_\Sigma = \int_\Sigma dx^1 dx^2 \sqrt{\tilde{E}_i^3 \tilde{E}^{3i}}.$$

According to the rules of canonical quantization we should promote the triads $\tilde{E}_i^3$ to quantum operators,

$$\hat{\tilde{E}}_i^3 \sim \frac{\delta}{\delta A_3^i}.$$

It turns out that the area A_Σ can be promoted to a well defined quantum operator despite the fact that we are dealing with product of two functional derivatives and worse we have a square-root to contend with as well.[13] Putting $N = 2J$, we talk of being in the J-th representation. We note that $\sum_i T^i T^i = J(J+1)1$. This quantity is important in the final formula for the area spectrum. We simply state the result below,

$$\hat{A}_\Sigma W_\gamma[A] = 8\pi \ell_{\text{Planck}}^2 \beta \sum_I \sqrt{j_I(j_I+1)} W_\gamma[A]$$

where the sum is over all edges I of the Wilson loop that pierce the surface Σ.

The formula for the volume of a region R is given by

$$V = \int_R d^3x \sqrt{\det(q)} = \tfrac{1}{6} \int_R dx^3 \sqrt{\epsilon_{abc}\epsilon^{ijk} \tilde{E}^a_i \tilde{E}^b_j \tilde{E}^c_k}.$$

The quantization of the volume proceeds the same way as with the area. As we take the derivative, and each time we do so we bring down the tangent vector $\dot{\gamma}^a$, when the volume operator acts on non-intersecting Wilson loops the result vanishes. Quantum states with non-zero volume must therefore involve intersections. Given that the anti-symmetric summation is taken over in the formula for the volume we would need at least intersections with three non-coplanar lines. Actually it turns out that one needs at least four-valent vertices for the volume operator to be non-vanishing.

We now consider Wilson loops with intersections. We assume the real representation where the gauge group is SU(2). Wilson loops are an over complete basis as there are identities relating different Wilson loops. These come about from the fact that Wilson loops are based on matrices (the holonomy) and these matrices satisfy identities. Given any two SU(2) matrices $\mathbb{A}$ and $\mathbb{B}$ it is easy to check that,

$$\mathrm{Tr}(\mathbb{A})\,\mathrm{Tr}(\mathbb{B}) = \mathrm{Tr}(\mathbb{A}\mathbb{B}) + \mathrm{Tr}(\mathbb{A}\mathbb{B}^{-1}).$$

This implies that given two loops γ and η that intersect, we will have,

$$W_\gamma[A] W_\eta[A] = W_{\gamma \circ \eta}[A] + W_{\gamma \circ \eta^{-1}}[A]$$

where by η^{-1} we mean the loop η traversed in the opposite direction and $\gamma \circ \eta$ means the loop obtained by going around the loop γ and then along η. See figure below. Given that the matrices are unitary one has that $W_\gamma[A] = W_{\gamma^{-1}}[A]$. Also given the cyclic property of the matrix traces (i.e. $Tr(\mathbb{A}\mathbb{B}) = Tr(\mathbb{B}\mathbb{A})$) one has that $W_{\gamma \circ \eta}[A] = W_{\eta \circ \gamma}[A]$. These identities can be combined with each other into further identities of increasing complexity adding more loops. These identities are the so-called Mandelstam identities. Spin networks certain are linear combinations of intersecting Wilson loops designed to address the over completeness introduced by the Mandelstam identities (for trivalent intersections they eliminate the over-completeness entirely) and actually constitute a basis for all gauge invariant functions.

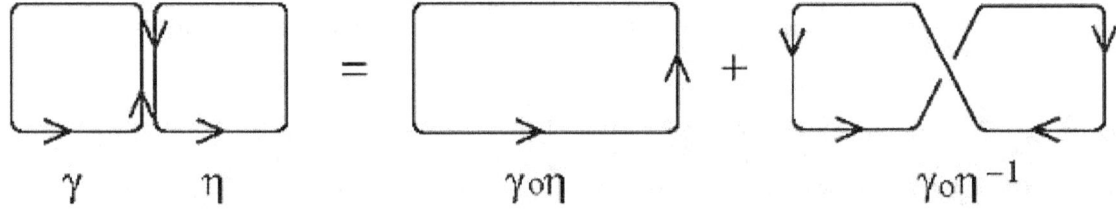

Graphical representation of the simplest non-trivial Mandelstam identity relating different Wilson loops.

As mentioned above the holonomy tells you how to propagate test spin half particles. A spin network state assigns an amplitude to a set of spin half particles tracing out a path in space, merging and splitting. These are described by spin networks γ: the edges are labelled by spins together with 'intertwiners' at the vertices which are prescription for how to sum over different ways the spins are rerouted. The sum over rerouting are chosen as such to make the form of the intertwiner invariant under Gauss gauge transformations.

31.4.5 Real variables, modern analysis and LQG

Main article: Hamiltonian constraint of LQG

Let us go into more detail about the technical difficulties associated with using Ashtekar's variables:

With Ashtekar's variables one uses a complex connection and so the relevant gauge group as actually SL(2, $\mathbb{C}$) and not SU(2). As SL(2, $\mathbb{C}$) is non-compact it creates serious problems for the rigorous construction of the necessary mathematical machinery. The group SU(2) is on the other hand is compact and the relevant constructions needed have been developed.

As mentioned above, because Ashtekar's variables are complex it results in complex general relativity. To recover the real theory one has to impose what are known as the reality conditions. These require that the densitized triad be real and that the real part of the Ashtekar connection equals the compatible spin connection (the compatibility condition being $\nabla_a e_b^I = 0$) determined by the desitized triad. The expression for compatible connection Γ_a^i is rather complicated and as such non-polynomial formula enters through the back door.

Before we state the next difficulty we should give a definition; a tensor density of weight W transforms like an ordinary tensor, except that in additional the W th power of the Jacobian,

$$J = \left| \frac{\partial x^a}{\partial x'^b} \right|$$

appears as a factor, i.e.

$$T'^{a...}_{b...} = J^W \frac{\partial x'^a}{\partial x^c} \cdots \frac{\partial x^d}{\partial x'^b} T^{c...}_{d...} .$$

It turns out that it is impossible, on general grounds, to construct a UV-finite, diffeomorphism non-violating operator corresponding to $\sqrt{\det(q)} H$. The reason is that the rescaled Hamiltonian constraint is a scalar density of weight two while it can be shown that only scalar densities of weight one have a chance to result in a well defined operator. Thus, one is forced to work with the original unrescaled, density one-valued, Hamiltonian constraint. However, this is non-polynomial and the whole virtue of the complex variables is questioned. In fact, all the solutions constructed for Ashtekar's Hamiltonian constraint only vanished for finite regularization (physics), however, this violates spatial diffeomorphism invariance.

Without the implementation and solution of the Hamiltonian constraint no progress can be made and no reliable predictions are possible!

To overcome the first problem one works with the configuration variable

$$A_a^i = \Gamma_a^i + \beta K_a^i$$

where β is real (as pointed out by Barbero, who introduced real variables some time after Ashtekar's variables[14][15]). The Guass law and the spatial diffeomorphism constraints are the same. In real Ashtekar variables the Hamiltonian is

$$H = \frac{\epsilon_{ijk} F_{ab}^k \tilde{E}_i^a \tilde{E}_j^b}{\sqrt{\det(q)}} + 2\frac{\beta^2+1}{\beta^2} \frac{(\tilde{E}_i^a \tilde{E}_j^b - \tilde{E}_j^a \tilde{E}_i^b)}{\sqrt{\det(q)}} (A_a^i - \Gamma_a^i)(A_b^j - \Gamma_b^j) = H_E + H' .$$

The complicated relationship between Γ_a^i and the desitized triads causes serious problems upon quantization. It is with the choice $\beta = \pm i$ that the second more complicated term is made to vanish. However, as mentioned above Γ_a^i reappears in the reality conditions. Also we still have the problem of the $1/\sqrt{\det(q)}$ factor.

Thiemann was able to make it work for real β. First he could simplify the troublesome $1/\sqrt{\det(q)}$ by using the identity

$$\{A_c^k, V\} = \frac{\epsilon_{abc} \epsilon^{ijk} \tilde{E}_i^a \tilde{E}_j^b}{\sqrt{\det(q)}}$$

where V is the volume. The A_c^k and V can be promoted to well defined operators in the loop representation and the Poisson bracket is replaced by a commutator upon quantization; this takes care of the first term. It turns out that a similar trick can be used to treat the second term. One introduces the quantity

$$K = \int d^3x K_a^i \tilde{E}_i^a$$

and notes that

$$K_a^i = \{A_a^i, K\} .$$

We are then able to write

$$A_a^i - \Gamma_a^i = \beta K_a^i = \beta \{A_a^i, K\} .$$

The reason the quantity K is easier to work with at the time of quantization is that it can be written as

$$K = -\{V, \int d^3x H_E\}$$

where we have used that the integrated densitized trace of the extrinsic curvature, K, is the "time derivative of the volume".

In the long history of canonical quantum gravity formulating the Hamiltonian constraint as a quantum operator

(Wheeler–DeWitt equation) in a mathematically rigorous manner has been a formidable problem. It was in the loop representation that a mathematically well defined Hamiltonian constraint was finally formulated in 1996.[9] We leave more details of its construction to the article Hamiltonian constraint of LQG. This together with the quantum versions of the Gauss law and spatial diffeomorphism constrains written in the loop representation are the central equations of LQG (modern canonical quantum General relativity).

Finding the states that are annihilated by these constraints (the physical states), and finding the corresponding physical inner product, and observables is the main goal of the technical side of LQG.

A very important aspect of the Hamiltonian operator is that it only acts at vertices (a consequence of this is that Thiemann's Hamiltonian operator, like Ashtekar's operator, annihilates non-intersecting loops except now it is not just formal and has rigorous mathematical meaning). More precisely, its action is non-zero on at least vertices of valence three and greater and results in a linear combination of new spin networks where the original graph has been modified by the addition of lines at each vertex together and a change in the labels of the adjacent links of the vertex.

31.4.6 Solving the quantum constraints

Main articles: spectrum, dual space and Rigged Hilbert space

We solve, at least approximately, all the quantum constraint equations and for the physical inner product to make physical predictions.

Before we move on to the constraints of LQG, lets us consider certain cases. We start with a kinematic Hilbert space $\mathcal{H}_{\text{Kin}}$ as so is equipped with an inner product—the kinematic inner product $\langle \phi, \psi \rangle_{\text{Kin}}$.

i) Say we have constraints $\hat{C}_I$ whose zero eigenvalues lie in their discrete spectrum. Solutions of the first constraint, $\hat{C}_1$, correspond to a subspace of the kinematic Hilbert space, $\mathcal{H}_1 \subset \mathcal{H}_{\text{Kin}}$. There will be a projection operator P_1 mapping $\mathcal{H}_{\text{Kin}}$ onto $\mathcal{H}_1$. The kinematic inner product structure is easily employed to provide the inner product structure after solving this first constraint; the new inner product $\langle \phi, \psi \rangle_1$ is simply

$$\langle \phi, \psi \rangle_1 = \langle P\phi, P\psi \rangle_{\text{Kin}}$$

They are based on the same inner product and are states normalizable with respect to it.

ii) The zero point is not contained in the point spectrum of all the $\hat{C}_I$, there is then no non-trivial solution $\Psi \in \mathcal{H}_{\text{Kin}}$ to the system of quantum constraint equations $\hat{C}_I \Psi = 0$ for all I.

For example the zero eigenvalue of the operator

$$\hat{C} = \left(i \frac{d}{dx} - k \right)$$

on $L_2(\mathbb{R}, dx)$ lies in the continuous spectrum $\mathbb{R}$ but the formal "eigenstate" $\exp(-ikx)$ is not normalizable in the kinematic inner product,

$$\int_{-\infty}^{\infty} dx \psi^*(x) \psi(x) = \int_{-\infty}^{\infty} dx e^{ikx} e^{-ikx} = \int_{-\infty}^{\infty} dx = \infty$$

and so does not belong to the kinematic Hilbert space $\mathcal{H}_{\text{Kin}}$. In these cases we take a dense subset $\mathcal{S}$ of $\mathcal{H}_{\text{Kin}}$ (intuitively this means either any point in $\mathcal{S}$ is either in $\mathcal{H}_{\text{Kin}}$ or arbitrarily close to a point in $\mathcal{H}_{\text{Kin}}$) with very good convergence properties and consider its dual space $\mathcal{S}'$ (intuitively these map elements of $\mathcal{S}$ onto finite complex numbers in a linear manner), then $\mathcal{S} \subset \mathcal{H}_{\text{Kin}} \subset \mathcal{S}'$ (as $\mathcal{S}'$ contains distributional functions). The constraint operator is then implemented on this larger dual space, which contains distributional functions, under the adjoint action on the operator. One looks for solutions on this larger space. This comes at the price that the solutions must be given a new Hilbert space inner product with respect to which they are normalizable (see article on rigged Hilbert space). In this case we have a generalized projection operator on the new space of states. We cannot use the above formula for the new inner product as it diverges, instead the new inner product is given by the simply modification of the above,

$$\langle \phi, \psi \rangle_1 = \langle P\phi, \psi \rangle_{\text{Kin}}.$$

The generalized projector P is known as a rigging map.

Let us move to LQG, additional complications will arise from the fact the constraint algebra is not a Lie algebra due to the bracket between two Hamiltonian constraints.

The Gauss law is solved by the use of spin network states. They provide a basis for the Kinematic Hilbert space $\mathcal{H}_{\text{Kin}}$. The spatial diffeomorphism constraint has been solved. The induced inner product on $\mathcal{H}_{\text{Diff}}$ (we do not pursue the details) has a very simple description in terms of spin network states; given two spin networks s and s', with associated spin network states ψ_s and $\psi_{s'}$, the inner product is 1 if s and s' are related to each other by a spatial diffeomorphism and zero otherwise.

The Hamiltonian constraint maps diffeomorphism invariant states onto non-diffeomorphism invaiant states as so does not preserve the diffeomorphism Hilbert space $\mathcal{H}_{\text{Diff}}$. This is an unavoidable consequence of the operator algebra, in particular the commutator:

$[\hat{C}(\vec{N}), \hat{H}(M)] \propto \hat{H}(\mathcal{L}_{\vec{N}} M)$

as can be seen by applying this to $\psi_s \in \mathcal{H}_{Diff}$,

$(\vec{C}(\vec{N})\hat{H}(M) - \hat{H}(M)\vec{C}(\vec{N}))\psi_s \propto \hat{H}(\mathcal{L}_{\vec{N}} M)\psi_s$

and using $\vec{C}(\vec{N})\psi_s = 0$ to obtain

$\vec{C}(\vec{N})[\hat{H}(M)\psi_s] \propto \hat{H}(\mathcal{L}_{\vec{N}} M)\psi_s \neq 0$

and so $\hat{H}(M)\psi_s$ is not in $\mathcal{H}_{Diff}$.

This means that you can't just solve the diffeomorphism constraint and then the Hamiltonian constraint. This problem can be circumvented by the introduction of the Master constraint, with its trivial operator algebra, one is then able in principle to construct the physical inner product from $\mathcal{H}_{\text{Diff}}$.

31.5 Spin foams

Main articles: spin network, spin foam, BF model and Barrett–Crane model

In loop quantum gravity (LQG), a spin network represents a "quantum state" of the gravitational field on a 3-dimensional hypersurface. The set of all possible spin networks (or, more accurately, "s-knots" - that is, equivalence classes of spin networks under diffeomorphisms) is countable; it constitutes a basis of LQG Hilbert space.

In physics, a spin foam is a topological structure made out of two-dimensional faces that represents one of the configurations that must be summed to obtain a Feynman's path integral (functional integration) description of quantum gravity. It is closely related to loop quantum gravity.

31.5.1 Spin foam derived from the Hamiltonian constraint operator

The Hamiltonian constraint generates `time' evolution. Solving the Hamiltonian constraint should tell us how quantum states evolve in `time' from an initial spin network state to a final spin network state. One approach to solving the Hamiltonian constraint starts with what is called the Dirac delta function. This is a rather singular function of the real line, denoted $\delta(x)$, that is zero everywhere except at $x = 0$ but whose integral is finite and nonzero. It can be represented as a Fourier integral,

$\delta(x) = \int e^{ikx} dk$.

One can employ the idea of the delta function to impose the condition that the Hamiltonian constraint should vanish. It is obvious that

$\prod_{x \in \Sigma} \delta(\hat{H}(x))$

is non-zero only when $\hat{H}(x) = 0$ for all x in Σ. Using this we can `project' out solutions to the Hamiltonian constraint. With analogy to the Fourier integral given above, this (generalized) projector can formally be written as

$\int [dN] e^{i \int d^3 x N(x) \hat{H}(x)}$.

Interestingly, this is formally spatially diffeomorphism-invariant. As such it can be applied at the spatially diffeomorphism-invariant level. Using this the physical inner product is formally given by

$\left\langle \int [dN] e^{i \int d^3 x N(x) \hat{H}(x)} s_{\text{int}} s_{\text{fin}} \right\rangle_{\text{Diff}}$

where s_{int} are the initial spin network and s_{fin} is the final spin network.

The exponential can be expanded

$$\left\langle \int [dN](1 + i\int d^3x N(x)\hat{H}(x) + \tfrac{i^2}{2!}[\int d^3x N(x)\hat{H}(x)][\int d^3x' N(x')\hat{H}(x')] + \ldots)s_{\text{int}}, s_{\text{fin}} \right\rangle_{\text{Diff}}$$

and each time a Hamiltonian operator acts it does so by adding a new edge at the vertex. The summation over different sequences of actions of $\hat{H}$ can be visualized as a summation over different histories of `interaction vertices' in the `time' evolution sending the initial spin network to the final spin network. This then naturally gives rise to the two-complex (a combinatorial set of faces that join along edges, which in turn join on vertices) underlying the spin foam description; we evolve forward an initial spin network sweeping out a surface, the action of the Hamiltonian constraint operator is to produce a new planar surface starting at the vertex. We are able to use the action of the Hamiltonian constraint on the vertex of a spin network state to associate an amplitude to each "interaction" (in analogy to Feynman diagrams). See figure below. This opens up a way of trying to directly link canonical LQG to a path integral description. Now just as a spin networks describe quantum space, each configuration contributing to these path integrals, or sums over history, describe `quantum space-time'. Because of their resemblance to soap foams and the way they are labeled John Baez gave these `quantum space-times' the name `spin foams'.

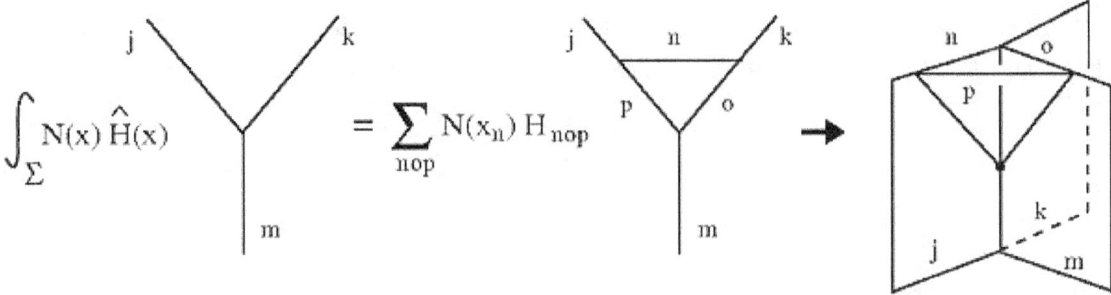

The action of the Hamiltonian constraint translated to the path integral or so-called spin foam description. A single node splits into three nodes, creating a spin foam vertex. $N(x_n)$ is the value of N at the vertex and H_{nop} are the matrix elements of the Hamiltonian constraint $\hat{H}$.

There are however severe difficulties with this particular approach, for example the Hamiltonian operator is not self-adjoint, in fact it is not even a normal operator (i.e. the operator does not commute with its adjoint) and so the spectral theorem cannot be used to define the exponential in general. The most serious problem is that the $\hat{H}(x)$'s are not mutually commuting, it can then be shown the formal quantity $\int [dN]e^{i\int d^3x N(x)\hat{H}(x)}$ cannot even define a (generalized) projector. The Master constraint (see below) does not suffer from these problems and as such offers a way of connecting the canonical theory to the path integral formulation.

31.5.2 Spin foams from BF theory

It turns out there are alternative routes to formulating the path integral, however their connection to the Hamiltonian formalism is less clear. One way is to start with the so-called BF theory. This is a simpler theory to general relativity. It has no local degrees of freedom and as such depends only on topological aspects of the fields. BF theory is what is known as a topological field theory. Surprisingly, it turns out that general relativity can be obtained from BF theory by imposing a constraint,[16] BF theory involves a field B_{ab}^{IJ} and if one chooses the field B to be the (anti-symmetric) product of two tetrads

$$B_{ab}^{IJ} = \tfrac{1}{2}(E_a^I E_b^J - E_b^I E_a^J)$$

(tetrads are like triads but in four spacetime dimensions), one recovers general relativity. The condition that the B field be given by the product of two tetrads is called the simplicity constraint. The spin foam dynamics of the topological field theory is well understood. Given the spin foam `interaction' amplitudes for this simple theory, one then tries to implement the simplicity conditions to obtain a path integral for general relativity. The non-trivial task of constructing a spin foam model is then reduced to the question of how this simplicity constraint should be imposed in the quantum theory. The first attempt at this was the famous Barrett–Crane model.[17] However this model was shown to be problematic, for example there did not seem to be enough degrees of freedom to ensure the correct classical limit.[18] It has been argued that the simplicity constraint was imposed too strongly at the quantum level and should only be imposed in the sense of expectation values just as with the Lorenz gauge condition $\partial_\mu \hat{A}^\mu$ in the Gupta–Bleuler formalism of quantum electrodynamics. New models have now been put forward, sometimes motivated by imposing the simplicity conditions in a weaker sense.

Another difficulty here is that spin foams are defined on a discretization of spacetime. While this presents no problems for a topological field theory as it has no local degrees of freedom, it presents problems for GR. This is known as the problem triangularization dependence.

31.5.3 Modern formulation of spin foams

Just as imposing the classical simplicity constraint recovers general relativity from BF theory, one expects an appropriate quantum simplicity constraint will recover quantum gravity from quantum BF theory.

Much progress has been made with regard to this issue by Engle, Pereira, and Rovelli[19] and Freidal and Krasnov[20] in defining spin foam interaction amplitudes with much better behaviour.

An attempt to make contact between EPRL-FK spin foam and the canonical formulation of LQG has been made.[21]

31.5.4 Spin foam derived from the Master constraint operator

See below.

31.5.5 Spin foams from consistent discretisations

31.6 The semi-classical limit

31.6.1 What is the semiclassical limit?

Main articles: Correspondence principle and classical limit

The **classical limit** or **correspondence limit** is the ability of a physical theory to approximate or "recover" classical mechanics when considered over special values of its parameters.[22] The classical limit is used with physical theories that predict non-classical behavior.

In physics, the **correspondence principle** states that the behavior of systems described by the theory of quantum mechanics (or by the old quantum theory) reproduces classical physics in the limit of large quantum numbers. In other words, it says that for large orbits and for large energies, quantum calculations must agree with classical calculations.[23]

The principle was formulated by Niels Bohr in 1920,[24] though he had previously made use of it as early as 1913 in developing his model of the atom.[25]

There are two basic requirements in establishing the semi-classical limit of any quantum theory:

i) reproduction of the Poisson brackets (of the diffeomorphism constraints in the case of general relativity). This is extremely important because, as noted above, the Poisson bracket algebra formed between the (smeared) constraints themselves completely determines the classical theory. This is analogous to establishing Ehrenfest's theorem;

ii) the specification of a complete set of classical observables whose corresponding operators (see complete set of commuting observables for the quantum mechanical definition of a complete set of observables) when acted on by appropriate semi-classical states reproduce the same classical variables with small quantum corrections (a subtle point is that states that are semi-classical for one class of observables may not be semi-classical for a different class of observables[26]).

This may be easily done, for example, in ordinary quantum mechanics for a particle but in general relativity this becomes a highly non-trivial problem as we will see below.

31.6.2 Why might LQG not have general relativity as its semiclassical limit?

Any candidate theory of quantum gravity must be able to reproduce Einstein's theory of general relativity as a classical limit of a quantum theory. This is not guaranteed because of a feature of quantum field theories which is that they have different sectors, these are analogous to the different phases that come about in the thermodynamical limit of statistical systems. Just as different phases are physically different, so are different sectors of a quantum field theory. It may turn out that LQG belongs to an unphysical sector - one in which you do not recover general relativity in the semi classical limit (in fact there might not be any physical sector at all).

Theorems establishing the uniqueness of the loop representation as defined by Ashtekar et al. (i.e. a certain concrete realization of a Hilbert space and associated operators reproducing the correct loop algebra - the realization that everybody was using) have been given by two groups (Lewandowski, Okolow, Sahlmann and Thiemann)[27] and (Christian Fleischhack).[28] Before this result was established it was not known whether there could be other examples of Hilbert spaces with operators invoking the same loop algebra, other realizations, not equivalent to the one that had been used so far. These uniqueness theorems imply no others exist and so if LQG does not have the correct semiclassical limit then this would mean the end of the loop representation of quantum gravity altogether.

31.6.3 Difficulties checking the semiclassical limit of LQG

There are difficulties in trying to establish LQG gives Einstein's theory of general relativity in the semi classical limit. There are a number of particular difficulties in establishing the semi-classical limit

1. There is no operator corresponding to infinitesimal spacial diffeomorphisms (it is not surprising that the theory has no generator of infinitesimal spatial 'translations' as it predicts spatial geometry has a discrete nature, compare to the situation in condensed matter). Instead it must be approximated by finite spatial diffeomorphisms and so the Poisson bracket structure of the classical theory is not exactly reproduced. This problem can be circumvented with the introduction of the so-called Master constraint (see below)[29]

2. There is the problem of reconciling the discrete combinatorial nature of the quantum states with the continuous nature of the fields of the classical theory.

3. There are serious difficulties arising from the structure of the Poisson brackets involving the spatial diffeomorphism and Hamiltonian constraints. In particular, the algebra of (smeared) Hamiltonian constraints does not close, it is proportional to a sum over infinitesimal spatial diffeomorphisms (which, as we have just noted, does not exist in the quantum theory) where the coefficients of proportionality are not constants but have non-trivial phase space dependence - as such it does not form a Lie algebra. However, the situation is much improved by the introduction of the Master constraint.[29]

4. The semi-classical machinery developed so far is only appropriate to non-graph-changing operators, however, Thiemann's Hamiltonian constraint is a graph-changing operator - the new graph it generates has degrees of freedom upon which the coherent state does not depend and so their quantum fluctuations are not suppressed. There is also the restriction, so far, that these coherent states are only defined at the Kimematic level, and now one has to lift them to the level of $\mathcal{H}_{Diff}$ and $\mathcal{H}_{Phys}$. It can be shown that Thiemann's Hamiltonian constraint is required to be graph changing in order to resolve problem 3 in some sense. The Master constraint algebra however is trivial and so the requirement that it be graph changing can be lifted and indeed non-graph changing Master constraint operators have been defined.

5. Formulating observables for classical general relativity is a formidable problem by itself because of its non-linear nature and space-time diffeomorphism invariance. In fact a systematic approximation scheme to calculate observables has only been recently developed.[30][31]

Difficulties in trying to examine the semi classical limit of the theory should not be confused with it having the wrong semi classical limit.

31.6.4 Progress in demonstrating LQG has the correct semiclassical limit

Much details here to be written up...

Concerning issue number 2 above one can consider so-called weave states. Ordinary measurements of geometric quantities are macroscopic, and planckian discreteness is smoothed out. The fabric of a T-shirt is analogous. At a distance it is a smooth curved two-dimensional surface. But a closer inspection we see that it is actually composed of thousands of one-dimensional linked threads. The image of space given in LQG is similar, consider a very large spin network formed by a very large number of nodes and links, each of Planck scale. But probed at a macroscopic scale, it appears as a three-dimensional continuous metric geometry.

As far as the editor knows problem 4 of having semi-classical machinery for non-graph changing operators is as the moment still out of reach.

To make contact with familiar low energy physics it is mandatory to have to develop approximation schemes both for the physical inner product and for Dirac observables.

The spin foam models have been intensively studied can be viewed as avenues toward approximation schemes for the physical inner product.

Markopoulou et al. adopted the idea of noiseless subsystems in an attempt to solve the problem of the low energy limit in background independent quantum gravity theories[32][33][34] The idea has even led to the intriguing possibility of matter of the standard model being identified with emergent degrees of freedom from some versions of LQG (see section below: *LQG and related research programs*).

As Wightman emphasized in the 1950s, in Minkowski QFTs the $n-$ point functions

$$W(x_1,\ldots,x_n) = \langle 0|\phi(x_n)\ldots\phi(x_1)|0\rangle \;,$$

completely determine the theory. In particular, one can calculate the scattering amplitudes from these quantities. As explained below in the section on the *Background independent scattering amplitudes*, in the background-independent context, the $n-$ point functions refer to a state and in gravity that state can naturally encode information about a specific geometry which can then appear in the expressions of these quantities. To leading order LQG calculations have been shown to agree in an appropriate sense with the $n-$ point functions calculated in the effective low energy quantum general relativity.

31.7 Improved dynamics and the Master constraint

Main articles: Hamiltonian (quantum mechanics) and Hamiltonian constraint of LQG

31.7.1 The Master constraint

Thiemann's Master constraint should not be confused with the Master equation to do with random processes. The Master Constraint Programme for Loop Quantum Gravity (LQG) was proposed as a classically equivalent way to impose the infinite number of Hamiltonian constraint equations

$$H(x) = 0$$

(x being a continuous index) in terms of a single Master constraint,

$$M = \int d^3x \frac{[H(x)]^2}{\sqrt{\det(q(x))}} \;.$$

which involves the square of the constraints in question. Note that $H(x)$ were infinitely many whereas the Master constraint is only one. It is clear that if M vanishes then so do the infinitely many $H(x)$'s. Conversely, if all the $H(x)$'s vanish then so does M, therefore they are equivalent. The Master constraint M involves an appropriate averaging over all space and so is invariant under spatial diffeomorphisms (it is invariant under spatial "shifts" as it is a summation over all such spatial "shifts" of a quantity that transforms as a scalar). Hence its Poisson bracket with the (smeared) spacial diffeomorphism constraint, $C(\vec{N})$, is simple:

$$\{M, C(\vec{N})\} = 0 \;.$$

(it is $su(2)$ invariant as well). Also, obviously as any quantity Poisson commutes with itself, and the Master constraint being a single constraint, it satisfies

$$\{M, M\} = 0 \;.$$

We also have the usual algebra between spatial diffeomorphisms. This represents a dramatic simplification of the Poisson bracket structure, and raises new hope in understanding the dynamics and establishing the semi-classical limit.[35]

An initial objection to the use of the Master constraint was that on first sight it did not seem to encode information about the observables; because the Mater constraint is quadratic in the constraint, when you compute its Poisson bracket with any quantity, the result is proportional to the constraint, therefore it always vanishes when the constraints are imposed and as such does not select out particular phase space functions. However, it was realized that the condition

$$\{\{M, O\}, O\}_{M=0} = 0$$

is equivalent to O being a Dirac observable. So the Master constraint does capture information about the observables. Because of its significance this is known as the Master equation.[35]

That the Master constraint Poisson algebra is an honest Lie algebra opens up the possibility of using a certain method, know as group averaging, in order to construct solutions of the infinite number of Hamiltonian constraints, a physical inner product thereon and Dirac observables via what is known as refined algebraic quantization RAQ[36]

31.7.2 Testing the Master constraint

The constraints in their primitive form are rather singular, this was the reason for integrating them over test functions to obtain smeared constraints. However, it would appear that the equation for the Master constraint, given above, is even more singular involving the product of two primitive constraints (although integrated over space). Squaring the constraint is dangerous as it could lead to worsened ultraviolent behaviour of the corresponding operator and hence the Master constraint programme must be approached with due care.

In doing so the Master constraint programme has been satisfactorily tested in a number of model systems with non-trivial constraint algebras, free and interacting field theories.[37][38][39][40][41] The Master constraint for LQG was established as a genuine positive self-adjoint operator and the physical Hilbert space of LQG was shown to be non-empty,[42] an obvious consistency test LQG must pass to be a viable theory of quantum General relativity.

31.7.3 Applications of the Master constraint

The Master constraint has been employed in attempts to approximate the physical inner product and define more rigorous path integrals.[43][44][45][46]

The Consistent Discretizations approach to LQG,[47][48] is an application of the master constraint program to construct the physical Hilbert space of the canonical theory.

31.7.4 Spin foam from the Master constraint

It turns out that the Master constraint is easily generalized to incorporate the other constraints. It is then referred to as the extended Master constraint, denoted M_E. We can define the extended Master constraint which imposes both the Hamiltonian constraint and spatial diffeomorphism constraint as a single operator,

$$M_E = \int_\Sigma d^3x \frac{H(x)^2 - q^{ab}V_a(x)V_b(x)}{\sqrt{det(q)}}.$$

Setting this single constraint to zero is equivalent to $H(x) = 0$ and $V_a(x) = 0$ for all x in Σ. This constraint implements the spatial diffeomorphism and Hamiltonian constraint at the same time on the Kinematic Hilbert space. The physical inner product is then defined as

$$\langle \phi, \psi \rangle_{\text{Phys}} = \lim_{T \to \infty} \left\langle \phi, \int_{-T}^{T} dt e^{it\hat{M}_E} \psi \right\rangle$$

(as $\delta(\hat{M}_E) = \lim_{T \to \infty} \int_{-T}^{T} dt e^{it\hat{M}_E}$). A spin foam representation of this expression is obtained by splitting the t-parameter in discrete steps and writing

$$e^{it\hat{M}_E} = \lim_{n \to \infty} [e^{it\hat{M}_E/n}]^n = \lim_{n \to \infty} [1 + it\hat{M}_E/n]^n.$$

The spin foam description then follows from the application of $[1 + it\hat{M}_E/n]$ on a spin network resulting in a linear combination of new spin networks whose graph and labels have been modified. Obviously an approximation is made by truncating the value of n to some finite integer. An advantage of the extended Master constraint is that we are working at the kinematic level and so far it is only here we have access semi-classical coherent states. Moreover, one can find none graph changing versions of this Master constraint operator, which are the only type of operators appropriate for these coherent states.

31.7.5 Algebraic quantum gravity

The Master constraint programme has evolved into a fully combinatorial treatment of gravity known as Algebraic Quantum Gravity (AQG).[49] While AQG is inspired by LQG, it differs drastically from it because in AQG there is fundamentally no topology or differential structure - it is background independent in a more generalized sense and could possibly have something to say about topology change. In this new formulation of quantum gravity existing semiclassical machinery, which is only viable for non-graph changing operators, can be employed, and progress has been made in establishing it has the correct semiclassical limit and providing contact with familiar low energy physics.[50][51] See Thiemann's book for details.

31.8 Physical applications of LQG

31.8.1 Black hole entropy

Main articles: Black hole thermodynamics, Isolated horizon and Immirzi parameter

The Immirzi parameter (also known as the Barbero-Immirzi parameter) is a numerical coefficient appearing in loop quantum gravity. It may take real or imaginary values.

An artist depiction of two black holes merging, a process in which the laws of thermodynamics are upheld.

Black hole thermodynamics is the area of study that seeks to reconcile the laws of thermodynamics with the existence of black hole event horizons. The no hair conjecture of general relativity states that a black hole is characterized only by its mass, its charge, and its angular momentum; hence, it has no entropy. It appears, then, that one can violate the second law of thermodynamics by dropping an object with nonzero entropy into a black hole.[52] Work by Stephen Hawking and Jacob Bekenstein showed that one can preserve the second law of thermodynamics by assigning to each black hole a *black-hole entropy*

$$S_{\text{BH}} = \frac{k_{\text{B}} A}{4\ell_{\text{P}}^2},$$

where A is the area of the hole's event horizon, k_{B} is the Boltzmann constant, and $\ell_{\text{P}} = \sqrt{G\hbar/c^3}$ is the Planck length.[53] The fact that the black hole entropy is also the maximal entropy that can be obtained by the Bekenstein bound (wherein the Bekenstein bound becomes an equality) was the main observation that led to the holographic principle.[52]

An oversight in the application of the no-hair theorem is the assumption that the relevant degrees of freedom accounting for the entropy of the black hole must be classical in nature; what if they were purely quantum mechanical instead and had non-zero entropy? Actually, this is what is realized in the LQG derivation of black hole entropy, and can be seen as a consequence of its background-independence - the classical black hole spacetime comes about from the semi-classical limit of the quantum state of the gravitational field, but there are many quantum states that have the same semiclasical limit. Specifically, in LQG[54] it is possible to associate a quantum geometrical interpretation to the microstates: These are the quantum geometries of the horizon which are consistent with the area, A, of the black hole and the topology of the horizon (i.e. spherical). LQG offers a geometric explanation of the finiteness of the entropy and of the proportionality of the area of the horizon.[55][56] These calculations have been generalized to rotating black holes.[57]

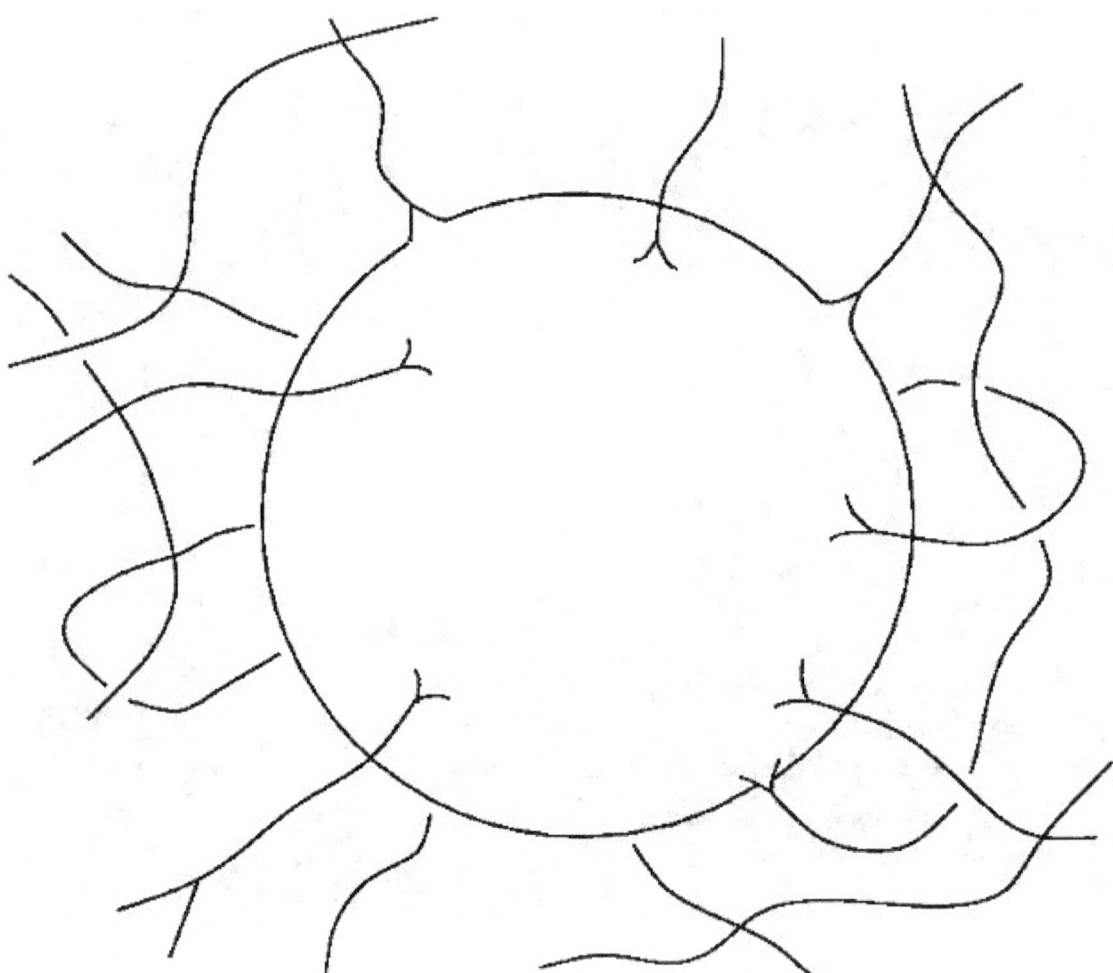

Representation of quantum geometries of the horizon. Polymer excitations in the bulk puncture the horizon, endowing it with quantized area. Intrinsically the horizon is flat except at punctures where it acquires a quantized deficit angle or quantized amount of curvature. These deficit angles add up to 4π.

It is possible to derive, from the covariant formulation of full quantum theory (Spinfoam) the correct relation between energy and area (1st law), the Unruh temperature and the distribution that yields Hawking entropy.[58] The calculation makes use of the notion of dynamical horizon and is done for non-extremal black holes.

A recent success of the theory in this direction is the computation of the entropy of all non singular black holes directly from theory and independent of Immirzi parameter.[59] The result is the expected formula $S = A/4$, where S is the entropy and A the area of the black hole, derived by Bekenstein and Hawking on heuristic grounds. This is the only known derivation of this formula from a fundamental theory, for the case of generic non singular black holes.

Older attempts at this calculation had difficulties. The problem was that although Loop quantum gravity predicted that the entropy of a black hole is proportional to the area of the event horizon, the result depended on a crucial free parameter in the theory, the above-mentioned Immirzi parameter. However, there is no known computation of the Immirzi parameter, so it had to be fixed by demanding agreement with Bekenstein and Hawking's calculation of the black hole entropy.

31.8.2 Loop quantum cosmology

Main articles: loop quantum cosmology, Big bounce and inflation (cosmology)

The popular and technical literature makes extensive references to LQG-related topic of loop quantum cosmology. LQC was mainly developed by Martin Bojowald, it was popularized Loop quantum cosmology in *Scientific American* for predicting a Big Bounce prior to the Big Bang. Loop quantum cosmology (LQC) is a symmetry-reduced model of classical general relativity quantized using methods that mimic those of loop quantum gravity (LQG) that predicts a "quantum bridge" between contracting and expanding cosmological branches.

Achievements of LQC have been the resolution of the big bang singularity, the prediction of a Big Bounce, and a natural mechanism for inflation (cosmology).

LQC models share features of LQG and so is a useful toy model. However, the results obtained are subject to the usual restriction that a truncated classical theory, then quantized, might not display the true behaviour of the full theory due to artificial suppression of degrees of freedom that might have large quantum fluctuations in the full theory. It has been argued that singularity avoidance in LQC are by mechanisms only available in these restrictive models and that singularity avoidance in the full theory can still be obtained but by a more subtle feature of LQG.[60][61]

31.8.3 Loop Quantum Gravity phenomenology

Quantum gravity effects are notoriously difficult to measure because the Planck length is so incredibly small. However recently physicists have started to consider the possibility of measuring quantum gravity effects, mostly from astrophysical observations and gravitational wave detectors.

31.8.4 Background independent scattering amplitudes

Loop quantum gravity is formulated in a background-independent language. No spacetime is assumed a priori, but rather it is built up by the states of theory themselves - however scattering amplitudes are derived from n -point functions (Correlation function (quantum field theory)) and these, formulated in conventional quantum field theory, are functions of points of a background space-time. The relation between the background-independent formalism and the conventional formalism of quantum field theory on a given spacetime is far from obvious, and it is far from obvious how to recover low-energy quantities from the full background-independent theory. One would like to derive the n -point functions of the theory from the background-independent formalism, in order to compare them with the standard perturbative expansion of quantum general relativity and therefore check that loop quantum gravity yields the correct low-energy limit.

A strategy for addressing this problem has been suggested;[62] the idea is to study the boundary amplitude, namely a path integral over a finite space-time region, seen as a function of the boundary value of the field.[63] In conventional quantum field theory, this boundary amplitude is well–defined[64][65] and codes the physical information of the theory; it does so in quantum gravity as well, but in a fully background–independent manner.[66] A generally covariant definition of n -point functions can then be based on the idea that the distance between physical points –arguments of the n -point function is determined by the state of the gravitational field on the boundary of the spacetime region considered.

Progress has been made in calculating background independent scattering amplitudes this way with the use of spin foams. This is a way to extract physical information from the theory. Claims to have reproduced the correct behaviour for graviton scattering amplitudes and to have recovered classical gravity have been made. "We have calculated Newton's law starting from a world with no space and no time." - Carlo Rovelli.

31.8.5 planck stars

Carlo Rovelli has written a paper claiming inside a black hole is a planck star, that if correct, would resolve the black hole firewall and black hole information paradox.

31.9 Gravitons, string theory, super symmetry, extra dimensions in LQG

Main articles: graviton, string theory, supersymmetry, Kaluza–Klein theory and supergravity

Some quantum theories of gravity posit a spin-2 quantum field that is quantized, giving rise to gravitons. In string theory one generally starts with quantized excitations on top of a classically fixed background. This theory is thus described as background dependent. Particles like photons as well as changes in the spacetime geometry (gravitons) are both described as excitations on the string worldsheet. While string theory is "background dependent", the choice of background, like a gauge fixing, does not affect the physical predictions. This is not the case, however, for quantum field theories, which give different predictions for different backgrounds. In contrast, loop quantum gravity, like general relativity, is manifestly background independent, eliminating the (in some sense) "redundant" background required in string theory. Loop quantum gravity, like string theory, also aims to overcome the nonrenormalizable divergences of quantum field theories.

LQG never introduces a background and excitations living on this background, so LQG does not use gravitons as building blocks. Instead one expects that one may recover a kind of semiclassical limit or weak field limit where something like "gravitons" will show up again. In contrast, gravitons play a key role in string theory where they are among the first (massless) level of excitations of a superstring.

LQG differs from string theory in that it is formulated in 3 and 4 dimensions and without supersymmetry or Kaluza-Klein extra dimensions, while the latter requires both to be true. There is no experimental evidence to date that confirms string theory's predictions of supersymmetry and Kaluza–Klein extra dimensions. In a 2003 paper A dialog on quantum gravity,[67] Carlo Rovelli regards the fact LQG is formulated in 4 dimensions and without supersymmetry as a strength of the theory as it represents the most parsimonious explanation, consistent with current experimental results, over its rival string/M-theory. Proponents of string theory will often point to the fact that, among other things, it demonstrably reproduces the established theories of general relativity and quantum field theory in the appropriate limits, which Loop Quantum Gravity has struggled to do. In that sense string theory's connection to established physics may be considered more reliable and less speculative, at the mathematical level. Peter Woit in Not Even Wrong and Lee Smolin in The Trouble with Physics regard string/M-theory to be in conflict with current known experimental results.

Since LQG has been formulated in 4 dimensions (with and without supersymmetry), and M-theory requires supersymmetry and 11 dimensions, a direct comparison between the two has not been possible. It is possible to extend mainstream LQG formalism to higher-dimensional supergravity, general relativity with supersymmetry and Kaluza–Klein extra dimensions should experimental evidence establish their existence. It would therefore be desirable to have higher-dimensional Supergravity loop quantizations at one's disposal in order to compare these approaches. In fact a series of recent papers have been published attempting just this.[68][69][70][71][72][73][74][75] Most recently, Thiemann at el have made progress toward calculating black hole entropy for supergravity in higher dimensions. It will be interesting to compare these results to the corresponding super string calculations.[76][77]

As of April 2013 LHC has failed to find evidence of supersymmetry or Kaluza–Klein extra dimensions, which has encouraged LQG researchers. Shaposhnikov in his paper "Is there a new physics between electroweak and Planck scales?" has proposed the neutrino minimal standard model,[78] which claims the most parsimonious theory is a standard model extended with neutrinos, plus gravity, and that extra dimensions, GUT physics, and supersymmetry, string/M-theory physics are unrealized in nature, and that any theory of quantum gravity must be four dimensional, like loop quantum gravity.

31.10 LQG and related research programs

Main articles: noncommutative geometry, twistor theory, entropic gravity, Sundance Bilson-Thompson, Asymptotic safety in quantum gravity, Causal dynamical triangulation, group field theory and consistent discretizations

Several research groups have attempted to combine LQG with other research programs: Johannes Aastrup, Jesper M. Grimstrup et al. research combines noncommutative geometry with loop quantum gravity,[79] Laurent Freidel, Simone Speziale, et al., spinors and twistor theory with loop quantum gravity,[80] and Lee Smolin et al. with Verlinde entropic gravity and loop gravity.[81] Stephon Alexander, Antonino Marciano and Lee Smolin have attempted to explain the origins of weak force chirality in terms of Ashketar's variables, which describe gravity as chiral,[82] and LQG with Yang–Mills theory fields [83] in four dimensions. Sundance Bilson-Thompson, Hackett et al.,[84][85] has attempted to introduce standard model via LQG"s degrees of freedom as an emergent property (by employing the idea noiseless subsystems a useful notion introduced in more general situation for constrained systems by Fotini Markopoulou-Kalamara et al.[86]) LQG has also drawn philosophical comparisons with Causal dynamical triangulation [87] and asymptotically safe gravity,[88] and the spinfoam with group field theory and AdS/CFT correspondence.[89] Smolin and Wen have suggested combining LQG with String-net liquid, tensors, and Smolin and Fotini Markopoulou-Kalamara Quantum Graphity. There is the consistent discretizations approach. In addition to what has already mentioned above, Pullin and Gambini provide a framework to connect the path integral and canonical approaches to quantum gravity. They may help reconcile the spin foam and canonical loop representation approaches. Recent research by Chris Duston and Matilde Marcolli introduces topology change via topspin networks.[90]

31.11 Problems and comparisons with alternative approaches

Main article: List of unsolved problems in physics

Some of the major unsolved problems in physics are theoretical, meaning that existing theories seem incapable of explaining a certain observed phenomenon or experimental result. The others are experimental, meaning that there is a difficulty in creating an experiment to test a proposed theory or investigate a phenomenon in greater detail.

Can quantum mechanics and general relativity be realized as a fully consistent theory (perhaps as a quantum field theory)?[7] Is spacetime fundamentally continuous or discrete? Would a consistent theory involve a force mediated by a hypothetical graviton, or be a product of a discrete structure of spacetime itself (as in loop quantum gravity)? Are there deviations from the predictions of general relativity at very small or very large scales or in other extreme circumstances that flow from a quantum gravity theory?

The theory of LQG is one possible solution to the problem of quantum gravity, as is string theory. There are substantial differences however. For example, string theory also addresses unification, the understanding of all known forces and particles as manifestations of a single entity, by postulating extra dimensions and so-far unobserved additional particles and symmetries. Contrary to this, LQG is based only on quantum theory and general relativity and its scope is limited to understanding the quantum aspects of the gravitational interaction. On the other hand, the consequences of LQG are radical, because they fundamentally change the nature of space and time and provide a tentative but detailed physical and mathematical picture of quantum spacetime.

Presently, no semiclassical limit recovering general relativity has been shown to exist. This means it remains unproven that LQG's description of spacetime at the Planck scale has the right continuum limit (described by general relativity with possible quantum corrections). Specifically, the dynamics of the theory is encoded in the Hamiltonian constraint, but there is no candidate Hamiltonian.[91] Other technical problems include finding off-shell closure of the constraint algebra and physical inner product vector space, coupling to matter fields of Quantum field theory, fate of the renormalization of the graviton in perturbation theory that lead to ultraviolet divergence beyond 2-loops (see One-loop Feynman diagram in Feynman diagram).[91]

While there has been a recent proposal relating to observation of naked singularities,[92] and doubly special relativity as a part of a program called loop quantum cosmology, there is no experimental observation for which loop quantum gravity makes a prediction not made by the Standard Model or general relativity (a problem that plagues all current theories of quantum gravity). Because of the above-mentioned lack of a semiclassical limit, LQG has not yet even reproduced the predictions made by general relativity.

An alternative criticism is that general relativity may be an effective field theory, and therefore quantization ignores the fundamental degrees of freedom.

31.12 See also

31.13 Notes

[1] Rovelli, Carlo (August 2008). "Loop Quantum Gravity". *CERN*. Retrieved 14 September 2014.

[2] Rovelli, C. (2011). "Zakopane lectures on loop gravity". arXiv:1102.3660 [gr-qc].

[3] Muxin, H. (2011). "Cosmological constant in loop quantum gravity vertex amplitude". *Physical Review D* **84** (6): 064010. arXiv:1105.2212. Bibcode:2011PhRvD..84f4010H. doi:10.1103/PhysRevD.84.064010.

[4] Fairbairn, W. J.; Meusburger, C. (2011). "q-Deformation of Lorentzian spin foam models". arXiv:1112.2511 [gr-qc].

[5] Rovelli, C. (2004). *Quantum Gravity*. Cambridge Monographs on Mathematical Physics. p. 71. ISBN 978-0-521-83733-0.

[6] Kauffman, S.; Smolin, L. (7 April 1997). "A Possible Solution For The Problem Of Time In Quantum Cosmology". *Edge.org*. Retrieved 2014-08-20.

[7] Smolin, L. (2006). "The Case for Background Independence". In Rickles, D.; French, S.; Saatsi, J. T. "The Structural Foundations of Quantum Gravity". Clarendon Press. pp. 196*ff*. ISBN 978-0-19-926969-3.

[8] Rovelli, C. (2004). *Quantum Gravity*. Cambridge Monographs on Mathematical Physics. p. 13ff. ISBN 978-0-521-83733-0.

[9] Thiemann, T. (1996). "Anomaly-free formulation of non-perturbative, four-dimensional Lorentzian quantum gravity". *Physics Letters B* **380**: 257–264. arXiv:gr-qc/9606088. Bibcode:1996PhLB..380..257T. doi:10.1016/0370-2693(96)00532-1.

[10] Baez, J.; de Muniain, J. P. (1994). *Gauge Fields, Knots and Quantum Gravity*. Series on Knots and Everything. Vol. 4. World Scientific. Part III, chapter 4. ISBN 978-981-02-1729-7.

[11] Thiemann, T. (2003). "Lectures on Loop Quantum Gravity". *Lecture Notes in Physics* **631**: 41–135. doi:10.1007/978-3-540-45230-0_3.

[12] Rovelli, C.; Smolin, L. (1988). "Knot Theory and Quantum Gravity". *Physical Review Letters* **61** (10): 1155–1958. Bibcode:1988PhRvL..61.1155R. doi:10.1103/PhysRevLett.61.1155.

[13] Gambini, R.; Pullin, J. (2011). *A First Course in Loop Quantum Gravity*. Oxford University Press. Section 8.2. ISBN 978-0-19-959075-9.

[14] Fernando, J.; Barbero, G. (1995). "Reality Conditions and Ashtekar Variables: A Different Perspective". *Physical Review D* **51**: 5498–5506. arXiv:gr-qc/9410013. Bibcode:1995PhRvD..51.5498B. doi:10.1103/PhysRevD.51.5498.

[15] Fernando, J.; Barbero, G. (1995). "Real Ashtekar Variables for Lorentzian Signature Space-times". *Physical Review D* **51**: 5507–5520. arXiv:gr-qc/9410014. Bibcode:1995PhRvD..51.5507B. doi:10.1103/PhysRevD.51.5507.

[16] Bojowald, M.; Alejandro, P.. "Spin Foam Quantization and Anomalies". arXiv:gr-qc/0303026 [gr-qc].

[17] Barrett, J.; Crane, L. (2000). *Classical and Quantum Gravity* **17**: 3101. arXiv:gr-qc/9904025. Bibcode:2000CQGra..17.3101B. doi:10.1088/0264-9381/17/16/302..

[18] Rovelli, C.; Alesci, E. (2007). "The complete LQG propagator I. Difficulties with the Barrett–Crane vertex". *Physical Review D* **76**: 104012. arXiv:hep-th/0703074. Bibcode:2007PhRvD..76b4012B. doi:10.1103/PhysRevD.76.024012.

[19] Engle, J.; Pereira, R.; Rovelli, C. (2009). *Physical Review Letters* **99**: 161301.

[20] Freidal, L.; Krasnov, K. (2008). *Classical and Quantum Gravity* **25**: 125018.

[21] Alesci, E.; Thiemann, T.; Zipfel, A. (2011). "Linking covariant and canonical LQG: new solutions to the Euclidean Scalar Constraint". arXiv:1109.1290.

[22] Bohm, D. (1989). *Quantum Theory*. Dover Publications. ISBN 978-0-486-65969-5.

[23] Tipler, P.; Llewellyn, R. (2008). *Modern Physics* (5th ed.). W. H. Freeman and Co. pp. 160–161. ISBN 978-0-7167-7550-8.

[24] Bohr, N. (1920). "Über die Serienspektra der Element". *Zeitschrift für Physik* **2** (5): 423–478. Bibcode:1920ZPhy....2..423B. doi:10.1007/BF01329978. (English translation in Bohr 1976, pp. 241–282)

[25] Jammer, M. (1989). *The Conceptual Development of Quantum Mechanics* (2nd ed.). Tomash Publishers. Section 3.2. ISBN 978-0-88318-617-6.

[26] Ashtekar, A.; Bombelli, L.; Corichi, A. (2005). "Semiclassical States for Constrained Systems". *Physical Review D* **72**: 025008. arXiv:hep-ph/0504114. Bibcode:2005PhRvD..72a5008C. doi:10.1103/PhysRevD.72.015008.

[27] Lewandowski, J.; Okołów, A.; Sahlmann, H.; Thiemann, T. (2005). "Uniqueness of Diffeomorphism Invariant States on Holonomy-Flux Algebras". *Communications in Mathematical Physics* **267**: 703–733. doi:10.1007/s00220-006-0100-7.

[28] Fleischhack, C. (2006). "Irreducibility of the Weyl algebra in loop quantum gravity". *Physical Review Letters* **97**: 061302. doi:10.1103/physrevlett.97.061302.

[29] Thiemann, T. (2008). *Modern Canonical General Relativity*. Cambridge Monographs on Mathematical Physics. Cambridge University Press. Section 10.6. ISBN 978-0-521-74187-3.

[30] "Partial and Complete Observables for Hamiltonian Constrained Systems". *General Relativity and Gravitation* **39**: 1891–1927. 2007. arXiv:gr-qc/0411013. Bibcode:2007GReGr..39.1891D. doi:10.1007/s10714-007-0495-2.

[31] "Partial and Complete Observables for Canonical General Relativity". *Classical and Quantum Gravity* **23**: 6155–6184. arXiv:gr-qc/0507106. Bibcode:2006CQGra..23.6155D. doi:10.1088/0264-9381/23/22/006.

[32] Dreyer, O.; Markopoulou, f.; Smolin, L. (2006). "Symmetry and entropy of black hole horizons". *Nuclear Physics B* **774**: 1–13.

[33] Kribs, D. W.; Markopoulou, F.. "Geometry from quantum particles". arXiv:gr-qc/0510052.

[34] Markopoulou, F.; Poulin, D. "Noiseless subsystems and the low energy limit of spin foam models" (unpublished). Missing or empty |title= (help)

[35] *The Phoenix Project: Master Constraint Programme for Loop Quantum Gravity*, Class.Quant.Grav.23:2211-2248,2006 or http://fr.arxiv.org/pdf/gr-qc/0305080

[36] *Modern Canonical Quantum General Relativity* by Thomas Thiemann

[37] *Testing the Master Constraint Programme for Loop Quantum Gravity I. General Framework*, Bianca Dittrich, Thomas Thiemann, Class.Quant.Grav. 23 (2006) 1025-1066.

[38] *Testing the Master Constraint Programme for Loop Quantum Gravity II. Finite Dimensional Systems*, Bianca Dittrich, Thomas Thiemann, Class.Quant.Grav. 23 (2006) 1067-1088.

[39] *Testing the Master Constraint Programme for Loop Quantum Gravity III. SL(2,R) Models*, Bianca Dittrich, Thomas Thiemann, Class.Quant.Grav. 23 (2006) 1089-1120.

[40] *Testing the Master Constraint Programme for Loop Quantum Gravity IV. Free Field Theories*, Bianca Dittrich, Thomas Thiemann, Class.Quant.Grav. 23 (2006) 1121-1142.

[41] *Testing the Master Constraint Programme for Loop Quantum Gravity V. Interacting Field Theories*, Bianca Dittrich, Thomas Thiemann, Class.Quant.Grav. 23 (2006) 1143-1162.

[42] *Quantum Spin Dynamics VIII. The Master Constraint*, Thomas Thiemann, Class.Quant.Grav. 23 (2006) 2249-2266.

[43] *Approximating the physical inner product of Loop Quantum Cosmology*, Benjamin Bahr, Thomas Thiemann, Class.Quant.Grav.24:2109-2138,2007.

[44] *On the Relation between Operator Constraint --, Master Constraint --, Reduced Phase Space --, and Path Integral Quantisation*, Muxin Han, Thomas Thiemann, Class.Quant.Grav.27:225019,2010.

[45] *On the Relation between Rigging Inner Product and Master Constraint Direct Integral Decomposition*, Muxin Han, Thomas Thiemann, J.Math.Phys.51:092501,2010.

[46] *A Path-integral for the Master Constraint of Loop Quantum Gravity*, Muxin Han, Class.Quant.Grav.27:215009,2010

[47] *Emergent diffeomorphism invariance in a discrete loop quantum gravity model*, Rodolfo Gambini, Jorge Pullin, Class.Quant.Grav.26:035002,2009

[48] Section 10.2.2 *A First Course in Loop quantum Gravity*, Rodolfo Gambinni, Jorge Pullin, Oxford University Press, first published 2011.

[49] *Algebraic Quantum Gravity (AQG) I. Conceptual Setup*, K. Giesel, T. Thiemann, Class.Quant.Grav.24:2465-2498,2007.

[50] *Algebraic Quantum Gravity (AQG) II. Semiclassical Analysis*, K. Giesel, T. Thiemann, Class.Quant.Grav.24:2499-2564,2007.

[51] *Algebraic Quantum Gravity (AQG) III. Semiclassical Perturbation Theory*, K. Giesel, T. Thiemann, Class.Quant.Grav.24:2565-2588,2007.

[52] Bousso, Raphael (2002). "The Holographic Principle". *Reviews of Modern Physics* **74** (3): 825–874. arXiv:hep-th/0203101. Bibcode:2002RvMP...74..825B. doi:10.1103/RevModPhys.74.825.

[53] Majumdar, Parthasarathi (1998). "Black Hole Entropy and Quantum Gravity". *ArXiv: General Relativity and Quantum Cosmology* **73**: 147. arXiv:gr-qc/9807045. Bibcode:1999InJPB..73..147M.

[54] See List of loop quantum gravity researchers

[55] Rovelli, Carlo (1996). "Black Hole Entropy from Loop Quantum Gravity". *Physical Review Letters* **77** (16): 3288–3291. arXiv:gr-qc/9603063. Bibcode:1996PhRvL..77.3288R. doi:10.1103/PhysRevLett.77.3288.

[56] Ashtekar, Abhay; Baez, John; Corichi, Alejandro; Krasnov, Kirill (1998). "Quantum Geometry and Black Hole Entropy". *Physical Review Letters* **80** (5): 904–907. arXiv:gr-qc/9710007. Bibcode:1998PhRvL..80..904A. doi:10.1103/PhysRevLett.80.904.

[57] *Quantum horizons and black hole entropy: Inclusion of distortion and rotation*, Abhay Ashtekar, Jonathan Engle, Chris Van Den Broeck, Class.Quant.Grav.22:L27-L34, 2005.

[58] Bianchi, Eugenio (2012). "Entropy of Non-Extremal Black Holes from Loop Gravity". arXiv:gr-qc/1204.5122.

[59] http://inspirehep.net/record/940357?ln=en. http://inspirehep.net/record/1111991.

[60] *On (Cosmological) Singularity Avoidance in Loop Quantum Gravity*, Johannes Brunnemann, Thomas Thiemann, Class.Quant.Grav. 23 (2006) 1395-1428.

[61] *Unboundedness of Triad -- Like Operators in Loop Quantum Gravity*, Johannes Brunnemann, Thomas Thiemann, Class.Quant.Grav. 23 (2006) 1429-1484.

[62] L. Modesto, C. Rovelli:*Particle scattering in loop quantum gravity*, Phys Rev Lett 95 (2005) 191301

[63] R Oeckl, *A 'general boundary' formulation for quantum mechanics and quantum gravity*, Phys Lett B575 (2003) 318-324 ; *Schrodinger's cat and the clock: lessons for quantum gravity*, Class Quant Grav 20 (2003) 5371-5380l

[64] F. Conrady, C. Rovelli *Generalized Schrodinger equation in Euclidean field theory*", Int J Mod Phys A 19, (2004) 1-32.

[65] L Doplicher, *Generalized Tomonaga-Schwinger equation from the Hadamard formula*, Phys Rev D70 (2004) 064037

[66] F. Conrady, L. Doplicher, R. Oeckl, C. Rovelli, M. Testa, *Minkowski vacuum in background independent quantum gravity*, Phys Rev D69 (2004) 064019.

[67] http://arxiv.org/abs/arXiv:hep-th/0310077

[68] *New Variables for Classical and Quantum Gravity in all Dimensions I. Hamiltonian Analysis*, Norbert Bodendorfer, Thomas Thiemann, Andreas Thurn, Class. Quantum Grav. 30 (2013) 045001

[69] *New Variables for Classical and Quantum Gravity in all Dimensions II. Lagrangian Analysis*, Norbert Bodendorfer, Thomas Thiemann, Andreas Thurn, Quantum Grav. 30 (2013) 045002

[70] *New Variables for Classical and Quantum Gravity in all Dimensions III. Quantum Theory*, Norbert Bodendorfer, Thomas Thiemann, Andreas Thurn, Class. Quantum Grav. 30 (2013) 045003

[71] *New Variables for Classical and Quantum Gravity in all Dimensions IV. Matter Coupling*, Norbert Bodendorfer, Thomas Thiemann, Andreas Thurn, Class. Quantum Grav. 30 (2013) 045004

[72] *On the Implementation of the Canonical Quantum Simplicity Constraint*, Norbert Bodendorfer, Thomas Thiemann, Andreas Thurn, Class. Quantum Grav. 30 (2013) 045005

[73] *Towards Loop Quantum Supergravity (LQSG) I. Rarita-Schwinger Sector*, Norbert Bodendorfer, Thomas Thiemann, Andreas Thurn, Class. Quantum Grav. 30 (2013) 045006

[74] *Towards Loop Quantum Supergravity (LQSG) II. p-Form Sector*, Norbert Bodendorfer, Thomas Thiemann, Andreas Thurn, Class. Quantum Grav. 30 (2013) 045007

[75] *Towards Loop Quantum Supergravity (LQSG)*, Norbert Bodendorfer, Thomas Thiemann, Andreas Thurn, Phys. Lett. B 711: 205-211 (2012)

[76] *New Variables for Classical and Quantum Gravity in all Dimensions V. Isolated Horizon Boundary Degrees of Freedom*, Norbert Bodendorfer, Thomas Thiemann, Andreas Thurn, http://uk.arxiv.org/pdf/1304.2679.

[77] *Black hole entropy from loop quantum gravity in higher dimensions*, Norbert Bodendorfer http://uk.arxiv.org/pdf/1307.5029

[78] http://arxiv.org/abs/0708.3550

[79] http://arxiv.org/abs/1203.6164

[80] http://arxiv.org/abs/1006.0199

[81] http://arxiv.org/abs/1001.3668

[82] http://arxiv.org/abs/1212.5246

[83] http://arxiv.org/abs/1105.3480

[84] *Quantum gravity and the standard model*, Sundance O. Bilson-Thompson, Fotini Markopoulou, Lee Smolin, Class.Quant.Grav.24:3975-3994,2007.

[85] For a precise review and outlook of this research see: *Emergent Braided Matter of Quantum Geometry*, Sundance Bilson-Thompson, Jonathan Hackett, Louis Kauffman, Yidun Wan, SIGMA 8 (2012), 014, 43 pages.

[86] *Constrained Mechanics and Noiseless Subsystems*, Tomasz Konopka, Fotini Markopoulou, arXiv:gr-qc/0601028.

[87] http://www.perimeterinstitute.ca/people/renate-loll

[88] wwnpqft.inln.cnrs.fr/pdf/Bianchi.pdf

[89] http://arxiv.org/abs/0804.0632

[90] http://arxiv.org/abs/1308.2934

[91] Nicolai, Hermann; Peeters, Kasper; Zamaklar, Marija (2005). "Loop quantum gravity: an outside view". *Classical and Quantum Gravity* **22** (19): R193–R247. arXiv:hep-th/0501114. Bibcode:2005CQGra..22R.193N. doi:10.1088/0264-9381/22/19/R01.

[92] Goswami et al.; Joshi, Pankaj S.; Singh, Parampreet (2006). "Quantum evaporation of a naked singularity". *Physical Review Letters* **96** (3): 31302. arXiv:gr-qc/0506129. Bibcode:2006PhRvL..96c1302G. doi:10.1103/PhysRevLett.96.031302.

31.14 References

- Topical Reviews

    - Rovelli, Carlo (2011). "Zakopane lectures on loop gravity". arXiv:1102.3660.

    - Rovelli, Carlo (1998). "Loop Quantum Gravity". *Living Reviews in Relativity* **1**. Retrieved 2008-03-13.

    - Thiemann, Thomas (2003). "Lectures on Loop Quantum Gravity". *Lectures Notes in Physics*. Lecture Notes in Physics **631**: 41–135. arXiv:gr-qc/0210094. Bibcode:2003LNP...631...41T. doi:10.1007/978-3-540-45230-0_3. ISBN 978-3-540-40810-9.

    - Ashtekar, Abhay; Lewandowski, Jerzy (2004). "Background Independent Quantum Gravity: A Status Report". *Classical and Quantum Gravity* **21** (15): R53–R152. arXiv:gr-qc/0404018. Bibcode:2004CQGra..21R..53A. doi:10.1088/0264-9381/21/15/R01.

    - Carlo Rovelli and Marcus Gaul, *Loop Quantum Gravity and the Meaning of Diffeomorphism Invariance*, e-print available as gr-qc/9910079.

    - Lee Smolin, *The case for background independence*, e-print available as hep-th/0507235.

    - Alejandro Corichi, *Loop Quantum Geometry: A primer*, e-print available as .

    - Alejandro Perez, *Introduction to loop quantum gravity and spin foams*, e-print available as .

    - Hermann Nicolai and Kasper Peeters *Loop and spin foam quantum gravity: A Brief guide for beginners.*, e-print available as .

- Popular books:

    - Lee Smolin, *Three Roads to Quantum Gravity*

    - Carlo Rovelli, *Che cos'è il tempo? Che cos'è lo spazio?*, Di Renzo Editore, Roma, 2004. French translation: *Qu'est ce que le temps? Qu'est ce que l'espace?*, Bernard Gilson ed, Brussel, 2006. English translation: *What is Time? What is space?*, Di Renzo Editore, Roma, 2006.

- Julian Barbour, *The End of Time: The Next Revolution in Our Understanding of the Universe*
- Musser, George (2008). "The Complete Idiot's Guide to String Theory". *The Physics Teacher* (Indianapolis: Alpha) **47** (2): 368. Bibcode:2009PhTea..47Q.128H. doi:10.1119/1.3072469. ISBN 978-1-59257-702-6. – Focuses on string theory but has an extended discussion of loop gravity as well.

- Magazine articles:
  - Lee Smolin, "Atoms of Space and Time", *Scientific American*, January 2004
  - Martin Bojowald, "Following the Bouncing Universe", *Scientific American*, October 2008

- Easier introductory, expository or critical works:
  - Abhay Ashtekar, *Gravity and the quantum*, e-print available as gr-qc/0410054 (2004)
  - John C. Baez and Javier Perez de Muniain, *Gauge Fields, Knots and Quantum Gravity*, World Scientific (1994)
  - Carlo Rovelli, *A Dialog on Quantum Gravity*, e-print available as hep-th/0310077 (2003)
  - Rodolfo Gambini and Jorge Pullin, *A First Course in Loop Quantum Gravity*, Oxford (2011)
  - Carlo Rovelli and Francesca Vidotto, *Covariant Loop Quantum Gravity*, Cambridge (2014); draft available online

- More advanced introductory/expository works:
  - Carlo Rovelli, *Quantum Gravity*, Cambridge University Press (2004); draft available online
  - Thomas Thiemann, *Introduction to modern canonical quantum general relativity*, e-print available as gr-qc/0110034
  - Thomas Thiemann, *Introduction to Modern Canonical Quantum General Relativity*, Cambridge University Press (2007)
  - Abhay Ashtekar, *New Perspectives in Canonical Gravity*, Bibliopolis (1988).
  - Abhay Ashtekar, *Lectures on Non-Perturbative Canonical Gravity*, World Scientific (1991)
  - Rodolfo Gambini and Jorge Pullin, *Loops, Knots, Gauge Theories and Quantum Gravity*, Cambridge University Press (1996)
  - Hermann Nicolai, Kasper Peeters, Marija Zamaklar, *Loop quantum gravity: an outside view*, e-print available as hep-th/0501114
  - H. Nicolai and K. Peeters, *Loop and Spin Foam Quantum Gravity: A Brief Guide for Beginners*, e-print available as hep-th/0601129
  - T. Thiemann The LQG – String: Loop Quantum Gravity Quantization of String Theory (2004)

- Conference proceedings:
  - John C. Baez (ed.), *Knots and Quantum Gravity*

- Fundamental research papers:
  - Ashtekar, Abhay (1986). "New variables for classical and quantum gravity". *Physical Review Letters* **57** (18): 2244–2247. Bibcode:1986PhRvL..57.2244A. doi:10.1103/PhysRevLett.57.2244. PMID 10033673
  - Ashtekar, Abhay (1987). "New Hamiltonian formulation of general relativity". *Physical Review D* **36** (6): 1587–1602. Bibcode:1987PhRvD..36.1587A. doi:10.1103/PhysRevD.36.1587
  - Roger Penrose, *Angular momentum: an approach to combinatorial space-time* in *Quantum Theory and Beyond*, ed. Ted Bastin, Cambridge University Press, 1971
  - Rovelli, Carlo; Smolin, Lee (1988). "Knot theory and quantum gravity". *Physical Review Letters* **61** (10): 1155. Bibcode:1988PhRvL..61.1155R. doi:10.1103/PhysRevLett.61.1155.
  - Rovelli, Carlo; Smolin, Lee (1990). "Loop space representation of quantum general relativity". *Nuclear Physics* **B331**: 80–152.
  - Carlo Rovelli and Lee Smolin, *Discreteness of area and volume in quantum gravity*, Nucl. Phys., **B442** (1995) 593-622, e-print available as gr-qc/9411005

- Kuchař, Karel (1973). "Canonical Quantization of Gravity". In Israel, Werner. *Relativity, Astrophysics and Cosmology*. D. Reidel. pp. 237–288. ISBN 90-277-0369-8.
- Thiemann, Thomas (2006). "Loop Quantum Gravity: An Inside View". *Approaches to Fundamental Physics*. Lecture Notes in Physics **721**: 185. arXiv:hep-th/0608210. Bibcode:2007LNP...721..185T. doi:10.1007/978-3-540-71117-9_10. ISBN 978-3-540-71115-5.

31.15 External links

- "Loop Quantum Gravity" by Carlo Rovelli Physics World, November 2003
- Quantum Foam and Loop Quantum Gravity
- Abhay Ashtekar: Semi-Popular Articles . Some excellent popular articles suitable for beginners about space, time, GR, and LQG.
- Loop Quantum Gravity: Lee Smolin.
- Loop Quantum Gravity on arxiv.org
- A list of LQG references catered to fresh graduates
- Loop Quantum Gravity Lectures Online by Lee Smolin
- Spin networks, spin foams and loop quantum gravity
- Wired magazine, News: *Moving Beyond String Theory*
- April 2006 Scientific American Special Issue, *A Matter of Time*, has Lee Smolin LQG Article *Atoms of Space and Time*
- September 2006, The Economist, article *Looping the loop*
- Gamma-ray Large Area Space Telescope: http://glast.gsfc.nasa.gov/
- Zeno meets modern science. Article from Acta Physica Polonica B by Z.K. Silagadze.
- Did pre-big bang universe leave its mark on the sky? - According to a model based on "loop quantum gravity" theory, a parent universe that existed before ours may have left an imprint (*New Scientist*, 10 April 2008)

Chapter 32

Causal dynamical triangulation

Causal dynamical triangulation (abbreviated as **CDT**) invented by Renate Loll, Jan Ambjørn and Jerzy Jurkiewicz, and popularized by Fotini Markopoulou and Lee Smolin, is an approach to quantum gravity that like loop quantum gravity is background independent. This means that it does not assume any pre-existing arena (dimensional space), but rather attempts to show how the spacetime fabric itself evolves. The Loops '05 conference, hosted by many loop quantum gravity theorists, included several presentations which discussed CDT in great depth, and revealed it to be a pivotal insight for theorists. It has sparked considerable interest as it appears to have a good semi-classical description. At large scales, it re-creates the familiar 4-dimensional spacetime, but it shows spacetime to be 2-d near the Planck scale, and reveals a fractal structure on slices of constant time. These interesting results agree with the findings of Lauscher and Reuter, who use an approach called Quantum Einstein Gravity, and with other recent theoretical work. A brief article appeared in the February 2007 issue of *Scientific American*, which gives an overview of the theory, explained why some physicists are excited about it, and put it in historical perspective. The same publication gives CDT, and its primary authors, a feature article in its July 2008 issue.

32.1 Introduction

Near the Planck scale, the structure of spacetime itself is constantly changing, due to quantum fluctuations. CDT theory uses a triangulation process which varies dynamically and follows deterministic rules, to map out how this can evolve into dimensional spaces similar to that of our universe. The results of researchers suggest that this is a good way to model the early universe, and describe its evolution. Using a structure called a simplex, it divides spacetime into tiny triangular sections. A simplex is the multidimensional analogue of a triangle; a 3-simplex is usually called a tetrahedron, while the 4-simplex, which is the basic building block in this theory, is also known as the pentachoron. Each simplex is geometrically flat, but simplices can be "glued" together in a variety of ways to create curved spacetimes. Where previous attempts at triangulation of quantum spaces have produced jumbled universes with far too many dimensions, or minimal universes with too few, CDT avoids this problem by allowing only those configurations where cause precedes any event. In other words, the timelines of all joined edges of simplices must agree.

32.2 Derivation

CDT is a modification of quantum Regge calculus where spacetime is discretized by approximating it with a piecewise linear manifold in a process called triangulation. In this process, a d-dimensional spacetime is considered as formed by space slices that are labeled by a discrete time variable t. Each space slice is approximated by a simplicial manifold composed by regular $(d-1)$-dimensional simplices and the connection between these slices is made by a piecewise linear manifold of d-simplices. In place of a smooth manifold there is a network of triangulation nodes, where space is locally flat (within each simplex) but globally curved, as with the individual faces and the overall surface of a geodesic dome. The line segments which make up each triangle can represent either a space-like or time-like extent, depending on whether they lie on a given time slice, or connect a vertex at time t with one at time $t+1$. The crucial development, which makes this a relatively successful theory, is that the network of simplices is constrained to evolve in a way that preserves causality. This allows a path integral to be calculated non-perturbatively, by summation of all possible (allowed) configurations of the simplices, and correspondingly, of all possible spatial geometries.

Simply put, each individual simplex is like a building block of spacetime, but the edges that have a time arrow must agree in direction, wherever the edges are joined. This rule preserves causality, a feature missing from previous theories. When simplexes are joined in this way, the manifold evolves in an orderly fashion, and eventually creates the observed framework of dimensions. CDT builds upon the earlier work of Barrett and Crane, and Baez and Barret, which demonstrates the feasibility and utility of this approach, but by introducing the causality constraint as a fundamental rule (influencing the process from the very start) Loll, Ambjørn, and Jurkiewicz created something different. Where others had regarded causality as an emergent property, they made it one of the primary ingredients.

32.3 Advantages and Disadvantages

CDT derives the observed nature and properties of spacetime from a small set of assumptions, without adjusting factors. The idea of deriving what is observed from first principles is very attractive to physicists. CDT models the character of spacetime both in the ultra-microscopic realm near the Planck scale, and at the scale of the cosmos, so CDT may provide insights into the nature of reality.

Evaluation of the observable implications of CDT relies heavily on Monte Carlo simulation by computer. Some feel that this makes CDT an inelegant quantum gravity theory. Also, it has been argued that discrete time-slicing may not accurately reproduce all possible modes of a dynamical system. However, research by Markopoulou and Smolin demonstrates that the cause for those concerns may be limited. Therefore, many physicists still regard this line of reasoning as promising.

32.4 Related theories

CDT has some similarities with loop quantum gravity, especially with its spin foam formulations. For example, the Lorentzian Barrett–Crane model is essentially a non-perturbative prescription for computing path integrals, just like CDT. There are important differences, however. Spin foam formulations of quantum gravity use different degrees of freedom and different Lagrangians. For example, in CDT, the distance, or "the interval", between any two points in a given triangulation can be calculated exactly (triangulations are eigenstates of the distance operator). This is not true for spin foams or loop quantum gravity in general.

Another approach to quantum gravity that is closely related to causal dynamical triangulation is called causal sets. Both CDT and causal sets attempt to model the spacetime with a discrete causal structure. The main difference between the two is that the causal set approach is relatively general, whereas CDT assumes a more specific relationship between the lattice of spacetime events and geometry. Consequently, the Lagrangian of CDT is constrained by the initial assumptions to the extent that it can be written down explicitly and analyzed (see, for example, hep-th/0505154, page 5), whereas there is more freedom in how one might write down an action for causal-set theory.

32.5 See also

- Asymptotic safety in quantum gravity
- Causal sets
- Fractal cosmology
- Loop quantum gravity
- 5-cell
- Planck scale
- Quantum gravity
- Regge calculus
- Simplex

- Simplicial manifold
- Spin foam

32.6 References

- Quantum gravity: progress from an unexpected direction
- Jan Ambjørn, Jerzy Jurkiewicz, and Renate Loll - "The Self-Organizing Quantum Universe", Scientific American, July 2008
- Alpert, Mark "The Triangular Universe" Scientific American page 24, February 2007
- Ambjørn, J.; Jurkiewicz, J.; Loll, R. - Quantum Gravity or the Art of Building Spacetime
- Loll, R.; Ambjørn, J.; Jurkiewicz, J. - The Universe from Scratch - a less technical recent overview
- Loll, R.; Ambjørn, J.; Jurkiewicz, J. - Reconstructing the Universe - a technically detailed overview
- Markopoulou, Fotini; Smolin, Lee - Gauge Fixing in Causal Dynamical Triangulations - shows that varying the time-slice gives similar results

Early papers on the subject:

- R. Loll, *Discrete Lorentzian Quantum Gravity*, arXiv:hep-th/0011194v1 21 Nov 2000
- J Ambjørn, A. Dasgupta, J. Jurkiewicz, and R. Loll, *A Lorentzian cure for Euclidean troubles*, arXiv:hep-th/0201104 v1 14 Jan 2002
- Causal dynamical triangulation on arxiv.org

32.7 External links

- Renate Loll's talk at Loops '05
- John Baez' talk at Loops '05
- Pentatope: from MathWorld
- Simplex: from MathWorld
- Tetrahedron: from MathWorld
- (Re-)Constructing the Universe from Renate Loll's homepage
- Renate Loll on the Quantum Origins of Space and Time as broadcast by TVO

Chapter 33

Lie algebra

"Lie bracket" redirects here. For the operation on vector fields, see Lie bracket of vector fields.

In mathematics, a **Lie algebra** (/'liː/, not /'laɪ/) is a vector space together with a non-associative multiplication called "Lie bracket" $[x, y]$. It was introduced to study the concept of infinitesimal transformations. The term "Lie algebra" (after Sophus Lie) was introduced by Hermann Weyl in the 1930s. In older texts, the name "**infinitesimal group**" is used.

Lie algebras are closely related to Lie groups which are groups that are also smooth manifolds, with the property that the group operations of multiplication and inversion are smooth maps. Any Lie group gives rise to a Lie algebra. Conversely, to any finite-dimensional Lie algebra over real or complex numbers, there is a corresponding connected Lie group unique up to covering (Lie's third theorem). This correspondence between Lie groups and Lie algebras allows one to study Lie groups in terms of Lie algebras.

33.1 Definitions

A **Lie algebra** is a vector space $\mathfrak{g}$ over some field F together with a binary operation $[\cdot, \cdot] : \mathfrak{g} \times \mathfrak{g} \to \mathfrak{g}$ called the **Lie bracket**, which satisfies the following axioms:

- Bilinearity:

$$[ax + by, z] = a[x, z] + b[y, z], \quad [z, ax + by] = a[z, x] + b[z, y]$$

  for all scalars a, b in F and all elements x, y, z in $\mathfrak{g}$.

- Alternating on $\mathfrak{g}$:

$$[x, x] = 0$$

  for all x in $\mathfrak{g}$.

- The Jacobi identity:

$$[x, [y, z]] + [z, [x, y]] + [y, [z, x]] = 0$$

for all x, y, z in $\mathfrak{g}$.

Note that the bilinearity and alternating properties imply anticommutativity, i.e., $[x,y] = -[y,x]$, for all elements x, y in $\mathfrak{g}$, while anticommutativity only implies the alternating property if the field's characteristic is not 2.[1]

It is customary to express a Lie algebra in lower-case fraktur, like $\mathfrak{g}$. If a Lie algebra is associated with a Lie group, then the spelling of the Lie algebra is the same as that Lie group. For example, the Lie algebra of $SU(n)$ is written as $\mathfrak{su}(n)$.

33.1.1 Generators and dimension

Elements of a Lie algebra $\mathfrak{g}$ are said to be **generators** of the Lie algebra if the smallest subalgebra of $\mathfrak{g}$ containing them is $\mathfrak{g}$ itself. The **dimension** of a Lie algebra is its dimension as a vector space over F. The cardinality of a minimal generating set of a Lie algebra is always less than or equal to its dimension.

33.1.2 Homomorphisms, subalgebras, and ideals

The Lie bracket is not associative in general, meaning that $[[x, y], z]$ need not equal $[x, [y, z]]$. Nonetheless, much of the terminology that was developed in the theory of associative rings or associative algebras is commonly applied to Lie algebras. A subspace $\mathfrak{h} \subseteq \mathfrak{g}$ that is closed under the Lie bracket is called a **Lie subalgebra**. If a subspace $I \subseteq \mathfrak{g}$ satisfies a stronger condition that

$$[\mathfrak{g}, I] \subseteq I,$$

then I is called an **ideal** in the Lie algebra $\mathfrak{g}$.[2] A **homomorphism** between two Lie algebras (over the same base field) is a linear map that is compatible with the respective commutators:

$$f : \mathfrak{g} \to \mathfrak{g}', \quad f([x, y]) = [f(x), f(y)],$$

for all elements x and y in $\mathfrak{g}$. As in the theory of associative rings, ideals are precisely the kernels of homomorphisms, given a Lie algebra $\mathfrak{g}$ and an ideal I in it, one constructs the **factor algebra** $\mathfrak{g}/I$, and the first isomorphism theorem holds for Lie algebras.

Let S be a subset of $\mathfrak{g}$. The set of elements x such that $[x, s] = 0$ for all s in S forms a subalgebra called the centralizer of S. The centralizer of $\mathfrak{g}$ itself is called the center of $\mathfrak{g}$. Similar to centralizers, if S is a subspace,[3] then the set of x such that $[x, s]$ is in S for all s in S forms a subalgebra called the normalizer of S.

33.1.3 Direct sum and semidirect product

Given two Lie algebras $\mathfrak{g}$ and $\mathfrak{g}'$, their direct sum is the Lie algebra consisting of the vector space $\mathfrak{g} \oplus \mathfrak{g}'$, of the pairs $(x, x'), x \in \mathfrak{g}, x' \in \mathfrak{g}'$, with the operation

$$[(x, x'), (y, y')] = ([x, y], [x', y']), \quad x, y \in \mathfrak{g}, \ x', y' \in \mathfrak{g}'.$$

Let $\mathfrak{g}$ be a Lie algebra and $\mathfrak{i}$ its ideal. If the canonical map $\mathfrak{g} \to \mathfrak{g}/\mathfrak{i}$ splits (i.e., admits a section), then $\mathfrak{g}$ is said to be a semidirect product of $\mathfrak{i}$ and $\mathfrak{g}/\mathfrak{i}$.

Levi's theorem says that a finite-dimensional Lie algebra is a semidirect product of its radical and the complementary subalgebra (Levi subalgebra).

33.2 Properties

33.2.1 Admits an enveloping algebra

See also: Universal enveloping algebra

For any associative algebra A with multiplication $*$, one can construct a Lie algebra $L(A)$. As a vector space, $L(A)$ is the same as A. The Lie bracket of two elements of $L(A)$ is defined to be their commutator in A:

$$[a, b] = a * b - b * a.$$

The associativity of the multiplication * in A implies the Jacobi identity of the commutator in $L(A)$. For example, the associative algebra of $n \times n$ matrices over a field F gives rise to the general linear Lie algebra $\mathfrak{gl}_n(F)$. The associative algebra A is called an **enveloping algebra** of the Lie algebra $L(A)$. Every Lie algebra can be embedded into one that arises from an associative algebra in this fashion; see universal enveloping algebra.

33.2.2 Representation

Given a vector space V, let $\mathfrak{gl}(V)$ denote the Lie algebra enveloped by the associative algebra of all linear endomorphisms of V. A representation of a Lie algebra $\mathfrak{g}$ on V is a Lie algebra homomorphism

$$\pi : \mathfrak{g} \to \mathfrak{gl}(V).$$

A representation is said to be faithful if its kernel is trivial. Every finite-dimensional Lie algebra has a faithful representation on a finite-dimensional vector space (Ado's theorem).[4]

For example,

$$\mathrm{ad} : \mathfrak{g} \to \mathfrak{gl}(\mathfrak{g})$$

given by $\mathrm{ad}(x)(y) = [x, y]$ is a representation of $\mathfrak{g}$ on the vector space $\mathfrak{g}$ called the adjoint representation. A derivation on the Lie algebra $\mathfrak{g}$ (in fact on any non-associative algebra) is a linear map $\delta : \mathfrak{g} \to \mathfrak{g}$ that obeys the Leibniz' law, that is,

$$\delta([x, y]) = [\delta(x), y] + [x, \delta(y)]$$

for all x and y in the algebra. For any x, $\mathrm{ad}(x)$ is a derivation; a consequence of the Jacobi identity. Thus, the image of ad lies in the subalgebra of $\mathfrak{gl}(\mathfrak{g})$ consisting of derivations on $\mathfrak{g}$. A derivation that happens to be in the image of ad is called an inner derivation. If $\mathfrak{g}$ is semisimple, every derivation on $\mathfrak{g}$ is inner.

33.3 Examples

33.3.1 Vector spaces

- Any vector space V endowed with the identically zero Lie bracket becomes a Lie algebra. Such Lie algebras are called abelian, cf. below. Any one-dimensional Lie algebra over a field is abelian, by the antisymmetry of the Lie bracket.

- The real vector space of all $n \times n$ skew-hermitian matrices is closed under the commutator and forms a real Lie algebra denoted $\mathfrak{u}(n)$. This is the Lie algebra of the unitary group $U(n)$.

33.3.2 Subspaces

- The subspace of the general linear Lie algebra $\mathfrak{gl}_n(F)$ consisting of matrices of trace zero is a subalgebra,[5] the special linear Lie algebra, denoted $\mathfrak{sl}_n(F)$.

33.3.3 Real matrix groups

- Any Lie group G defines an associated real Lie algebra $\mathfrak{g}$ =Lie(G). The definition in general is somewhat technical, but in the case of real matrix groups, it can be formulated via the exponential map, or the matrix exponent. The Lie algebra $\mathfrak{g}$ consists of those matrices X for which $\exp(tX) \in G$, ∀ real numbers t.

   The Lie bracket of $\mathfrak{g}$ is given by the commutator of matrices. As a concrete example, consider the special linear group SL(n,**R**), consisting of all $n \times n$ matrices with real entries and determinant 1. This is a matrix Lie group, and its Lie algebra consists of all $n \times n$ matrices with real entries and trace 0.

33.3.4 Three dimensions

- The three-dimensional Euclidean space $\mathbf{R}^3$ with the Lie bracket given by the cross product of vectors becomes a three-dimensional Lie algebra.

- The Heisenberg algebra $H_3(\mathrm{R})$ is a three-dimensional Lie algebra generated by elements x, y and z with Lie brackets

$$[x, y] = z, \quad [x, z] = 0, \quad [y, z] = 0$$

It is explicitly realized as the space of 3×3 strictly upper-triangular matrices, with the Lie bracket given by the matrix commutator,

$$x = \begin{pmatrix} 0 & 1 & 0 \\ 0 & 0 & 0 \\ 0 & 0 & 0 \end{pmatrix}, \quad y = \begin{pmatrix} 0 & 0 & 0 \\ 0 & 0 & 1 \\ 0 & 0 & 0 \end{pmatrix}, \quad z = \begin{pmatrix} 0 & 0 & 1 \\ 0 & 0 & 0 \\ 0 & 0 & 0 \end{pmatrix}.$$

Any element of the Heisenberg group is thus representable as a product of group generators, i.e., matrix exponentials of these Lie algebra generators,

$$\begin{pmatrix} 1 & a & c \\ 0 & 1 & b \\ 0 & 0 & 1 \end{pmatrix} = e^{by} e^{cz} e^{ax}.$$

- The commutation relations between the x, y, and z components of the angular momentum operator in quantum mechanics are the same as those of $\mathfrak{su}(2)$ and $\mathfrak{so}(3)$,

$$[L_x, L_y] = i\hbar L_z$$
$$[L_y, L_z] = i\hbar L_x$$
$$[L_z, L_x] = i\hbar L_y$$

(The physicist convention for Lie algebras is used in the above equations, hence the factor of i.) The Lie algebra formed by these operators have, in fact, representations of all finite dimensions.

33.3.5 Infinite dimensions

- An important class of infinite-dimensional real Lie algebras arises in differential topology. The space of smooth vector fields on a differentiable manifold M forms a Lie algebra, where the Lie bracket is defined to be the commutator of vector fields. One way of expressing the Lie bracket is through the formalism of Lie derivatives, which identifies a vector field X with a first order partial differential operator LX acting on smooth functions by letting $LX(f)$ be the directional derivative of the function f in the direction of X. The Lie bracket $[X,Y]$ of two vector fields is the vector field defined through its action on functions by the formula:

$$L_{[X,Y]}f = L_X(L_Y f) - L_Y(L_X f).$$

- A Kac–Moody algebra is an example of an infinite-dimensional Lie algebra.

- The Moyal algebra is an infinite-dimensional Lie algebra which contains all classical Lie algebras as subalgebras.

33.4 Structure theory and classification

Lie algebras can be classified to some extent. In particular, this has an application to the classification of Lie groups.

33.4.1 Abelian, nilpotent, and solvable

Analogously to abelian, nilpotent, and solvable groups, defined in terms of the derived subgroups, one can define abelian, nilpotent, and solvable Lie algebras.

A Lie algebra $\mathfrak{g}$ is **abelian** if the Lie bracket vanishes, i.e. $[x,y] = 0$, for all x and y in $\mathfrak{g}$. Abelian Lie algebras correspond to commutative (or abelian) connected Lie groups such as vector spaces K^n or tori T^n, and are all of the form $\mathfrak{k}^n$, meaning an n-dimensional vector space with the trivial Lie bracket.

A more general class of Lie algebras is defined by the vanishing of all commutators of given length. A Lie algebra $\mathfrak{g}$ is **nilpotent** if the lower central series

$$\mathfrak{g} > [\mathfrak{g},\mathfrak{g}] > [[\mathfrak{g},\mathfrak{g}],\mathfrak{g}] > [[[\mathfrak{g},\mathfrak{g}],\mathfrak{g}],\mathfrak{g}] > \cdots$$

becomes zero eventually. By Engel's theorem, a Lie algebra is nilpotent if and only if for every u in $\mathfrak{g}$ the adjoint endomorphism

$$\mathrm{ad}(u) : \mathfrak{g} \to \mathfrak{g}, \quad \mathrm{ad}(u)v = [u,v]$$

is nilpotent.

More generally still, a Lie algebra $\mathfrak{g}$ is said to be **solvable** if the derived series:

$$\mathfrak{g} > [\mathfrak{g},\mathfrak{g}] > [[\mathfrak{g},\mathfrak{g}],[\mathfrak{g},\mathfrak{g}]] > [[[\mathfrak{g},\mathfrak{g}],[\mathfrak{g},\mathfrak{g}]],[[\mathfrak{g},\mathfrak{g}],[\mathfrak{g},\mathfrak{g}]]] > \cdots$$

becomes zero eventually.

Every finite-dimensional Lie algebra has a unique maximal solvable ideal, called its radical. Under the Lie correspondence, nilpotent (respectively, solvable) connected Lie groups correspond to nilpotent (respectively, solvable) Lie algebras.

33.4.2 Simple and semisimple

A Lie algebra is "simple" if it has no non-trivial ideals and is not abelian. A Lie algebra $\mathfrak{g}$ is called **semisimple** if its radical is zero. Equivalently, $\mathfrak{g}$ is semisimple if it does not contain any non-zero abelian ideals. In particular, a simple Lie algebra is semisimple. Conversely, it can be proven that any semisimple Lie algebra is the direct sum of its minimal ideals, which are canonically determined simple Lie algebras.

The concept of semisimplicity for Lie algebras is closely related with the complete reducibility (semisimplicity) of their representations. When the ground field F has characteristic zero, any finite-dimensional representation of a semisimple Lie algebra is semisimple (i.e., direct sum of irreducible representations.) In general, a Lie algebra is called reductive if the adjoint representation is semisimple. Thus, a semisimple Lie algebra is reductive.

33.4.3 Cartan's criterion

Cartan's criterion gives conditions for a Lie algebra to be nilpotent, solvable, or semisimple. It is based on the notion of the Killing form, a symmetric bilinear form on $\mathfrak{g}$ defined by the formula

$$K(u, v) = \mathrm{tr}(\mathrm{ad}(u)\,\mathrm{ad}(v)),$$

where tr denotes the trace of a linear operator. A Lie algebra $\mathfrak{g}$ is semisimple if and only if the Killing form is nondegenerate. A Lie algebra $\mathfrak{g}$ is solvable if and only if $K(\mathfrak{g}, [\mathfrak{g}, \mathfrak{g}]) = 0$.

33.4.4 Classification

The Levi decomposition expresses an arbitrary Lie algebra as a semidirect sum of its solvable radical and a semisimple Lie algebra, almost in a canonical way. Furthermore, semisimple Lie algebras over an algebraically closed field have been completely classified through their root systems. However, the classification of solvable Lie algebras is a 'wild' problem, and cannot be accomplished in general.

33.5 Relation to Lie groups

See also: Lie group–Lie algebra correspondence

Although Lie algebras are often studied in their own right, historically they arose as a means to study Lie groups.

Lie's fundamental theorems describe a relation between Lie groups and Lie algebras. In particular, any Lie group gives rise to a canonically determined Lie algebra (concretely, *the tangent space at the identity*); and, conversely, for any Lie algebra there is a corresponding connected Lie group (Lie's third theorem; see the Baker–Campbell–Hausdorff formula). This Lie group is not determined uniquely; however, any two connected Lie groups with the same Lie algebra are *locally isomorphic*, and in particular, have the same universal cover. For instance, the special orthogonal group SO(3) and the special unitary group SU(2) give rise to the same Lie algebra, which is isomorphic to $\mathbf{R}^3$ with the cross-product, while SU(2) is a simply-connected twofold cover of SO(3).

Given a Lie group, a Lie algebra can be associated to it either by endowing the tangent space to the identity with the differential of the adjoint map, or by considering the left-invariant vector fields as mentioned in the examples. In the case of real matrix groups, the Lie algebra $\mathfrak{g}$ consists of those matrices X for which $\exp(tX) \in G$ for all real numbers t, where exp is the exponential map.

Some examples of Lie algebras corresponding to Lie groups are the following:

- The Lie algebra $\mathfrak{gl}_n(\mathbb{C})$ for the group $\mathrm{GL}_n(\mathbb{C})$ is the algebra of complex $n \times n$ matrices
- The Lie algebra $\mathfrak{sl}_n(\mathbb{C})$ for the group $\mathrm{SL}_n(\mathbb{C})$ is the algebra of complex $n \times n$ matrices with trace 0
- The Lie algebras $\mathfrak{o}(n)$ for the group $\mathrm{O}(n)$ and $\mathfrak{so}(n)$ for $\mathrm{SO}(n)$ are both the algebra of real anti-symmetric $n \times n$ matrices (See Antisymmetric matrix: Infinitesimal rotations for a discussion)
- The Lie algebra $\mathfrak{u}(n)$ for the group $\mathrm{U}(n)$ is the algebra of skew-Hermitian complex $n \times n$ matrices while the Lie algebra $\mathfrak{su}(n)$ for $\mathrm{SU}(n)$ is the algebra of skew-Hermitian, traceless complex $n \times n$ matrices.

In the above examples, the Lie bracket $[X, Y]$ (for X and Y matrices in the Lie algebra) is defined as $[X, Y] = XY - YX$.

Given a set of generators T^a, the **structure constants** f^{abc} express the Lie brackets of pairs of generators as linear combinations of generators from the set, i.e., $[T^a, T^b] = f^{abc} T^c$. The structure constants determine the Lie brackets of elements of the Lie algebra, and consequently nearly completely determine the group structure of the Lie group. The structure of the Lie group near the identity element is displayed explicitly by the Baker–Campbell–Hausdorff formula, an expansion in Lie algebra elements X, Y and their Lie brackets, all nested together within a single exponent, $\exp(tX)\exp(tY) = \exp(tX + tY + \tfrac{1}{2} t^2 [X, Y] + O(t^3))$.

The mapping from Lie groups to Lie algebras is functorial, which implies that homomorphisms of Lie groups lift to homomorphisms of Lie algebras, and various properties are satisfied by this lifting: it commutes with composition, it maps Lie subgroups, kernels, quotients and cokernels of Lie groups to subalgebras, kernels, quotients and cokernels of Lie algebras, respectively.

The functor **L** which takes each Lie group to its Lie algebra and each homomorphism to its differential is faithful and exact. It is however not an equivalence of categories: different Lie groups may have isomorphic Lie algebras (for example SO(3) and SU(2)), and there are (infinite dimensional) Lie algebras that are not associated to any Lie group.[6]

However, when the Lie algebra $\mathfrak{g}$ is finite-dimensional, one can associate to it a simply connected Lie group having $\mathfrak{g}$ as its Lie algebra. More precisely, the Lie algebra functor **L** has a left adjoint functor **Γ** from finite-dimensional (real) Lie algebras to Lie groups, factoring through the full subcategory of simply connected Lie groups.[7] In other words, there is a natural isomorphism of bifunctors

$$\mathrm{Hom}(\Gamma(\mathfrak{g}), H) \cong \mathrm{Hom}(\mathfrak{g}, \mathrm{L}(H)).$$

The adjunction $\mathfrak{g} \to \mathrm{L}(\Gamma(\mathfrak{g}))$ (corresponding to the identity on $\Gamma(\mathfrak{g})$) is an isomorphism, and the other adjunction $\Gamma(\mathrm{L}(H)) \to H$ is the projection homomorphism from the universal cover group of the identity component of H to H. It follows immediately that if G is simply connected, then the Lie algebra functor establishes a bijective correspondence between Lie group homomorphisms $G \to H$ and Lie algebra homomorphisms $\mathbf{L}(G) \to \mathbf{L}(H)$.

The universal cover group above can be constructed as the image of the Lie algebra under the exponential map. More generally, we have that the Lie algebra is homeomorphic to a neighborhood of the identity. But globally, if the Lie group is compact, the exponential will not be injective, and if the Lie group is not connected, simply connected or compact, the exponential map need not be surjective.

If the Lie algebra is infinite-dimensional, the issue is more subtle. In many instances, the exponential map is not even locally a homeomorphism (for example, in Diff($\mathbf{S}^1$), one may find diffeomorphisms arbitrarily close to the identity that are not in the image of exp). Furthermore, some infinite-dimensional Lie algebras are not the Lie algebra of any group.

The correspondence between Lie algebras and Lie groups is used in several ways, including in the classification of Lie groups and the related matter of the representation theory of Lie groups. Every representation of a Lie algebra lifts uniquely to a representation of the corresponding connected, simply connected Lie group, and conversely every representation of any Lie group induces a representation of the group's Lie algebra; the representations are in one to one correspondence. Therefore, knowing the representations of a Lie algebra settles the question of representations of the group.

As for classification, it can be shown that any connected Lie group with a given Lie algebra is isomorphic to the universal cover mod a discrete central subgroup. So classifying Lie groups becomes simply a matter of counting the discrete subgroups of the center, once the classification of Lie algebras is known (solved by Cartan et al. in the semisimple case).

33.6 Category theoretic definition

Using the language of category theory, a **Lie algebra** can be defined as an object A in **Vec**k, the category of vector spaces over a field k of characteristic not 2, together with a morphism $[.,.]: A \otimes A \to A$, where $\otimes$ refers to the monoidal product of **Vec**k, such that

- $[\cdot, \cdot] \circ (\mathrm{id} + \tau_{A,A}) = 0$

- $[\cdot, \cdot] \circ ([\cdot, \cdot] \otimes \mathrm{id}) \circ (\mathrm{id} + \sigma + \sigma^2) = 0$

where $\tau (a \otimes b) := b \otimes a$ and σ is the cyclic permutation braiding $(\mathrm{id} \otimes \tau_{A,A}) \circ (\tau_{A,A} \otimes \mathrm{id})$. In diagrammatic form:

33.7 See also

33.8 Notes

[1] Humphreys p. 1

[2] Due to the anticommutativity of the commutator, the notions of a left and right ideal in a Lie algebra coincide.

[3] Jacobson 1962, pg. 28

[4] Jacobson 1962, Ch. VI

[5] Humphreys p.2

[6] Beltita 2005, pg. 75

[7] Adjoint property is discussed in more general context in Hofman & Morris (2007) (e.g., page 130) but is a straightforward consequence of, e.g., Bourbaki (1989) Theorem 1 of page 305 and Theorem 3 of page 310.

33.9 References

- Beltita, Daniel. *Smooth Homogeneous Structures in Operator Theory*, CRC Press, 2005. ISBN 978-1-4200-3480-6

- Boza, Luis; Fedriani, Eugenio M. & Núñez, Juan. *A new method for classifying complex filiform Lie algebras*, Applied Mathematics and Computation, 121 (2-3): 169–175, 2001

- Bourbaki, Nicolas. "Lie Groups and Lie Algebras - Chapters 1-3", Springer, 1989, ISBN 3-540-64242-0

- Erdmann, Karin & Wildon, Mark. *Introduction to Lie Algebras*, 1st edition, Springer, 2006. ISBN 1-84628-040-0

- Hall, Brian C. *Lie Groups, Lie Algebras, and Representations: An Elementary Introduction*, Springer, 2003. ISBN 0-387-40122-9

- Hofman, Karl & Morris, Sidney. "The Lie Theory of Connected Pro-Lie Groups", European Mathematical Society, 2007, ISBN 978-3-03719-032-6

- Humphreys, James E. *Introduction to Lie Algebras and Representation Theory*, Second printing, revised. Graduate Texts in Mathematics, 9. Springer-Verlag, New York, 1978. ISBN 0-387-90053-5

- Jacobson, Nathan, *Lie algebras*, Republication of the 1962 original. Dover Publications, Inc., New York, 1979. ISBN 0-486-63832-4

- Kac, Victor G. et al. *Course notes for MIT 18.745: Introduction to Lie Algebras*, math.mit.edu

- O'Connor, J.J. & Robertson, E.F. Biography of Sophus Lie, MacTutor History of Mathematics Archive, www-history.mcs.st-andrews.ac.uk

- O'Connor, J.J. & Robertson, E.F. Biography of Wilhelm Killing, MacTutor History of Mathematics Archive, www-history.mcs.st-andrews.ac.uk

- Serre, Jean-Pierre. "Lie Algebras and Lie Groups", 2nd edition, Springer, 2006. ISBN 3-540-55008-9

- Steeb, W.-H. *Continuous Symmetries, Lie Algebras, Differential Equations and Computer Algebra*, second edition, World Scientific, 2007, ISBN 978-981-270-809-0

- Varadarajan, V.S. *Lie Groups, Lie Algebras, and Their Representations*, 1st edition, Springer, 2004. ISBN 0-387-90969-9.

33.10 External links

- Hazewinkel, Michiel, ed. (2001), "Lie algebra", *Encyclopedia of Mathematics*, Springer, ISBN 978-1-55608-010-4

Chapter 34

Lie group

In mathematics, a **Lie group** /ˈliː/ is a group that is also a differentiable manifold, with the property that the group operations are compatible with the smooth structure. Lie groups are named after Sophus Lie, who laid the foundations of the theory of continuous transformation groups. The term *groupes de Lie* first appeared in French in 1893 in the thesis of Lie's student Arthur Tresse, page 3.[1]

Lie groups represent the best-developed theory of continuous symmetry of mathematical objects and structures, which makes them indispensable tools for many parts of contemporary mathematics, as well as for modern theoretical physics. They provide a natural framework for analysing the continuous symmetries of differential equations (differential Galois theory), in much the same way as permutation groups are used in Galois theory for analysing the discrete symmetries of algebraic equations. An extension of Galois theory to the case of continuous symmetry groups was one of Lie's principal motivations.

34.1 Overview

Lie groups are smooth[Note 1] differentiable manifolds and as such can be studied using differential calculus, in contrast with the case of more general topological groups. One of the key ideas in the theory of Lie groups is to replace the *global* object, the group, with its *local* or linearized version, which Lie himself called its "infinitesimal group" and which has since become known as its Lie algebra.

Lie groups play an enormous role in modern geometry, on several different levels. Felix Klein argued in his Erlangen program that one can consider various "geometries" by specifying an appropriate transformation group that leaves certain geometric properties invariant. Thus Euclidean geometry corresponds to the choice of the group E(3) of distance-preserving transformations of the Euclidean space $\mathbf{R}^3$, conformal geometry corresponds to enlarging the group to the conformal group, whereas in projective geometry one is interested in the properties invariant under the projective group. This idea later led to the notion of a G-structure, where *G* is a Lie group of "local" symmetries of a manifold. On a "global" level, whenever a Lie group acts on a geometric object, such as a Riemannian or a symplectic manifold, this action provides a measure of rigidity and yields a rich algebraic structure. The presence of continuous symmetries expressed via a Lie group action on a manifold places strong constraints on its geometry and facilitates analysis on the manifold. Linear actions of Lie groups are especially important, and are studied in representation theory.

In the 1940s–1950s, Ellis Kolchin, Armand Borel, and Claude Chevalley realised that many foundational results concerning Lie groups can be developed completely algebraically, giving rise to the theory of algebraic groups defined over an arbitrary field. This insight opened new possibilities in pure algebra, by providing a uniform construction for most finite simple groups, as well as in algebraic geometry. The theory of automorphic forms, an important branch of modern number theory, deals extensively with analogues of Lie groups over adele rings; p-adic Lie groups play an important role, via their connections with Galois representations in number theory.

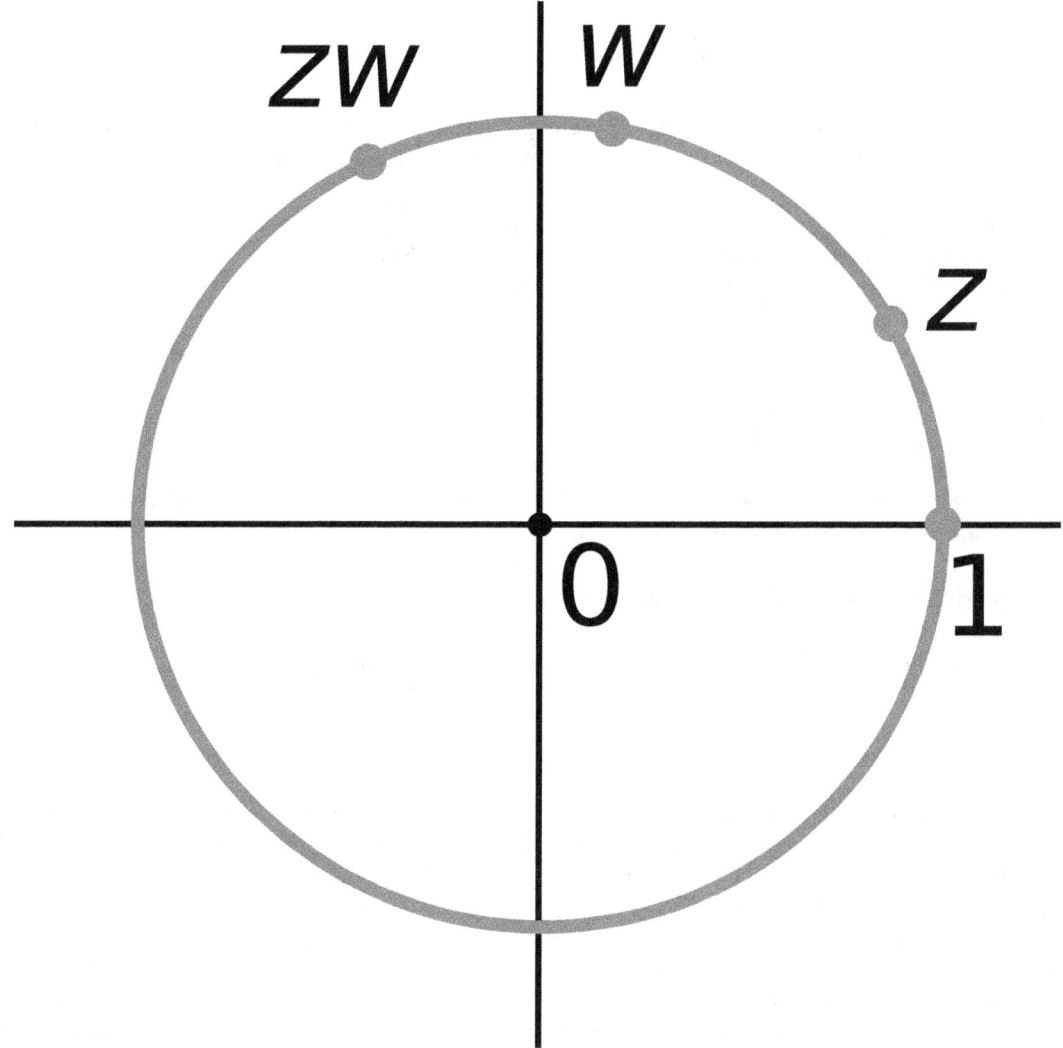

The circle of center 0 and radius 1 in the complex plane is a Lie group with complex multiplication.

34.2 Definitions and examples

A **real Lie group** is a group that is also a finite-dimensional real smooth manifold, in which the group operations of multiplication and inversion are smooth maps. Smoothness of the group multiplication

$$\mu : G \times G \to G \quad \mu(x,y) = xy$$

means that μ is a smooth mapping of the product manifold $G \times G$ into G. These two requirements can be combined to the single requirement that the mapping

$$(x,y) \mapsto x^{-1}y$$

be a smooth mapping of the product manifold into G.

34.2.1 First examples

- The 2×2 real invertible matrices form a group under multiplication, denoted by GL(2, **R**) or by GL2(**R**):

$$\mathrm{GL}(2, \mathbf{R}) = \left\{ A = \begin{pmatrix} a & b \\ c & d \end{pmatrix} : \det A = ad - bc \neq 0 \right\}.$$

This is a four-dimensional noncompact real Lie group. This group is disconnected; it has two connected components corresponding to the positive and negative values of the determinant.

- The rotation matrices form a subgroup of GL(2, **R**), denoted by SO(2, **R**). It is a Lie group in its own right: specifically, a one-dimensional compact connected Lie group which is diffeomorphic to the circle. Using the rotation angle φ as a parameter, this group can be parametrized as follows:

$$\mathrm{SO}(2, \mathbf{R}) = \left\{ \begin{pmatrix} \cos\varphi & -\sin\varphi \\ \sin\varphi & \cos\varphi \end{pmatrix} : \varphi \in \mathbf{R}/2\pi\mathbf{Z} \right\}.$$

Addition of the angles corresponds to multiplication of the elements of SO(2, **R**), and taking the opposite angle corresponds to inversion. Thus both multiplication and inversion are differentiable maps.

- The orthogonal group also forms an interesting example of a Lie group.

All of the previous examples of Lie groups fall within the class of classical groups.

34.2.2 Related concepts

A **complex Lie group** is defined in the same way using complex manifolds rather than real ones (example: SL(2, **C**)), and similarly, using an alternate metric completion of **Q**, one can define a *p*-adic Lie group over the *p*-adic numbers. Hilbert's fifth problem asked whether replacing differentiable manifolds with topological or analytic ones can yield new examples. The answer to this question turned out to be negative: in 1952, Gleason, Montgomery and Zippin showed that if G is a topological manifold with continuous group operations, then there exists exactly one analytic structure on G which turns it into a Lie group (see also Hilbert–Smith conjecture). If the underlying manifold is allowed to be infinite-dimensional (for example, a Hilbert manifold), then one arrives at the notion of an infinite-dimensional Lie group. It is possible to define analogues of many Lie groups over finite fields, and these give most of the examples of finite simple groups.

The language of category theory provides a concise definition for Lie groups: a Lie group is a group object in the category of smooth manifolds. This is important, because it allows generalization of the notion of a Lie group to Lie supergroups.

34.3 More examples of Lie groups

See also: Table of Lie groups and List of simple Lie groups

Lie groups occur in abundance throughout mathematics and physics. Matrix groups or algebraic groups are (roughly) groups of matrices (for example, orthogonal and symplectic groups), and these give most of the more common examples of Lie groups.

34.3.1 Examples with a specific number of dimensions

- The circle group $\mathbf{S}^1$ consisting of angles mod 2π under addition or, alternatively, the complex numbers with absolute value 1 under multiplication. This is a one-dimensional compact connected abelian Lie group.

- The 3-sphere $\mathbf{S}^3$ forms a Lie group by identification with the set of quaternions of unit norm, called versors. The only other spheres that admit the structure of a Lie group are the 0-sphere $\mathbf{S}^0$ (real numbers with absolute value 1) and the circle $\mathbf{S}^1$ (complex numbers with absolute value 1). For example, for even $n > 1$, $\mathbf{S}^n$ is not a

34.3. MORE EXAMPLES OF LIE GROUPS

Lie group because it does not admit a nonvanishing vector field and so *a fortiori* cannot be parallelizable as a differentiable manifold. Of the spheres only S^0, S^1, S^3, and S^7 are parallelizable. The last carries the structure of a Lie quasigroup (a nonassociative group), which can be identified with the set of unit octonions.

- The (3-dimensional) metaplectic group is a double cover of SL(2, **R**) playing an important role in the theory of modular forms. It is a connected Lie group that cannot be faithfully represented by matrices of finite size, i.e., a nonlinear group.

- The Heisenberg group is a connected nilpotent Lie group of dimension 3, playing a key role in quantum mechanics.

- The Lorentz group is a 6-dimensional Lie group of linear isometries of the Minkowski space.

- The Poincaré group is a 10-dimensional Lie group of affine isometries of the Minkowski space.

- The group U(1)×SU(2)×SU(3) is a Lie group of dimension 1+3+8=12 that is the gauge group of the Standard Model in particle physics. The dimensions of the factors correspond to the 1 photon + 3 vector bosons + 8 gluons of the standard model

- The exceptional Lie groups of types G_2, F_4, E_6, E_7, E_8 have dimensions 14, 52, 78, 133, and 248. Along with the A-B-C-D series of simple Lie groups, the exceptional groups complete the list of simple Lie groups. There is also a Lie group named $E_7\frac{1}{2}$ of dimension 190, but it is not a *simple* Lie group.

34.3.2 Examples with n dimensions

- Euclidean space **R**^{n} with ordinary vector addition as the group operation becomes an n-dimensional noncompact abelian Lie group.

- The Euclidean group E(n, **R**) is the Lie group of all Euclidean motions, i.e., isometric affine maps, of n-dimensional Euclidean space **R**^{n}.

- The orthogonal group O(n, **R**), consisting of all $n \times n$ orthogonal matrices with real entries is an $n(n-1)/2$-dimensional Lie group. This group is disconnected, but it has a connected subgroup SO(n, **R**) of the same dimension consisting of orthogonal matrices of determinant 1, called the special orthogonal group (for $n = 3$, the rotation group SO(3)).

- The unitary group U(n) consisting of $n \times n$ unitary matrices (with complex entries) is a compact connected Lie group of dimension n^2. Unitary matrices of determinant 1 form a closed connected subgroup of dimension $n^2 - 1$ denoted SU(n), the special unitary group.

- Spin groups are double covers of the special orthogonal groups, used for studying fermions in quantum field theory (among other things).

- The group GL(n, **R**) of invertible matrices (under matrix multiplication) is a Lie group of dimension n^2, called the general linear group. It has a closed connected subgroup SL(n, **R**), the special linear group, consisting of matrices of determinant 1 which is also a Lie group.

- The symplectic group Sp($2n$, **R**) consists of all $2n \times 2n$ matrices preserving a *symplectic form* on **R**^{2n}. It is a connected Lie group of dimension $2n^2 + n$.

- The group of invertible upper triangular n by n matrices is a solvable Lie group of dimension $n(n+1)/2$. (cf. Borel subgroup)

- The A-series, B-series, C-series and D-series, whose elements are denoted by An, Bn, Cn, and Dn, are infinite families of simple Lie groups.

34.3.3 Constructions

There are several standard ways to form new Lie groups from old ones:

- The product of two Lie groups is a Lie group.

- Any topologically closed subgroup of a Lie group is a Lie group. This is known as the Closed subgroup theorem or **Cartan's theorem**.

- The quotient of a Lie group by a closed normal subgroup is a Lie group.

- The universal cover of a connected Lie group is a Lie group. For example, the group **R** is the universal cover of the circle group S^1. In fact any covering of a differentiable manifold is also a differentiable manifold, but by specifying *universal* cover, one guarantees a group structure (compatible with its other structures).

34.3.4 Related notions

Some examples of groups that are *not* Lie groups (except in the trivial sense that any group can be viewed as a 0-dimensional Lie group, with the discrete topology), are:

- Infinite-dimensional groups, such as the additive group of an infinite-dimensional real vector space. These are not Lie groups as they are not *finite-dimensional* manifolds

- Some totally disconnected groups, such as the Galois group of an infinite extension of fields, or the additive group of the *p*-adic numbers. These are not Lie groups because their underlying spaces are not real manifolds. (Some of these groups are "*p*-adic Lie groups"). In general, only topological groups having similar local properties to R^n for some positive integer n can be Lie groups (of course they must also have a differentiable structure)

34.4 Early history

According to the most authoritative source on the early history of Lie groups (Hawkins, p. 1), Sophus Lie himself considered the winter of 1873–1874 as the birth date of his theory of continuous groups. Hawkins, however, suggests that it was "Lie's prodigious research activity during the four-year period from the fall of 1869 to the fall of 1873" that led to the theory's creation (*ibid*). Some of Lie's early ideas were developed in close collaboration with Felix Klein. Lie met with Klein every day from October 1869 through 1872: in Berlin from the end of October 1869 to the end of February 1870, and in Paris, Göttingen and Erlangen in the subsequent two years (*ibid*, p. 2). Lie stated that all of the principal results were obtained by 1884. But during the 1870s all his papers (except the very first note) were published in Norwegian journals, which impeded recognition of the work throughout the rest of Europe (*ibid*, p. 76). In 1884 a young German mathematician, Friedrich Engel, came to work with Lie on a systematic treatise to expose his theory of continuous groups. From this effort resulted the three-volume *Theorie der Transformationsgruppen*, published in 1888, 1890, and 1893.

Lie's ideas did not stand in isolation from the rest of mathematics. In fact, his interest in the geometry of differential equations was first motivated by the work of Carl Gustav Jacobi, on the theory of partial differential equations of first order and on the equations of classical mechanics. Much of Jacobi's work was published posthumously in the 1860s, generating enormous interest in France and Germany (Hawkins, p. 43). Lie's *idée fixe* was to develop a theory of symmetries of differential equations that would accomplish for them what Évariste Galois had done for algebraic equations: namely, to classify them in terms of group theory. Lie and other mathematicians showed that the most important equations for special functions and orthogonal polynomials tend to arise from group theoretical symmetries. In Lie's early work, the idea was to construct a theory of *continuous groups*, to complement the theory of discrete groups that had developed in the theory of modular forms, in the hands of Felix Klein and Henri Poincaré. The initial application that Lie had in mind was to the theory of differential equations. On the model of Galois theory and polynomial equations, the driving conception was of a theory capable of unifying, by the study of symmetry, the whole area of ordinary differential equations. However, the hope that Lie Theory would unify the entire field of ordinary differential equations was not fulfilled. Symmetry methods for ODEs continue to be studied, but do not dominate the subject. There is a differential Galois theory, but it was developed by others, such as Picard and Vessiot, and it provides a theory of quadratures, the indefinite integrals required to express solutions.

Additional impetus to consider continuous groups came from ideas of Bernhard Riemann, on the foundations of geometry, and their further development in the hands of Klein. Thus three major themes in 19th century mathematics were combined by Lie in creating his new theory: the idea of symmetry, as exemplified by Galois through the algebraic notion of a group; geometric theory and the explicit solutions of differential equations of mechanics, worked out by

Poisson and Jacobi; and the new understanding of geometry that emerged in the works of Plücker, Möbius, Grassmann and others, and culminated in Riemann's revolutionary vision of the subject.

Although today Sophus Lie is rightfully recognized as the creator of the theory of continuous groups, a major stride in the development of their structure theory, which was to have a profound influence on subsequent development of mathematics, was made by Wilhelm Killing, who in 1888 published the first paper in a series entitled *Die Zusammensetzung der stetigen endlichen Transformationsgruppen* (*The composition of continuous finite transformation groups*) (Hawkins, p. 100). The work of Killing, later refined and generalized by Élie Cartan, led to classification of semisimple Lie algebras, Cartan's theory of symmetric spaces, and Hermann Weyl's description of representations of compact and semisimple Lie groups using highest weights.

In 1900 David Hilbert challenged Lie theorists with his Fifth Problem presented at the International Congress of Mathematicians in Paris.

Weyl brought the early period of the development of the theory of Lie groups to fruition, for not only did he classify irreducible representations of semisimple Lie groups and connect the theory of groups with quantum mechanics, but he also put Lie's theory itself on firmer footing by clearly enunciating the distinction between Lie's *infinitesimal groups* (i.e., Lie algebras) and the Lie groups proper, and began investigations of topology of Lie groups.[2] The theory of Lie groups was systematically reworked in modern mathematical language in a monograph by Claude Chevalley.

34.5 The concept of a Lie group, and possibilities of classification

Lie groups may be thought of as smoothly varying families of symmetries. Examples of symmetries include rotation about an axis. What must be understood is the nature of 'small' transformations, e.g., rotations through tiny angles, that link nearby transformations. The mathematical object capturing this structure is called a Lie algebra (Lie himself called them "infinitesimal groups"). It can be defined because Lie groups are manifolds, so have tangent spaces at each point.

The Lie algebra of any compact Lie group (very roughly: one for which the symmetries form a bounded set) can be decomposed as a direct sum of an abelian Lie algebra and some number of simple ones. The structure of an abelian Lie algebra is mathematically uninteresting (since the Lie bracket is identically zero); the interest is in the simple summands. Hence the question arises: what are the simple Lie algebras of compact groups? It turns out that they mostly fall into four infinite families, the "classical Lie algebras" A_n, B_n, C_n and D_n, which have simple descriptions in terms of symmetries of Euclidean space. But there are also just five "exceptional Lie algebras" that do not fall into any of these families. E_8 is the largest of these.

34.6 Properties

- The diffeomorphism group of a Lie group acts transitively on the Lie group
- Every Lie group is parallelizable, and hence an orientable manifold (there is a bundle isomorphism between its tangent bundle and the product of itself with the tangent space at the identity)

34.7 Types of Lie groups and structure theory

Lie groups are classified according to their algebraic properties (simple, semisimple, solvable, nilpotent, abelian), their connectedness (connected or simply connected) and their compactness.

- Compact Lie groups are all known: they are finite central quotients of a product of copies of the circle group S^1 and simple compact Lie groups (which correspond to connected Dynkin diagrams).
- Any simply connected solvable Lie group is isomorphic to a closed subgroup of the group of invertible upper triangular matrices of some rank, and any finite-dimensional irreducible representation of such a group is 1-dimensional. Solvable groups are too messy to classify except in a few small dimensions.
- Any simply connected nilpotent Lie group is isomorphic to a closed subgroup of the group of invertible upper triangular matrices with 1's on the diagonal of some rank, and any finite-dimensional irreducible representation

of such a group is 1-dimensional. Like solvable groups, nilpotent groups are too messy to classify except in a few small dimensions.

- Simple Lie groups are sometimes defined to be those that are simple as abstract groups, and sometimes defined to be connected Lie groups with a simple Lie algebra. For example, SL(2, **R**) is simple according to the second definition but not according to the first. They have all been classified (for either definition).

- Semisimple Lie groups are Lie groups whose Lie algebra is a product of simple Lie algebras.[3] They are central extensions of products of simple Lie groups.

The identity component of any Lie group is an open normal subgroup, and the quotient group is a discrete group. The universal cover of any connected Lie group is a simply connected Lie group, and conversely any connected Lie group is a quotient of a simply connected Lie group by a discrete normal subgroup of the center. Any Lie group G can be decomposed into discrete, simple, and abelian groups in a canonical way as follows. Write

G_{con} for the connected component of the identity

G_{sol} for the largest connected normal solvable subgroup

G_{nil} for the largest connected normal nilpotent subgroup

so that we have a sequence of normal subgroups

$1 \subseteq G_{nil} \subseteq G_{sol} \subseteq G_{con} \subseteq G.$

Then

G/G_{con} is discrete

G_{con}/G_{sol} is a central extension of a product of simple connected Lie groups.

G_{sol}/G_{nil} is abelian. A connected abelian Lie group is isomorphic to a product of copies of **R** and the circle group S^1.

$G_{nil}/1$ is nilpotent, and therefore its ascending central series has all quotients abelian.

This can be used to reduce some problems about Lie groups (such as finding their unitary representations) to the same problems for connected simple groups and nilpotent and solvable subgroups of smaller dimension.

34.8 The Lie algebra associated with a Lie group

Main article: Lie group–Lie algebra correspondence

To every Lie group we can associate a Lie algebra whose underlying vector space is the tangent space of the Lie group at the identity element and which completely captures the local structure of the group. Informally we can think of elements of the Lie algebra as elements of the group that are "infinitesimally close" to the identity, and the Lie bracket is related to the commutator of two such infinitesimal elements. Before giving the abstract definition we give a few examples:

- The Lie algebra of the vector space $\mathbf{R}^n$ is just $\mathbf{R}^n$ with the Lie bracket given by
  $[A, B] = 0.$
  (In general the Lie bracket of a connected Lie group is always 0 if and only if the Lie group is abelian.)

- The Lie algebra of the general linear group GL(n, **R**) of invertible matrices is the vector space M(n, **R**) of square matrices with the Lie bracket given by
  $[A, B] = AB - BA.$
  If G is a closed subgroup of GL(n, **R**) then the Lie algebra of G can be thought of informally as the matrices m of M(n, **R**) such that $1 + \varepsilon m$ is in G, where ε is an infinitesimal positive number with $\varepsilon^2 = 0$ (of course, no such real number ε exists). For example, the orthogonal group O(n, **R**) consists of matrices A with $AA^T = 1$, so the Lie algebra consists of the matrices m with $(1 + \varepsilon m)(1 + \varepsilon m)^T = 1$, which is equivalent to $m + m^T = 0$ because $\varepsilon^2 = 0$.

- Formally, when working over the reals, as here, this is accomplished by considering the limit as $\varepsilon \to 0$; but the "infinitesimal" language generalizes directly to Lie groups over general rings.

The concrete definition given above is easy to work with, but has some minor problems: to use it we first need to represent a Lie group as a group of matrices, but not all Lie groups can be represented in this way, and it is not obvious that the Lie algebra is independent of the representation we use. To get around these problems we give the general definition of the Lie algebra of a Lie group (in 4 steps):

1. Vector fields on any smooth manifold M can be thought of as derivations X of the ring of smooth functions on the manifold, and therefore form a Lie algebra under the Lie bracket $[X, Y] = XY - YX$, because the Lie bracket of any two derivations is a derivation.

2. If G is any group acting smoothly on the manifold M, then it acts on the vector fields, and the vector space of vector fields fixed by the group is closed under the Lie bracket and therefore also forms a Lie algebra.

3. We apply this construction to the case when the manifold M is the underlying space of a Lie group G, with G acting on $G = M$ by left translations $L_g(h) = gh$. This shows that the space of left invariant vector fields (vector fields satisfying $L_g*X_h = X_{gh}$ for every h in G, where L_g* denotes the differential of L_g) on a Lie group is a Lie algebra under the Lie bracket of vector fields.

4. Any tangent vector at the identity of a Lie group can be extended to a left invariant vector field by left translating the tangent vector to other points of the manifold. Specifically, the left invariant extension of an element v of the tangent space at the identity is the vector field defined by $v^\wedge g = L_g*v$. This identifies the tangent space T_eG at the identity with the space of left invariant vector fields, and therefore makes the tangent space at the identity into a Lie algebra, called the Lie algebra of G, usually denoted by a Fraktur $\mathfrak{g}$. Thus the Lie bracket on $\mathfrak{g}$ is given explicitly by $[v, w] = [v^\wedge, w^\wedge]e$.

This Lie algebra $\mathfrak{g}$ is finite-dimensional and it has the same dimension as the manifold G. The Lie algebra of G determines G up to "local isomorphism", where two Lie groups are called **locally isomorphic** if they look the same near the identity element. Problems about Lie groups are often solved by first solving the corresponding problem for the Lie algebras, and the result for groups then usually follows easily. For example, simple Lie groups are usually classified by first classifying the corresponding Lie algebras.

We could also define a Lie algebra structure on T_e using right invariant vector fields instead of left invariant vector fields. This leads to the same Lie algebra, because the inverse map on G can be used to identify left invariant vector fields with right invariant vector fields, and acts as -1 on the tangent space T_e.

The Lie algebra structure on T_e can also be described as follows: the commutator operation

$$(x, y) \to xyx^{-1}y^{-1}$$

on $G \times G$ sends (e, e) to e, so its derivative yields a bilinear operation on T_eG. This bilinear operation is actually the zero map, but the second derivative, under the proper identification of tangent spaces, yields an operation that satisfies the axioms of a Lie bracket, and it is equal to twice the one defined through left-invariant vector fields.

34.9 Homomorphisms and isomorphisms

If G and H are Lie groups, then a Lie group homomorphism $f : G \to H$ is a smooth group homomorphism. In the case of complex Lie groups, such a homomorphism is required to be a holomorphic map. However, these requirements are a bit stringent; over real or complex numbers, every continuous homomorphism between Lie groups turns out to be (real or complex) analytic.

The composition of two Lie homomorphisms is again a homomorphism, and the class of all Lie groups, together with these morphisms, forms a category. Moreover, every Lie group homomorphism induces a homomorphism between the corresponding Lie algebras. Let $\phi: G \to H$ be a Lie group homomorphism and let ϕ_* be its derivative at the identity. If we identify the Lie algebras of G and H with their tangent spaces at the identity elements then ϕ_* is a map between the corresponding Lie algebras:

$$\phi_* : \mathfrak{g} \to \mathfrak{h}$$

One can show that ϕ_* is actually a Lie algebra homomorphism (meaning that it is a linear map which preserves the Lie bracket). In the language of category theory, we then have a covariant functor from the category of Lie groups to the category of Lie algebras which sends a Lie group to its Lie algebra and a Lie group homomorphism to its derivative at the identity.

Two Lie groups are called *isomorphic* if there exists a bijective homomorphism between them whose inverse is also a Lie group homomorphism. Equivalently, it is a diffeomorphism which is also a group homomorphism.

Ado's theorem says every finite-dimensional Lie algebra is isomorphic to a matrix Lie algebra. For every finite-dimensional matrix Lie algebra, there is a linear group (matrix Lie group) with this algebra as its Lie algebra. So every abstract Lie algebra is the Lie algebra of some (linear) Lie group.

The *global structure* of a Lie group is not determined by its Lie algebra; for example, if Z is any discrete subgroup of the center of G then G and G/Z have the same Lie algebra (see the table of Lie groups for examples). A *connected* Lie group is simple, semisimple, solvable, nilpotent, or abelian if and only if its Lie algebra has the corresponding property.

If we require that the Lie group be simply connected, then the global structure is determined by its Lie algebra: for every finite-dimensional Lie algebra $\mathfrak{g}$ over $\mathbf{F}$ there is a simply connected Lie group G with $\mathfrak{g}$ as Lie algebra, unique up to isomorphism. Moreover every homomorphism between Lie algebras lifts to a unique homomorphism between the corresponding simply connected Lie groups.

34.10 The exponential map

Main article: Exponential map

The exponential map from the Lie algebra $M(n, \mathbf{R})$ of the general linear group $GL(n, \mathbf{R})$ to $GL(n, \mathbf{R})$ is defined by the usual power series:

$$\exp(A) = 1 + A + \frac{A^2}{2!} + \frac{A^3}{3!} + \cdots$$

for matrices A. If G is any subgroup of $GL(n, \mathbf{R})$, then the exponential map takes the Lie algebra of G into G, so we have an exponential map for all matrix groups.

The definition above is easy to use, but it is not defined for Lie groups that are not matrix groups, and it is not clear that the exponential map of a Lie group does not depend on its representation as a matrix group. We can solve both problems using a more abstract definition of the exponential map that works for all Lie groups, as follows.

Every vector v in $\mathfrak{g}$ determines a linear map from $\mathbf{R}$ to $\mathfrak{g}$ taking 1 to v, which can be thought of as a Lie algebra homomorphism. Because $\mathbf{R}$ is the Lie algebra of the simply connected Lie group $\mathbf{R}$, this induces a Lie group homomorphism $c : \mathbf{R} \to G$ so that

$$c(s+t) = c(s)c(t)$$

for all s and t. The operation on the right hand side is the group multiplication in G. The formal similarity of this formula with the one valid for the exponential function justifies the definition

$$\exp(v) = c(1).$$

This is called the *exponential map*, and it maps the Lie algebra $\mathfrak{g}$ into the Lie group G. It provides a diffeomorphism between a neighborhood of 0 in $\mathfrak{g}$ and a neighborhood of e in G. This exponential map is a generalization of the exponential function for real numbers (because $\mathbf{R}$ is the Lie algebra of the Lie group of positive real numbers with multiplication), for complex numbers (because $\mathbf{C}$ is the Lie algebra of the Lie group of non-zero complex numbers with multiplication) and for matrices (because $M(n, \mathbf{R})$ with the regular commutator is the Lie algebra of the Lie group $GL(n, \mathbf{R})$ of all invertible matrices).

Because the exponential map is surjective on some neighbourhood N of e, it is common to call elements of the Lie algebra **infinitesimal generators** of the group G. The subgroup of G generated by N is the identity component of G.

34.10. THE EXPONENTIAL MAP

The exponential map and the Lie algebra determine the *local group structure* of every connected Lie group, because of the Baker–Campbell–Hausdorff formula: there exists a neighborhood U of the zero element of $\mathfrak{g}$, such that for u, v in U we have

$$\exp(u)\exp(v) = \exp\left(u + v + \tfrac{1}{2}[u,v] + \tfrac{1}{12}[[u,v],v] - \tfrac{1}{12}[[u,v],u] - \cdots\right),$$

where the omitted terms are known and involve Lie brackets of four or more elements. In case u and v commute, this formula reduces to the familiar exponential law $\exp(u)\exp(v) = \exp(u+v)$.

The exponential map relates Lie group homomorphisms. That is, if $\phi : G \to H$ is a Lie group homomorphism and $\phi_* : \mathfrak{g} \to \mathfrak{h}$ the induced map on the corresponding Lie algebras, then for all $x \in \mathfrak{g}$ we have

$$\phi(\exp(x)) = \exp(\phi_*(x)).$$

In other words the following diagram commutes:

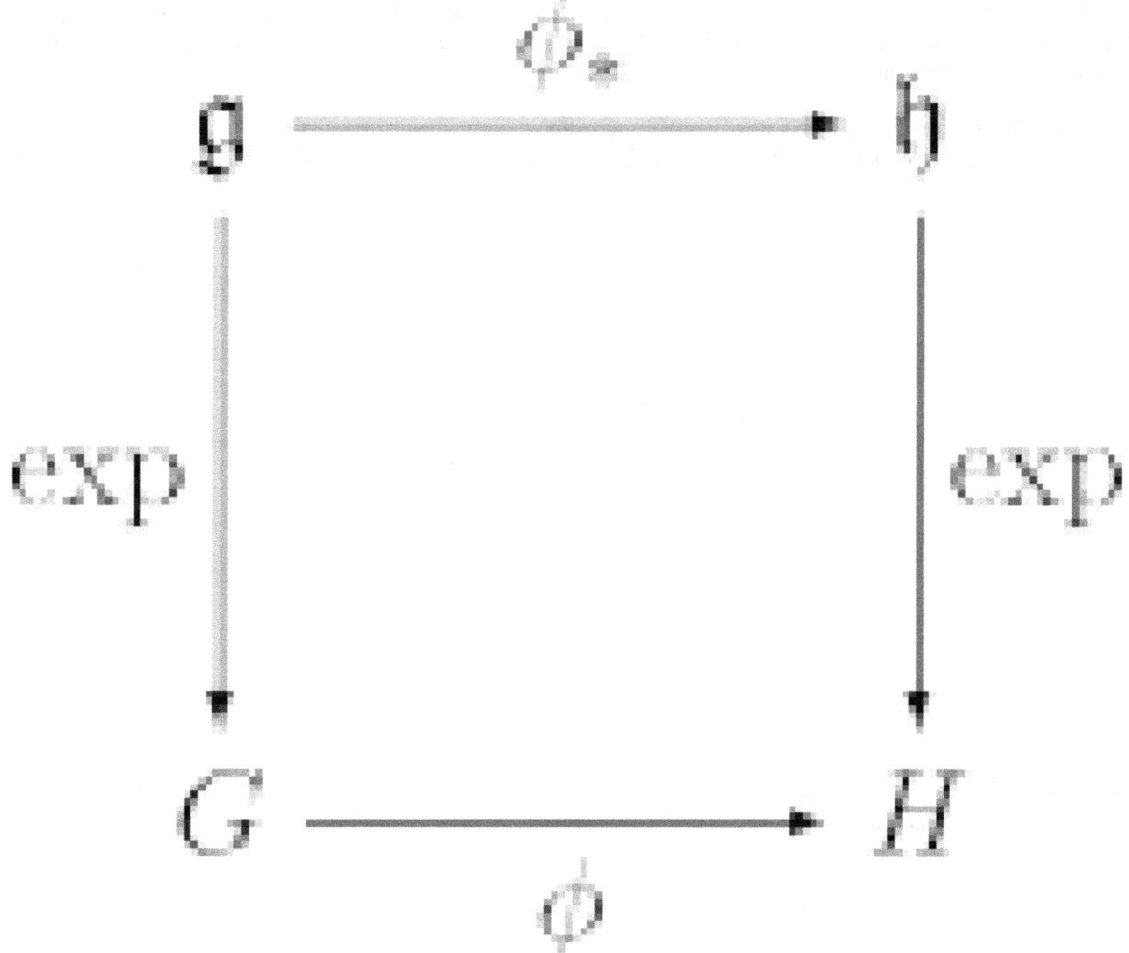

[4] (In short, exp is a natural transformation from the functor Lie to the identity functor on the category of Lie groups.)

The exponential map from the Lie algebra to the Lie group is not always onto, even if the group is connected (though it does map onto the Lie group for connected groups that are either compact or nilpotent). For example, the exponential map of SL(2, **R**) is not surjective. Also, exponential map is not surjective nor injective for infinite-dimensional (see below) Lie groups modelled on C^∞ Fréchet space, even from arbitrary small neighborhood of 0 to corresponding neighborhood of 1.

34.11 Infinite-dimensional Lie groups

Lie groups are often defined to be finite-dimensional, but there are many groups that resemble Lie groups, except for being infinite-dimensional. The simplest way to define infinite-dimensional Lie groups is to model them on Banach spaces, and in this case much of the basic theory is similar to that of finite-dimensional Lie groups. However this is inadequate for many applications, because many natural examples of infinite-dimensional Lie groups are not Banach manifolds. Instead one needs to define Lie groups modeled on more general locally convex topological vector spaces. In this case the relation between the Lie algebra and the Lie group becomes rather subtle, and several results about finite-dimensional Lie groups no longer hold.

Some of the examples that have been studied include:

- The group of diffeomorphisms of a manifold. Quite a lot is known about the group of diffeomorphisms of the circle. Its Lie algebra is (more or less) the Witt algebra, which has a central extension called the Virasoro algebra, used in string theory and conformal field theory. Very little is known about the diffeomorphism groups of manifolds of larger dimension. The diffeomorphism group of spacetime sometimes appears in attempts to quantize gravity.

- The group of smooth maps from a manifold to a finite-dimensional Lie group is an example of a gauge group (with operation of pointwise multiplication), and is used in quantum field theory and Donaldson theory. If the manifold is a circle these are called loop groups, and have central extensions whose Lie algebras are (more or less) Kac–Moody algebras.

- There are infinite-dimensional analogues of general linear groups, orthogonal groups, and so on. One important aspect is that these may have *simpler* topological properties: see for example Kuiper's theorem. In M-Theory theory, for example, a 10 dimensional SU(N) gauge theory becomes an 11 dimensional theory when N becomes infinite.

- A specific example is that $SU(\infty)$ is equal to the group of area preserving diffeomorphisms of a torus.

34.12 See also

- Lie subgroup
- E_8
- Adjoint representation of a Lie group
- Adjoint endomorphism
- Haar measure
- Homogeneous space
- List of Lie group topics
- List of simple Lie groups
- Moufang polygon
- Riemannian manifold
- Representations of Lie groups
- Table of Lie groups
- Lie algebra
- Symmetry in quantum mechanics
- Lie group action

34.13 References

[1] Arthur Tresse (1893). "Sur les invariants différentiels des groupes continus de transformations". *Acta Mathematica* **18**: 1–88. doi:10.1007/bf02418270.

[2] Borel (2001).

[3] Helgason, Sigurdur (1978). *Differential Geometry, Lie Groups, and Symmetric Spaces*. New York: Academic Press. p. 131. ISBN 0-12-338460-5.

[4] http://www.math.sunysb.edu/~{}vkiritch/MAT552/ProblemSet1.pdf

34.14 Notes

[1] having derivatives of all orders

34.15 References

- Adams, John Frank (1969), *Lectures on Lie Groups*, Chicago Lectures in Mathematics, Chicago: Univ. of Chicago Press, ISBN 0-226-00527-5.

- Borel, Armand (2001), *Essays in the history of Lie groups and algebraic groups*, History of Mathematics **21**, Providence, R.I.: American Mathematical Society, ISBN 978-0-8218-0288-5, MR 1847105

- Bourbaki, Nicolas, *Elements of mathematics: Lie groups and Lie algebras.* Chapters 1–3 ISBN 3-540-64242-0, Chapters 4–6 ISBN 3-540-42650-7, Chapters 7–9 ISBN 3-540-43405-4

- Chevalley, Claude (1946), *Theory of Lie groups*, Princeton: Princeton University Press, ISBN 0-691-04990-4.

- P. M. Cohn (1957) *Lie Groups*, Cambridge Tracts in Mathematical Physics.

- J. L. Coolidge (1940) *A History of Geometrical Methods*, pp 304–17, Oxford University Press (Dover Publications 2003).

- Fulton, William; Harris, Joe (1991), *Representation theory. A first course*, Graduate Texts in Mathematics, Readings in Mathematics **129**, New York: Springer-Verlag, ISBN 978-0-387-97495-8, MR 1153249, ISBN 978-0-387-97527-6

- Robert Gilmore (2008) *Lie groups, physics, and geometry: an introduction for physicists, engineers and chemists*, Cambridge University Press ISBN 9780521884006 .

- Hall, Brian C. (2003), *Lie Groups, Lie Algebras, and Representations: An Elementary Introduction*, Springer, ISBN 0-387-40122-9.

- F. Reese Harvey (1990) *Spinors and calibrations*, Academic Press, ISBN 0-12-329650-1 .

- Hawkins, Thomas (2000), *Emergence of the theory of Lie groups*, Sources and Studies in the History of Mathematics and Physical Sciences, Berlin, New York: Springer-Verlag, ISBN 978-0-387-98963-1, MR 1771134 Borel's review

- Knapp, Anthony W. (2002), *Lie Groups Beyond an Introduction*, Progress in Mathematics **140** (2nd ed.), Boston: Birkhäuser, ISBN 0-8176-4259-5.

- Nijenhuis, Albert (1959). "Review: *Lie groups*, by P. M. Cohn". *Bulletin of the American Mathematical Society* **65** (6): 338–341. doi:10.1090/s0002-9904-1959-10358-x.

- Rossmann, Wulf (2001), *Lie Groups: An Introduction Through Linear Groups*, Oxford Graduate Texts in Mathematics, Oxford University Press, ISBN 978-0-19-859683-7. The 2003 reprint corrects several typographical mistakes.

- Sattinger, David H.; Weaver, O. L. (1986). *Lie groups and algebras with applications to physics, geometry, and mechanics*. Springer-Verlag. ISBN 3-540-96240-9.

- Serre, Jean-Pierre (1965), *Lie Algebras and Lie Groups: 1964 Lectures given at Harvard University*, Lecture notes in mathematics **1500**, Springer, ISBN 3-540-55008-9.

- Stillwell, John (2008). *Naive Lie Theory*. Springer. ISBN 0-387-98289-2.

- Heldermann Verlag Journal of Lie Theory

- Steeb, Willi-Hans (2007), *Continuous Symmetries, Lie algebras, Differential Equations and Computer Algebra: second edition*, World Scientific Publishing, ISBN 981-270-809-X.

- Lie Groups. Representation Theory and Symmetric Spaces Wolfgang Ziller, Vorlesung 2010

Chapter 35

Heterotic string theory

This article is about string theory. For heterosis in biology, see Heterosis.

In string theory, a **heterotic string** is a closed string (or loop) which is a hybrid ('heterotic') of a superstring and a bosonic string. There are two kinds of heterotic string, the Heterotic type SO(32) and the Heterotic type $E_8 \times E_8$, abbreviated to HO and HE. Heterotic string theory was first developed in 1985 by David Gross, Jeffrey Harvey, Emil Martinec, and Ryan Rohm (the so-called "Princeton String Quartet"[1]), in one of the key papers that fueled the first superstring revolution.

In string theory, the left-moving and the right-moving excitations are completely decoupled,[2] and it is possible to construct a string theory whose left-moving (counter-clockwise) excitations are treated as a bosonic string propagating in $D = 26$ dimensions, while the right-moving (clock-wise) excitations are treated as a superstring in $D = 10$ dimensions.

The mismatched 16 dimensions must be compactified on an even, self-dual lattice (a discrete subgroup of a linear space). There are two possible even self-dual lattices in 16 dimensions, and it leads to two types of the heterotic string. They differ by the gauge group in 10 dimensions. One gauge group is SO(32) (the HO string) while the other is $E_8 \times E_8$ (the HE string).[3]

These two gauge groups also turned out to be the only two anomaly-free gauge groups that can be coupled to the $N = 1$ supergravity in 10 dimensions other than $U(1)^{496}$ and $E_8 \times U(1)^{248}$, which is suspected to lie in the swampland.

Every heterotic string must be a closed string, not an open string; it is not possible to define any boundary conditions that would relate the left-moving and the right-moving excitations because they have a different character.

A heterotic string is embedded in the membrane that creates harmonics on the string which translate into mass and energy through mechanisms discussed above.

35.1 String duality

String duality is a class of symmetries in physics that link different string theories. In the 1990s, it was realized that the strong coupling limit of the HO theory is type I string theory — a theory that also contains open strings; this relation is called S-duality. The HO and HE theories are also related by T-duality.

Because the various superstring theories were shown to be related by dualities, it was proposed that that each type of string was a different aspect of a single underlying theory called M-theory.

35.2 References

[1] D. Overbye, "String theory, at 20, explains it all (or not)". *NY Times*, 2004-12-07

[2] "String Theory and M-Theory", Becker, Becker and Schwarz, p.253

[3] Polchinski, Joseph. (1998). *String Theory: Volume 2*, p. 45.

Chapter 36

Topological defect

Also see topological excitations and the base concepts: topology, differential equations, quantum mechanics & condensed matter physics.

In mathematics and physics, a **topological soliton** or a **topological defect** is a solution of a system of partial differential equations or of a quantum field theory homotopically distinct from the vacuum solution; it can be proven to exist because the boundary conditions entail the existence of homotopically distinct solutions. Typically, this occurs because the boundary on which the boundary conditions are specified has a non-trivial homotopy group which is preserved in differential equations; the solutions to the differential equations are then topologically distinct, and are classified by their homotopy class. Topological defects are not only stable against small perturbations, but cannot decay or be undone or be de-tangled, precisely because there is no continuous transformation that will map them (homotopically) to a uniform or "trivial" solution.

Examples include the soliton or solitary wave which occurs in many exactly solvable models, the screw dislocations in crystalline materials, the skyrmion and the Wess–Zumino–Witten model in quantum field theory.

Topological defects are believed to drive phase transitions in condensed matter physics. Notable examples of topological defects are observed in lambda transition universality class systems including: screw/edge-dislocations in liquid crystals, magnetic flux tubes in superconductors and vortices in superfluids.

The authenticity of a topological defect depends on the authenticity of the vacuum in which the system will tend towards if infinite time elapses; false and true topological defects can be distinguished if the defect is in a false vacuum and a true vacuum, respectively.

36.1 Cosmology

Certain grand unified theories predict topological defects to have formed in the early universe. According to the Big Bang theory, the universe cooled from an initial hot, dense state triggering a series of phase transitions much like what happens in condensed-matter systems.

In physical cosmology, a topological defect is an (often) stable configuration of matter predicted by some theories to form at phase transitions in the very early universe.

36.1.1 Symmetry breakdown

Depending on the nature of symmetry breakdown, various solitons are believed to have formed in the early universe according to the Higgs–Kibble mechanism. The well-known topological defects are magnetic monopoles, cosmic strings, domain walls, skyrmions and textures.

As the universe expanded and cooled, symmetries in the laws of physics began breaking down in regions that spread at the speed of light; topological defects occur where different regions came into contact with each other. The matter in these defects is in the original symmetric phase, which persists after a phase transition to the new asymmetric phase is completed.

36.1.2 Types of topological defects

Various different types of topological defects are possible, with the type of defect formed being determined by the symmetry properties of the matter and the nature of the phase transition. They include:

- Domain walls, two-dimensional membranes that form when a discrete symmetry is broken at a phase transition. These walls resemble the walls of a closed-cell foam, dividing the universe into discrete cells.

- Cosmic strings are one-dimensional lines that form when an axial or cylindrical symmetry is broken.

- Monopoles, cube-like defects that form when a spherical symmetry is broken, are predicted to have magnetic charge, either north or south (and so are commonly called "magnetic monopoles").

- Textures form when larger, more complicated symmetry groups are completely broken. They are not as localized as the other defects, and are unstable. Other more complex hybrids of these defect types are also possible.

- Extra dimensions and higher dimensions.

36.1.3 Observation

Topological defects, of the cosmological type, are extremely high-energy phenomena and are likely impossible to produce in artificial Earth-bound physics experiments, but topological defects that formed during the universe's formation could theoretically be observed.

No topological defects of any type have yet been observed by astronomers, however, and certain types are not compatible with current observations; in particular, if domain walls and monopoles were present in the observable universe, they would result in significant deviations from what astronomers can see. Because of these observations, the formation of these structures *within the observable universe* is highly constrained, requiring special circumstances (see: *inflation*). On the other hand, cosmic strings have been suggested as providing the initial 'seed'-gravity around which the large-scale structure of the cosmos of matter has condensed. Textures are similarly benign. In late 2007, a cold spot in the cosmic microwave background was interpreted as possibly being a sign of a texture lying in that direction.[1]

36.2 Condensed matter

In condensed matter physics, the theory of homotopy groups provides a natural setting for description and classification of defects in ordered systems.[2] Topological methods have been used in several problems of condensed matter theory. Poénaru and Toulouse used topological methods to obtain a condition for line (string) defects in liquid crystals can cross each other without entanglement. It was a non-trivial application of topology that first led to the discovery of peculiar hydrodynamic behavior in the A-phase of superfluid helium-3.[2]

36.2.1 Classification

An *ordered medium* is defined as a region of space described by a function $f(r)$ that assigns to every point in the region an *order parameter*, and the possible values of the order parameter space constitute an *order parameter space*. The homotopy theory of defects uses the fundamental group of the order parameter space of a medium to discuss the existence, stability and classifications of topological defects in that medium.[2]

Suppose R is the order parameter space for a medium, and let G be a Lie group of transformations on R. Let H be the symmetry subgroup of G for the medium. Then, the order parameter space can be written as the Lie group quotient[3] $R=G/H$.

If G is a universal cover for G/H then, it can be shown[3] that $\pi n\ (G/H)=\pi n\text{-}1\ (H)$, where πi denotes the i-th homotopy group.

Various types of defects in the medium can be characterized by elements of various homotopy groups of the order parameter space. For example, (in three dimensions), line defects correspond to elements of $\pi 1\ (R)$, point defects correspond to elements of $\pi 2\ (R)$, textures correspond to elements of $\pi 3\ (R)$. However, defects which belong to the

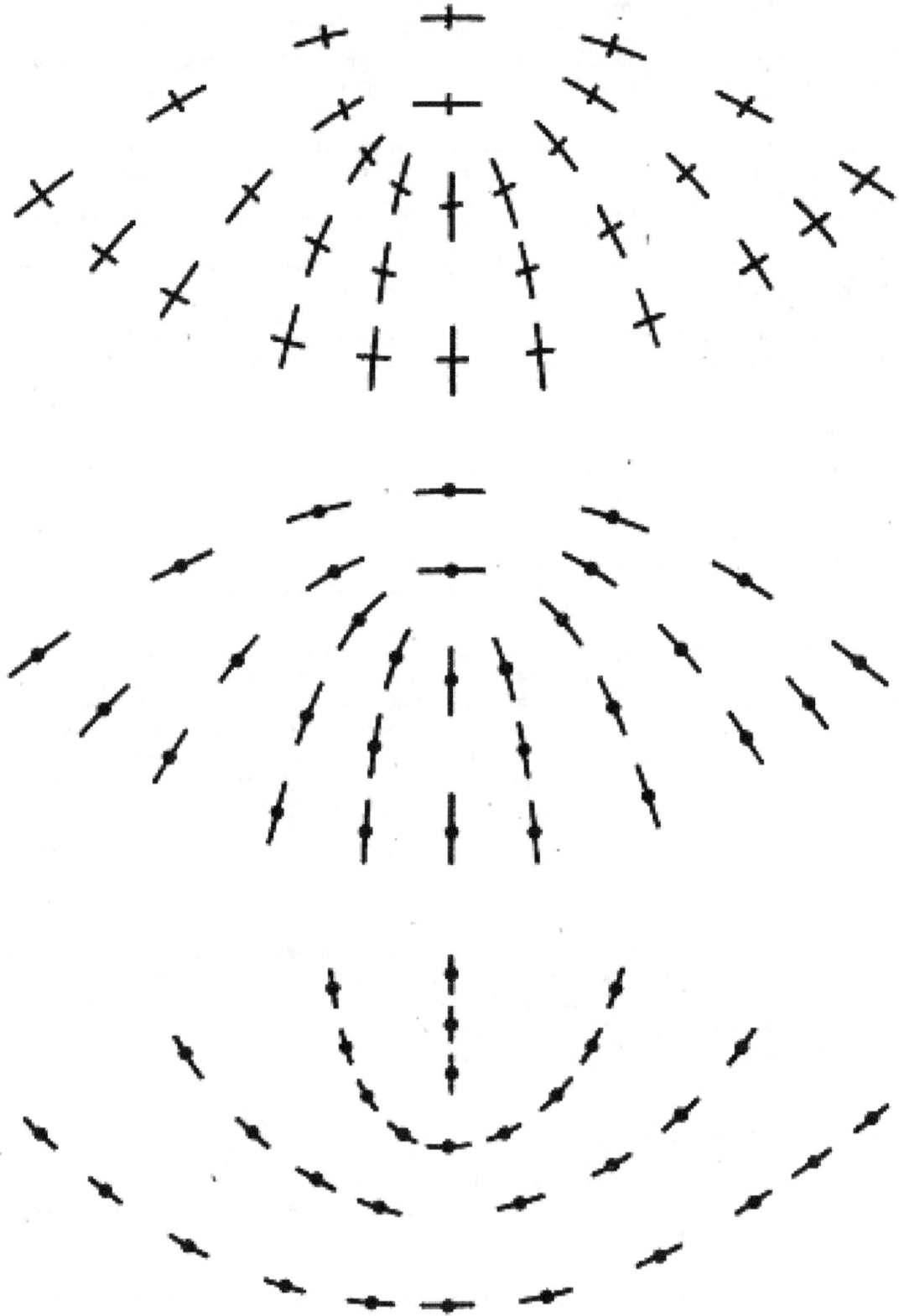

Classes of stable defects in biaxial nematics

same conjugacy class of $\pi 1\ (R)$ can be deformed continuously to each other,[2] and hence, distinct defects correspond to distinct conjugacy classes.

Poénaru and Toulouse showed that[4] crossing defects get entangled if and only if they are members of separate conjugacy classes of $\pi 1\ (R)$.

36.2.2 Stable defects

The homotopy theory is deeply related to the stability of topological defects. In the case of line defect, if the closed path can be continuously deformed into one point, the defect is not stable, and otherwise, it is stable.

Unlike in cosmology and field theory, topological defects in condensed matter can be experimentally observed.[5] Ferromagnetic materials have regions of magnetic alignment separated by domain walls. Nematic and bi-axial nematic liquid crystals display a variety of defects including monopoles, strings, textures etc.[2] Defects can also been found in biochemistry, notably in the process of protein folding.

36.3 Images

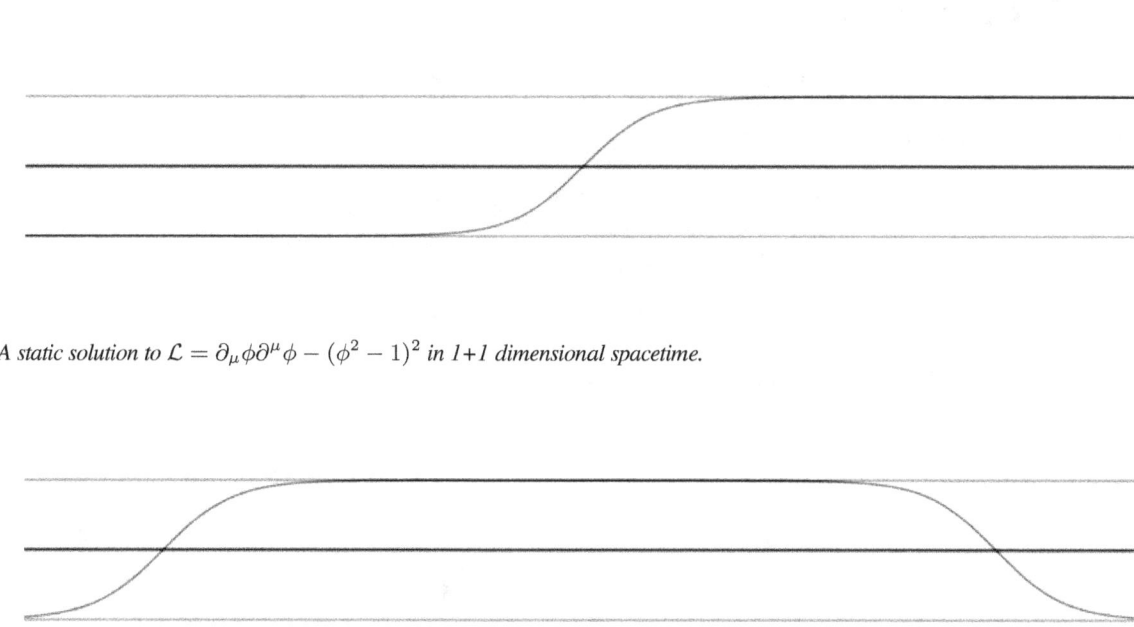

A static solution to $\mathcal{L} = \partial_\mu \phi \partial^\mu \phi - (\phi^2 - 1)^2$ in 1+1 dimensional spacetime.

A soliton and an antisoliton colliding with velocities $\pm sinh(0.05)$ and annihilating.

36.4 See also

- Quantum vortex
- Dislocation
- Vector soliton

- Quantum topology
- Topological entropy in physics
- Topological order
- Topological quantum field theory
- Topological quantum number
- Topological string theory

36.5 References

[1] Cruz, M.; N. Turok; P. Vielva; E. Martínez-González; M. Hobson (2007). "A Cosmic Microwave Background Feature Consistent with a Cosmic Texture". *Science* **318** (5856): 1612–4. arXiv:0710.5737. Bibcode:2007Sci...318.1612C. doi:10.1126/science.1148694. PMID 17962521. Retrieved 2007-10-25.

[2] Mermin, N. D. (1979). "The topological theory of defects in ordered media". *Reviews of Modern Physics* **51** (3): 591. Bibcode:1979RvMP...51..591M. doi:10.1103/RevModPhys.51.591.

[3] Nakahara, Mikio (2003). *Geometry, Topology and Physics*. Taylor & Francis. ISBN 0-7503-0606-8.

[4] Poénaru, V.; Toulouse, G. (1977). "The crossing of defects in ordered media and the topology of 3-manifolds". *Le Journal de Physique* **38** (8).

[5] "Topological defects". Cambridge cosmology.

36.6 External links

- Cosmic Strings & other Topological Defects
- http://demonstrations.wolfram.com/SeparationOfTopologicalSingularities/

Chapter 37

Magnetic monopole

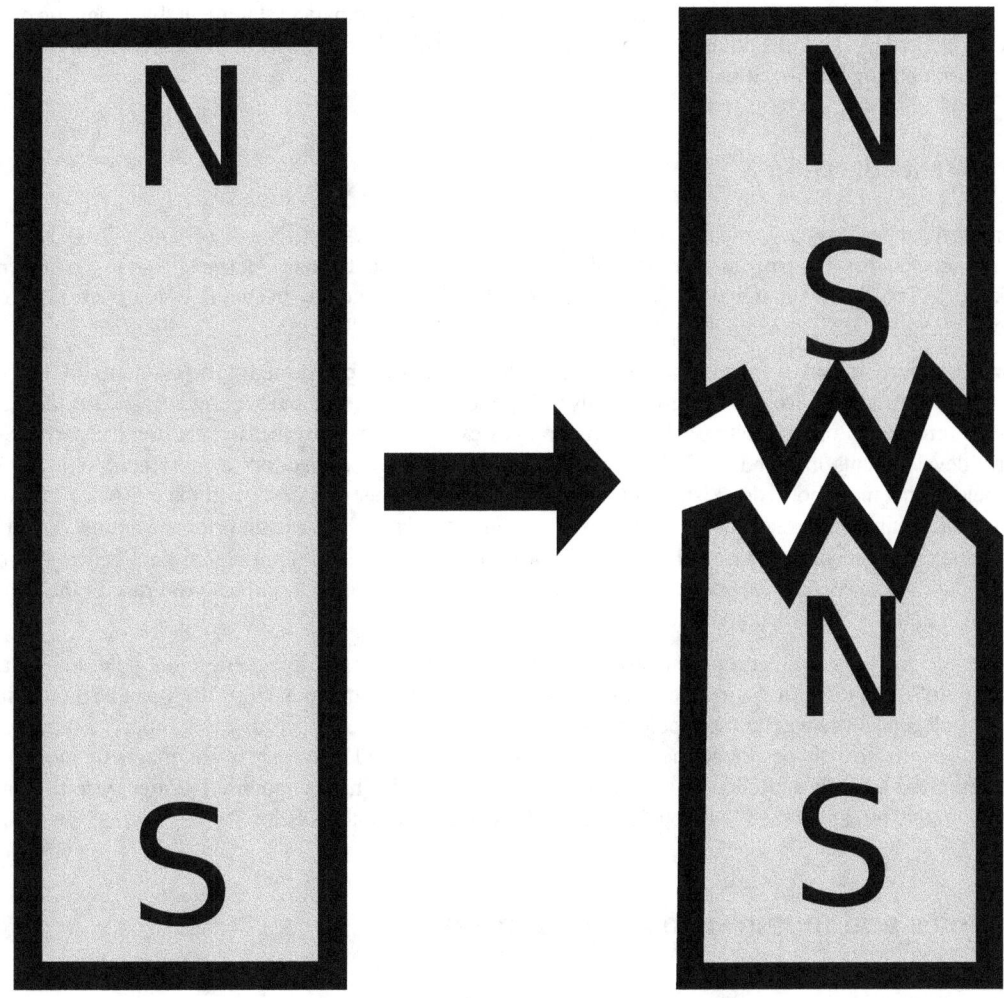

*It is impossible to make **magnetic monopoles** from a bar magnet. If a bar magnet is cut in half, it is not the case that one half has the north pole and the other half has the south pole. Instead, each piece has its own north and south poles. A magnetic monopole cannot be created from normal matter such as atoms and electrons, but would instead be a new elementary particle.*

A **magnetic monopole** is a hypothetical elementary particle in particle physics that is an isolated magnet with only one magnetic pole (a north pole without a south pole or vice-versa).[1][2] In more technical terms, a magnetic monopole would have a net "magnetic charge". Modern interest in the concept stems from particle theories, notably the grand unified and superstring theories, which predict their existence.[3][4]

Magnetism in bar magnets and electromagnets does not arise from magnetic monopoles, and in fact there is no conclusive experimental evidence that magnetic monopoles exist at all in the universe.

Some condensed matter systems contain effective (non-isolated) magnetic monopole *quasi*-particles,[5] or contain phenomena that are mathematically analogous to magnetic monopoles.[6]

37.1 Historical background

37.1.1 Pre-twentieth century

Many early scientists attributed the magnetism of lodestones to two different "magnetic fluids" ("effluvia"), a north-pole fluid at one end and a south-pole fluid at the other, which attracted and repelled each other in analogy to positive and negative electric charge.[7][8] However, an improved understanding of electromagnetism in the nineteenth century showed that the magnetism of lodestones was properly explained by Ampère's circuital law, not magnetic monopole fluids. Gauss's law for magnetism, one of Maxwell's equations, is the mathematical statement that magnetic monopoles do not exist. Nevertheless, it was pointed out by Pierre Curie in 1894[9] that magnetic monopoles *could* conceivably exist, despite not having been seen so far.

37.1.2 Twentieth century

The *quantum* theory of magnetic charge started with a paper by the physicist Paul A.M. Dirac in 1931.[10] In this paper, Dirac showed that if *any* magnetic monopoles exist in the universe, then all electric charge in the universe must be quantized.[11] The electric charge *is*, in fact, quantized, which is consistent with (but does not prove) the existence of monopoles.[11]

Since Dirac's paper, several systematic monopole searches have been performed. Experiments in 1975[12] and 1982[13] produced candidate events that were initially interpreted as monopoles, but are now regarded as inconclusive.[14] Therefore, it remains an open question whether monopoles exist. Further advances in theoretical particle physics, particularly developments in grand unified theories and quantum gravity, have led to more compelling arguments (detailed below) that monopoles do exist. Joseph Polchinski, a string-theorist, described the existence of monopoles as "one of the safest bets that one can make about physics not yet seen".[15] These theories are not necessarily inconsistent with the experimental evidence. In some theoretical models, magnetic monopoles are unlikely to be observed, because they are too massive to be created in particle accelerators (see below), and also too rare in the Universe to enter a particle detector with much probability.[15]

Some condensed matter systems propose a structure superficially similar to a magnetic monopole, known as a flux tube. The ends of a flux tube form a magnetic dipole, but since they move independently, they can be treated for many purposes as independent magnetic monopole quasiparticles. Since 2009, numerous news reports from the popular media[16][17] have incorrectly described these systems as the long-awaited discovery of the magnetic monopoles, but the two phenomena are only superficially related to one another.[18][19] These condensed-matter systems continue to be an area of active research. (See "Monopoles" in condensed-matter systems below.)

37.2 Poles and magnetism in ordinary matter

Main article: Magnetism

All matter ever isolated to date—including every atom on the periodic table and every particle in the standard model—has zero magnetic monopole charge. Therefore, the ordinary phenomena of magnetism and magnets have nothing to do with magnetic monopoles.

Instead, magnetism in ordinary matter comes from two sources. First, electric currents create magnetic fields according to Ampère's law. Second, many elementary particles have an "intrinsic" magnetic moment, the most important of which is the electron magnetic dipole moment. (This magnetism is related to quantum-mechanical "spin".)

Mathematically, the magnetic field of an object is often described in terms of a multipole expansion. This is an expression of the field as the sum of component fields with specific mathematical forms. The first term in the expansion is called the "monopole" term, the second is called "dipole", then "quadrupole", then "octupole", and so on. Any of these terms can be present in the multipole expansion of an electric field, for example. However, in the multipole expansion of a *magnetic* field, the "monopole" term is always exactly zero (for ordinary matter). A magnetic monopole, if it exists, would have the defining property of producing a magnetic field whose "monopole" term is nonzero.

A magnetic dipole is something whose magnetic field is predominantly or exactly described by the magnetic dipole term of the multipole expansion. The term "dipole" means "two poles", corresponding to the fact that a dipole magnet typically contains a "north pole" on one side and a "south pole" on the other side. This is analogous to an electric dipole, which has positive charge on one side and negative charge on the other. However, an electric dipole and magnetic dipole are fundamentally quite different. In an electric dipole made of ordinary matter, the positive charge is made of protons and the negative charge is made of electrons, but a magnetic dipole does *not* have different types of matter creating the north pole and south pole. Instead, the two magnetic poles arise simultaneously from the aggregate effect of all the currents and intrinsic moments throughout the magnet. Because of this, the two poles of a magnetic dipole must always have equal and opposite strength, and the two poles cannot be separated from each other.

37.3 Maxwell's equations

Maxwell's equations of electromagnetism relate the electric and magnetic fields to each other and to the motions of electric charges. The standard equations provide for electric charges, but they posit no magnetic charges. Except for this difference, the equations are symmetric under the interchange of the electric and magnetic fields.[20] In fact, symmetric Maxwell's equations can be written when all charges (and hence electric currents) are zero, and this is how the electromagnetic wave equation is derived.

Fully symmetric Maxwell's equations can also be written if one allows for the possibility of "magnetic charges" analogous to electric charges.[21] With the inclusion of a variable for the density of these magnetic charges, say ϱ_m, there will also be a "magnetic current density" variable in the equations, $\mathbf{j}_m$.

If magnetic charges do not exist – or if they do exist but are not present in a region of space – then the new terms in Maxwell's equations are all zero, and the extended equations reduce to the conventional equations of electromagnetism such as $\nabla \cdot \mathbf{B} = 0$ (where $\nabla \cdot$ is divergence and $\mathbf{B}$ is the magnetic $\mathbf{B}$ field).

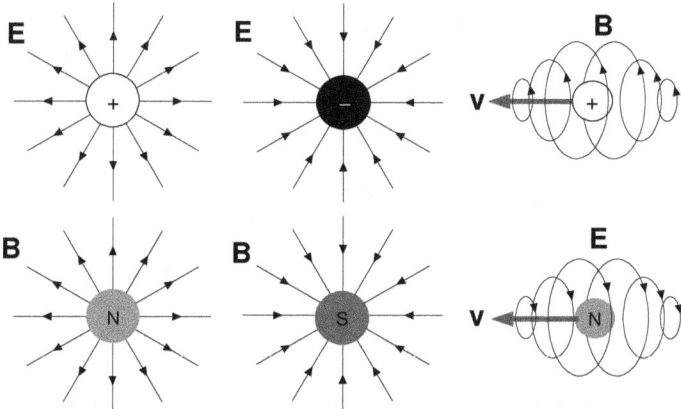

Left: Fields due to stationary electric and magnetic monopoles. **Right:** In motion (velocity **v**), an *electric* charge induces a **B** field while a *magnetic* charge induces an **E** field. Conventional current is used.

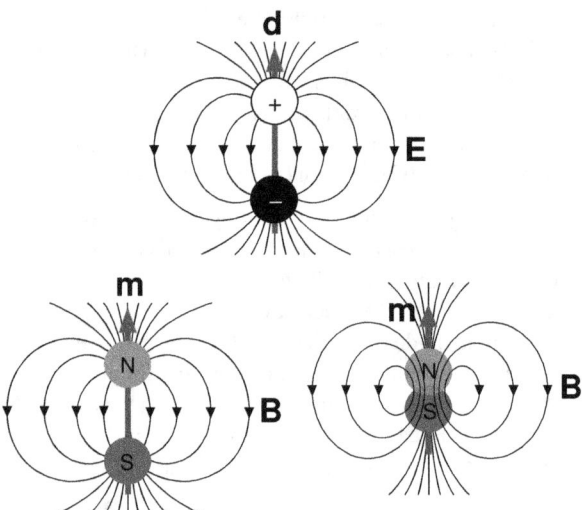

Top: **E** field due to an electric dipole moment **d**. **Bottom left:** **B** field due to a *mathematical* magnetic dipole **m** formed by two magnetic monopoles. **Bottom right:** **B** field due to a natural magnetic dipole moment **m** found in ordinary matter (*not* from monopoles).
The **E** fields and **B** fields due to electric charges (black/white) and magnetic poles (red/blue).[22][23]

37.3.1 In Gaussian cgs units

The extended Maxwell's equations are as follows, in Gaussian cgs units:[24]

In these equations ϱ_m is the *magnetic charge density*, $\mathbf{j}_m$ is the *magnetic current density*, and q_m is the *magnetic charge* of a test particle, all defined analogously to the related quantities of electric charge and current; **v** is the particle's velocity and c is the speed of light. For all other definitions and details, see Maxwell's equations. For the equations in nondimensionalized form, remove the factors of c.

37.3.2 In SI units

In SI units, there are two conflicting units in use for magnetic charge q_m: webers (Wb) and ampere·meters (A·m). The conversion between them is $q_m(\text{Wb}) = \mu_0 q_m(\text{A·m})$, since the units are 1 Wb = 1 H·A = (1 H·m^{-1})·(1 A·m) by dimensional analysis (H is the henry – the SI unit of inductance).

Maxwell's equations then take the following forms (using the same notation above):[26]

37.3.3 Tensor formulation

Maxwell's equations in the language of tensors makes Lorentz covariance clear. The generalized equations are:[27][28]

where

- F is the electromagnetic tensor, $\star$ denotes the Hodge dual, (so $\star F$ is the dual tensor to F),
- for a particle with electric charge q_e and magnetic charge q_m; v is the four-velocity and p the four-momentum,

- for an electric and magnetic charge distribution; $J_e = (\varrho_e, \mathbf{j}_e)$ is the electric four-current and $J_m = (\varrho_m, \mathbf{j}_m)$ the magnetic four-current.

For a particle having only electric charge, one can express its field using a four-potential, according to the standard covariant formulation of classical electromagnetism:

$$F_{\alpha\beta} = \partial_\alpha A_\beta - \partial_\beta A_\gamma$$

However, this formula is inadequate for a particle that has both electric and magnetic charge, and we must add a term involving another potential P.[29][30]

$$F_{\alpha\beta} = \partial_\alpha A_\beta - \partial_\beta A_\alpha + \partial^\mu(\epsilon_{\alpha\beta\mu\nu}P^\nu),$$

This formula for the fields is often called the Cabibbo-Ferrari relation, though Shanmugadhasan proposed it earlier.[30] The quantity $\varepsilon^{\alpha\beta\gamma\delta}$ is the Levi-Civita symbol, and the indices (as usual) behave according to the Einstein summation convention.

37.3.4 Duality transformation

The generalized Maxwell's equations possess a certain symmetry, called a *duality transformation*. One can choose any real angle ξ, and simultaneously change the fields and charges everywhere in the universe as follows (in Gaussian units):[31]

where the primed quantities are the charges and fields before the transformation, and the unprimed quantities are after the transformation. The fields and charges after this transformation still obey the same Maxwell's equations. The matrix is a two-dimensional rotation matrix.

Because of the duality transformation, one cannot uniquely decide whether a particle has an electric charge, a magnetic charge, or both, just by observing its behavior and comparing that to Maxwell's equations. For example, it is merely a convention, not a requirement of Maxwell's equations, that electrons have electric charge but not magnetic charge; after a $\xi = \pi/2$ transformation, it would be the other way around. The key empirical fact is that all particles ever observed have the same ratio of magnetic charge to electric charge.[31] Duality transformations can change the ratio to any arbitrary numerical value, but cannot change the fact that all particles have the same ratio. Since this is the case, a duality transformation can be made that sets this ratio to be zero, so that all particles have no magnetic charge. This choice underlies the "conventional" definitions of electricity and magnetism.[31]

37.4 Dirac's quantization

One of the defining advances in quantum theory was Paul Dirac's work on developing a relativistic quantum electromagnetism. Before his formulation, the presence of electric charge was simply "inserted" into the equations of quantum mechanics (QM), but in 1931 Dirac showed that a discrete charge naturally "falls out" of QM. That is to say, we can maintain the form of Maxwell's equations and still have magnetic charges.

Consider a system consisting of a single stationary electric monopole (an electron, say) and a single stationary magnetic monopole. Classically, the electromagnetic field surrounding them has a momentum density given by the Poynting vector, and it also has a total angular momentum, which is proportional to the product $q_e q_m$, and independent of the distance between them.

Quantum mechanics dictates, however, that angular momentum is quantized in units of $\hbar$, so therefore the product $q_e q_m$ must also be quantized. This means that if even a single magnetic monopole existed in the universe, and the form of Maxwell's equations is valid, all electric charges would then be quantized.

What are the units in which magnetic charge would be quantized? Although it would be possible simply to integrate over all space to find the total angular momentum in the above example, Dirac took a different approach. This led him to new ideas. He considered a point-like magnetic charge whose magnetic field behaves as q_m/r^2 and is directed in the radial direction, located at the origin. Because the divergence of **B** is equal to zero almost everywhere, except for the locus of the magnetic monopole at $r = 0$, one can locally define the vector potential such that the curl of the vector potential **A** equals the magnetic field **B**.

However, the vector potential cannot be defined globally precisely because the divergence of the magnetic field is proportional to the Dirac delta function at the origin. We must define one set of functions for the vector potential on the "northern hemisphere" (the half-space $z > 0$ above the particle), and another set of functions for the "southern hemisphere". These two vector potentials are matched at the "equator" (the plane $z = 0$ through the particle), and they differ by a gauge transformation. The wave function of an electrically-charged particle (a "probe charge") that orbits the "equator" generally changes by a phase, much like in the Aharonov–Bohm effect. This phase is proportional to the electric charge q_e of the probe, as well as to the magnetic charge q_m of the source. Dirac was originally considering an electron whose wave function is described by the Dirac equation.

Because the electron returns to the same point after the full trip around the equator, the phase φ of its wave function $e^{i\varphi}$ must be unchanged, which implies that the phase φ added to the wave function must be a multiple of 2π:

where ε_0 is the vacuum permittivity, $\hbar = h/2\pi$ is the reduced Planck's constant, c is the speed of light, and $\mathbb{Z}$ is the set of integers.

This is known as the **Dirac quantization condition**. The hypothetical existence of a magnetic monopole would imply that the electric charge must be quantized in certain units; also, the existence of the electric charges implies that the magnetic charges of the hypothetical magnetic monopoles, if they exist, must be quantized in units inversely proportional to the elementary electric charge.

At the time it was not clear if such a thing existed, or even had to. After all, another theory could come along that would explain charge quantization without need for the monopole. The concept remained something of a curiosity. However, in the time since the publication of this seminal work, no other widely accepted explanation of charge quantization has appeared. (The concept of local gauge invariance—see gauge theory below—provides a natural explanation of charge quantization, without invoking the need for magnetic monopoles; but only if the U(1) gauge group is compact, in which case we will have magnetic monopoles anyway.)

If we maximally extend the definition of the vector potential for the southern hemisphere, it will be defined everywhere except for a semi-infinite line stretched from the origin in the direction towards the northern pole. This semi-infinite line is called the Dirac string and its effect on the wave function is analogous to the effect of the solenoid in the Aharonov–Bohm effect. The quantization condition comes from the requirement that the phases around the Dirac string are trivial, which means that the Dirac string must be unphysical. The Dirac string is merely an artifact of the coordinate chart used and should not be taken seriously.

The Dirac monopole is a singular solution of Maxwell's equation (because it requires removing the worldline from spacetime); in more complicated theories, it is superseded by a smooth solution such as the 't Hooft–Polyakov monopole.

37.5 Topological interpretation

37.5.1 Dirac string

Main article: Dirac string

A gauge theory like electromagnetism is defined by a gauge field, which associates a group element to each path in space time. For infinitesimal paths, the group element is close to the identity, while for longer paths the group element is the successive product of the infinitesimal group elements along the way.

In electrodynamics, the group is U(1), unit complex numbers under multiplication. For infinitesimal paths, the group element is $1 + iA_\mu dx^\mu$ which implies that for finite paths parametrized by s, the group element is:

$$\prod_s \left(1 + ieA_\mu \frac{dx^\mu}{ds} ds\right) = \exp\left(ie \int A \cdot dx\right).$$

The map from paths to group elements is called the Wilson loop or the holonomy, and for a U(1) gauge group it is the phase factor which the wavefunction of a charged particle acquires as it traverses the path. For a loop:

$$e \oint_{\partial D} A \cdot dx = e \int_D (\nabla \times A) dS = e \int_D B \, dS.$$

So that the phase a charged particle gets when going in a loop is the magnetic flux through the loop. When a small solenoid has a magnetic flux, there are interference fringes for charged particles which go around the solenoid, or around different sides of the solenoid, which reveal its presence.

But if all particle charges are integer multiples of e, solenoids with a flux of $2\pi/e$ have no interference fringes, because the phase factor for any charged particle is $e^{2\pi i} = 1$. Such a solenoid, if thin enough, is quantum-mechanically invisible. If such a solenoid were to carry a flux of $2\pi/e$, when the flux leaked out from one of its ends it would be indistinguishable from a monopole.

Dirac's monopole solution in fact describes an infinitesimal line solenoid ending at a point, and the location of the solenoid is the singular part of the solution, the Dirac string. Dirac strings link monopoles and antimonopoles of opposite magnetic charge, although in Dirac's version, the string just goes off to infinity. The string is unobservable, so you can put it anywhere, and by using two coordinate patches, the field in each patch can be made nonsingular by sliding the string to where it cannot be seen.

37.5.2 Grand unified theories

Main article: 't Hooft–Polyakov monopole

In a U(1) gauge group with quantized charge, the group is a circle of radius $2\pi/e$. Such a U(1) gauge group is called compact. Any U(1) which comes from a Grand Unified Theory is compact – because only compact higher gauge groups make sense. The size of the gauge group is a measure of the inverse coupling constant, so that in the limit of a large-volume gauge group, the interaction of any fixed representation goes to zero.

The case of the U(1) gauge group is a special case because all its irreducible representations are of the same size – the charge is bigger by an integer amount, but the field is still just a complex number – so that in U(1) gauge field theory it is possible to take the decompactified limit with no contradiction. The quantum of charge becomes small, but each charged particle has a huge number of charge quanta so its charge stays finite. In a non-compact U(1) gauge group theory, the charges of particles are generically not integer multiples of a single unit. Since charge quantization is an experimental certainty, it is clear that the U(1) gauge group of electromagnetism is compact.

GUTs lead to compact U(1) gauge groups, so they explain charge quantization in a way that seems to be logically independent from magnetic monopoles. However, the explanation is essentially the same, because in any GUT which breaks down into a U(1) gauge group at long distances, there are magnetic monopoles.

The argument is topological:

1. The holonomy of a gauge field maps loops to elements of the gauge group. Infinitesimal loops are mapped to group elements infinitesimally close to the identity.

2. If you imagine a big sphere in space, you can deform an infinitesimal loop which starts and ends at the north pole as follows: stretch out the loop over the western hemisphere until it becomes a great circle (which still starts and ends at the north pole) then let it shrink back to a little loop while going over the eastern hemisphere. This is called *lassoing the sphere*.

3. Lassoing is a sequence of loops, so the holonomy maps it to a sequence of group elements, a continuous path in the gauge group. Since the loop at the beginning of the lassoing is the same as the loop at the end, the path in the group is closed.

4. If the group path associated to the lassoing procedure winds around the U(1), the sphere contains magnetic charge. During the lassoing, the holonomy changes by the amount of magnetic flux through the sphere.

5. Since the holonomy at the beginning and at the end is the identity, the total magnetic flux is quantized. The magnetic charge is proportional to the number of windings N, the magnetic flux through the sphere is equal to $2\pi N/e$. This is the Dirac quantization condition, and it is a topological condition which demands that the long distance U(1) gauge field configurations be consistent.

6. When the U(1) gauge group comes from breaking a compact Lie group, the path which winds around the U(1) group enough times is topologically trivial in the big group. In a non-U(1) compact Lie group, the covering space is a Lie group with the same Lie algebra, but where all closed loops are contractible. Lie groups are homogenous, so that any cycle in the group can be moved around so that it starts at the identity, then its lift to the covering group ends at P, which is a lift of the identity. Going around the loop twice gets you to P^2, three times to P^3, all lifts of the identity. But there are only finitely many lifts of the identity, because the lifts can't accumulate. This number of times one has to traverse the loop to make it contractible is small, for example if the GUT group is SO(3), the covering group is SU(2), and going around any loop twice is enough.

7. This means that there is a continuous gauge-field configuration in the GUT group allows the U(1) monopole configuration to unwind itself at short distances, at the cost of not staying in the U(1). In order to do this with as little energy as possible, you should leave only the U(1) gauge group in the neighborhood of one point, which is called the **core** of the monopole. Outside the core, the monopole has only magnetic field energy.

Hence, the Dirac monopole is a topological defect in a compact U(1) gauge theory. When there is no GUT, the defect is a singularity – the core shrinks to a point. But when there is some sort of short-distance regulator on space time, the monopoles have a finite mass. Monopoles occur in lattice U(1), and there the core size is the lattice size. In general, they are expected to occur whenever there is a short-distance regulator.

37.5.3 String theory

In our universe, quantum gravity provides the regulator. When gravity is included, the monopole singularity can be a black hole, and for large magnetic charge and mass, the black hole mass is equal to the black hole charge, so that the mass of the magnetic black hole is not infinite. If the black hole can decay completely by Hawking radiation, the lightest charged particles cannot be too heavy.[33] The lightest monopole should have a mass less than or comparable to its charge in natural units.

So in a consistent holographic theory, of which string theory is the only known example, there are always finite-mass monopoles. For ordinary electromagnetism, the mass bound is not very useful because it is about same size as the Planck mass.

37.5.4 Mathematical formulation

In mathematics, a (classical) gauge field is defined as a connection over a principal G-bundle over spacetime. G is the gauge group, and it acts on each fiber of the bundle separately.

A *connection* on a G bundle tells you how to glue fibers together at nearby points of M. It starts with a continuous symmetry group G which acts on the fiber F, and then it associates a group element with each infinitesimal path. Group multiplication along any path tells you how to move from one point on the bundle to another, by having the G element associated to a path act on the fiber F.

In mathematics, the definition of bundle is designed to emphasize topology, so the notion of connection is added on as an afterthought. In physics, the connection is the fundamental physical object. One of the fundamental observations in the theory of characteristic classes in algebraic topology is that many homotopical structures of nontrivial principal bundles may be expressed as an integral of some polynomial over **any** connection over it. Note that a connection over a trivial bundle can never give us a nontrivial principal bundle.

If space time is $\mathbf{R}^4$ the space of all possible connections of the G-bundle is connected. But consider what happens when we remove a timelike worldline from spacetime. The resulting spacetime is homotopically equivalent to the topological sphere S^2.

A principal G-bundle over S^2 is defined by covering S^2 by two charts, each homeomorphic to the open 2-ball such that their intersection is homeomorphic to the strip $S^1 \times I$. 2-balls are homotopically trivial and the strip is homotopically equivalent to the circle S^1. So a topological classification of the possible connections is reduced to classifying the transition functions. The transition function maps the strip to G, and the different ways of mapping a strip into G are given by the first homotopy group of G.

So in the G-bundle formulation, a gauge theory admits Dirac monopoles provided G is not simply connected, whenever there are paths that go around the group that cannot be deformed to a constant path (a path whose image consists of a single point). U(1), which has quantized charges, is not simply connected and can have Dirac monopoles while **R**, its universal covering group, **is** simply connected, doesn't have quantized charges and does not admit Dirac monopoles. The mathematical definition is equivalent to the physics definition provided that, following Dirac, gauge fields are allowed which are defined only patch-wise and the gauge field on different patches are glued after a gauge transformation.

The total magnetic flux is none other than the first Chern number of the principal bundle, and depends only upon the choice of the principal bundle, and not the specific connection over it. In other words, it's a topological invariant.

This argument for monopoles is a restatement of the lasso argument for a pure U(1) theory. It generalizes to $d + 1$ dimensions with $d \geq 2$ in several ways. One way is to extend everything into the extra dimensions, so that U(1) monopoles become sheets of dimension $d - 3$. Another way is to examine the type of topological singularity at a point with the homotopy group $\pi d_{-2}(G)$.

37.6 Grand unified theories

In more recent years, a new class of theories has also suggested the existence of magnetic monopoles.

During the early 1970s, the successes of quantum field theory and gauge theory in the development of electroweak theory and the mathematics of the strong nuclear force led many theorists to move on to attempt to combine them in a single theory known as a Grand Unified Theory (GUT). Several GUTs were proposed, most of which implied the presence of a real magnetic monopole particle. More accurately, GUTs predicted a range of particles known as dyons, of which the most basic state was a monopole. The charge on magnetic monopoles predicted by GUTs is either 1 or 2 gD, depending on the theory.

The majority of particles appearing in any quantum field theory are unstable, and they decay into other particles in a variety of reactions that must satisfy various conservation laws. Stable particles are stable because there are no lighter particles into which they can decay and still satisfy the conservation laws. For instance, the electron has a lepton number of one and an electric charge of one, and there are no lighter particles that conserve these values. On the other hand, the muon, essentially a heavy electron, can decay into the electron plus two quanta of energy, and hence it is not stable.

The dyons in these GUTs are also stable, but for an entirely different reason. The dyons are expected to exist as a side effect of the "freezing out" of the conditions of the early universe, or a symmetry breaking. In this scenario, the dyons arise due to the configuration of the vacuum in a particular area of the universe, according to the original Dirac theory. They remain stable not because of a conservation condition, but because there is no simpler *topological* state into which they can decay.

The length scale over which this special vacuum configuration exists is called the *correlation length* of the system. A correlation length cannot be larger than causality would allow, therefore the correlation length for making magnetic monopoles must be at least as big as the horizon size determined by the metric of the expanding universe. According to that logic, there should be at least one magnetic monopole per horizon volume as it was when the symmetry breaking took place.

Cosmological models of the events following the big bang make predictions about what the horizon volume was, which lead to predictions about present-day monopole density. Early models predicted an enormous density of monopoles, in clear contradiction to the experimental evidence.[34][35] This was called the "monopole problem". Its widely accepted resolution was not a change in the particle-physics prediction of monopoles, but rather in the cosmological models used to infer their present-day density. Specifically, more recent theories of cosmic inflation drastically reduce the predicted number of magnetic monopoles, to a density small enough to make it unsurprising that humans have never seen one.[36] This resolution of the "monopole problem" was regarded as a success of cosmic inflation theory. (However, of course, it is only a noteworthy success if the particle-physics monopole prediction is correct.[37]) For these reasons, monopoles became a major interest in the 1970s and 80s, along with the other "approachable" predictions of GUTs such as proton decay.

Many of the other particles predicted by these GUTs were beyond the abilities of current experiments to detect. For instance, a wide class of particles known as the X and Y bosons are predicted to mediate the coupling of the electroweak and strong forces, but these particles are extremely heavy and well beyond the capabilities of any reasonable particle accelerator to create.

37.7 Searches for magnetic monopoles

A number of attempts have been made to detect magnetic monopoles. One of the simpler ones is to use a loop of superconducting wire to look for even tiny magnetic sources, a so-called "superconducting quantum interference device", or SQUID. Given the predicted density, loops small enough to fit on a lab bench would expect to see about one monopole event per year. Although there have been tantalizing events recorded, in particular the event recorded by Blas Cabrera on the night of February 14, 1982 (thus, sometimes referred to as the "Valentine's Day Monopole"[38]), there has never been reproducible evidence for the existence of magnetic monopoles.[13] The lack of such events places a limit on the number of monopoles of about one monopole per 10^{29} nucleons.

Another experiment in 1975 resulted in the announcement of the detection of a moving magnetic monopole in cosmic rays by the team led by P. Buford Price.[12] Price later retracted his claim, and a possible alternative explanation was offered by Alvarez.[39] In his paper it was demonstrated that the path of the cosmic ray event that was claimed to be due to a magnetic monopole could be reproduced by the path followed by a platinum nucleus decaying first to osmium, and then to tantalum.

Other experiments rely on the strong coupling of monopoles with photons, as is the case for any electrically-charged particle as well. In experiments involving photon exchange in particle accelerators, monopoles should be produced in reasonable numbers, and detected due to their effect on the scattering of the photons. The probability of a particle being created in such experiments is related to their mass – with heavier particles being less likely to be created – so by examining the results of such experiments, limits on the mass of a magnetic monopole can be calculated. The most recent such experiments suggest that monopoles with masses below 600 GeV/c^2 do not exist, while upper limits on their mass due to the very existence of the universe – which would have collapsed by now if they were too heavy – are about 10^{17} GeV/c^2.

The MoEDAL experiment, installed at the Large Hadron Collider, is currently searching for magnetic monopoles and large supersymmetric particles using layers of special plastic sheets attached to the walls around LHCb's VELO detector. The particles it is looking for will damage the sheets along their path, with various identifying features.

The Russian astrophysicist Igor Novikov claims the fields of macroscopic black holes to be potential magnetic monopoles, representing the entrance to an Einstein–Rosen bridge.[40]

37.8 "Monopoles" in condensed-matter systems

Since around 2003, various condensed-matter physics groups have used the term "magnetic monopole" to describe a different and largely unrelated phenomenon.[18][19]

A true magnetic monopole would be a new elementary particle, and would violate the law $\nabla \cdot \mathbf{B} = 0$. A monopole of this kind, which would help to explain the law of charge quantization as formulated by Paul Dirac in 1931,[41] has never been observed in experiments.

The monopoles studied by condensed-matter groups have none of these properties. They are not a new elementary particle, but rather are an emergent phenomenon in systems of everyday particles (protons, neutrons, electrons, photons); in other words, they are quasi-particles. They are not sources for the **B**-field (i.e., they do not violate $\nabla \cdot \mathbf{B} = 0$); instead, they are sources for other fields, for example the **H**-field,[5] or the "B*-field" (related to superfluid vorticity)[6] They are not directly relevant to grand unified theories or other aspects of particle physics, and do not help explain charge quantization—except insofar as studies of analogous situations can help confirm that the mathematical analyses involved are sound.[42]

There are a number of examples in condensed-matter physics where collective behavior leads to emergent phenomena that resemble magnetic monopoles in certain respects,[17][43][44][45] including most prominently the spin ice materials.[5][46] While these should not be confused with hypothetical elementary monopoles existing in the vacuum, they nonetheless have similar properties and can be probed using similar techniques.

Some researchers use the term **magnetricity** to describe the manipulation of magnetic monopole quasiparticles in spin ice,[46][47] in analogy to the word "electricity".

One example of the work on magnetic monopole quasiparticles is a paper published in the journal *Science* in September 2009, in which researchers Jonathan Morris and Alan Tennant from the Helmholtz-Zentrum Berlin für Materialien und Energie (HZB) along with Santiago Grigera from Instituto de Física de Líquidos y Sistemas Biológicos (IFLYSIB, CONICET) and other colleagues from Dresden University of Technology, University of St. Andrews and Oxford University described the observation of quasiparticles resembling magnetic monopoles. A single crystal of the spin ice material dysprosium titanate was cooled to a temperature between 0.6 kelvin and 2.0 kelvin. Using observations of neutron scattering, the magnetic moments were shown to align into interwoven tubelike bundles resembling Dirac strings. At the defect formed by the end of each tube, the magnetic field looks like that of a monopole. Using an applied magnetic field to break the symmetry of the system, the researchers were able to control the density and orientation of these strings. A contribution to the heat capacity of the system from an effective gas of these quasiparticles was also described.[16][48]

This research went on to win the 2012 Europhysics Prize for condensed matter physics.

Another example is a paper in the February 11, 2011 issue of *Nature Physics* which describes creation and measurement of long-lived magnetic monopole quasiparticle currents in spin ice. By applying a magnetic-field pulse to crystal of dysprosium titanate at 0.36 K, the authors created a relaxing magnetic current that lasted for several minutes. They measured the current by means of the electromotive force it induced in a solenoid coupled to a sensitive amplifier, and quantitatively described it using a chemical kinetic model of point-like charges obeying the Onsager–Wien mechanism of carrier dissociation and recombination. They thus derived the microscopic parameters of monopole motion in spin ice and identified the distinct roles of free and bound magnetic charges.[47]

In superfluids, there is a field **B***, related to superfluid vorticity, which is mathematically analogous to the magnetic **B**-field. Because of the similarity, the field **B*** is called a "synthetic magnetic field". In January 2014, it was reported that monopole quasiparticles[49] for the **B*** field were created and studied in a spinor Bose–Einstein condensate.[6] This constitutes the first example of a magnetic monopole observed within a system governed by quantum field theory.[42]

37.9 Further descriptions in particle physics

In physics the phrase "magnetic monopole" usually denoted a Yang–Mills potential A and Higgs field ϕ whose equations of motion are determined by the Yang–Mills action

$$\int (F_A, F_A) + (D_A\phi, D_A\phi) - \lambda(1 - \|\phi\|^2)^2.$$

In mathematics, the phrase customarily refers to a static solution to these equations in the Bogomolny–Parasad–Sommerfeld limit $\lambda \to \phi$ which realizes, within topological class, the absolutes minimum of the functional

$$\int_{R^3} (F_A, F_A) + (D_A\phi, D_A\phi).$$

This means that it in a connection A on a principal G-bundle over $\mathbf{R}^3$ (c.f. also Connections on a manifold; principal G-object) and a section ϕ of the associated adjoint bundle of Lie algebras such that the curvature FA and covariant derivative $DA\ \phi$ satisfy the Bogomolny equations

$$F_A = *D_A\phi$$

and the boundary conditions.

$$\|\phi\| = 1 - \frac{m}{r} + \theta(r^2), \quad \|D_A\phi\| = \mathcal{O}(r^2)$$

Pure mathematical advances in the theory of monopoles from the 1980s onwards have often proceeded on the basis of physically motived questions.

The equations themselves are invariant under gauge transformation and orientation-preserving symmetries. When γ is large, $\phi/\|\phi\|$ defines a mapping from a 2-sphere of radius γ in $\mathbf{R}^3$ to an adjoint orbit G/k and the homotopy class of this mapping is called the magnetic charge. Most work has been done in the case G = SU(2), where the charge is a positive integer k. The absolute minimum value of the functional is then $8\pi k$ and the coefficient m in the asymptotic expansion of $\phi/\|\phi\|$ is $k/2$.

The first SU(2) solution was found by E. B. Bogomolny, J. K. Parasad and C. M. Sommerfield in 1975. It is spherically symmetric of charge 1 and has the form

$$A = \left(\frac{1}{\sinh\gamma} - \frac{1}{\gamma}\right) \epsilon_{ijk}\frac{x_j}{\gamma}\sigma_k\, dx_i,$$

$$\phi = \left(\frac{1}{\tanh\gamma} - \frac{1}{\gamma}\right) \frac{x_j}{\gamma}\sigma_i$$

In 1980, C.H.Taubes[50] showed by a gluing construction that there exist solutions for all large k and soon after explicit axially-symmetric solutions were found. The first exact solution in the general case was given in 1981 by R.S.Ward for $k = 2$ in terms of elliptic function.

There are two ways of solving the Bogomolny equations. The first is by twistor methods. In the formulation of N.J. Hitchin,[51] an arbitrary solution corresponds to a holomorphic vector bundle over the complex surface TP^1, the tangent bundle of the projective line. This is naturally isomorphic to the space of oriented straight lines in $\mathbf{R}^3$.

The boundary condition show that the holomorphic bundle is an extension of line bundles determined by a compact algebraic curve of genus $(k-1)^2$ (the spectral curve) in TP^1, satisfying certain constraints.

The second method, due to W.Nahm,[52] involves solving an eigen value problem for the coupled Dirac operator and transforming the equations with their boundary conditions into a system of ordinary differential equations, the Nahm equations.

$$\frac{dT_1}{ds} = [T_2, T_3],\quad \frac{dT_2}{ds} = [T_3, T_1],\quad \frac{dT_3}{ds} = [T_1, T_2]$$

where $Ti(s)$ is a $k \times k$-matrix valued function on (0,2).

Both constructions are based on analogous procedures for instantons, the key observation due to N.S.Manton being of the self-dual Yang–Mills equations (c.f. also Yang–Mills field) in $\mathbf{R}^4$.

The equivalence of the two methods for SU(2) and their general applicability was established in[53] (see also[54]). Explicit formulas for A and ϕ are difficult to obtain by either method, despite some exact solutions of Nahm's equations in symmetric situations.[55]

The case of a more general Lie group G, where the stabilizer of ϕ at infinity is a maximal torus, was treated by M.K.Murray[56] from the twistor point of view, where the single spectral curve of an SU(2)-monopole is replaced by a collection of curves indexed by the vortices of the Dynkin diagram of G. The corresponding Nahm construction was designed by J.Hustubise and Murray.[57]

The moduli space (c.f. also Moduli theory) of all SU(2) monopoles of charge k up to gauge equivalence was shown by Taubes[58] to be a smooth non-compact manifold of dimension $4k - 1$. Restricting to gauge transformations that preserve the connection at infinity gives a $4k$-dimensional manifold Mk, which is a circle bundle over the true moduli space and carries a natural complete hyper-Kähler metric[59] (c.f. also Kähler–Einstein manifold). With suspected to any of the complex structures of the hyper-Kähler family, this manifold is holomorphically equivalent to the space of based rational mapping of degree k from P_1 to itself.[60]

The metric is known in twistor terms,[59] and its Kähler potential can be written using the Riemann theta functions of the spectral curve,[54] but only the case $k = 2$ is known in a more conventional and usable form[59] (as of 2000). This Atiyah–Hitchin manifold, the Einstein Taub-NUT metric and $\mathbf{R}^4$ are the only 4-dimensional complete hyper-Kähler manifolds with a non-triholomorphic SU(2) action. Its geodesics have been studied and a programme of Manton concerning monopole dynamics put into effect. Further dynamical features have been elucidated by numerical and analytical techniques.

A cyclic k-fold conering of Mk splits isometrically is a product $Mk \times S^1 \times \mathbf{R}^3$, where Mk is the space of strongly centred monopoles. This space features in an application of S-duality in theoretical physics, and in[61] G.B.Segal and A.Selby studied its topology and the L^2 harmonic forms defined on it, partially confirming the physical prediction.

Magnetic monopole on hyperbolic three-space were investigated from the twistor point of view by M. F. Atiyah[62] (replacing the complex surface TP^1 by the complement of the anti-diagonal in $P^1 \times P^1$) and in terms of discrete Nahm equations by Murray and M. A. Singer.[63]

37.10   See also

- Bogomolny equations

- Dirac string
- Dyon
- Felix Ehrenhaft
- Gauss's law for magnetism
- Halbach array
- Instanton
- Meron
- Soliton
- 't Hooft–Polyakov monopole
- Wu–Yang monopole

37.11 Notes

[1] Dark Cosmos: In Search of Our Universe's Missing Mass and Energy, by Dan Hooper, p192

[2] Particle Data Group summary of magnetic monopole search

[3] Wen, Xiao-Gang; Witten, Edward, *Electric and magnetic charges in superstring models*, Nuclear Physics B, Volume 261, p. 651–677

[4] S. Coleman, *The Magnetic Monopole 50 years Later*, reprinted in *Aspects of Symmetry*

[5] C. Castelnovo, R. Moessner and S. L. Sondhi (January 3, 2008). "Magnetic monopoles in spin ice". *Nature* **451**: 42–45. arXiv:0710.5515. Bibcode:2008Natur.451...42C. doi:10.1038/nature06433.

[6] Ray, M.W.; Ruokokoski, E.; Kandel, S.; Möttönen, M.; Hall, D. S. (2014). "Observation of Dirac monopoles in a synthetic magnetic field". *Nature* **505** (7485): 657–660. arXiv:1408.3133. Bibcode:2014Natur.505..657R. doi:10.1038/nature12954. ISSN 0028-0836.

[7] The encyclopædia britannica, Volume 17, p352

[8] Principles of Physics by William Francis Magie, p424

[9] Pierre Curie, *Sur la possibilité d'existence de la conductibilité magnétique et du magnétisme libre* (*On the possible existence of magnetic conductivity and free magnetism*), Séances de la Société Française de Physique (Paris), p76 (1894). (French)Free access online copy.

[10] Paul Dirac, "Quantised Singularities in the Electromagnetic Field". Proc. Roy. Soc. (London) **A 133**, 60 (1931). Free web link.

[11] Lecture notes by Robert Littlejohn, University of California, Berkeley, 2007–8

[12] P. B. Price; E. K. Shirk; W. Z. Osborne; L. S. Pinsky (August 25, 1975). "Evidence for Detection of a Moving Magnetic Monopole". *Physical Review Letters* (American Physical Society) **35** (8): 487–490. Bibcode:1975PhRvL..35..487P. doi:10.1103/PhysRevLett.35.487.

[13] Blas Cabrera (May 17, 1982). "First Results from a Superconductive Detector for Moving Magnetic Monopoles". *Physical Review Letters* (American Physical Society) **48** (20): 1378–1381. Bibcode:1982PhRvL..48.1378C. doi:10.1103/PhysRevLett.48.1378.

[14] Milton p.60

[15] Polchinski, arXiv 2003

[16] "Magnetic Monopoles Detected in a Real Magnet for the First Time". Science Daily. September 4, 2009. Retrieved September 4, 2009.

[17] Making magnetic monopoles, and other exotica, in the lab, Symmetry Breaking, January 29, 2009. Retrieved January 31, 2009.

[18] Magnetic monopoles spotted in spin ices, September 3, 2009. "Oleg Tchernyshyov at Johns Hopkins University [a researcher in this field] cautions that the theory and experiments are specific to spin ices, and are not likely to shed light on magnetic monopoles as predicted by Dirac."

[19] Elizabeth Gibney (29 January 2014). "Quantum cloud simulates magnetic monopole". *Nature (news section)*. doi:10.1038/nature.2014.1461 "This is not the first time that physicists have created monopole analogues. In 2009, physicists observed magnetic monopoles in a crystalline material called spin ice, which, when cooled to near-absolute zero, seems to fill with atom-sized, classical monopoles. These are magnetic in a true sense, but cannot be studied individually. Similar analogues have also been seen in other materials, such as in superfluid helium.... Steven Bramwell, a physicist at University College London who pioneered work on monopoles in spin ices, says that the [2014 experiment led by David Hall] is impressive, but that what it observed is not a Dirac monopole in the way many people might understand it. "There's a mathematical analogy here, a neat and beautiful one. But they're not magnetic monopoles."

[20] The fact that the electric and magnetic fields can be written in a symmetric way is specific to the fact that space is three-dimensional. When the equations of electromagnetism are extrapolated to other dimensions, the magnetic field is described as being a rank-two antisymmetric tensor, whereas the electric field remains a true vector. In dimensions other than three, these two mathematical objects do not have the same number of components.

[21] http://www.ieeeghn.org/wiki/index.php/STARS:Maxwell%27s_Equations

[22] Parker, C.B. (1994). *McGraw-Hill Encyclopaedia of Physics* (2nd ed.). McGraw-Hill. ISBN 0-07-051400-3.

[23] M. Mansfield, C. O'Sullivan (2011). *Understanding Physics* (4th ed.). John Wiley & Sons. ISBN 978-0-47-0746370.

[24] F. Moulin (2001). "Magnetic monopoles and Lorentz force". *Nuovo Cimento B* **116** (8): 869–877. arXiv:math-ph/0203043. Bibcode:2001NCimB.116..869M.

[25] Wolfgang Rindler (November 1989). "Relativity and electromagnetism: The force on a magnetic monopole". *American Journal of Physics* (American Journal of Physics) **57** (11): 993–994. Bibcode:1989AmJPh..57..993R. doi:10.1119/1.15782.

[26] For the convention where magnetic charge has units of webers, see Jackson 1999. In particular, for Maxwell's equations, see section 6.11, equation (6.150), page 273, and for the Lorentz force law, see page 290, exercise 6.17(a). For the convention where magnetic charge has units of ampere-meters, see (for example) arXiv:physics/0508099v1, eqn (4).

[27] J.A. Heras, G. Baez (2009). "The covariant formulation of Maxwell's equations expressed in a form independent of specific units". arXiv:0901.0194.

[28] F. Moulin (2002). "Magnetic monopoles and Lorentz force". arXiv:math-ph/0203043.

[29] Shanmugadhasan, S. "The Dynamical Theory of Magnetic Monopoles", *Canadian Journal of Physics* Vol. 30, p. 218. (1952).

[30] Fryberger, D. "On Generalized Electromagnetism and Dirac Algebra", *Foundations of Physics* Vol. 19, p. 125 (1989).

[31] Jackson 1999, section 6.11.

[32] Jackson 1999, section 6.11, equation (6.153), page 275

[33] Nima Arkani-Hamed, Lubos Motl, Alberto Nicolis, Cumrun Vafa: The String Landscape, Black Holes and Gravity as the Weakest Force(arXiv:hep-th/0601001, JHEP 0706:060,2007)

[34] Zel'dovich, Ya.; Khlopov; M. Yu. Khlopov (1978). "On the concentration of relic monopoles in the universe". *Phys. Lett.* **B79** (3): 239–41. Bibcode:1978PhLB...79..239Z. doi:10.1016/0370-2693(78)90232-0.

[35] Preskill, John (1979). "Cosmological production of superheavy magnetic monopoles". *Phys. Rev. Lett.* **43** (19): 1365. Bibcode:1979PhRvL..43.1365P. doi:10.1103/PhysRevLett.43.1365.

[36] Preskill, John (1984). "Magnetic Monopoles". *Ann. Rev. Nucl. Part. Sci.* **34**: 461. Bibcode:1984ARNPS..34..461P. doi:10.1146/annurev.ns.34.120184.002333.

[37] Rees, Martin. (1998). *Before the Beginning* (New York: Basic Books) p. 185 ISBN 0-201-15142-1

[38] http://www.nature.com/nature/journal/v429/n6987/full/429010a.html

[39] Alvarez, Luis W. "Analysis of a Reported Magnetic Monopole". In ed. Kirk, W. T. "Proceedings of the 1975 international symposium on lepton and photon interactions at high energies". International symposium on lepton and photon interactions at high energies, Aug 21, 1975. p. 967.

[40] „If the structures of the magnetic fields appear to be magnetic monopoles, that are macroscopic in size, then this is a wormhole." Taken from All About Space, issue No. 24, April 2014, item „Could wormholes really exist?"

[41] "Quantised Singularities in the Electromagnetic Field" Paul Dirac, *Proceedings of the Royal Society*, May 29, 1931. Retrieved February 1, 2014.

[42] Elizabeth Gibney (29 January 2014). "Quantum cloud simulates magnetic monopole". *Nature (news section)*. doi:10.1038/nature.2014.14612.

[43] Zhong, Fang; Naoto Nagosa, Mei S. Takahashi, Atsushi Asamitsu, Roland Mathieu, Takeshi Ogasawara, Hiroyuki Yamada, Masashi Kawasaki, Yoshinori Tokura, Kiyoyuki Terakura (October 3, 2003). "The Anomalous Hall Effect and Magnetic Monopoles in Momentum Space". *Science* 302 (5642) 92–95. doi:10.1126/science.1089408. ISSN 1095-9203. http://www.sciencemag.org/cgi/content/abstract/302/5642/92. Retrieved August 2, 2007.

[44] Inducing a Magnetic Monopole with Topological Surface States, American Association for the Advancement of Science (AAAS) *Science Express* magazine, Xiao-Liang Qi, Rundong Li, Jiadong Zang, Shou-Cheng Zhang, January 29, 2009. Retrieved January 31, 2009.

[45] *Artificial Magnetic Monopoles Discovered*

[46] S. T. Bramwell, S. R. Giblin, S. Calder, R. Aldus, D. Prabhakaran, T. Fennell (15 October 2009). "Measurement of the charge and current of magnetic monopoles in spin ice". *Nature* **461**: 956–959. arXiv:0907.0956. Bibcode:2009Natur.461..956B. doi:10.1038/nature08500. PMID 19829376.

[47] S. R. Giblin, S. T. Bramwell, P. C. W. Holdsworth, D. Prabhakaran & I. Terry (February 13, 2011). "Creation and measurement of long-lived magnetic monopole currents in spin ice" **7** (3). Nature Physics. Bibcode:2011NatPh...7..252G. doi:10.1038/nphys1896. Retrieved February 28, 2011.

[48] D.J.P. Morris, D.A. Tennant, S.A. Grigera, B. Klemke, C. Castelnovo, R. Moessner, C. Czter-nasty, M. Meissner, K.C. Rule, J.-U. Hoffmann, K. Kiefer, S. Gerischer, D. Slobinsky, and R.S. Perry (September 3, 2009) [2009-07-09]. "Dirac Strings and Magnetic Monopoles in Spin Ice $Dy_2Ti_2O_7$". *Science* **326** (5951): 411–4. arXiv:1011.1174. Bibcode:2009Sci...326..411M. doi:10.1126/science.1178868. PMID 19729617.

[49] Ville Pietilä, Mikko Möttönen, *Creation of Dirac Monopoles in Spinor Bose–Einstein Condensates*, Phys. Rev. Lett. 103, 030401 (2009)

[50] A.Jaffe, C.H.Taubes (1980). *Vortices and monopoles*.

[51] N.J. Hitchin (1982). *Monopoles and geodesics*.

[52] W.Nahm (1982). *The construction of all self-dual monopoles by the ADHM method*.

[53] N.J. Hitchin (1983). *On the construction of monopoles*.

[54] N.J. Hitchin (1999). *Integrable sustems in Riemannian geometry* (K.Uhlenbeck ed.). C-L.Terng (ed.).

[55] N.J. Hitchin, N.S. Manton, M.K. Murray (1995). *Symmetric Monopoles*.

[56] M.K.Murray (1983). *Monopoles and spectral curves for arbitrary Lie groups*.

[57] J.Hurtubise, M.K.Murray (1989). *On the construction of Monopoles for the classical groups*.

[58] C.H.Taubes (1983). *Stability in Yang–Mills theories*.

[59] M.F. Atiyah; N.J. Hitchin (1988). *The geometry and dynamics of magnetic monopoles*. Princeton Univ.Press.

[60] S.K.Donaldson (1984). *Nahm's equations and the classification of monopoles*.

[61] G.B.Segal, A.Selby (1996). *The cohomology of the space of magnetic monopoles*.

[62] M.F.Atiyah (1987). *Magnetic monopoles in hyperbolic space, Vector bundles on algebraic varieties*. Oxford University Press.

[63] M.K.Murray (2000). *On the complete integrability of the discrete Nahm equations*.

37.12 References

- Brau, Charles A. (2004). *Modern Problems in Classical Electrodynamics*. Oxford University Press. ISBN 0-19-514665-4.

- Hitchin, N.J.; Murray, M.K. (1988). *Spectral curves and the ADHM method*.

- Jackson, John David (1999). *Classical Electrodynamics* (3rd ed.). New York: Wiley. ISBN 0-471-30932-X.

- Milton, Kimball A. (June 2006). "Theoretical and experimental status of magnetic monopoles". *Reports on Progress in Physics* **69** (6): 1637–1711. arXiv:hep-ex/0602040. Bibcode:2006RPPh...69.1637M. doi:10.1088/0034-4885/69/6/R02.

- Shnir, Yakov M. (2005). *Magnetic Monopoles*. Springer-Verlag. ISBN 3-540-25277-0.

- Sutcliffe, P.M. (1997). *BPS monopoles*.

37.13 External links

- Magnetic Monopole Searches (lecture notes)

- Particle Data Group summary of magnetic monopole search

- 'Race for the Pole' Dr David Milstead Freeview 'Snapshot' video by the Vega Science Trust and the BBC/OU.

- Interview with Jonathan Morris about magnetic monopoles and magnetic monopole quasiparticles. Drillingsraum, April 16, 2010

- *Nature*, 2009

- *Sciencedaily*, 2009

- H. Kadowaki, N. Doi, Y. Aoki, Y.Tabata, T.J. Sato, J.W. Lynn, K. Matsuhira, Z. Hiroi (2009). "Observation of Magnetic Monopoles in Spin Ice". arXiv:0908.3568.

- *Video of lecture by Paul Dirac on magnetic monopoles*, 1975 on YouTube

This article incorporates material from N. Hitchin (2001), "Magnetic Monopole", in Hazewinkel, Michiel, Encyclopedia of Mathematics, Springer, ISBN 978-1-55608-010-4, which is licensed under the Creative Commons Attribution/Share-Alike License and GNU Free Documentation License.

Chapter 38

Cosmic string

Not to be confused with string in string theory.

Cosmic strings are hypothetical 1-dimensional topological defects which may have formed during a symmetry breaking phase transition in the early universe when the topology of the vacuum manifold associated to this symmetry breaking was not simply connected. It is expected that at least one string per Hubble volume is formed. Their existence was first contemplated by the theoretical physicist Tom Kibble in the 1970s.

The formation of cosmic strings is somewhat analogous to the imperfections that form between crystal grains in solidifying liquids, or the cracks that form when water freezes into ice. The phase transitions leading to the production of cosmic strings are likely to have occurred during the earliest moments of the universe's evolution, just after cosmological inflation, and are a fairly generic prediction in both Quantum field theory and String theory models of the Early universe.

38.1 Theories containing cosmic strings

In string theory the role of cosmic strings can be played by the fundamental strings (or F-strings) themselves that define the theory perturbatively, by D-strings which are related to the F-strings by weak-strong or so called S-duality, or higher-dimensional D-, NS- or M-branes that are partially wrapped on compact cycles associated to extra spacetime dimensions so that only one non-compact dimension remains.[1]

The prototypical example of a quantum field theory with cosmic strings is the Abelian Higgs model. The quantum field theory and string theory cosmic strings are expected to have many properties in common, but more research is needed to determine the precise distinguishing features. The F-strings for instance are fully quantum-mechanical and do not have a classical definition, whereas the field theory cosmic strings are almost exclusively treated classically.

38.2 Dimensions

Cosmic strings, if they exist, would be extremely thin with diameters of the same order of magnitude as that of a proton, i.e. ~ 1 fm, or smaller. Given that this scale is much smaller than any cosmological scale these strings are often studied in the zero width, or Nambu-Goto approximation. Under this assumption strings behave as one-dimensional objects and obey the Nambu-Goto action, which is classically equivalent to the Polyakov action that defines the bosonic sector of superstring theory.

In field theory, the string width is set by the scale of the symmetry breaking phase transition. In string theory, the string width is set (in the simplest cases) by the fundamental string scale, warp factors (associated to the spacetime curvature of an internal six-dimensional spacetime manifold) and/or the size of internal compact dimensions. (In string theory, the universe is either 10- or 11-dimensional, depending on the strength of interactions and the curvature of spacetime.)

38.3 Gravitation

A string is a geometrical deviation from Euclidean geometry in spacetime characterized by an angular deficit: a circle around the outside of a string would comprise a total angle less than 360°. From the general theory of relativity such a geometrical defect must be in tension, and would be manifested by mass. Even though cosmic strings are thought to be extremely thin, they would have immense density, and so would represent significant gravitational wave sources. A cosmic string about a kilometer in length may be more massive than the Earth.

However general relativity predicts that the gravitational potential of a straight string vanishes: there is no gravitational force on static surrounding matter. The only gravitational effect of a straight cosmic string is a relative deflection of matter (or light) passing the string on opposite sides (a purely topological effect). A closed cosmic string gravitates in a more conventional way.

During the expansion of the universe, cosmic strings would form a network of loops, and in the past it was thought that their gravity could have been responsible for the original clumping of matter into galactic superclusters. It is now calculated that their contribution to the structure formation in the universe is less than 10%.

38.3.1 Negative Mass Cosmic String

The standard model of cosmic string is a geometrical structure with an angle deficit, which thus is in tension and hence has positive mass. In 1995, Visser *et al.* proposed that cosmic string could theoretically also exist with an angle excess, and thus negative tension and hence negative mass. The stability of such exotic matter string is problematic; however, they suggested that if a negative mass string were to be wrapped around a wormhole in the early universe, such a wormhole could be stabilized sufficiently to exist in the present day.[2][3]

38.4 Observational evidence

It was once thought that the gravitational influence of cosmic strings might contribute to the large-scale clumping of matter in the universe, but all that is known today through galaxy surveys and precision measurements of the cosmic microwave background (CMB) fits an evolution out of random, gaussian fluctuations. These precise observations therefore tend to rule out a significant role for cosmic strings and currently it is known that the contribution of cosmic strings to the CMB cannot be more than 10%.

The violent oscillations of cosmic strings generically lead to the formation of cusps and kinks. These in turn cause parts of the string to pinch off into isolated loops. These loops have a finite lifespan and decay (primarily) via gravitational radiation. This radiation which leads to the strongest signal from cosmic strings may in turn be detectable in gravitational wave experiments, such as LIGO and LISA. An important open question is to what extent do the pinched off loops backreact or change the initial state of the emitting cosmic string—such backreaction effects are almost always neglected in computations and are known to be important, even for order of magnitude estimates.

Gravitational lensing of a galaxy by a straight section of a cosmic string would produce two identical, undistorted images of the galaxy. In 2003 a group led by Mikhail Sazhin reported the accidental discovery of two seemingly identical galaxies very close together in the sky, leading to speculation that a cosmic string had been found.[4] However, observations by the Hubble Space Telescope in January 2005 showed them to be a pair of similar galaxies, not two images of the same galaxy.[5][6] A cosmic string would produce a similar duplicate image of fluctuations in the cosmic microwave background, which might be detectable by the Planck Surveyor mission.[7] A 2013 analysis of data from Planck mission failed to find any evidence of Cosmic strings [8]

A second piece of evidence supporting cosmic string theory is a phenomenon observed in observations of the "double quasar" called Q0957+561A,B. Originally discovered by Dennis Walsh, Bob Carswell, and Ray Weymann in 1979, the double image of this quasar is caused by a galaxy positioned between it and the Earth. The gravitational lens effect of this intermediate galaxy bends the quasar's light so that it follows two paths of different lengths to Earth. The result is that we see two images of the same quasar, one arriving a short time after the other (about 417.1 days later). However, a team of astronomers at the Harvard-Smithsonian Center for Astrophysics led by Rudolph Schild studied the quasar and found that during the period between September 1994 and July 1995 the two images appeared to have no time delay; changes in the brightness of the two images occurred simultaneously on four separate occasions. Schild and his team believe that the only explanation for this observation is that a cosmic string passed between the Earth and the quasar during that time period traveling at very high speed and oscillating with a period of about 100 days.[9]

The earthbound Laser Interferometer Gravitational-Wave Observatory (LIGO) and especially the space-based gravitational wave detector Laser Interferometer Space Antenna (LISA) will search for gravitational waves and are likely to be sensitive enough to detect signals from cosmic strings, provided the relevant cosmic string tensions are not too small.

38.5 String theory and cosmic strings

During the early days of string theory both string theorists and cosmic string theorists believed that there was no direct connection between superstrings and cosmic strings (the names were chosen independently by analogy with ordinary string). The possibility of cosmic strings being produced in the early universe was first envisioned by quantum field theorist Tom Kibble in 1976, and this sprouted the first flurry of interest in the field. In 1985, during the first superstring revolution, Edward Witten contemplated on the possibility of fundamental superstrings having been produced in the early universe and stretched to macroscopic scales, in which case (following the nomenclature of Tom Kibble) they would then be referred to as cosmic superstrings. He concluded that had they been produced they would have either disintegrated into smaller strings before ever reaching macroscopic scales (in the case of Type I superstring theory), they would always appear as boundaries of domain walls whose tension would force the strings to collapse rather than grow to cosmic scales (in the context of Heterotic superstring theory), or having a characteristic energy scale close to the Planck energy they would be produced before cosmological inflation and hence be diluted away with the expansion of the universe and not be observable.

Much has changed since these early days, primarily due to the second superstring revolution. It is now known that string theory in addition to the fundamental strings which define the theory perturbatively also contains other one-dimensional objects, such as D-strings, and higher-dimensional objects such as D-branes, NS-branes and M-branes partially wrapped on compact internal spacetime dimensions, while being spatially extended in one non-compact dimension. The possibility of large compact dimensions and large warp factors allows strings with tension much lower than the Planck scale. Furthermore, various dualities that have been discovered point to the conclusion that actually all these apparently different types of string are just the same object as it appears in different regions of parameter space. These new developments have largely revived interest in cosmic strings, starting in the early 2000s.

In 2002, Henry Tye and collaborators predicted the production of cosmic superstrings during the last stages of brane inflation,[10] a string theory construction of the early universe that gives leads to an expanding universe and cosmological inflation. It was subsequently realized by string theorist Joseph Polchinski that the expanding Universe could have stretched a "fundamental" string (the sort which superstring theory considers) until it was of intergalactic size. Such a stretched string would exhibit many of the properties of the old "cosmic" string variety, making the older calculations useful again. As theorist Tom Kibble remarks, "string theory cosmologists have discovered cosmic strings lurking everywhere in the undergrowth". Older proposals for detecting cosmic strings could now be used to investigate superstring theory.

Superstrings, D-strings or the other stringy objects mentioned above stretched to intergalactic scales would radiate gravitational waves, which could be detected using experiments like LIGO and especially the space-based gravitational wave experiment LISA. They might also cause slight irregularities in the cosmic microwave background, too subtle to have been detected yet but possibly within the realm of future observability.

Note that most of these proposals depend, however, on the appropriate cosmological fundamentals (strings, branes, etc.), and no convincing experimental verification of these has been confirmed to date. Cosmic strings nevertheless provide a window into string theory. If cosmic strings are observed which is a real possibility for a wide range of cosmological string models this would provide the first experimental evidence of a string theory model underlying the structure of spacetime.

38.6 See also

- 0-dimensional topological defect: magnetic monopole
- 1-dimensional topological defect: cosmic string
- 2-dimensional topological defect: domain wall
- cosmic string loop stabilised by a fermionic supercurrent: vorton

38.7 References

[1] Copeland, Edmund J; Myers, Robert C; Polchinski, Joseph (2004). "Cosmic F- and D-strings". *Journal of High Energy Physics* **2004** (6): 013. arXiv:hep-th/0312067. Bibcode:2004JHEP...06..013C. doi:10.1088/1126-6708/2004/06/013.

[2] Cramer, John; Forward, Robert; Morris, Michael; Visser, Matt; Benford, Gregory; Landis, Geoffrey (1995). "Natural wormholes as gravitational lenses". *Physical Review D* **51** (6): 3117. arXiv:astro-ph/9409051. Bibcode:1995PhRvD..51.3117C. doi:10.1103/PhysRevD.51.3117.

[3] "Searching for a 'Subway to the Stars'" (Press release).

[4] Sazhin, M.; Longo, G.; Capaccioli, M.; Alcala, J. M.; Silvotti, R.; Covone, G.; Khovanskaya, O.; Pavlov, M.; Pannella, M. et al. (2003). "CSL-1: Chance projection effect or serendipitous discovery of a gravitational lens induced by a cosmic string?". *Monthly Notices of the Royal Astronomical Society* **343** (2): 353. arXiv:astro-ph/0302547. Bibcode:2003MNRAS.343..353S. doi:10.1046/j.1365-8711.2003.06568.x.

[5] Agol, Eric; Hogan, Craig; Plotkin, Richard (2006). "Hubble imaging excludes cosmic string lens". *Physical Review D* **73** (8): 87302. arXiv:astro-ph/0603838. Bibcode:2006PhRvD..73h7302A. doi:10.1103/PhysRevD.73.087302.

[6] Sazhin, M. V.; Capaccioli, M.; Longo, G.; Paolillo, M.; Khovanskaya, O. S.; Grogin, N. A.; Schreier, E. J.; Covone, G. (2006). "The true nature of CSL-1". arXiv:0601494 [astro-ph].

[7] Fraisse, Aurélien; Ringeval, Christophe; Spergel, David; Bouchet, François (2008). "Small-angle CMB temperature anisotropies induced by cosmic strings". *Physical Review D* **78** (4): 43535. arXiv:0708.1162. Bibcode:2008PhRvD..78d3535F. doi:10.1103/PhysRevD.78.043535.

[8] Planck Collaboration; Ade, P. A. R.; Aghanim, N.; Armitage-Caplan, C.; Arnaud, M.; Ashdown, M.; Atrio-Barandela, F.; Aumont, J. et al. (2013). "Planck 2013 results. XXV. Searches for cosmic strings and other topological defects". arXiv:1303.5085 [astro-ph.CO].

[9] Schild, R.; Masnyak, I. S.; Hnatyk, B. I.; Zhdanov, V. I. (2004). "Anomalous fluctuations in observations of Q0957+561 A,B: Smoking gun of a cosmic string?". *Astronomy and Astrophysics* **422** (2): 477. arXiv:astro-ph/0406434. Bibcode:2004A&A...422..477S. doi:10.1051/0004-6361:20040274.

[10] Sarangi, Saswat; Tye, S.-H.Henry (2002). "Cosmic string production towards the end of brane inflation". *Physics Letters B* **536** (3–4): 185. arXiv:hep-th/0204074. Bibcode:2002PhLB..536..185S. doi:10.1016/S0370-2693(02)01824-5.

- Dr. Kip Thorne, ITP & CalTech. *Spacetime Warps and the Quantum: A Glimpse of the Future.* Lecture slides and audio

38.8 External links

- An artistic perspective of Cosmic Strings

- A simulation of cosmic string

- http://www.damtp.cam.ac.uk/user/gr/public/cs_interact.html

- Sazhin, M.; Longo, G.; Capaccioli, M.; Alcala, J. M.; Silvotti, R.; Covone, G.; Khovanskaya, O.; Pavlov, M.; Pannella, M. et al. (2003). "CSL-1: Chance projection effect or serendipitous discovery of a gravitational lens induced by a cosmic string?". *Monthly Notices of the Royal Astronomical Society* **343** (2): 353. arXiv:astro-ph/0302547. Bibcode:2003MNRAS.343..353S. doi:10.1046/j.1365-8711.2003.06568.x.

- Schild, R.; Masnyak, I. S.; Hnatyk, B. I.; Zhdanov, V. I. (2004). "Anomalous fluctuations in observations of Q0957+561 A,B: Smoking gun of a cosmic string?". *Astronomy and Astrophysics* **422** (2): 477. arXiv:astro-ph/0406434. Bibcode:2004A&A...422..477S. doi:10.1051/0004-6361:20040274.

- Kibble, T. W. B. (2004). "Cosmic strings reborn?". arXiv:0410073 [astro-ph].

- Lo, Amy S.; Wright, Edward L. (2005). "Signatures of Cosmic Strings in the Cosmic Microwave Background". arXiv:0503120 [astro-ph].

- Sazhin, M.; Capaccioli, M.; Longo, G.; Paolillo, M.; Khovanskaya, O. (2006). "Further Spectroscopic Observations of the CSL 1 Object". *The Astrophysical Journal* **636**: L5. arXiv:astro-ph/0506400. Bibcode:2006ApJ...636L...5S. doi:10.1086/499429.

- Agol, Eric; Hogan, Craig; Plotkin, Richard (2006). "Hubble imaging excludes cosmic string lens". *Physical Review D* **73** (8): 87302. arXiv:astro-ph/0603838. Bibcode:2006PhRvD..73h7302A. doi:10.1103/PhysRevD.73.087302.

- Cosmic strings and superstrings on arxiv.org

Chapter 39

Domain wall (string theory)

For other uses, see Domain wall (disambiguation).

In physics, a **domain wall** is any of several similar things in string theory, magnetism, or optics. These phenomena can all be generically described as topological solitons that occur whenever a discrete symmetry is spontaneously broken.[1]

39.1 String theory

In string theory, a domain wall is a theoretical 2-dimensional singularity. A domain wall is meant to represent an object of codimension one embedded into space (a defect in space localized in one spatial dimension). For example, D8-branes are domain walls in type II string theory. In M-theory, the existence of Horava–Witten domain walls, "ends of the world" that carry an E8 gauge theory, is important for various relations between superstring theory and M-theory.

If domain walls exist, it seems plausible that gravitational waves would be violently emitted if two such walls collided. As the Laser Interferometer Gravitational-Wave Observatory and future observatories of its kind will search for direct evidence of gravitational waves, this phenomenon would be included as well in such searches.

39.2 See also

- Topological defect
- Cosmic string
- Membrane (M-theory)
- Gravitational singularity

39.3 References

[1] S. Weinberg, *The Quantum Theory of Fields*, Vol. 2. Chap 23, Cambridge University Press (1995).

Chapter 40

Inflation (cosmology)

"Inflation model" and "Inflation theory" redirect here. For a general rise in the price level, see Inflation. For other uses, see Inflation (disambiguation).

In physical cosmology, **cosmic inflation**, **cosmological inflation**, or just **inflation** is the exponential expansion of

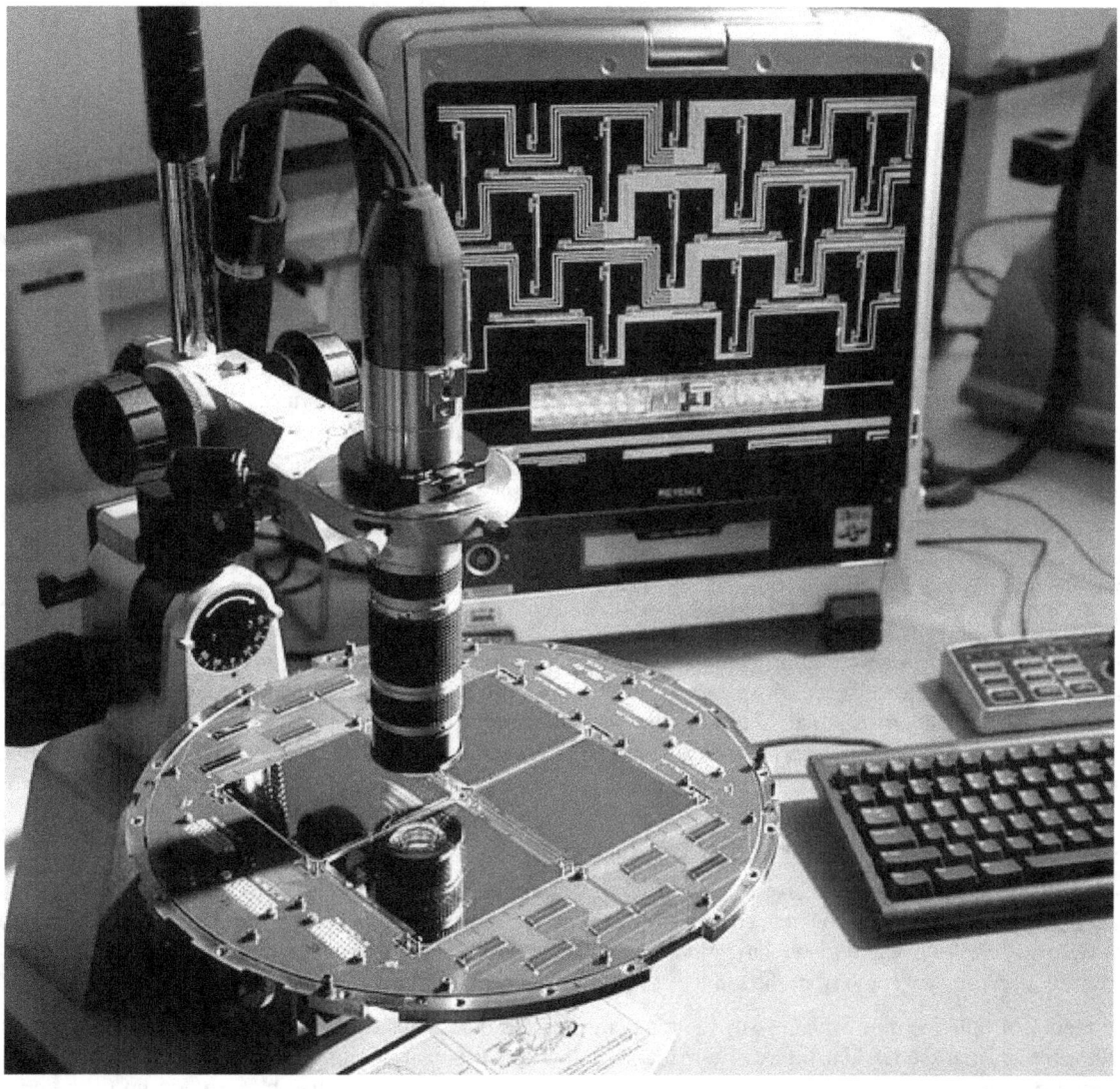

Evidence of gravitational waves in the infant universe may have been uncovered by the BICEP2 radio telescope.[1][2][3][4]

space in the early universe. The inflationary epoch lasted from 10^{-36} seconds after the Big Bang to sometime between

10^{-33} and 10^{-32} seconds. Following the inflationary period, the universe continues to expand, but at a less accelerated rate.

The inflationary hypothesis was developed in the 1980s by physicists Alan Guth and Andrei Linde.[5]

Inflation explains the origin of the large-scale structure of the cosmos. Quantum fluctuations in the microscopic inflationary region, magnified to cosmic size, become the seeds for the growth of structure in the universe (see galaxy formation and evolution and structure formation).[6] Many physicists also believe that inflation explains why the Universe appears to be the same in all directions (isotropic), why the cosmic microwave background radiation is distributed evenly, why the universe is flat, and why no magnetic monopoles have been observed.

While the detailed particle physics mechanism responsible for inflation is not known, the basic picture makes a number of predictions that have been confirmed by observation.[7][8] The hypothetical field thought to be responsible for inflation is called the inflaton.[9]

On 17 March 2014, astrophysicists of the BICEP2 collaboration announced the detection of inflationary gravitational waves in the B-mode power spectrum, which if confirmed, would provide clear experimental evidence for the theory of inflation.[1][2][3][4][10][11] However, on 19 June 2014, lowered confidence in confirming the findings was reported;[10][12][13] and on 19 September 2014, even more lowered confidence.[14][15]

40.1 Overview

Main article: Metric expansion of space

An expanding universe generally has a cosmological horizon, which, by analogy with the more familiar horizon caused by the curvature of the Earth's surface, marks the boundary of the part of the universe that an observer can see. Light (or other radiation) emitted by objects beyond the cosmological horizon never reaches the observer, because the space in between the observer and the object is expanding too rapidly.

The observable universe is one *causal patch* of a much larger unobservable universe; there are parts of the universe that cannot communicate with us yet. These parts of the universe are outside our current cosmological horizon. In the standard hot big bang model, without inflation, the cosmological horizon moves out, bringing new regions into view. Yet as a local observer sees these regions for the first time, they look no different from any other region of space the local observer has already seen: they have a background radiation that is at nearly exactly the same temperature as the background radiation of other regions, and their space-time curvature is evolving lock-step with ours. This presents a mystery: how did these new regions know what temperature and curvature they were supposed to have? They couldn't have learned it by getting signals, because they were not in communication with our past light cone before.[16][17]

Inflation answers this question by postulating that all the regions come from an earlier era with a big vacuum energy, or cosmological constant. A space with a cosmological constant is qualitatively different: instead of moving outward, the cosmological horizon stays put. For any one observer, the distance to the cosmological horizon is constant. With exponentially expanding space, two nearby observers are separated very quickly; so much so, that the distance between them quickly exceeds the limits of communications. The spatial slices are expanding very fast to cover huge volumes. Things are constantly moving beyond the cosmological horizon, which is a fixed distance away, and everything becomes homogeneous very quickly.

As the inflationary field slowly relaxes to the vacuum, the cosmological constant goes to zero, and space begins to expand normally. The new regions that come into view during the normal expansion phase are exactly the same regions that were pushed out of the horizon during inflation, and so they are necessarily at nearly the same temperature and curvature, because they come from the same little patch of space.

The theory of inflation thus explains why the temperatures and curvatures of different regions are so nearly equal. It also predicts that the total curvature of a space-slice at constant global time is zero. This prediction implies that the total ordinary matter, dark matter, and residual vacuum energy in the universe have to add up to the critical density, and the evidence strongly supports this. More strikingly, inflation allows physicists to calculate the minute differences in temperature of different regions from quantum fluctuations during the inflationary era, and many of these quantitative predictions have been confirmed.[18][19]

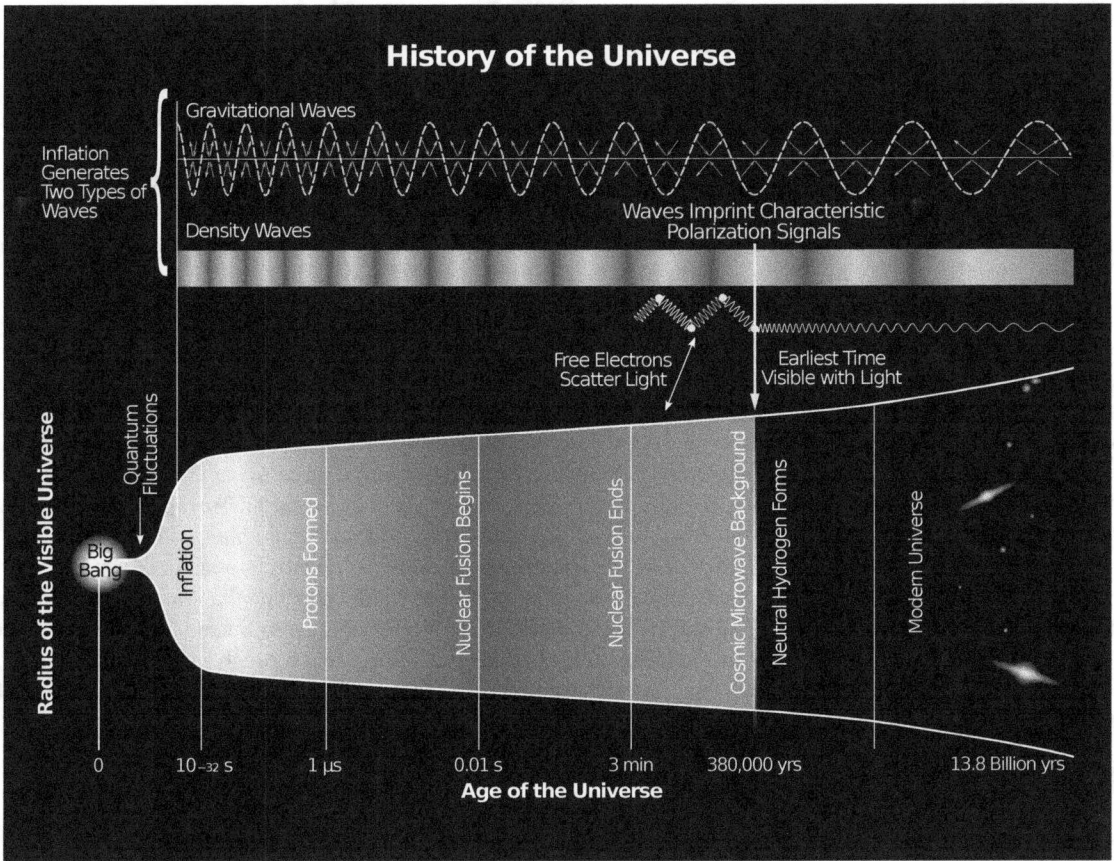

History of the Universe - gravitational waves are hypothesized to arise from cosmic inflation, a faster-than-light expansion just after the Big Bang (17 March 2014).[1][2][3]

40.1.1 Space expands

To say that space expands exponentially means that two inertial observers are moving farther apart with accelerating velocity. In stationary coordinates for one observer, a patch of an inflating universe has the following polar metric:[20][21]

$$ds^2 = -\left(1 - \Lambda r^2\right) dt^2 + \frac{1}{1 - \Lambda r^2}\, dr^2 + r^2\, d\Omega^2.$$

This is just like an inside-out black hole metric—it has a zero in the dt component on a fixed radius sphere called the cosmological horizon. Objects are drawn away from the observer at $r = 0$ towards the cosmological horizon, which they cross in a finite proper time. This means that any inhomogeneities are smoothed out, just as any bumps or matter on the surface of a black hole horizon are swallowed and disappear.

Since the space–time metric has no explicit time dependence, once an observer has crossed the cosmological horizon, observers closer in take its place. This process of falling outward and replacement points closer in are always steadily replacing points further out—an exponential expansion of space–time.

This steady-state exponentially expanding spacetime is called a de Sitter space, and to sustain it there must be a cosmological constant, a vacuum energy proportional to Λ everywhere. In this case, the equation of state is $p = -\rho$. The physical conditions from one moment to the next are stable: the rate of expansion, called the Hubble parameter, is nearly constant, and the scale factor of the universe is proportional to e^{Ht}. Inflation is often called a period of *accelerated expansion* because the distance between two fixed observers is increasing exponentially (i.e. at an accelerating rate as they move apart), while Λ can stay approximately constant (see deceleration parameter).

40.1.2 Few inhomogeneities remain

Cosmological inflation has the important effect of smoothing out inhomogeneities, anisotropies and the curvature of space. This pushes the universe into a very simple state, in which it is completely dominated by the inflaton field, the source of the cosmological constant, and the only significant inhomogeneities are the tiny quantum fluctuations in the inflaton. Inflation also dilutes exotic heavy particles, such as the magnetic monopoles predicted by many extensions to the Standard Model of particle physics. If the universe was only hot enough to form such particles *before* a period of inflation, they would not be observed in nature, as they would be so rare that it is quite likely that there are none in the observable universe. Together, these effects are called the inflationary "no-hair theorem"[22] by analogy with the no hair theorem for black holes.

The "no-hair" theorem works essentially because the cosmological horizon is no different from a black-hole horizon, except for philosophical disagreements about what is on the other side. The interpretation of the no-hair theorem is that the universe (observable and unobservable) expands by an enormous factor during inflation. In an expanding universe, energy densities generally fall, or get diluted, as the volume of the universe increases. For example, the density of ordinary "cold" matter (dust) goes down as the inverse of the volume: when linear dimensions double, the energy density goes down by a factor of eight; the radiation energy density goes down even more rapidly as the universe expands since the wavelength of each photon is stretched (redshifted), in addition to the photons being dispersed by the expansion. When linear dimensions are doubled, the energy density in radiation falls by a factor of sixteen (see the solution of the energy density continuity equation for an ultra-relativistic fluid). During inflation, the energy density in the inflaton field is roughly constant. However, the energy density in everything else, including inhomogeneities, curvature, anisotropies, exotic particles, and standard-model particles is falling, and through sufficient inflation these all become negligible. This leaves the universe flat and symmetric, and (apart from the homogeneous inflaton field) mostly empty, at the moment inflation ends and reheating begins.[23]

40.1.3 Key requirement

A key requirement is that inflation must continue long enough to produce the present observable universe from a single, small inflationary Hubble volume. This is necessary to ensure that the universe appears flat, homogeneous and isotropic at the largest observable scales. This requirement is generally thought to be satisfied if the universe expanded by a factor of at least 10^{26} during inflation.[24]

40.1.4 Reheating

Inflation is a period of supercooled expansion, when the temperature drops by a factor of 100,000 or so. (The exact drop is model dependent, but in the first models it was typically from 10^{27}K down to 10^{22}K.[25]) This relatively low temperature is maintained during the inflationary phase. When inflation ends the temperature returns to the pre-inflationary temperature; this is called *reheating* or thermalization because the large potential energy of the inflaton field decays into particles and fills the universe with Standard Model particles, including electromagnetic radiation, starting the radiation dominated phase of the Universe. Because the nature of the inflation is not known, this process is still poorly understood, although it is believed to take place through a parametric resonance.[26][27]

40.2 Motivations

Inflation resolves several problems in the Big Bang cosmology that were discovered in the 1970s.[28] Inflation was first discovered by Guth while investigating the problem of why no magnetic monopoles are seen today; he found that a positive-energy false vacuum would, according to general relativity, generate an exponential expansion of space. It was very quickly realised that such an expansion would resolve many other long-standing problems. These problems arise from the observation that to look like it does *today*, the universe would have to have started from very finely tuned, or "special" initial conditions at the Big Bang. Inflation attempts to resolve these problems by providing a dynamical mechanism that drives the universe to this special state, thus making a universe like ours much more likely in the context of the Big Bang theory.

40.2.1 Horizon problem

Main article: Horizon problem

The horizon problem is the problem of determining why the universe appears statistically homogeneous and isotropic in accordance with the cosmological principle.[29][30][31] For example, molecules in a canister of gas are distributed homogeneously and isotropically because they are in thermal equilibrium: gas throughout the canister has had enough time to interact to dissipate inhomogeneities and anisotropies. The situation is quite different in the big bang model without inflation, because gravitational expansion does not give the early universe enough time to equilibrate. In a big bang with only the matter and radiation known in the Standard Model, two widely separated regions of the observable universe cannot have equilibrated because they move apart from each other faster than the speed of light—thus have never come into causal contact: in the history of the universe, back to the earliest times, it has not been possible to send a light signal between the two regions. Because they have no interaction, it is difficult to explain why they have the same temperature (are thermally equilibrated). This is because the Hubble radius in a radiation or matter-dominated universe expands much more quickly than physical lengths and so points that are out of communication are coming into communication. Historically, two proposed solutions were the *Phoenix universe* of Georges Lemaître[32] and the related oscillatory universe of Richard Chase Tolman,[33] and the Mixmaster universe of Charles Misner.[30][34] Lemaître and Tolman proposed that a universe undergoing a number of cycles of contraction and expansion could come into thermal equilibrium. Their models failed, however, because of the buildup of entropy over several cycles. Misner made the (ultimately incorrect) conjecture that the Mixmaster mechanism, which made the universe *more* chaotic, could lead to statistical homogeneity and isotropy.

40.2.2 Flatness problem

Main article: Flatness problem

Another problem is the flatness problem (which is sometimes called one of the Dicke coincidences, with the other being the cosmological constant problem).[35][36] It had been known in the 1960s that the density of matter in the universe was comparable to the critical density necessary for a flat universe (that is, a universe whose large scale geometry is the usual Euclidean geometry, rather than a non-Euclidean hyperbolic or spherical geometry).[37]:61

Therefore, regardless of the shape of the universe the contribution of spatial curvature to the expansion of the universe could not be much greater than the contribution of matter. But as the universe expands, the curvature redshifts away more slowly than matter and radiation. Extrapolated into the past, this presents a fine-tuning problem because the contribution of curvature to the universe must be exponentially small (sixteen orders of magnitude less than the density of radiation at big bang nucleosynthesis, for example). This problem is exacerbated by recent observations of the cosmic microwave background that have demonstrated that the universe is flat to the accuracy of a few percent.[38]

40.2.3 Magnetic-monopole problem

The magnetic monopole problem (sometimes called the exotic-relics problem) says that if the early universe were very hot, a large number of very heavy, stable magnetic monopoles would be produced. This is a problem with Grand Unified Theories, which proposes that at high temperatures (such as in the early universe) the electromagnetic force, strong, and weak nuclear forces are not actually fundamental forces but arise due to spontaneous symmetry breaking from a single gauge theory.[39] These theories predict a number of heavy, stable particles that have not yet been observed in nature. The most notorious is the magnetic monopole, a kind of stable, heavy "knot" in the magnetic field.[40][41] Monopoles are expected to be copiously produced in Grand Unified Theories at high temperature,[42][43] and they should have persisted to the present day, to such an extent that they would become the primary constituent of the universe.[44][45] Not only is that not the case, but all searches for them have failed, placing stringent limits on the density of relic magnetic monopoles in the universe.[46] A period of inflation that occurs below the temperature where magnetic monopoles can be produced would offer a possible resolution of this problem: monopoles would be separated from each other as the universe around them expands, potentially lowering their observed density by many orders of magnitude. Though, as cosmologist Martin Rees has written, "Skeptics about exotic physics might not be hugely impressed by a theoretical argument to explain the absence of particles that are themselves only hypothetical. Preventive medicine can readily seem 100 percent effective against a disease that doesn't exist!"[47]

40.3 History

40.3.1 Precursors

In the early days of General Relativity, Albert Einstein introduced the cosmological constant to allow a static solution, which was a three-dimensional sphere with a uniform density of matter. A little later, Willem de Sitter found a highly symmetric inflating universe, which described a universe with a cosmological constant that is otherwise empty.[48] It was discovered that Einstein's solution is unstable, and if there are small fluctuations, it eventually either collapses or turns into de Sitter's.

In the early 1970s Zeldovich noticed the serious flatness and horizon problems of big bang cosmology; before his work, cosmology was presumed to be symmetrical on purely philosophical grounds. In the Soviet Union, this and other considerations led Belinski and Khalatnikov to analyze the chaotic BKL singularity in General Relativity. Misner's Mixmaster universe attempted to use this chaotic behavior to solve the cosmological problems, with limited success.

In the late 1970s, Sidney Coleman applied the instanton techniques developed by Alexander Polyakov and collaborators to study the fate of the false vacuum in quantum field theory. Like a metastable phase in statistical mechanics—water below the freezing temperature or above the boiling point—a quantum field would need to nucleate a large enough bubble of the new vacuum, the new phase, in order to make a transition. Coleman found the most likely decay pathway for vacuum decay and calculated the inverse lifetime per unit volume. He eventually noted that gravitational effects would be significant, but he did not calculate these effects and did not apply the results to cosmology.

In the Soviet Union, Alexei Starobinsky noted that quantum corrections to general relativity should be important in the early universe. These generically lead to curvature-squared corrections to the Einstein–Hilbert action and a form of $f(R)$ modified gravity. The solution to Einstein's equations in the presence of curvature squared terms, when the curvatures are large, leads to an effective cosmological constant. Therefore, he proposed that the early universe went through a de Sitter phase, an inflationary era.[49] This resolved the problems of cosmology, and led to specific predictions for the corrections to the microwave background radiation, corrections that were calculated in detail shortly afterwards.

In 1978, Zeldovich noted the monopole problem, which was an unambiguous quantitative version of the horizon problem, this time in a fashionable subfield of particle physics, which led to several speculative attempts to resolve it. In 1980, working in the west, Alan Guth realized that false vacuum decay in the early universe would solve the problem, leading him to propose scalar driven inflation. Starobinsky's and Guth's scenarios both predicted an initial deSitter phase, differing only in the details of the mechanism.

40.3.2 Early inflationary models

According to Andrei Linde, the earliest theory of inflation was proposed by Erast Gliner (1965) but the theory was not taken seriously except by Andrei Sakharov, 'who made an attempt to calculate density perturbations produced in this scenario." [50] Independently, inflation was proposed in January 1980 by Alan Guth as a mechanism to explain the nonexistence of magnetic monopoles;[51][52] it was Guth who coined the term "inflation".[5] At the same time, Starobinsky argued that quantum corrections to gravity would replace the initial singularity of the universe with an exponentially expanding deSitter phase.[53] In October 1980, Demosthenes Kazanas suggested that exponential expansion could eliminate the particle horizon and perhaps solve the horizon problem,[54] while Sato suggested that an exponential expansion could eliminate domain walls (another kind of exotic relic).[55] In 1981 Einhorn and Sato[56] published a model similar to Guth's and showed that it would resolve the puzzle of the magnetic monopole abundance in Grand Unified Theories. Like Guth, they concluded that such a model not only required fine tuning of the cosmological constant, but also would very likely lead to a much too granular universe, i.e., to large density variations resulting from bubble wall collisions.

Guth proposed that as the early universe cooled, it was trapped in a false vacuum with a high energy density, which is much like a cosmological constant. As the very early universe cooled it was trapped in a metastable state (it was supercooled), which it could only decay out of through the process of bubble nucleation via quantum tunneling. Bubbles of true vacuum spontaneously form in the sea of false vacuum and rapidly begin expanding at the speed of light. Guth recognized that this model was problematic because the model did not reheat properly: when the bubbles nucleated, they did not generate any radiation. Radiation could only be generated in collisions between bubble walls. But if inflation lasted long enough to solve the initial conditions problems, collisions between bubbles became exceedingly rare. In any one causal patch it is likely that only one bubble will nucleate.

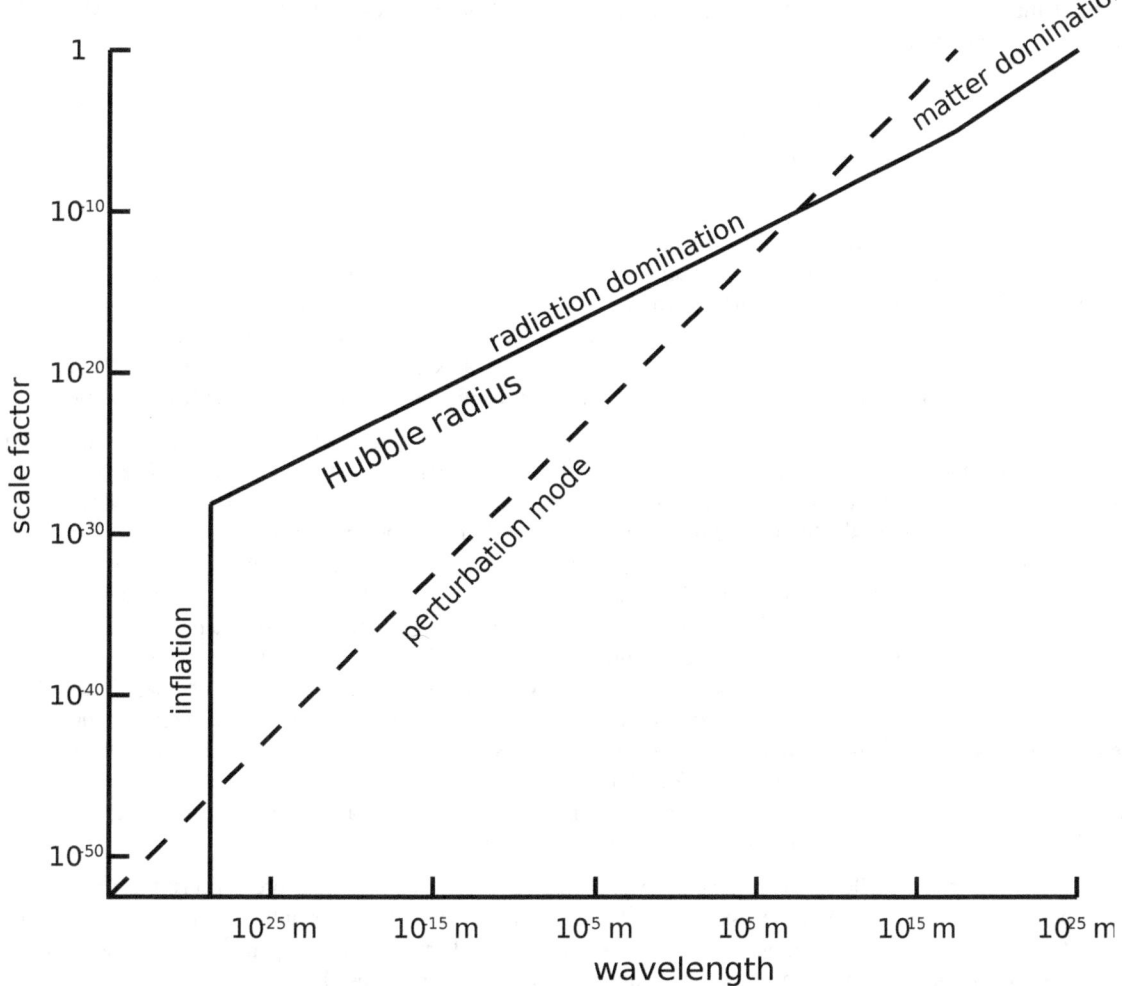

The physical size of the Hubble radius (solid line) as a function of the linear expansion (scale factor) of the universe. During cosmological inflation, the Hubble radius is constant. The physical wavelength of a perturbation mode (dashed line) is also shown. The plot illustrates how the perturbation mode grows larger than the horizon during cosmological inflation before coming back inside the horizon, which grows rapidly during radiation domination. If cosmological inflation had never happened, and radiation domination continued back until a gravitational singularity, then the mode would never have been outside the horizon in the very early universe, and no causal mechanism could have ensured that the universe was homogeneous on the scale of the perturbation mode.

40.3.3 Slow-roll inflation

The bubble collision problem was solved by Andrei Linde[57] and independently by Andreas Albrecht and Paul Steinhardt[58] in a model named *new inflation* or *slow-roll inflation* (Guth's model then became known as *old inflation*). In this model, instead of tunneling out of a false vacuum state, inflation occurred by a scalar field rolling down a potential energy hill. When the field rolls very slowly compared to the expansion of the universe, inflation occurs. However, when the hill becomes steeper, inflation ends and reheating can occur.

40.3.4 Effects of asymmetries

Eventually, it was shown that new inflation does not produce a perfectly symmetric universe, but that tiny quantum fluctuations in the inflaton are created. These tiny fluctuations form the primordial seeds for all structure created in the later universe.[59] These fluctuations were first calculated by Viatcheslav Mukhanov and G. V. Chibisov in the Soviet Union in analyzing Starobinsky's similar model.[60][61][62] In the context of inflation, they were worked out independently of the work of Mukhanov and Chibisov at the three-week 1982 Nuffield Workshop on the Very Early Universe at Cambridge University.[63] The fluctuations were calculated by four groups working separately over the course of the workshop: Stephen Hawking;[64] Starobinsky;[65] Guth and So-Young Pi;[66] and James M. Bardeen, Paul Steinhardt and Michael Turner.[67]

40.4 Observational status

Inflation is a mechanism for realizing the cosmological principle, which is the basis of the standard model of physical cosmology: it accounts for the homogeneity and isotropy of the observable universe. In addition, it accounts for the observed flatness and absence of magnetic monopoles. Since Guth's early work, each of these observations has received further confirmation, most impressively by the detailed observations of the cosmic microwave background made by the Wilkinson Microwave Anisotropy Probe (WMAP) spacecraft.[18] This analysis shows that the universe is flat to an accuracy of at least a few percent, and that it is homogeneous and isotropic to a part in 100,000.

In addition, inflation predicts that the structures visible in the universe today formed through the gravitational collapse of perturbations that were formed as quantum mechanical fluctuations in the inflationary epoch. The detailed form of the spectrum of perturbations called a nearly-scale-invariant Gaussian random field (or Harrison–Zel'dovich spectrum) is very specific and has only two free parameters, the amplitude of the spectrum and the *spectral index*, which measures the slight deviation from scale invariance predicted by inflation (perfect scale invariance corresponds to the idealized de Sitter universe).[68] Inflation predicts that the observed perturbations should be in thermal equilibrium with each other (these are called *adiabatic* or *isentropic* perturbations). This structure for the perturbations has been confirmed by the WMAP spacecraft and other cosmic microwave background experiments,[18] and galaxy surveys, especially the ongoing Sloan Digital Sky Survey.[69] These experiments have shown that the one part in 100,000 inhomogeneities observed have exactly the form predicted by theory. Moreover, there is evidence for a slight deviation from scale invariance. The *spectral index*, n_s is equal to one for a scale-invariant spectrum. The simplest models of inflation predict that this quantity is between 0.92 and 0.98.[70][71][72][73] From the data taken by the WMAP spacecraft it can be inferred that $n_s = 0.963 \pm 0.012$,[74] implying that it differs from one at the level of two standard deviations (2σ). This is considered an important confirmation of the theory of inflation.[18]

A number of theories of inflation have been proposed that make radically different predictions, but they generally have much more fine tuning than is necessary.[70][71] As a physical model, however, inflation is most valuable in that it robustly predicts the initial conditions of the universe based on only two adjustable parameters: the spectral index (that can only change in a small range) and the amplitude of the perturbations. Except in contrived models, this is true regardless of how inflation is realized in particle physics.

Occasionally, effects are observed that appear to contradict the simplest models of inflation. The first-year WMAP data suggested that the spectrum might not be nearly scale-invariant, but might instead have a slight curvature.[75] However, the third-year data revealed that the effect was a statistical anomaly.[18] Another effect has been remarked upon since the first cosmic microwave background satellite, the Cosmic Background Explorer: the amplitude of the quadrupole moment of the cosmic microwave background is unexpectedly low and the other low multipoles appear to be preferentially aligned with the ecliptic plane. Some have claimed that this is a signature of non-Gaussianity and thus contradicts the simplest models of inflation. Others have suggested that the effect may be due to other new physics, foreground contamination, or even publication bias.[76]

An experimental program is underway to further test inflation with more precise measurements of the cosmic microwave background. In particular, high precision measurements of the so-called "B-modes" of the polarization of the background radiation could provide evidence of the gravitational radiation produced by inflation, and could also show whether the energy scale of inflation predicted by the simplest models (10^{15}–10^{16} GeV) is correct.[71][72] In March 2014, it was announced that B-mode polarization of the background radiation consistent with that predicted from inflation had been demonstrated by a South Pole experiment, a collaboration led by four principal investigators from the California Institute of Technology, Harvard University, Stanford University, and the University of Minnesota BICEP2.[1][2] Other potentially corroborating measurements are expected to be performed by the Planck spacecraft, although it is unclear if the signal will be visible, or if contamination from foreground sources will interfere with these measurements.[77] Other forthcoming measurements, such as those of 21 centimeter radiation (radiation emitted and absorbed from neutral hydrogen before the first stars turned on), may measure the power spectrum with even greater resolution than the cosmic microwave background and galaxy surveys, although it is not known if these measurements will be possible or if interference with radio sources on earth and in the galaxy will be too great.[78]

Dark energy is broadly similar to inflation, and is thought to be causing the expansion of the present-day universe to accelerate. However, the energy scale of dark energy is much lower, 10^{-12} GeV, roughly 27 orders of magnitude less than the scale of inflation.

40.5 Theoretical status

In the early proposal of Guth, it was thought that the inflaton was the Higgs field, the field that explains the mass of the elementary particles.[52] It is now believed by some that the inflaton cannot be the Higgs field[79] although the recent discovery of the Higgs boson has increased the number of works considering the Higgs field as inflaton.[80] One problem of this identification is the current tension with experimental data at the electroweak scale,[81] which is currently under study at the Large Hadron Collider (LHC). Other models of inflation relied on the properties of grand unified theories.[58] Since the simplest models of grand unification have failed, it is now thought by many physicists that inflation will be included in a supersymmetric theory like string theory or a supersymmetric grand unified theory. At present, while inflation is understood principally by its detailed predictions of the initial conditions for the hot early universe, the particle physics is largely *ad hoc* modelling. As such, though predictions of inflation have been consistent with the results of observational tests, there are many open questions about the theory.

40.5.1 Fine-tuning problem

One of the most severe challenges for inflation arises from the need for fine tuning in inflationary theories. In new inflation, the *slow-roll conditions* must be satisfied for inflation to occur. The slow-roll conditions say that the inflaton potential must be flat (compared to the large vacuum energy) and that the inflaton particles must have a small mass.[82] In order for the new inflation theory of Linde, Albrecht and Steinhardt to be successful, therefore, it seemed that the universe must have a scalar field with an especially flat potential and special initial conditions. However, there are ways to explain these fine-tunings. For example, classically scale invariant field theories, where scale invariance is broken by quantum effects, provide an explanation of the flatness of inflationary potentials, as long as the theory can be studied through perturbation theory.[83]

Andrei Linde

Andrei Linde proposed a theory known as *chaotic inflation* in which he suggested that the conditions for inflation are actually satisfied quite generically and inflation will occur in virtually any universe that begins in a chaotic, high energy state and has a scalar field with unbounded potential energy.[84] However, in his model the inflaton field necessarily takes values larger than one Planck unit: for this reason, these are often called *large field* models and the competing new inflation models are called *small field* models. In this situation, the predictions of effective field theory are thought to be invalid, as renormalization should cause large corrections that could prevent inflation.[85] This problem has not yet been resolved and some cosmologists argue that the small field models, in which inflation can occur at a much lower energy scale, are better models of inflation.[86] While inflation depends on quantum field theory (and the semiclassical approximation to quantum gravity) in an important way, it has not been completely reconciled with these theories.

Robert Brandenberger has commented on fine-tuning in another situation.[87] The amplitude of the primordial inhomogeneities produced in inflation is directly tied to the energy scale of inflation. There are strong suggestions that this scale is around 10^{16} GeV or 10^{-3} times the Planck energy. The natural scale is naïvely the Planck scale so this small value could be seen as another form of fine-tuning (called a hierarchy problem): the energy density given by the scalar potential is down by 10^{-12} compared to the Planck density. This is not usually considered to be a critical problem, however, because the scale of inflation corresponds naturally to the scale of gauge unification.

40.5.2 Eternal inflation

Main article: Eternal inflation

In many models of inflation, the inflationary phase of the universe's expansion lasts forever in at least some regions of the universe. This occurs because inflating regions expand very rapidly, reproducing themselves. Unless the rate of decay to the non-inflating phase is sufficiently fast, new inflating regions are produced more rapidly than non-inflating regions. In such models most of the volume of the universe at any given time is inflating. All models of eternal inflation produce an infinite multiverse, typically a fractal.

Although new inflation is classically rolling down the potential, quantum fluctuations can sometimes bring it back up to previous levels. These regions in which the inflaton fluctuates upwards expand much faster than regions in which the inflaton has a lower potential energy, and tend to dominate in terms of physical volume. This steady state, which first developed by Vilenkin,[88] is called "eternal inflation". It has been shown that any inflationary theory with an unbounded potential is eternal.[89] It is a popular conclusion among physicists that this steady state cannot continue forever into the past.[90][91][92] The inflationary spacetime, which is similar to de Sitter space, is incomplete without a contracting region. However, unlike de Sitter space, fluctuations in a contracting inflationary space will collapse to form a gravitational singularity, a point where densities become infinite. Therefore, it is necessary to have a theory for the universe's initial conditions. Linde, however, believes inflation may be past eternal.[93]

In eternal inflation, regions with inflation have an exponentially growing volume, while regions that are not inflating don't. This suggests that the volume of the inflating part of the universe in the global picture is always unimaginably larger than the part that has stopped inflating, even though inflation eventually ends as seen by any single pre-inflationary observer. Scientists disagree about how to assign a probability distribution to this hypothetical anthropic landscape. If the probability of different regions is counted by volume, one should expect that inflation will never end, or applying boundary conditions that a local observer exists to observe it, that inflation will end as late as possible. Some physicists believe this paradox can be resolved by weighting observers by their pre-inflationary volume.

40.5.3 Initial conditions

Some physicists have tried to avoid the initial conditions problem by proposing models for an eternally inflating universe with no origin.[94][95][96][97] These models propose that while the universe, on the largest scales, expands exponentially it was, is and always will be, spatially infinite and has existed, and will exist, forever.

Other proposals attempt to describe the ex nihilo creation of the universe based on quantum cosmology and the following inflation. Vilenkin put forth one such scenario.[88] Hartle and Hawking offered the no-boundary proposal for the initial creation of the universe in which inflation comes about naturally.[98]

Alan Guth has described the inflationary universe as the "ultimate free lunch":[99][100] new universes, similar to our own, are continually produced in a vast inflating background. Gravitational interactions, in this case, circumvent (but do not violate) the first law of thermodynamics (energy conservation) and the second law of thermodynamics (entropy and the arrow of time problem). However, while there is consensus that this solves the initial conditions problem, some have disputed this, as it is much more likely that the universe came about by a quantum fluctuation. Donald Page was an outspoken critic of inflation because of this anomaly.[101] He stressed that the thermodynamic arrow of time necessitates low entropy initial conditions, which would be highly unlikely. According to them, rather than solving this problem, the inflation theory further aggravates it – the reheating at the end of the inflation era increases entropy, making it necessary for the initial state of the Universe to be even more orderly than in other Big Bang theories with no inflation phase.

Hawking and Page later found ambiguous results when they attempted to compute the probability of inflation in the Hartle-Hawking initial state.[102] Other authors have argued that, since inflation is eternal, the probability doesn't matter as long as it is not precisely zero: once it starts, inflation perpetuates itself and quickly dominates the universe.[103][104]:223–225 However, Albrecht and Lorenzo Sorbo have argued that the probability of an inflationary cosmos, consistent with today's observations, emerging by a random fluctuation from some pre-existent state, *compared* with a non-inflationary cosmos overwhelmingly favours the inflationary scenario, simply because the "seed" amount of non-gravitational energy required for the inflationary cosmos is so much less than any required for a non-inflationary alternative, which outweighs any entropic considerations.[105]

Another problem that has occasionally been mentioned is the trans-Planckian problem or trans-Planckian effects.[106] Since the energy scale of inflation and the Planck scale are relatively close, some of the quantum fluctuations that have made up the structure in our universe were smaller than the Planck length before inflation. Therefore, there ought to be corrections from Planck-scale physics, in particular the unknown quantum theory of gravity. There has been some disagreement about the magnitude of this effect: about whether it is just on the threshold of detectability or completely undetectable.[107]

40.5.4 Hybrid inflation

Another kind of inflation, called *hybrid inflation*, is an extension of new inflation. It introduces additional scalar fields, so that while one of the scalar fields is responsible for normal slow roll inflation, another triggers the end of inflation: when inflation has continued for sufficiently long, it becomes favorable to the second field to decay into a much lower energy state.[108]

In hybrid inflation, one of the scalar fields is responsible for most of the energy density (thus determining the rate of expansion), while the other is responsible for the slow roll (thus determining the period of inflation and its termination). Thus fluctuations in the former inflaton would not affect inflation termination, while fluctuations in the latter would not affect the rate of expansion. Therefore hybrid inflation is not eternal.[109][110] When the second (slow-rolling) inflaton reaches the bottom of its potential, it changes the location of the minimum of the first inflaton's potential, which leads to a fast roll of the inflaton down its potential, leading to termination of inflation.

40.5.5 Inflation and string cosmology

The discovery of flux compactifications have opened the way for reconciling inflation and string theory.[111] A new theory, called *brane inflation* suggests that inflation arises from the motion of D-branes[112] in the compactified geometry, usually towards a stack of anti-D-branes. This theory, governed by the *Dirac-Born-Infeld action*, is very different from ordinary inflation. The dynamics are not completely understood. It appears that special conditions are necessary since inflation occurs in tunneling between two vacua in the string landscape. The process of tunneling between two vacua is a form of old inflation, but new inflation must then occur by some other mechanism.

40.5.6 Inflation and loop quantum gravity

When investigating the effects the theory of loop quantum gravity would have on cosmology, a loop quantum cosmology model has evolved that provides a possible mechanism for cosmological inflation. Loop quantum gravity assumes a quantized spacetime. If the energy density is larger than can be held by the quantized spacetime, it is thought to bounce back.

40.5.7 Inflation and generalized uncertainty principle (GUP)

The effects of generalized uncertainty principle (GUP) on the inflationary dynamics and the thermodynamics of the early Universe are studied.[113] Using the GUP approach, Tawfik et al. evaluated the tensorial and scalar density fluctuations in the inflation era and compared them with the standard case. They found a good agreement with the Wilkinson Microwave Anisotropy Probe data. Assuming that a quantum gas of scalar particles is confined within a thin layer near the apparent horizon of the Friedmann-Lemaitre-Robertson-Walker Universe that satisfies the boundary condition, Tawfik et al. calculated the number and entropy densities and the free energy arising from the quantum states using the GUP approach. Furthermore, a qualitative estimation for effects of the quantum gravity on all these thermodynamic quantities was introduced.

40.6 Alternatives to inflation

The flatness and horizon problems are naturally solved in the Einstein-Cartan-Sciama-Kibble theory of gravity, without needing an exotic form of matter and introducing free parameters.[114][115] This theory extends general relativity by removing a constraint of the symmetry of the affine connection and regarding its antisymmetric part, the torsion tensor, as a dynamical variable. The minimal coupling between torsion and Dirac spinors generates a spin-spin interaction that is significant in fermionic matter at extremely high densities. Such an interaction averts the unphysical Big Bang singularity, replacing it with a cusp-like bounce at a finite minimum scale factor, before which the Universe was contracting. The rapid expansion immediately after the Big Bounce explains why the present Universe at largest scales appears spatially flat, homogeneous and isotropic. As the density of the Universe decreases, the effects of torsion weaken and the Universe smoothly enters the radiation-dominated era.

There are models that explain some of the observations explained by inflation. However none of these "alternatives" has the same breadth of explanation as inflation, and still require inflation for a more complete fit with observation; they should therefore be regarded as adjuncts to inflation, rather than as alternatives.

String theory requires that, in addition to the three observable spatial dimensions, there exist additional dimensions that are curled up or compactified (see also Kaluza–Klein theory). Extra dimensions appear as a frequent component of supergravity models and other approaches to quantum gravity. This raised the contingent question of why four space-time dimensions became large and the rest became unobservably small. An attempt to address this question, called *string gas cosmology*, was proposed by Robert Brandenberger and Cumrun Vafa.[116] This model focuses on the dynamics of the early universe considered as a hot gas of strings. Brandenberger and Vafa show that a dimension of spacetime can only expand if the strings that wind around it can efficiently annihilate each other. Each string is a one-dimensional object, and the largest number of dimensions in which two strings will generically intersect (and, presumably, annihilate) is three. Therefore, one argues that the most likely number of non-compact (large) spatial dimensions is three. Current work on this model centers on whether it can succeed in stabilizing the size of the compactified dimensions and produce the correct spectrum of primordial density perturbations. For a recent review, see[117] The authors admits that their model "does not solve the entropy and flatness problems of standard cosmology ..... and we can provide no explanation for why the current universe is so close to being spatially flat".[118]

The ekpyrotic and cyclic models are also considered adjuncts to inflation. These models solve the horizon problem through an expanding epoch well *before* the Big Bang, and then generate the required spectrum of primordial density perturbations during a contracting phase leading to a Big Crunch. The universe passes through the Big Crunch and emerges in a hot Big Bang phase. In this sense they are reminiscent of the oscillatory universe proposed by Richard Chace Tolman: however in Tolman's model the total age of the universe is necessarily finite, while in these models this is not necessarily so. Whether the correct spectrum of density fluctuations can be produced, and whether the universe can successfully navigate the Big Bang/Big Crunch transition, remains a topic of controversy and current research. Ekpyrotic models avoid the magnetic monopole problem as long as the temperature at the Big Crunch/Big Bang transition remains below the Grand Unified Scale, as this is the temperature required to produce magnetic monopoles in the first place. As things stand, there is no evidence of any 'slowing down' of the expansion, but this is not surprising as each cycle is expected to last on the order of a trillion years.

Another adjunct, the varying speed of light model has also been theorized by Jean-Pierre Petit in 1988, John Moffat in 1992 as well Andreas Albrecht and João Magueijo in 1999, instead of superluminal expansion the speed of light was 60 orders of magnitude faster than its current value solving the horizon and homogeneity problems in the early universe.

40.7 Criticisms

Since its introduction by Alan Guth in 1980, the inflationary paradigm has become widely accepted. Nevertheless, many physicists, mathematicians, and philosophers of science have voiced criticisms, claiming untestable predictions and an alleged lack of serious empirical support.[103] In 1999, John Earman and Jesús Mosterín published a thorough critical review of inflationary cosmology, concluding, "we do not think that there are, as yet, good grounds for admitting any of the models of inflation into the standard core of cosmology."[119]

In order to work, and as pointed out by Roger Penrose from 1986 on, inflation requires extremely specific initial conditions of its own, so that the problem (or pseudo-problem) of initial conditions is not solved: "There is something fundamentally misconceived about trying to explain the uniformity of the early universe as resulting from a thermalization process. [...] For, if the thermalization is actually doing anything [...] then it represents a definite increasing of the entropy. Thus, the universe would have been even more special before the thermalization than after."[120] The problem of specific or "fine-tuned" initial conditions would not have been solved; it would have gotten worse.

A recurrent criticism of inflation is that the invoked inflation field does not correspond to any known physical field, and that its potential energy curve seems to be an ad hoc contrivance to accommodate almost any data obtainable. Paul J. Steinhardt, one of the founding fathers of inflationary cosmology, has recently become one of its sharpest critics. He calls 'bad inflation' a period of accelerated expansion whose outcome conflicts with observations, and 'good inflation' one compatible with them: "Not only is bad inflation more likely than good inflation, but no inflation is more likely than either.... Roger Penrose considered all the possible configurations of the inflaton and gravitational fields. Some of these configurations lead to inflation ... Other configurations lead to a uniform, flat universe directly – without inflation. Obtaining a flat universe is unlikely overall. Penrose's shocking conclusion, though, was that obtaining a flat universe without inflation is much more likely than with inflation – by a factor of 10 to the googol (10 to the 100) power!"[103][104]

40.8 See also

- Brane cosmology
- Conservation of angular momentum
- Cosmology

- Dark flow
- Doughnut theory of the universe
- Hubble's law
- Non-minimally coupled inflation
- Nonlinear optics
- Varying speed of light
- Warm inflation

40.9 Notes

[1] Staff (17 March 2014). "BICEP2 2014 Results Release". *National Science Foundation*. Retrieved 18 March 2014.

[2] Clavin, Whitney (17 March 2014). "NASA Technology Views Birth of the Universe". *NASA*. Retrieved 17 March 2014.

[3] Overbye, Dennis (17 March 2014). "Space Ripples Reveal Big Bang's Smoking Gun". *The New York Times*. Retrieved 17 March 2014.

[4] Overbye, Dennis (24 March 2014). "Ripples From the Big Bang". *New York Times*. Retrieved 24 March 2014.

[5] Chapter 17 of Peebles (1993).

[6] Tyson, Neil deGrasse and Donald Goldsmith (2004), *Origins: Fourteen Billion Years of Cosmic Evolution*, W. W. Norton & Co., pp. 84–5.

[7] Steinhardt, Paul J. (2011). "The inflation debate: Is the theory at the heart of modern cosmology deeply flawed?" (*Scientific American*, April; pp. 18-25). "...inflationary theory is powerfully predictive. For example, numerous observations of the cosmic microwave background radiation and the distribution of galaxies have confirmed that the spatial variations in energy in the early universe were nearly scale-invariant."

[8] Tsujikawa, Shinji (28 Apr 2003). "Introductory review of cosmic inflation". arXiv:hep-ph/0304257. "In fact temperature anisotropies observed by the COBE satellite in 1992 exhibit nearly scale-invariant spectra as predicted by the inflationary paradigm. Recent observations of WMAP also show strong evidence for inflation."

[9] Guth, Alan H. (1997). *The Inflationary Universe: The Quest for a New Theory of Cosmic Origins*. Basic Books. pp. 233–234. ISBN 0201328402.

[10] Ade, P.A.R. (BICEP2 Collaboration) et al. (19 June 2014). "Detection of B-Mode Polarization at Degree Angular Scales by BICEP2". *Physical Review Letters* **112** (24): 241101. arXiv:1403.3985. Bibcode:2014PhRvL.112x1101A. doi:10.1103/PhysRevLett.112.241101.

[11] http://www.math.columbia.edu/~{}woit/wordpress/?p=6865

[12] Overbye, Dennis (19 June 2014). "Astronomers Hedge on Big Bang Detection Claim". *New York Times*. Retrieved 20 June 2014.

[13] Amos, Jonathan (19 June 2014). "Cosmic inflation: Confidence lowered for Big Bang signal". *BBC News*. Retrieved 20 June 2014.

[14] Planck Collaboration Team (19 September 2014). "Planck intermediate results. XXX. The angular power spectrum of polarized dust emission at intermediate and high Galactic latitudes". *ArXiv*. arXiv:1409.5738. Retrieved 22 September 2014.

[15] Overbye, Dennis (22 September 2014). "Study Confirms Criticism of Big Bang Finding". *New York Times*. Retrieved 22 September 2014.

[16] Using Tiny Particles To Answer Giant Questions. Science Friday, 3 April 2009.

[17] See also Faster than light#Universal expansion.

[18] Spergel, D.N. (2006). "Three-year Wilkinson Microwave Anisotropy Probe (WMAP) observations: Implications for cosmology". "WMAP... confirms the basic tenets of the inflationary paradigm..."

[19] Our Baby Universe Likely Expanded Rapidly, Study Suggests

[20] Melia, Fulvio (2007). "The Cosmic Horizon"., *MNRAS*, **382**: 1917–1921. Bibcode:2007MNRAS.382.1917M. doi:10.1111/j.1365-2966.2007.12499.x.

[21] Melia, Fulvio et al. (2009). "The Cosmological Spacetime". *IJMP-D* **18**: 1889–1901. doi:10.1142/s0218271809015746.

[22] Kolb and Turner (1988).

[23] Barbara Sue Ryden (2003). *Introduction to cosmology*. Addison-Wesley. ISBN 978-0-8053-8912-8. "Not only is inflation very effective at driving down the number density of magnetic monopoles, it is also effective at driving down the number density of every other type of particle, including photons.":202–207

[24] This is usually quoted as 60 *e*-folds of expansion, where $e^{60} \approx 10^{26}$. It is equal to the amount of expansion since reheating, which is roughly $E_{\text{inflation}}/T_0$, where $T_0 = 2.7$ K is the temperature of the cosmic microwave background today. See, *e.g.* Kolb and Turner (1998) or Liddle and Lyth (2000).

[25] Guth, *Phase transitions in the very early universe*, in *The Very Early Universe*, ISBN 0-521-31677-4 eds Hawking, Gibbon & Siklos

[26] See Kolb and Turner (1988) or Mukhanov (2005).

[27] Kofman, Lev; Linde, Andrei; Starobinsky, Alexei (1994). "Reheating after inflation". *Physical Review Letters* **73** (5): 3195–3198. arXiv:hep-th/9405187. Bibcode:1986CQGra...3..811K. doi:10.1088/0264-9381/3/5/011.

[28] Much of the historical context is explained in chapters 15–17 of Peebles (1993).

[29] Misner, Charles W.; Coley, A A; Ellis, G F R; Hancock, M (1968). "The isotropy of the universe". *Astrophysical Journal* **151** (2): 431. Bibcode:1998CQGra..15..331W. doi:10.1088/0264-9381/15/2/008.

[30] Misner, Charles; Thorne, Kip S. and Wheeler, John Archibald (1973). *Gravitation*. San Francisco: W. H. Freeman. pp. 489–490, 525–526. ISBN 0-7167-0344-0.

[31] Weinberg, Steven (1971). *Gravitation and Cosmology*. John Wiley. pp. 740, 815. ISBN 0-471-92567-5.

[32] Lemaître, Georges (1933). "The expanding universe". *Annales de la Société Scientifique de Bruxelles* **47A**: 49., English in *Gen. Rel. Grav.* **29**:641–680, 1997.

[33] R. C. Tolman (1934). *Relativity, Thermodynamics, and Cosmology*. Oxford: Clarendon Press. ISBN 0-486-65383-8. LCCN 34032023. Reissued (1987) New York: Dover ISBN 0-486-65383-8.

[34] Misner, Charles W.; Leach, P G L (1969). "Mixmaster universe". *Physical Review Letters* **22** (15): 1071–74. Bibcode:2008JPhA...41o5201A. doi:10.1088/1751-8113/41/15/155201.

[35] Dicke, Robert H. (1970). *Gravitation and the Universe*. Philadelphia: American Philosopical Society.

[36] Dicke, Robert H.; P. J. E. Peebles (1979). "The big bang cosmology – enigmas and nostrums". In ed. S. W. Hawking and W. Israel. "General Relativity: an Einstein Centenary Survey". Cambridge University Press.

[37] Alan P. Lightman (1 January 1993). *Ancient Light: Our Changing View of the Universe*. Harvard University Press. ISBN 978-0-674-03363-4.

[38] What is the Universe Made Of?

[39] Since supersymmetric Grand Unified Theory is built into string theory, it is still a triumph for inflation that it is able to deal with these magnetic relics. See, *e.g.* Kolb and Turner (1988) and Template:Cite arxiv conference

[40] 't Hooft, Gerard (1974). "Magnetic monopoles in Unified Gauge Theories". *Nuclear Physics B* **79** (2): 276–84. Bibcode:1974NuPhB..79..276T. doi:10.1016/0550-3213(74)90486-6.

[41] Polyakov, Alexander M. (1974). "Particle spectrum in quantum field theory". *JETP Letters* **20**: 194–5. Bibcode:1974JETPL..20..194P.

[42] Guth, Alan; Tye, S. (1980). "Phase Transitions and Magnetic Monopole Production in the Very Early Universe". *Physical Review Letters* **44** (10): 631–635; Erratum *ibid.*,**44**:963, 1980. Bibcode:1980PhRvL..44..631G. doi:10.1103/PhysRevLett.44.631.

[43] Einhorn, Martin B; Stein, D. L.; Toussaint, Doug (1980). "Are Grand Unified Theories Compatible with Standard Cosmology?". *Physical Review D* **21** (12): 3295–3298. Bibcode:1980PhRvD..21.3295E. doi:10.1103/PhysRevD.21.3295.

[44] Zel'dovich, Ya.; Khlopov; M. Yu. Khlopov (1978). "On the concentration of relic monopoles in the universe". *Physics Letters B* **79** (3): 239–41. Bibcode:1978PhLB...79..239Z. doi:10.1016/0370-2693(78)90232-0.

[45] Preskill, John (1979). "Cosmological production of superheavy magnetic monopoles". *Physical Review Letters* **43** (19): 1365. Bibcode:1979PhRvL..43.1365P. doi:10.1103/PhysRevLett.43.1365.

[46] See, *e.g.* Yao, W.-M.; Amsler, C.; Asner, D.; Barnett, R. M.; Beringer, J.; Burchat, P. R.; Carone, C. D.; Caso, C.; Dahl, O.; d'Ambrosio, G.; De Gouvea, A.; Doser, M.; Eidelman, S.; Feng, J. L.; Gherghetta, T.; Goodman, M.; Grab, C.; Groom, D. E.; Gurtu, A.; Hagiwara, K.; Hayes, K. G.; Hernández-Rey, J. J.; Hikasa, K.; Jawahery, H.; Kolda, C.; Kwon, Y.; Mangano, M. L.; Manohar, A. V.; Masoni, A. et al. (2006). "Review of Particle Physics". *J. Phys. G* **33** (1): 1. arXiv:astro-ph/0601168. Bibcode:2006JPhG...33....1Y. doi:10.1088/0954-3899/33/1/001.

[47] Rees, Martin. (1998). *Before the Beginning* (New York: Basic Books) p. 185 ISBN 0-201-15142-1

[48] de Sitter, Willem (1917). "Einstein's theory of gravitation and its astronomical consequences. Third paper". *Monthly Notices of the Royal Astronomical Society* **78**: 3–28. Bibcode:1917MNRAS..78....3D. doi:10.1093/mnras/78.1.3.

[49] Starobinsky, A. A. (1979). "Spectrum Of Relict Gravitational Radiation And The Early State Of The Universe". *JETP Letters* **30**: 682. Bibcode:1979JETPL..30..682S.; *Pisma Zh. Eksp. Teor. Fiz.* **30**: 719. 1979. Bibcode:bibcode=1979ZhPmR..30..719S.

[50] Linde, Andrei. "Lectures in Inflationary Cosmology".

[51] SLAC seminar, "10^{-35} seconds after the Big Bang", 23 January 1980. see Guth (1997), pg 186

[52] Guth, Alan H. (1981). "Inflationary universe: A possible solution to the horizon and flatness problems" (PDF). *Physical Review D* **23** (2): 347–356. Bibcode:1981PhRvD..23..347G. doi:10.1103/PhysRevD.23.347.

[53] Starobinsky, Alexei A. (1980). "A new type of isotropic cosmological models without singularity". *Physics Letters B* **91**: 99–102. Bibcode:1980PhLB...91...99S. doi:10.1016/0370-2693(80)90670-X.

[54] Kazanas, D. (1980). "Dynamics of the universe and spontaneous symmetry breaking". *Astrophysical Journal* **241**: L59–63. Bibcode:1980ApJ...241L..59K. doi:10.1086/183361.

[55] Sato, K. (1981). "Cosmological baryon number domain structure and the first order phase transition of a vacuum". *Physics Letters B* **33**: 66–70. Bibcode:1981PhLB...99...66S. doi:10.1016/0370-2693(81)90805-4.

[56] Einhorn, Martin B; Sato, Katsuhiko (1981). "Monopole Production In The Very Early Universe In A First Order Phase Transition". *Nuclear Physics B* **180** (3): 385–404. Bibcode:1981NuPhB.180..385E. doi:10.1016/0550-3213(81)90057-2.

[57] Linde, A (1982). "A new inflationary universe scenario: A possible solution of the horizon, flatness, homogeneity, isotropy and primordial monopole problems". *Physics Letters B* **108** (6): 389–393. Bibcode:1982PhLB..108..389L. doi:10.1016/0370-2693(82)91219-9.

[58] Albrecht, Andreas; Steinhardt, Paul (1982). "Cosmology for Grand Unified Theories with Radiatively Induced Symmetry Breaking" (PDF). *Physical Review Letters* **48** (17): 1220–1223. Bibcode:1982PhRvL..48.1220A. doi:10.1103/PhysRevLett.48.1220.

[59] J.B. Hartle (2003). *Gravity: An Introduction to Einstein's General Relativity* (1st ed.). Addison Wesley. p. 411. ISBN 0-8053-8662-9

[60] See Linde (1990) and Mukhanov (2005).

[61] Chibisov, Viatcheslav F.; Chibisov, G. V. (1981). "Quantum fluctuation and "nonsingular" universe". *JETP Letters* **33**: 532–5. Bibcode:1981JETPL..33..532M.

[62] Mukhanov, Viatcheslav F. (1982). "The vacuum energy and large scale structure of the universe". *Soviet Physics JETP* **56**: 258–65.

[63] See Guth (1997) for a popular description of the workshop, or *The Very Early Universe*, ISBN 0-521-31677-4 eds Hawking, Gibbon & Siklos for a more detailed report

[64] Hawking, S.W. (1982). "The development of irregularities in a single bubble inflationary universe". *Physics Letters B* **115** (4): 295. Bibcode:1982PhLB..115..295H. doi:10.1016/0370-2693(82)90373-2.

[65] Starobinsky, Alexei A. (1982). "Dynamics of phase transition in the new inflationary universe scenario and generation of perturbations". *Physics Letters B* **117** (3–4): 175–8. Bibcode:1982PhLB..117..175S. doi:10.1016/0370-2693(82)90541-X.

[66] Guth, A.H. (1982). "Fluctuations in the new inflationary universe". *Physical Review Letters* **49** (15): 1110–3. Bibcode:1982PhRvL..49.1110G. doi:10.1103/PhysRevLett.49.1110.

[67] Bardeen, James M.; Steinhardt, Paul J.; Turner, Michael S. (1983). "Spontaneous creation Of almost scale-free density perturbations in an inflationary universe". *Physical Review D* **28** (4): 679. Bibcode:1983PhRvD..28..679B. doi:10.1103/PhysRevD.28.679.

[68] Perturbations can be represented by Fourier modes of a wavelength. Each Fourier mode is normally distributed (usually called Gaussian) with mean zero. Different Fourier components are uncorrelated. The variance of a mode depends only on its wavelength in such a way that within any given volume each wavelength contributes an equal amount of power to the spectrum of perturbations. Since the Fourier transform is in three dimensions, this means that the variance of a mode goes as k^{-3} to compensate for the fact that within any volume, the number of modes with a given wavenumber k goes as k^3.

[69] Tegmark, M.; Eisenstein, Daniel J.; Strauss, Michael A.; Weinberg, David H.; Blanton, Michael R.; Frieman, Joshua A.; Fukugita, Masataka; Gunn, James E. et al. (August 2006). "Cosmological constraints from the SDSS luminous red galaxies". *Physical Review D* **74** (12). arXiv:astro-ph/0608632. Bibcode:2006PhRvD..74l3507T. doi:10.1103/PhysRevD.74.123507. |first10= missing |last10= in Authors list (help); |first11= missing |last11= in Authors list (help); |first12= missing |last12= in Authors list (help); |first13= missing |last13= in Authors list (help); |first14= missing |last14= in Authors list (help); |first15= missing |last15= in Authors list (help); |first16= missing |last16= in Authors list (help); |first17= missing |last17= in Authors list (help); |first18= missing |last18= in Authors list (help); |first19= missing |last19= in Authors list (help); |first20= missing |last20= in Authors list (help); |first21= missing |last21= in Authors list (help); |first22= missing |last22= in Authors list (help); |first23= missing |last23= in Authors list (help); |first24= missing |last24= in Authors list (help); |first25= missing |last25= in Authors list (help); |first26= missing |last26= in Authors list (help); |first27= missing |last27= in Authors list (help); |first28= missing |last28= in Authors list (help); |first29= missing |last29= in Authors list (help); |first30= missing |last30= in Authors list (help)

[70] Steinhardt, Paul J. (2004). "Cosmological perturbations: Myths and facts". *Modern Physics Letters A* **19** (13 & 16): 967–82. Bibcode:2004MPLA...19..967S. doi:10.1142/S0217732304014252.

[71] Boyle, Latham A.; Steinhardt, PJ; Turok, N (2006). "Inflationary predictions for scalar and tensor fluctuations reconsidered". *Physical Review Letters* **96** (11): 111301. arXiv:astro-ph/0507455. Bibcode:2006PhRvL..96k1301B. doi:10.1103/PhysRevLett.96.1113 PMID 16605810.

[72] Tegmark, Max (2005). "What does inflation really predict?". *JCAP* **0504** (4): 001. arXiv:astro-ph/0410281. Bibcode:2005JCAP...04..001T. doi:10.1088/1475-7516/2005/04/001.

[73] This is known as a "red" spectrum, in analogy to redshift, because the spectrum has more power at longer wavelengths.

[74] Komatsu, E.; Smith, K. M.; Dunkley, J.; Bennett, C. L.; Gold, B.; Hinshaw, G.; Jarosik, N.; Larson, D. et al. (January 2010). "Seven-Year Wilkinson Microwave Anisotropy Probe (WMAP) Observations: Cosmological Interpretation". *The Astrophysical Journal Supplement Series* **192** (2): 18. arXiv:1001.4538. Bibcode:2011ApJS..192...18K. doi:10.1088/0067-0049/192/2/18.

[75] Spergel, D. N.; Verde, L.; Peiris, H. V.; Komatsu, E.; Nolta, M. R.; Bennett, C. L.; Halpern, M.; Hinshaw, G. et al. (2003). "First year Wilkinson Microwave Anisotropy Probe (WMAP) observations: determination of cosmological parameters". *Astrophysical Journal Supplement Series* **148** (1): 175. arXiv:astro-ph/0302209. Bibcode:2003ApJS..148..175S. doi:10.1086/377226. |first10= missing |last10= in Authors list (help); |first11= missing |last11= in Authors list (help); |first12= missing |last12= in Authors list (help); |first13= missing |last13= in Authors list (help); |first14= missing |last14= in Authors list (help); |first15= missing |last15= in Authors list (help); |first16= missing |last16= in Authors list (help); |first17= missing |last17= in Authors list (help)

[76] See cosmic microwave background#Low multipoles for details and references.

[77] Rosset, C.; (PLANCK-HFI collaboration) (2005). "Systematic effects in CMB polarization measurements". "Exploring the universe: Contents and structures of the universe (XXXIXth Rencontres de Moriond)".

[78] Loeb, A.; Zaldarriaga, M (2004). "Measuring the small-scale power spectrum of cosmic density fluctuations through 21 cm tomography prior to the epoch of structure formation". *Physical Review Letters* **92** (21): 211301. arXiv:astro-ph/0312134. Bibcode:2004PhRvL..92u1301L. doi:10.1103/PhysRevLett.92.211301. PMID 15245272.

[79] Guth, Alan (1997). *The Inflationary Universe*. Addison–Wesley. ISBN 0-201-14942-7.

[80] Choi, Charles (Jun 29, 2012). "Could the Large Hadron Collider Discover the Particle Underlying Both Mass and Cosmic Inflation?". Scientific American. Retrieved Jun 25, 2014."The virtue of so-called Higgs inflation models is that they might explain inflation within the current Standard Model of particle physics, which successfully describes how most known particles and forces behave. Interest in the Higgs is running hot this summer because CERN, the lab in Geneva, Switzerland, that runs the LHC, has said it will announce highly anticipated findings regarding the particle in early July."

[81] Salvio, Alberto (2013-08-09). "Higgs Inflation at NNLO after the Boson Discovery". *Phys.Lett. B727 (2013) 234-239*. arXiv:1308.2244. Bibcode:2013PhLB..727..234S. doi:10.1016/j.physletb.2013.10.042.

[82] Technically, these conditions are that the logarithmic derivative of the potential, $\epsilon = (1/2)(V'/V)^2$ and second derivative $\eta = V''/V$ are small, where V is the potential and the equations are written in reduced Planck units. See, *e.g.* Liddle and Lyth (2000), pg 42-43.

[83] Salvio, Strumia (2014-03-17). "Agravity". *JHEP 1406 (2014) 080*. arXiv:1403.4226. Bibcode:2014JHEP...06..080S. doi:10.1007/JHEP06(2014)080.

[84] Linde, Andrei D. (1983). "Chaotic inflation". *Physics Letters B* **129** (3): 171–81. Bibcode:1983PhLB..129..177L. doi:10.1016/0370-2693(83)90837-7.

[85] Technically, this is because the inflaton potential is expressed as a Taylor series in φ/mP_l, where φ is the inflaton and mP_l is the Planck mass. While for a single term, such as the mass term $m_\varphi^4(\varphi/mP_l)^2$, the slow roll conditions can be satisfied for φ much greater than mP_l, this is precisely the situation in effective field theory in which higher order terms would be expected to contribute and destroy the conditions for inflation. The absence of these higher order corrections can be seen as another sort of fine tuning. See *e.g.* Alabidi, Laila; Lyth, David H (2006). "Inflation models and observation". *JCAP* **0605** (5): 016. arXiv:astro-ph/0510441. Bibcode:2006JCAP...05..016A. doi:10.1088/1475-7516/2006/05/016.

[86] See, *e.g.* Lyth, David H. (1997). "What would we learn by detecting a gravitational wave signal in the cosmic microwave background anisotropy?". *Physical Review Letters* **78** (10): 1861–3. arXiv:hep-ph/9606387. Bibcode:1997PhRvL..78.1861L. doi:10.1103/PhysRevLett.78.1861.

[87] Template:Cite arxiv conference

[88] Vilenkin, Alexander (1983). "The birth of inflationary universes". *Physical Review D* **27** (12): 2848. Bibcode:1983PhRvD..27.2848V. doi:10.1103/PhysRevD.27.2848.

[89] A. Linde (1986). "Eternal chaotic inflation". *Modern Physics Letters A* **1** (2): 81. Bibcode:1986MPLA....1...81L. doi:10.1142/S0217732386000 A. Linde (1986). "Eternally existing self-reproducing chaotic inflationary universe". *Physics Letters B* **175** (4): 395–400. Bibcode:1986PhLB..175..395L. doi:10.1016/0370-2693(86)90611-8.

[90] A. Borde, A. Guth and A. Vilenkin (2003). "Inflationary space-times are incomplete in past directions". *Physical Review Letters* **90** (15): 151301. arXiv:gr-qc/0110012. Bibcode:2003PhRvL..90o1301B. doi:10.1103/PhysRevLett.90.151301. PMID 12732026.

[91] A. Borde (1994). "Open and closed universes, initial singularities and inflation". *Physical Review D* **50** (6): 3692–702. arXiv:gr-qc/9403049. Bibcode:1994PhRvD..50.3692B. doi:10.1103/PhysRevD.50.3692.

[92] A. Borde and A. Vilenkin (1994). "Eternal inflation and the initial singularity". *Physical Review Letters* **72** (21): 3305–9. arXiv:gr-qc/9312022. Bibcode:1994PhRvL..72.3305B. doi:10.1103/PhysRevLett.72.3305.

[93] Linde (2005, §V).

[94] Carroll, Sean M.; Chen, Jennifer (2005). "Does inflation provide natural initial conditions for the universe?". *Gen. Rel. Grav.* **37** (10): 1671–4. arXiv:gr-qc/0505037. Bibcode:2005GReGr..37.1671C. doi:10.1007/s10714-005-0148-2.

[95] Carroll, Sean M.; Jennifer Chen (2004). "Spontaneous inflation and the origin of the arrow of time". arXiv:hep-th/0410270 [hep-th]. Bibcode 2004hep.th...10270C.

[96] Aguirre, Anthony; Gratton, Steven (2003). "Inflation without a beginning: A null boundary proposal". *Physical Review D* **67** (8): 083515. arXiv:gr-qc/0301042. Bibcode:2003PhRvD..67h3515A. doi:10.1103/PhysRevD.67.083515.

[97] Aguirre, Anthony; Gratton, Steven (2002). "Steady-State Eternal Inflation". *Physical Review D* **65** (8): 083507. arXiv:astro-ph/0111191. Bibcode:2002PhRvD..65h3507A. doi:10.1103/PhysRevD.65.083507.

[98] Hartle, J.; Hawking, S. (1983). "Wave function of the universe". *Physical Review D* **28** (12): 2960. Bibcode:1983PhRvD..28.2960H. doi:10.1103/PhysRevD.28.2960.; See also Hawking (1998).

[99] Hawking (1998), p. 129.

[100] Wikiquote

[101] Page, Don N. (1983). "Inflation does not explain time asymmetry". *Nature* **304** (5921): 39. Bibcode:1983Natur.304...39P. doi:10.1038/304039a0.; see also Roger Penrose's book The Road to Reality: A Complete Guide to the Laws of the Universe.

[102] Hawking, S. W.; Page, Don N. (1988). "How probable is inflation?". *Nuclear Physics B* **298** (4): 789. Bibcode:1988NuPhB.298..789H. doi:10.1016/0550-3213(88)90008-9.

[103] Steinhardt, Paul J. (2011). "The inflation debate: Is the theory at the heart of modern cosmology deeply flawed?" (*Scientific American*, April; pp. 18-25).

[104] Paul J. Steinhardt; Neil Turok (2007). *Endless Universe: Beyond the Big Bang*. Broadway Books. ISBN 978-0-7679-1501-4.

[105] Albrecht, Andreas; Sorbo, Lorenzo (2004). "Can the universe afford inflation?". *Physical Review D* **70** (6): 063528. arXiv:hep-th/0405270. Bibcode:2004PhRvD..70f3528A. doi:10.1103/PhysRevD.70.063528.

[106] Martin, Jerome; Brandenberger, Robert (2001). "The trans-Planckian problem of inflationary cosmology". *Physical Review D* **63** (12): 123501. arXiv:hep-th/0005209. Bibcode:2001PhRvD..63l3501M. doi:10.1103/PhysRevD.63.123501.

[107] Martin, Jerome; Ringeval, Christophe (2004). "Superimposed Oscillations in the WMAP Data?". *Physical Review D* **69** (8): 083515. arXiv:astro-ph/0310382. Bibcode:2004PhRvD..69h3515M. doi:10.1103/PhysRevD.69.083515.

[108] Robert H. Brandenberger, "A Status Review of Inflationary Cosmology", proceedings Journal-ref: BROWN-HET-1256 (2001), (available from arXiv:hep-ph/0101119v1 11 January 2001)

[109] Andrei Linde, "Prospects of Inflation", *Physica Scripta Online* (2004) (available from arXiv:hep-th/0402051)

[110] Blanco-Pillado et al., "Racetrack inflation", (2004) (available from arXiv:hep-th/0406230)

[111] Kachru, Shamit; Kallosh, Renata; Linde, Andrei; Maldacena, Juan; McAllister, Liam; Trivedi, Sandip P (2003). "Towards inflation in string theory". *JCAP* **0310** (10): 013. arXiv:hep-th/0308055. Bibcode:2003JCAP...10..013K. doi:10.1088/1475-7516/2003/10/013.

[112] G. R. Dvali, S. H. Henry Tye, *Brane inflation, Phys.Lett.* **B450**, 72-82 (1999), arXiv:hep-ph/9812483.

[113] A. Tawfik, H. Magdy and A. Farag Ali, Gen.Rel.Grav. 45 (2013) 1227-1246

[114] Poplawski, N. J. (2010). "Cosmology with torsion: An alternative to cosmic inflation". *Physics Letters B* **694** (3): 181–185. arXiv:1007.0587. Bibcode:2010PhLB..694..181P. doi:10.1016/j.physletb.2010.09.056.

[115] Poplawski, N. (2012). "Nonsingular, big-bounce cosmology from spinor-torsion coupling". *Physical Review D* **85** (10): 107502. arXiv:1111.4595. Bibcode:2012PhRvD..85j7502P. doi:10.1103/PhysRevD.85.107502.

[116] Brandenberger, R; Vafa, C. (1989). "Superstrings in the early universe". *Nuclear Physics B* **316** (2): 391–410. Bibcode:1989NuPhB.316..391B. doi:10.1016/0550-3213(89)90037-0.

[117] Battefeld, Thorsten; Watson, Scott (2006). "String Gas Cosmology". *Reviews Modern Physics* **78** (2): 435–454. arXiv:hep-th/0510022. Bibcode:2006RvMP...78..435B. doi:10.1103/RevModPhys.78.435.

[118] Brandenberger, Robert H.; Nayeri, ALI; Patil, Subodh P.; Vafa, Cumrun (2007). "String Gas Cosmology and Structure Formation". *International Journal of Modern Physics A* **22** (21): 3621–3642. arXiv:hep-th/0608121. Bibcode:2007IJMPA..22.3621B. doi:10.1142/S0217751X07037159.

[119] Earman, John; Mosterín, Jesús (March 1999). "A Critical Look at Inflationary Cosmology". *Philosophy of Science* **66**: 1–49. doi:10.2307/188736 (inactive 2014-10-16). JSTOR 188736.

[120] Penrose, Roger (2004). *The Road to Reality: A Complete Guide to the Laws of the Universe*. London: Vintage Books, p. 755. See also Penrose, Roger (1989). "Difficulties with Inflationary Cosmology". *Annals of the New York Academy of Sciences* **271**: 249–264. Bibcode:1989NYASA.571..249P. doi:10.1111/j.1749-6632.1989.tb50513.x.

40.10 References

- Guth, Alan (1997). *The Inflationary Universe: The Quest for a New Theory of Cosmic Origins*. Perseus. ISBN 0-201-32840-2.

- Hawking, Stephen (1998). *A Brief History of Time*. Bantam. ISBN 0-553-38016-8.

- Hawking, Stephen; Gary Gibbons (1983). *The Very Early Universe*. Cambridge University Press. ISBN 0-521-31677-4.

- Kolb, Edward; Michael Turner (1988). *The Early Universe*. Addison-Wesley. ISBN 0-201-11604-9.

- Linde, Andrei (1990). *Particle Physics and Inflationary Cosmology*. Chur, Switzerland: Harwood. arXiv:hep-th/0503203. ISBN 3-7186-0490-6.

- Linde, Andrei (2005) "Inflation and String Cosmology", *eConf* **C040802** (2004) L024; *J. Phys. Conf. Ser.* **24** (2005) 151–60; arXiv:hep-th/0503195 v1 2005-03-24.

- Liddle, Andrew; David Lyth (2000). *Cosmological Inflation and Large-Scale Structure*. Cambridge. ISBN 0-521-57598-2.

- Lyth, David H.; Riotto, Antonio (1999). "Particle physics models of inflation and the cosmological density perturbation". *Phys. Rept.* **314** (1–2): 1–146. arXiv:hep-ph/9807278. Bibcode:1999PhR...314....1L. doi:10.1016/S0370-1573(98)00128-8.

- Mukhanov, Viatcheslav (2005). *Physical Foundations of Cosmology*. Cambridge University Press. ISBN 0-521-56398-4.

- Vilenkin, Alex (2006). *Many Worlds in One: The Search for Other Universes*. Hill and Wang. ISBN 0-8090-9523-8.

- Peebles, P. J. E. (1993). *Principles of Physical Cosmology*. Princeton University Press. ISBN 0-691-01933-9.

40.11 External links

- Was Cosmic Inflation The 'Bang' Of The Big Bang?, by Alan Guth, 1997

- An Introduction to Cosmological Inflation by Andrew Liddle, 1999

- update 2004 by Andrew Liddle

- hep-ph/0309238 Laura Covi: Status of observational cosmology and inflation

- hep-th/0311040 David H. Lyth: Which is the best inflation model?

- The Growth of Inflation *Symmetry*, December 2004

- Guth's logbook showing the original idea

- WMAP Bolsters Case for Cosmic Inflation, March 2006

- NASA March 2006 WMAP press release

- Max Tegmark's *Our Mathematical Universe* (2014), "Chapter 5: Inflation"

Chapter 41

Doublet–triplet splitting problem

In particle physics, the **doublet–triplet (splitting) problem** is a problem of *some* Grand Unified Theories, such as SU(5), SO(10), E_6. Grand unified theories predict Higgs bosons (doublets of $SU(2)$) arise from representations of the unified group that contain other states, in particular, states that are triplets of color. The primary problem with these color triplet Higgs, is that they can mediate proton decay in supersymmetric theories that are only suppressed by two powers of GUT scale (i.e. they are dimension 5 supersymmetric operators). In addition to mediating proton decay, they alter gauge coupling unification. The doublet–triplet problem is the question 'what keeps the doublets light while the triplets are heavy?'

41.1 Doublet–triplet splitting and the μ-problem

In 'minimal' SU(5), the way one accomplishes doublet–triplet splitting is through a combination of interactions

$\int d^2\theta\ \lambda H_{\bar{5}} \Sigma H_5 + \mu H_{\bar{5}} H_5$

where Σ is an adjoint of SU(5) and is traceless. When Σ acquires a vacuum expectation value

$\langle \Sigma \rangle = \text{diag}(2, 2, 2, -3, -3)f$

that breaks SU(5) to the Standard Model gauge symmetry the Higgs doublets and triplets acquire a mass

$\int d^2\theta\ (2\lambda f + \mu) H_{\bar{3}} H_3 + (-3\lambda f + \mu) H_{\bar{2}} H_2$

Since f is at the GUT scale (10^{16} GeV) and the Higgs doublets need to have a weak scale mass (100 GeV), this requires

$\mu \sim 3\lambda f \pm 100 \text{GeV}$.

So to solve this doublet–triplet splitting problem requires a tuning of the two terms to within one part in 10^{14}. This is also why the mu problem of the MSSM (i.e. why are the Higgs doublets so light) and doublet–triplet splitting are so closely intertwined.

41.1.1 Dimopoulos–Wilczek mechanism

In an SO(10) theory, there is a potential solution to the doublet–triplet splitting problem known as the 'Dimopoulos–Wilczek' mechanism. In SO(10), the adjoint field, Σ acquires a vacuum expectation value of the form

$\langle \Sigma \rangle = \text{diag}(i\sigma_2 f_3, i\sigma_2 f_3, i\sigma_2 f_3, i\sigma_2 f_2, i\sigma_2 f_2)$.

f_2 and f_3 give masses to the Higgs doublet and triplet, respectively, and are independent of each other, because Σ is traceless for any values they may have. If $f_2 = 0$, then the Higgs doublet remains massless. This is very similar to the way that doublet–triplet splitting is done in either higher-dimensional grand unified theories or string theory.

To arrange for the VEV to align along this direction (and still not mess up the other details of the model) often requires very contrived models, however.

41.2 Higgs representations in Grand Unified Theories

In SU(5):

$$5 \to (1,2)_{\frac{1}{2}} \oplus (3,1)_{-\frac{1}{3}}$$

$$\bar{5} \to (1,2)_{-\frac{1}{2}} \oplus (\bar{3},1)_{\frac{1}{3}}$$

In SO(10):

$$10 \to (1,2)_{\frac{1}{2}} \oplus (1,2)_{-\frac{1}{2}} \oplus (3,1)_{-\frac{1}{3}} \oplus (\bar{3},1)_{\frac{1}{3}}$$

41.3 Proton decay

Non-supersymmetric theories suffer from qratric radiative corrections to the mass squared of the electroweak Higgs boson (see hierarchy problem). In the presence of supersymmetry, the triplet Higgsino needs to be more massive than the GUT scale to prevent proton decay because it generates dimension 5 operators in MSSM; there it is not enough simply to require the triplet to have a GUT scale mass.

41.4 References

- 'Supersymmetry at Ordinary Energies. 1. Masses AND Conservation Laws.' Steven Weinberg. Published in Phys.Rev.D26:287,1982.

- 'Proton Decay in Supersymmetric Models.' Savas Dimopoulos, Stuart A. Raby, Frank Wilczek. Published in Phys.Lett.B112:133,1982.

- 'Incomplete Multiplets in Supersymmetric Unified Models.' Savas Dimopoulos, Frank Wilczek.

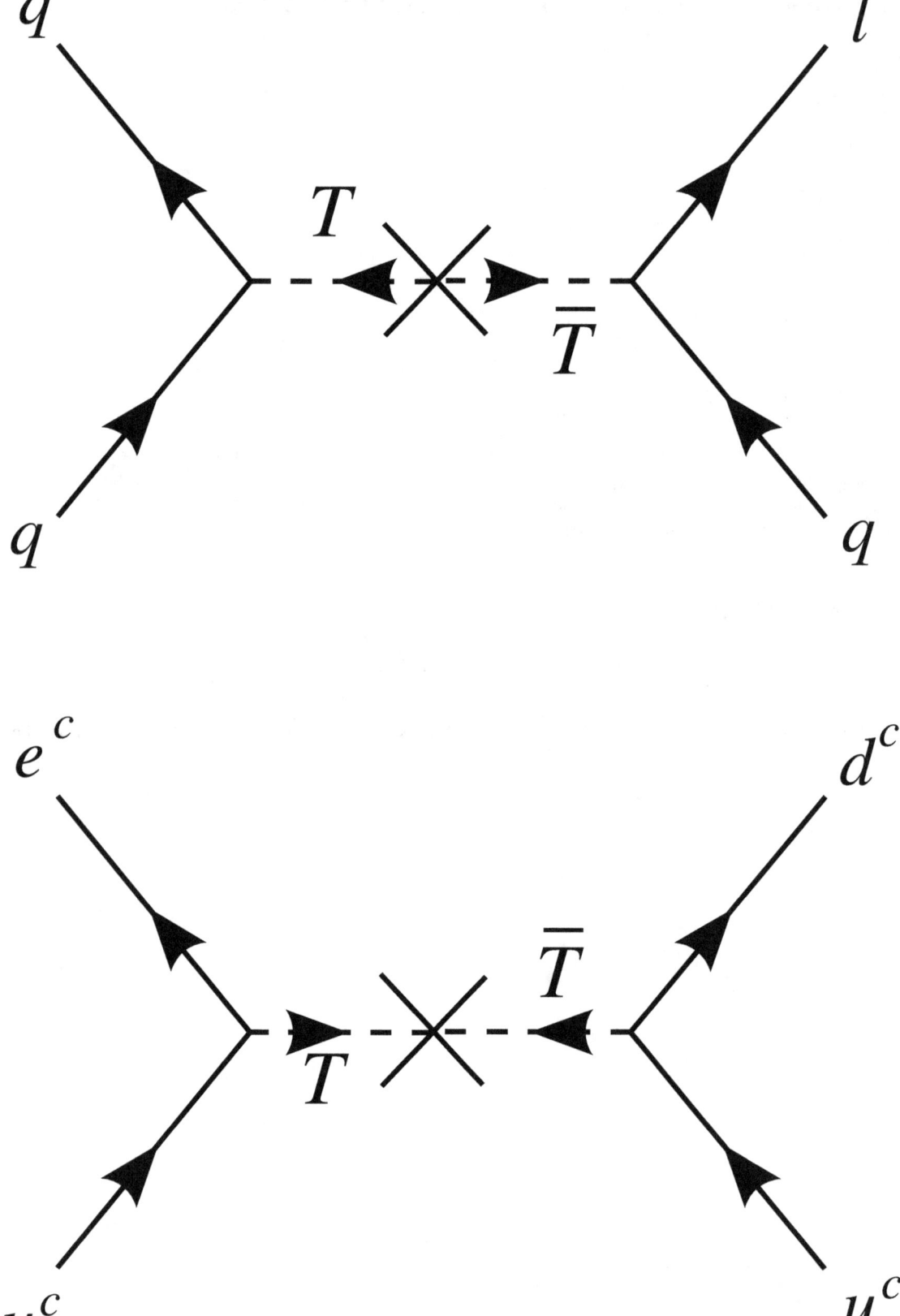

Dimension 6 proton decay mediated by the triplet Higgs $T(3,1)_{-\frac{1}{3}}$ and the anti-triplet Higgs $\bar{T}(\bar{3},1)_{\frac{1}{3}}$ in $SU(5)$ GUT

Chapter 42

Quantum chromodynamics

In theoretical physics, **quantum chromodynamics (QCD)** is the theory of strong interactions, a fundamental force describing the interactions between quarks and gluons which make up hadrons such as the proton, neutron and pion. QCD is a type of quantum field theory called a non-abelian gauge theory with symmetry group SU(3). The QCD analog of electric charge is a property called *color*. Gluons are the force carrier of the theory, like photons are for the electromagnetic force in quantum electrodynamics. The theory is an important part of the Standard Model of particle physics. A huge body of experimental evidence for QCD has been gathered over the years.

QCD enjoys two peculiar properties:

- **Confinement**, which means that the force between quarks does not diminish as they are separated. Because of this, when you do separate a quark from other quarks, the energy in the gluon field is enough to create another quark pair; they are thus forever bound into hadrons such as the proton and the neutron or the pion and kaon. Although analytically unproven, confinement is widely believed to be true because it explains the consistent failure of free quark searches, and it is easy to demonstrate in lattice QCD.

- **Asymptotic freedom**, which means that in very high-energy reactions, quarks and gluons interact very weakly creating a quark–gluon plasma. This prediction of QCD was first discovered in the early 1970s by David Politzer and by Frank Wilczek and David Gross. For this work they were awarded the 2004 Nobel Prize in Physics.

There is no known phase transition line separating these two properties; confinement is dominant in low-energy scales but, as energy increases, asymptotic freedom becomes dominant.

42.1 Terminology

The word *quark* was coined by American physicist Murray Gell-Mann (b. 1929) in its present sense. It originally comes from the phrase "Three quarks for Muster Mark" in *Finnegans Wake* by James Joyce. On June 27, 1978, Gell-Mann wrote a private letter to the editor of the *Oxford English Dictionary*, in which he related that he had been influenced by Joyce's words: "The allusion to three quarks seemed perfect." (Originally, only three quarks had been discovered.) Gell-Mann, however, wanted to pronounce the word to rhyme with "fork" rather than with "park", as Joyce seemed to indicate by rhyming words in the vicinity such as *Mark*. Gell-Mann got around that "by supposing that one ingredient of the line 'Three quarks for Muster Mark' was a cry of 'Three quarts for Mister ...' heard in H.C. Earwicker's pub", a plausible suggestion given the complex punning in Joyce's novel.[1]

The three kinds of charge in QCD (as opposed to one in quantum electrodynamics or QED) are usually referred to as "color charge" by loose analogy to the three kinds of color (red, green and blue) perceived by humans. Other than this nomenclature, the quantum parameter "color" is completely unrelated to the everyday, familiar phenomenon of color.

Since the theory of electric charge is dubbed "electrodynamics", the Greek word "chroma" Χρώμα (meaning color) is applied to the theory of color charge, "chromodynamics".

42.2 History

With the invention of bubble chambers and spark chambers in the 1950s, experimental particle physics discovered a large and ever-growing number of particles called hadrons. It seemed that such a large number of particles could not all be fundamental. First, the particles were classified by charge and isospin by Eugene Wigner and Werner Heisenberg; then, in 1953, according to strangeness by Murray Gell-Mann and Kazuhiko Nishijima. To gain greater insight, the hadrons were sorted into groups having similar properties and masses using the *eightfold way*, invented in 1961 by Gell-Mann and Yuval Ne'eman. Gell-Mann and George Zweig, correcting an earlier approach of Shoichi Sakata, went on to propose in 1963 that the structure of the groups could be explained by the existence of three flavors of smaller particles inside the hadrons: the quarks.

Perhaps the first remark that quarks should possess an additional quantum number was made[2] as a short footnote in the preprint of Boris Struminsky[3] in connection with Ω^- hyperon composed of three strange quarks with parallel spins (this situation was peculiar, because since quarks are fermions, such combination is forbidden by the Pauli exclusion principle):

> Three identical quarks cannot form an antisymmetric S-state. In order to realize an antisymmetric orbital S-state, it is necessary for the quark to have an additional quantum number.
> — B. V. Struminsky, *Magnetic moments of barions in the quark model*, JINR-Preprint P-1939, Dubna, Submitted on January 7, 1965

Boris Struminsky was a PhD student of Nikolay Bogolyubov. The problem considered in this preprint was suggested by Nikolay Bogolyubov, who advised Boris Struminsky in this research.[3] In the beginning of 1965, Nikolay Bogolyubov, Boris Struminsky and Albert Tavkhelidze wrote a preprint with a more detailed discussion of the additional quark quantum degree of freedom.[4] This work was also presented by Albert Tavchelidze without obtaining consent of his collaborators for doing so at an international conference in Trieste (Italy), in May 1965.[5][6]

A similar mysterious situation was with the Δ^{++} baryon; in the quark model, it is composed of three up quarks with parallel spins. In 1965, Moo-Young Han with Yoichiro Nambu and Oscar W. Greenberg independently resolved the problem by proposing that quarks possess an additional SU(3) gauge degree of freedom, later called color charge. Han and Nambu noted that quarks might interact via an octet of vector gauge bosons: the gluons.

Since free quark searches consistently failed to turn up any evidence for the new particles, and because an elementary particle back then was *defined* as a particle which could be separated and isolated, Gell-Mann often said that quarks were merely convenient mathematical constructs, not real particles. The meaning of this statement was usually clear in context: He meant quarks are confined, but he also was implying that the strong interactions could probably not be fully described by quantum field theory.

Richard Feynman argued that high energy experiments showed quarks are real particles: he called them *partons* (since they were parts of hadrons). By particles, Feynman meant objects which travel along paths, elementary particles in a field theory.

The difference between Feynman's and Gell-Mann's approaches reflected a deep split in the theoretical physics community. Feynman thought the quarks have a distribution of position or momentum, like any other particle, and he (correctly) believed that the diffusion of parton momentum explained diffractive scattering. Although Gell-Mann believed that certain quark charges could be localized, he was open to the possibility that the quarks themselves could not be localized because space and time break down. This was the more radical approach of S-matrix theory.

James Bjorken proposed that pointlike partons would imply certain relations should hold in deep inelastic scattering of electrons and protons, which were spectacularly verified in experiments at SLAC in 1969. This led physicists to abandon the S-matrix approach for the strong interactions.

The discovery of asymptotic freedom in the strong interactions by David Gross, David Politzer and Frank Wilczek allowed physicists to make precise predictions of the results of many high energy experiments using the quantum field theory technique of perturbation theory. Evidence of gluons was discovered in three-jet events at PETRA in 1979. These experiments became more and more precise, culminating in the verification of perturbative QCD at the level of a few percent at the LEP in CERN.

The other side of asymptotic freedom is confinement. Since the force between color charges does not decrease with distance, it is believed that quarks and gluons can never be liberated from hadrons. This aspect of the theory is verified within lattice QCD computations, but is not mathematically proven. One of the Millennium Prize Problems announced by the Clay Mathematics Institute requires a claimant to produce such a proof. Other aspects of non-perturbative QCD are the exploration of phases of quark matter, including the quark–gluon plasma.

42.3 Theory

42.3.1 Some definitions

Every field theory of particle physics is based on certain symmetries of nature whose existence is deduced from observations. These can be

- local symmetries, that is the symmetry acts independently at each point in spacetime. Each such symmetry is the basis of a gauge theory and requires the introduction of its own gauge bosons.
- global symmetries, which are symmetries whose operations must be simultaneously applied to all points of spacetime.

QCD is a gauge theory of the SU(3) gauge group obtained by taking the color charge to define a local symmetry.

Since the strong interaction does not discriminate between different flavors of quark, QCD has approximate **flavor symmetry**, which is broken by the differing masses of the quarks.

There are additional global symmetries whose definitions require the notion of chirality, discrimination between left and right-handed. If the spin of a particle has a positive projection on its direction of motion then it is called left-handed; otherwise, it is right-handed. Chirality and handedness are not the same, but become approximately equivalent at high energies.

- **Chiral** symmetries involve independent transformations of these two types of particle.
- **Vector** symmetries (also called diagonal symmetries) mean the same transformation is applied on the two chiralities.
- **Axial** symmetries are those in which one transformation is applied on left-handed particles and the inverse on the right-handed particles.

42.3.2 Additional remarks: duality

As mentioned, *asymptotic freedom* means that at large energy – this corresponds also to *short distances* – there is practically no interaction between the particles. This is in contrast – more precisely one would say *dual* – to what one is used to, since usually one connects the absence of interactions with *large* distances. However, as already mentioned in the original paper of Franz Wegner,[9] a solid state theorist who introduced 1971 simple gauge invariant lattice models, the high-temperature behaviour of the *original model*, e.g. the strong decay of correlations at large distances, corresponds to the low-temperature behaviour of the (usually ordered!) *dual model*, namely the asymptotic decay of non-trivial correlations, e.g. short-range deviations from almost perfect arrangements, for short distances. Here, in contrast to Wegner, we have only the dual model, which is that one described in this article.[10]

42.3.3 Symmetry groups

The color group SU(3) corresponds to the local symmetry whose gauging gives rise to QCD. The electric charge labels a representation of the local symmetry group U(1) which is gauged to give QED: this is an abelian group. If one considers a version of QCD with N_f flavors of massless quarks, then there is a global (chiral) flavor symmetry group $SU_L(N_f) \times SU_R(N_f) \times U_B(1) \times U_A(1)$. The chiral symmetry is spontaneously broken by the QCD vacuum to the vector (L+R) $SU_V(N_f)$ with the formation of a chiral condensate. The vector symmetry, $U_B(1)$ corresponds to the baryon number of quarks and is an exact symmetry. The axial symmetry $U_A(1)$ is exact in the classical theory, but broken in the quantum theory, an occurrence called an anomaly. Gluon field configurations called instantons are closely related to this anomaly.

There are two different types of SU(3) symmetry: there is the symmetry that acts on the different colors of quarks, and this is an exact gauge symmetry mediated by the gluons, and there is also a flavor symmetry which rotates different flavors of quarks to each other, or *flavor SU(3)*. Flavor SU(3) is an approximate symmetry of the vacuum of QCD, and is not a fundamental symmetry at all. It is an accidental consequence of the small mass of the three lightest quarks.

In the QCD vacuum there are vacuum condensates of all the quarks whose mass is less than the QCD scale. This includes the up and down quarks, and to a lesser extent the strange quark, but not any of the others. The vacuum is symmetric under SU(2) isospin rotations of up and down, and to a lesser extent under rotations of up, down and strange, or full flavor group SU(3), and the observed particles make isospin and SU(3) multiplets.

The approximate flavor symmetries do have associated gauge bosons, observed particles like the rho and the omega, but these particles are nothing like the gluons and they are not massless. They are emergent gauge bosons in an approximate string description of QCD.

42.3.4 Lagrangian

The dynamics of the quarks and gluons are controlled by the quantum chromodynamics Lagrangian. The gauge invariant QCD Lagrangian is

$$\mathcal{L}_{\text{QCD}} = \bar{\psi}_i \left(i (\gamma^\mu D_\mu)_{ij} - m\, \delta_{ij} \right) \psi_j - \frac{1}{4} G^a_{\mu\nu} G^{\mu\nu}_a$$

where $\psi_i(x)$ is the quark field, a dynamical function of spacetime, in the fundamental representation of the SU(3) gauge group, indexed by $i, j, \ldots$; $\mathcal{A}^a_\mu(x)$ are the gluon fields, also dynamical functions of spacetime, in the adjoint representation of the SU(3) gauge group, indexed by $a, b, \ldots$ The γ^μ are Dirac matrices connecting the spinor representation to the vector representation of the Lorentz group.

The symbol $G^a_{\mu\nu}$ represents the gauge invariant gluon field strength tensor, analogous to the electromagnetic field strength tensor, $F^{\mu\nu}$, in quantum electrodynamics. It is given by:[11]

$$G^a_{\mu\nu} = \partial_\mu \mathcal{A}^a_\nu - \partial_\nu \mathcal{A}^a_\mu + g f^{abc} \mathcal{A}^b_\mu \mathcal{A}^c_\nu,$$

where *fabc* are the structure constants of SU(3). Note that the rules to move-up or pull-down the *a*, *b*, or *c* indexes are *trivial*, (+, ..., +), so that $f^{abc} = f_{abc} = f^a{}_{bc}$ whereas for the μ or ν indexes one has the non-trivial *relativistic* rules, corresponding e.g. to the metric signature (+ − − −).

The constants *m* and *g* control the quark mass and coupling constants of the theory, subject to renormalization in the full quantum theory.

An important theoretical notion concerning the final term of the above Lagrangian is the *Wilson loop* variable. This loop variable plays a most important role in discretized forms of the QCD (see lattice QCD), and more generally, it distinguishes confined and deconfined states of a gauge theory. It was introduced by the Nobel prize winner Kenneth G. Wilson and is treated in a separate article.

42.3.5 Fields

Quarks are massive spin-1/2 fermions which carry a color charge whose gauging is the content of QCD. Quarks are represented by Dirac fields in the fundamental representation **3** of the gauge group SU(3). They also carry electric charge (either −1/3 or 2/3) and participate in weak interactions as part of weak isospin doublets. They carry global quantum numbers including the baryon number, which is 1/3 for each quark, hypercharge and one of the flavor quantum numbers.

Gluons are spin-1 bosons which also carry color charges, since they lie in the adjoint representation **8** of SU(3). They have no electric charge, do not participate in the weak interactions, and have no flavor. They lie in the singlet representation **1** of all these symmetry groups.

Every quark has its own antiquark. The charge of each antiquark is exactly the opposite of the corresponding quark.

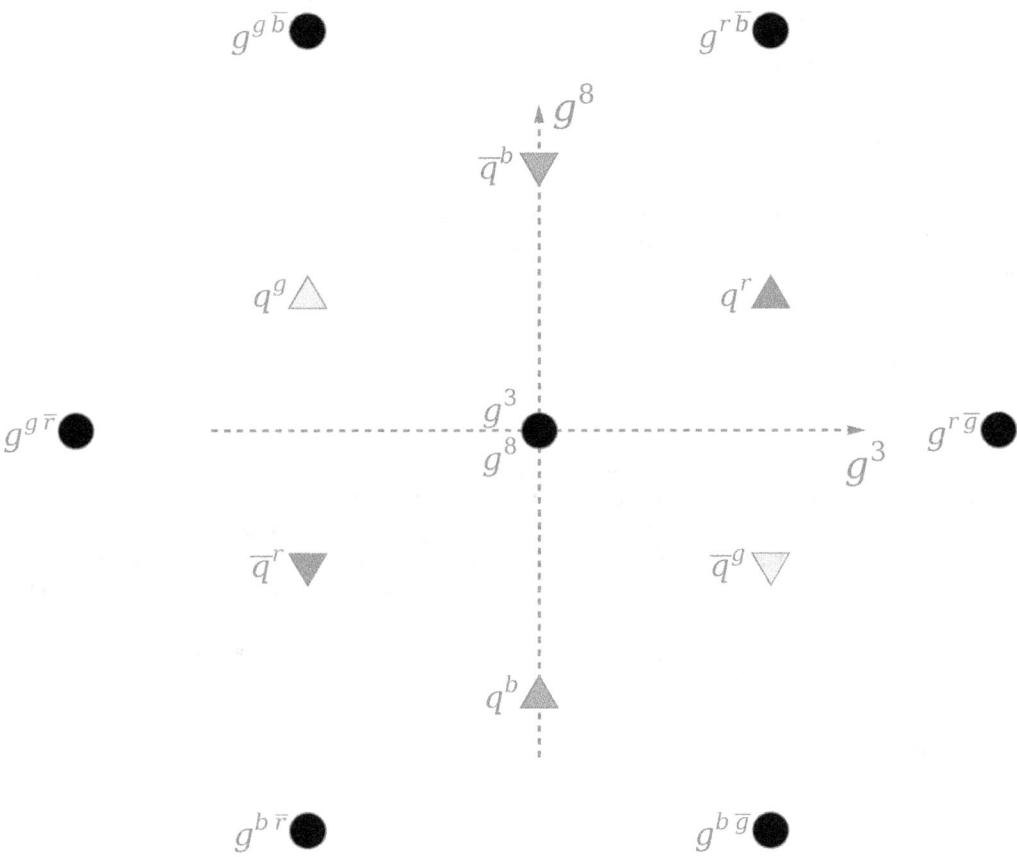

The pattern of strong charges for the three colors of quark, three antiquarks, and eight gluons (with two of zero charge overlapping).

42.3.6 Dynamics

According to the rules of quantum field theory, and the associated Feynman diagrams, the above theory gives rise to three basic interactions: a quark may emit (or absorb) a gluon, a gluon may emit (or absorb) a gluon, and two gluons may directly interact. This contrasts with QED, in which only the first kind of interaction occurs, since photons have no charge. Diagrams involving Faddeev–Popov ghosts must be considered too (except in the unitarity gauge).

42.3.7 Area law and confinement

Detailed computations with the above-mentioned Lagrangian[12] show that the effective potential between a quark and its anti-quark in a meson contains a term $\propto r$, which represents some kind of "stiffness" of the interaction between the particle and its anti-particle at large distances, similar to the entropic elasticity of a rubber band (see below). This leads to *confinement* [13] of the quarks to the interior of hadrons, i.e. mesons and nucleons, with typical radii R_c, corresponding to former "Bag models" of the hadrons[14]. The order of magnitude of the "bag radius" is 1 fm (= 10^{-15} m). Moreover, the above-mentioned stiffness is quantitatively related to the so-called "area law" behaviour of the expectation value of the Wilson loop product PW of the ordered coupling constants around a closed loop W; i.e. $\langle P_W \rangle$ is proportional to the *area* enclosed by the loop. For this behaviour the non-abelian behaviour of the gauge group is essential.

42.4 Methods

Further analysis of the content of the theory is complicated. Various techniques have been developed to work with QCD. Some of them are discussed briefly below.

42.4.1 Perturbative QCD

Main article: Perturbative QCD

This approach is based on asymptotic freedom, which allows perturbation theory to be used accurately in experiments performed at very high energies. Although limited in scope, this approach has resulted in the most precise tests of QCD to date.

42.4.2 Lattice QCD

Main article: Lattice QCD

Among non-perturbative approaches to QCD, the most well established one is lattice QCD. This approach uses a discrete set of spacetime points (called the lattice) to reduce the analytically intractable path integrals of the continuum theory to a very difficult numerical computation which is then carried out on supercomputers like the QCDOC which was constructed for precisely this purpose. While it is a slow and resource-intensive approach, it has wide applicability, giving insight into parts of the theory inaccessible by other means, in particular into the explicit forces acting between quarks and antiquarks in a meson. However, the numerical sign problem makes it difficult to use lattice methods to study QCD at high density and low temperature (e.g. nuclear matter or the interior of neutron stars).

42.4.3 1/N expansion

Main article: 1/N expansion

A well-known approximation scheme, the 1/N expansion, starts from the premise that the number of colors is infinite, and makes a series of corrections to account for the fact that it is not. Until now, it has been the source of qualitative insight rather than a method for quantitative predictions. Modern variants include the AdS/CFT approach.

42.4.4 Effective theories

For specific problems effective theories may be written down which give qualitatively correct results in certain limits. In the best of cases, these may then be obtained as systematic expansions in some parameter of the QCD Lagrangian. One such effective field theory is chiral perturbation theory or ChiPT, which is the QCD effective theory at low energies. More precisely, it is a low energy expansion based on the spontaneous chiral symmetry breaking of QCD, which is an exact symmetry when quark masses are equal to zero, but for the u,d and s quark, which have small mass, it is still a good approximate symmetry. Depending on the number of quarks which are treated as light, one uses either SU(2) ChiPT or SU(3) ChiPT . Other effective theories are heavy quark effective theory (which expands around heavy quark mass near infinity), and soft-collinear effective theory (which expands around large ratios of energy scales). In addition to effective theories, models like the Nambu–Jona-Lasinio model and the chiral model are often used when discussing general features.

42.4.5 QCD sum rules

Main article: QCD sum rules

Based on an Operator product expansion one can derive sets of relations that connect different observables with each other.

42.4.6 Nambu–Jona-Lasinio model

In one of his recent works, Kei-Ichi Kondo derived as a low-energy limit of QCD, a theory linked to the Nambu–Jona-Lasinio model since it is basically a particular non-local version of the Polyakov–Nambu–Jona-Lasinio model.[16] The later being in its local version, nothing but the Nambu–Jona-Lasinio model in which one has included the Polyakov loop effect, in order to describe a 'certain confinement'.

The Nambu–Jona-Lasinio model in itself is, among many other things, used because it is a 'relatively simple' model of chiral symmetry breaking, phenomenon present up to certain conditions (Chiral limit i.e. massless fermions) in QCD itself. In this model, however, there is no confinement. In particular, the energy of an isolated quark in the physical vacuum turns out well defined and finite.

42.5 Experimental tests

The notion of quark flavors was prompted by the necessity of explaining the properties of hadrons during the development of the quark model. The notion of color was necessitated by the puzzle of the Δ++. This has been dealt with in the section on the history of QCD.

The first evidence for quarks as real constituent elements of hadrons was obtained in deep inelastic scattering experiments at SLAC. The first evidence for gluons came in three jet events at PETRA.

Several good quantitative tests of perturbative QCD exist:

- The running of the QCD coupling as deduced from many observations
- Scaling violation in polarized and unpolarized deep inelastic scattering
- Vector boson production at colliders (this includes the Drell-Yan process)
- Jet cross sections in colliders
- Event shape observables at the LEP
- Heavy-quark production in colliders

Quantitative tests of non-perturbative QCD are fewer, because the predictions are harder to make. The best is probably the running of the QCD coupling as probed through lattice computations of heavy-quarkonium spectra. There is a recent claim about the mass of the heavy meson B_c . Other non-perturbative tests are currently at the level of 5% at best. Continuing work on masses and form factors of hadrons and their weak matrix elements are promising candidates for future quantitative tests. The whole subject of quark matter and the quark–gluon plasma is a non-perturbative test bed for QCD which still remains to be properly exploited.

One qualitative prediction of QCD is that there exist composite particles made solely of gluons called glueballs that have not yet been definitively observed experimentally. A definitive observation of a glueball with the properties predicted by QCD would strongly confirm the theory. In principle, if glueballs could be definitively ruled out, this would be a serious experimental blow to QCD. But, as of 2013, scientists are unable to confirm or deny the existence of glueballs definitively, despite the fact that particle accelerators have sufficient energy to generate them.

42.6 Cross-relations to solid state physics

There are unexpected cross-relations to solid state physics. For example, the notion of gauge invariance forms the basis of the well-known Mattis spin glasses,[17] which are systems with the usual spin degrees of freedom $s_i = \pm 1$ for $i = 1,...,N$, with the special fixed "random" couplings $J_{i,k} = \epsilon_i J_0 \epsilon_k$. Here the ϵ_i and ϵ_k quantities can independently and "randomly" take the values ±1, which corresponds to a most-simple gauge transformation ($s_i \to s_i \cdot \epsilon_i \quad J_{i,k} \to \epsilon_i J_{i,k} \epsilon_k \quad s_k \to s_k \cdot \epsilon_k$). This means that thermodynamic expectation values of measurable quantities, e.g. of the energy $\mathcal{H} := -\sum s_i J_{i,k} s_k$, are invariant.

However, here the *coupling degrees of freedom* $J_{i,k}$, which in the QCD correspond to the *gluons*, are "frozen" to fixed values (quenching). In contrast, in the QCD they "fluctuate" (annealing), and through the large number of gauge degrees of freedom the entropy plays an important role (see below).

For positive J_0 the thermodynamics of the Mattis spin glass corresponds in fact simply to a "ferromagnet in disguise", just because these systems have no "frustration" at all. This term is a basic measure in spin glass theory.[18] Quantitatively it is identical with the loop product $P_W := J_{i,k}J_{k,l}...J_{n,m}J_{m,i}$ along a closed loop W. However, for a Mattis spin glass – in contrast to "genuine" spin glasses – the quantity PW never becomes negative.

The basic notion "frustration" of the spin-glass is actually similar to the Wilson loop quantity of the QCD. The only difference is again that in the QCD one is dealing with SU(3) matrices, and that one is dealing with a "fluctuating" quantity. Energetically, perfect absence of frustration should be non-favorable and atypical for a spin glass, which means that one should add the loop product to the Hamiltonian, by some kind of term representing a "punishment". In the QCD the Wilson loop is essential for the Lagrangian rightaway.

The relation between the QCD and "disordered magnetic systems" (the spin glasses belong to them) were additionally stressed in a paper by Fradkin, Huberman und Shenker,[19] which also stresses the notion of duality.

A further analogy consists in the already mentioned similarity to polymer physics, where, analogously to Wilson Loops, so-called "entangled nets" appear, which are important for the formation of the entropy-elasticity (force proportional to the length) of a rubber band. The non-abelian character of the SU(3) corresponds thereby to the non-trivial "chemical links", which glue different loop segments together, and "asymptotic freedom" means in the polymer analogy simply the fact that in the short-wave limit, i.e. for $0 \leftarrow \lambda_w \ll R_c$ (where Rc is a characteristic correlation length for the glued loops, corresponding to the above-mentioned "bag radius", while λ_w is the wavelength of an excitation) any non-trivial correlation vanishes totally, as if the system had crystallized.[20]

There is also a correspondence between confinement in QCD – the fact that the color field is only different from zero in the interior of hadrons – and the behaviour of the usual magnetic field in the theory of type-II superconductors: there the magnetism is confined to the interiour of the Abrikosov flux-line lattice,[21] i.e., the London penetration depth λ of that theory is analogous to the confinement radius Rc of quantum chromodynamics. Mathematically, this correspondendence is supported by the second term, $\propto gG_\mu^a \bar\psi_i \gamma^\mu T_{ij}^a \psi_j$, on the r.h.s. of the Lagrangian.

42.7 See also

- For overviews, see Standard Model, its field theoretical formulation, strong interactions, quarks and gluons, hadrons, confinement, QCD matter, or quark–gluon plasma.

- For details, see gauge theory, quantization procedure including BRST quantization and Faddeev–Popov ghosts. A more general category is quantum field theory.

- For techniques, see Lattice QCD, 1/N expansion, perturbative QCD, Soft-collinear effective theory, heavy quark effective theory, chiral models, and the Nambu and Jona-Lasinio model.

- For experiments, see quark search experiments, deep inelastic scattering, jet physics, quark–gluon plasma.

- Symmetry in quantum mechanics

42.8 References

[1] Gell-Mann, Murray (1995). *The Quark and the Jaguar*. Owl Books. ISBN 978-0-8050-7253-2.

[2] Fyodor Tkachov (2009). "A contribution to the history of quarks: Boris Struminsky's 1965 JINR publication". arXiv:0904.0343 [physics.hist-ph].

[3] B. V. Struminsky, Magnetic moments of barions in the quark model. JINR-Preprint P-1939, Dubna, Russia. Submitted on January 7, 1965.

[4] N. Bogolubov, B. Struminsky, A. Tavkhelidze. On composite models in the theory of elementary particles. JINR Preprint D-1968, Dubna 1965.

[5] A. Tavkhelidze. Proc. Seminar on High Energy Physics and Elementary Particles, Trieste, 1965, Vienna IAEA, 1965, p. 763.

[6] V. A. Matveev and A. N. Tavkhelidze (INR, RAS, Moscow) The quantum number color, colored quarks and QCD (Dedicated to the 40th Anniversary of the Discovery of the Quantum Number Color). Report presented at the 99th Session of the JINR Scientific Council, Dubna, 19–20 January 2006.

[7] J. Polchinski, M. Strassler (2002). "Hard Scattering and Gauge/String duality". *Physical Review Letters* **88** (3): 31601. arXiv:hep-th/0109174. Bibcode:2002PhRvL..88c1601P. doi:10.1103/PhysRevLett.88.031601. PMID 11801052.

[8] Brower, Richard C.; Mathur, Samir D.; Chung-I Tan (2000). "Glueball Spectrum for QCD from AdS Supergravity Duality". *Nuclear Physics B* **587**: 249–276. arXiv:hep-th/0003115. Bibcode:2000NuPhB.587..249B. doi:10.1016/S0550-3213(00)00435-1.

[9] F. Wegner, *Duality in Generalized Ising Models and Phase Transitions without Local Order Parameter*, J. Math. Phys. **12** (1971) 2259–2272.

> Reprinted in Claudio Rebbi (ed.), *Lattice Gauge Theories and Monte Carlo Simulations*, World Scientific, Singapore (1983), p. 60–73. Abstract:

[10] Perhaps one can guess that in the "original" model mainly the quarks would fluctuate, whereas in the present one, the "dual" model, mainly the gluons do.

[11] M. Eidemüller, H.G. Dosch, M. Jamin (1999). "The field strength correlator from QCD sum rules". *Nucl.Phys.Proc.Suppl.86:421–425,2000* (Heidelberg, Germany). arXiv:hep-ph/9908318.

[12] See all standard textbooks on the QCD, e.g., those noted above

[13] Only at extremely large pressures and or temperatures, e.g. for $T \cong 5 \cdot 10^{12}$ K or larger, *confinement* gives way to a quark–gluon plasma.

[14] Kenneth A. Johnson, "The bag model of quark confinement", Scientific American, July 1979

[15] M. Cardoso et al., "Lattice QCD computation of the colour fields for the static hybrid quark–gluon–antiquark system, and microscopic study of the Casimir scaling", Phys. Rev. D 81, 034504 (2010)).

[16] Kei-Ichi Kondo (2010). "Toward a first-principle derivation of confinement and chiral-symmetry-breaking crossover transitions in QCD". *Physical Review D* **82** (6): 065024. arXiv:1005.0314v2. Bibcode:2010PhRvD..82f5024K. doi:10.1103/PhysRevD.82.065024.

[17] D.C. Mattis, Phys. Lett. 56a (1976) 421

[18] J. Vanninemus and G. Toulouse, J. Phys. C 10 (1977) 537

[19] E. Fradkin, B.A. Huberman, S. Shenker, *Gauge Symmetries in random magnetic systems*, Phys. Rev. B 18 (1978) 4783–4794,

[20] A. Bergmann, A. Owen, "Dielectric relaxation spectroscopy of poly[(R)−3-Hydroxybutyrate] (PHD) during crystallization", Polymer International 53 (7) (2004) 863–868,

[21] Mathematically, the flux-line lattices are described by Emil Artin's braid group, which is nonabelian, since one braid can wind around another one.

42.9 Further reading

- Greiner, Walter;Schäfer, Andreas (1994). *Quantum Chromodynamics*. Springer. ISBN 0-387-57103-5.
- Halzen, Francis; Martin, Alan (1984). *Quarks & Leptons: An Introductory Course in Modern Particle Physics*. John Wiley & Sons. ISBN 0-471-88741-2.
- Creutz, Michael (1985). *Quarks, Gluons and Lattices*. Cambridge University Press. ISBN 978-0-521-31535-7.

42.10 External links

- Particle data group
- The millennium prize for proving confinement
- Ab Initio Determination of Light Hadron Masses
- Andreas S Kronfeld *The Weight of the World Is Quantum Chromodynamics*
- Andreas S Kronfeld *Quantum chromodynamics with advanced computing*
- Standard model gets right answer
- Quantum Chromodynamics

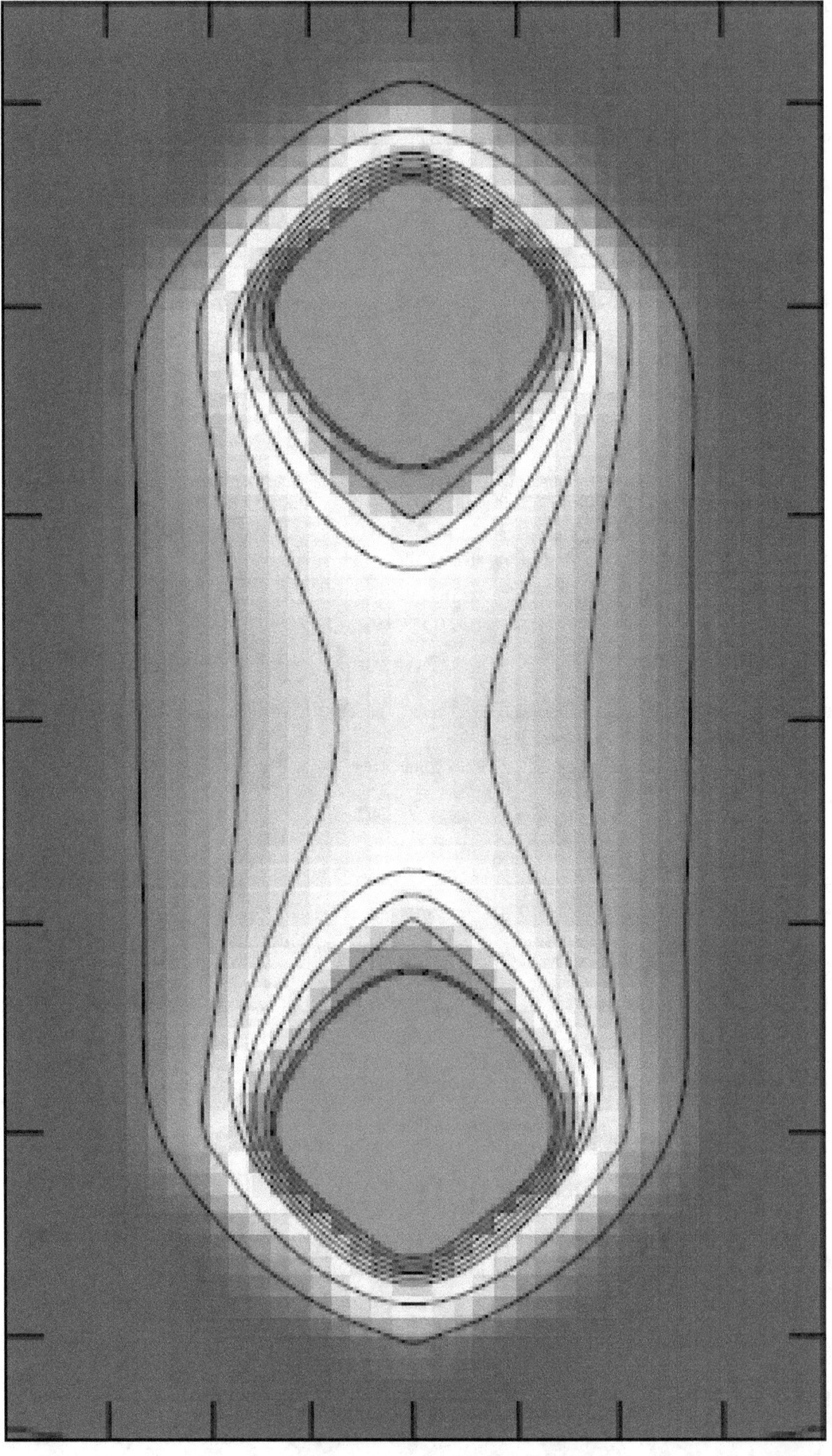

Chapter 43

Classical unified field theories

Since the 19th century, some physicists have attempted to develop a single theoretical framework that can account for the fundamental forces of nature – a unified field theory. **Classical unified field theories** are attempts to create a unified field theory based on classical physics. In particular, unification of gravitation and electromagnetism was actively pursued by several physicists and mathematicians in the years between World War I and World War II. This work spurred the purely mathematical development of differential geometry. Albert Einstein is the best known of the many physicists who attempted to develop a classical unified field theory.

This article describes various attempts at a classical (non-quantum), relativistic unified field theory. For a survey of classical relativistic field theories of gravitation that have been motivated by theoretical concerns other than unification, see Classical theories of gravitation. For a survey of current work toward creating a quantum theory of gravitation, see quantum gravity.

43.1 Overview

The early attempts at creating a unified field theory began with the Riemannian geometry of general relativity, and attempted to incorporate electromagnetic fields into a more general geometry, since ordinary Riemannian geometry seemed incapable of expressing the properties of the electromagnetic field. Einstein was not alone in his attempts to unify electromagnetism and gravity; a large number of mathematicians and physicists, including Hermann Weyl, Arthur Eddington, Theodor Kaluza, and R. Bach also attempted to develop approaches that could unify these interactions.[1][2] These scientists pursued several avenues of generalization, including extending the foundations of geometry and adding an extra spatial dimension.

43.2 Early work

The first attempts to provide a unified theory were by G. Mie in 1912 and Ernst Reichenbacher in 1916.[3][4] However, these theories were unsatisfactory, as they did not incorporate general relativity because general relativity had yet to be formulated. These efforts, along with those of Forster, involved making the metric tensor (which had previously been assumed to be symmetric and real-valued) into an asymmetric and/or complex-valued tensor, and they also attempted to create a field theory for matter as well.

43.3 Differential geometry and field theory

From 1918 until 1923, there were three distinct approaches to field theory: the gauge theory of Weyl, Kaluza's five-dimensional theory, and Eddington's development of affine geometry. Einstein corresponded with these researchers, and collaborated with Kaluza, but was not yet fully involved in the unification effort.

43.4 Weyl's infinitesimal geometry

In order to include electromagnetism into the geometry of general relativity, Hermann Weyl worked to generalize the Riemannian geometry upon which general relativity is based. His idea was to create a more general infinitesimal geometry. He noted that in addition to a metric field there could be additional degrees of freedom along a path between two points in a manifold, and he tried to exploit this by introducing a basic method for comparison of local size measures along such a path, in terms of a gauge field. This geometry generalized Riemannian geometry in that there was a vector field Q, in addition to the metric g, which together gave rise to both the electromagnetic and gravitational fields. This theory was mathematically sound, albeit complicated, resulting in difficult and high-order field equations. The critical mathematical ingredients in this theory, the Lagrangians and curvature tensor, were worked out by Weyl and colleagues. Then Weyl carried out an extensive correspondence with Einstein and others as to its physical validity, and the theory was ultimately found to be physically unreasonable. However, Weyl's principle of gauge invariance was later applied in a modified form to quantum field theory.

43.5 Kaluza's fifth dimension

Kaluza's approach to unification was to embed space-time into a five-dimensional cylindrical world; one of four space dimensions and one of time. Unlike Weyl's approach, Riemannian geometry was maintained, and the extra dimension allowed for the incorporation of the electromagnetic field vector into the geometry. Despite the relative mathematical elegance of this approach, in collaboration with Einstein and Einstein's aide Grommer it was determined that this theory did not admit a non-singular, static, spherically symmetric solution. This theory did have some influence on Einstein's later work and was further developed later by Klein in an attempt to incorporate relativity into quantum theory, in what is now known as Kaluza–Klein theory.

43.6 Eddington's affine geometry

Sir Arthur Stanley Eddington was a noted astronomer who became an enthusiastic and influential promoter of Einstein's general theory of relativity. He was among the first to propose an extension of the gravitational theory based on the affine connection as the fundamental structure field rather than the metric tensor which was the original focus of general relativity. Affine connection is the basis for *parallel transport* of vectors from one space-time point to another; Eddington assumed the affine connection to be symmetric in its covariant indices, because it seemed plausible that the result of parallel-transporting one infinitesimal vector along another should produce the same result as transporting the second along the first. (Later workers revisited this assumption.)

Eddington emphasized what he considered to be epistemological considerations; for example, he thought that the cosmological constant version of the general-relativistic field equation expressed the property that the universe was "self-gauging". Since the simplest cosmological model (the De Sitter universe) that solves that equation is a spherically symmetric, stationary, closed universe (exhibiting a cosmological red shift, which is more conventionally interpreted as due to expansion), it seemed to explain the overall form of the universe.

Like many other classical unified field theorists, Eddington considered that in the Einstein field equations for general relativity the stress–energy tensor $T_{\mu\nu}$, which represents matter/energy, was merely provisional, and that in a truly unified theory the source term would automatically arise as some aspect of the free-space field equations. He also shared the hope that an improved fundamental theory would explain why the two elementary particles then known (proton and electron) have quite different masses.

The Dirac equation for the relativistic quantum electron caused Eddington to rethink his previous conviction that fundamental physical theory had to be based on tensors. He subsequently devoted his efforts into development of a "Fundamental Theory" based largely on algebraic notions (which he called "E-frames"). Unfortunately his descriptions of this theory were sketchy and difficult to understand, so very few physicists followed up on his work.[5]

43.7 Einstein's geometric approaches

When the equivalent of Maxwell's equations for electromagnetism is formulated within the framework of Einstein's theory of general relativity, the electromagnetic field energy (being equivalent to mass as one would expect from

Einstein's famous equation E=mc^2) contributes to the stress tensor and thus to the curvature of space-time, which is the general-relativistic representation of the gravitational field; or putting it another way, certain configurations of curved space-time *incorporate* effects of an electromagnetic field. This suggests that a purely geometric theory ought to treat these two fields as different aspects of the same basic phenomenon. However, ordinary Riemannian geometry is unable to describe the properties of the electromagnetic field as a purely geometric phenomenon.

Einstein tried to form a generalized theory of gravitation that would unify the gravitational and electromagnetic forces (and perhaps others), guided by a belief in a single origin for the entire set of physical laws. These attempts initially concentrated on additional geometric notions such as vierbeins and "distant parallelism", but eventually centered around treating both the metric tensor and the affine connection as fundamental fields. (Because they are not independent, the metric-affine theory was somewhat complicated.) In general relativity, these fields are symmetric (in the matrix sense), but since antisymmetry seemed essential for electromagnetism, the symmetry requirement was relaxed for one or both fields. Einstein's proposed unified-field equations (fundamental laws of physics) were generally derived from a variational principle expressed in terms of the Riemann curvature tensor for the presumed space-time manifold.[6]

In field theories of this kind, particles appear as limited regions in space-time in which the field strength or the energy density are particularly high. Einstein and coworker Leopold Infeld managed to demonstrate that, in Einstein's final theory of the unified field, true singularities of the field did have trajectories resembling point particles. However, singularities are places where the equations break down, and Einstein believed that in an ultimate theory the laws should apply *everywhere*, with particles being soliton-like solutions to the (highly nonlinear) field equations. Further, the large-scale topology of the universe should impose restrictions on the solutions, such as quantization or discrete symmetries.

The degree of abstraction, combined with a relative lack of good mathematical tools for analyzing nonlinear equation systems, make it hard to connect such theories with the physical phenomena that they might describe. For example, it has been suggested that the torsion (antisymmetric part of the affine connection) might be related to isospin rather than electromagnetism; this is related to a discrete (or *"internal"*) symmetry known to Einstein as "displacement field duality".

Einstein became increasingly isolated in his research on a generalized theory of gravitation, and most physicists consider his attempts ultimately unsuccessful. In particular, his pursuit of a unification of the fundamental forces ignored developments in quantum physics (and vice versa), most notably the discovery of the strong nuclear force and weak nuclear force.[7]

43.8 Schrödinger's pure-affine theory

Inspired by Einstein's approach to a unified field theory and Eddington's idea of the affine connection as the sole basis for differential geometric structure for space-time, Erwin Schrödinger from 1940 to 1951 thoroughly investigated pure-affine formulations of generalized gravitational theory. Although he initially assumed a symmetric affine connection, like Einstein he later considered the nonsymmetric field.

Schrödinger's most striking discovery during this work was that the metric tensor was *induced* upon the manifold via a simple construction from the Riemann curvature tensor, which was in turn formed entirely from the affine connection. Further, taking this approach with the simplest feasible basis for the variational principle resulted in a field equation having the form of Einstein's general-relativistic field equation with a cosmological term arising *automatically*.[8]

Skepticism from Einstein and published criticisms from other physicists discouraged Schrödinger, and his work in this area has been largely ignored.

43.9 Later work

After the 1930s, progressively fewer scientists worked on classical unification, due to the continual development of quantum theory and the difficulties encountered in developing a quantum theory of gravity. Einstein continued to work on unified field theories of gravity and electromagnetism, but he became increasingly isolated in this research, which he pursued until his death. Despite the publicity of this work due to Einstein's celebrity status, it never resulted in a resounding success.

Most scientists, though not Einstein, eventually abandoned classical theories. Current mainstream research on unified field theories focuses on the problem of creating a quantum theory of gravity and unifying such a theory with the other fundamental theories in physics, which are quantum theories. (Some programs, most notably string theory, attempt to solve both of these problems at once.) With four fundamental forces now identified, gravity remains the one force whose unification proves problematic.

Although new "classical" unified field theories continue to be proposed from time to time, often involving non-traditional elements such as spinors, none has been generally accepted by physicists.

43.10 See also

- Classical field theories
- Fundamental equation of unified field theory

43.11 References

[1] Weyl, H. (1918). "Gravitation und Elektrizität". *Sitz. Preuss. Akad. Wiss.*: 465.

[2] Eddington, A. S. (1924). *The Mathematical Theory of Relativity, 2nd ed.* Cambridge Univ. Press.

[3] Mie, G. (1912). "Grundlagen einer Theorie der Materie". *Ann. Phys.* **37** (3): 511–534. Bibcode:1912AnP...342..511M. doi:10.1002/andp.19123420306.

[4] Reichenbächer, E. (1917). "Grundzüge zu einer Theorie der Elektrizität und der Gravitation". *Ann. Phys.* **52** (2): 134–173. Bibcode:1917AnP...357..134R. doi:10.1002/andp.19173570203.

[5] Kilmister, C. W. (1994). *Eddington's search for a fundamental theory*. Cambridge Univ. Press.

[6] Einstein, A. (1956). *The Meaning of Relativity. 5th ed.* Princeton Univ. Press.

[7] Gönner, Hubert F. M. "On the History of Unified Field Theories". *Living Reviews in Relativity.* Retrieved August 10, 2005.

[8] Schrödinger, E. (1950). *Space-Time Structure*. Cambridge Univ. Press.

Chapter 44

Riemannian geometry

Elliptic geometry is also sometimes called "Riemannian geometry".

Riemannian geometry is the branch of differential geometry that studies Riemannian manifolds, smooth manifolds with a *Riemannian metric*, i.e. with an inner product on the tangent space at each point that varies smoothly from point to point. This gives, in particular, local notions of angle, length of curves, surface area, and volume. From those some other global quantities can be derived by integrating local contributions.

Riemannian geometry originated with the vision of Bernhard Riemann expressed in his inaugural lecture *Ueber die Hypothesen, welche der Geometrie zu Grunde liegen* (*On the Hypotheses which lie at the Bases of Geometry*). It is a very broad and abstract generalization of the differential geometry of surfaces in $\mathbf{R}^3$. Development of Riemannian geometry resulted in synthesis of diverse results concerning the geometry of surfaces and the behavior of geodesics on them, with techniques that can be applied to the study of differentiable manifolds of higher dimensions. It enabled Einstein's general relativity theory, made profound impact on group theory and representation theory, as well as analysis, and spurred the development of algebraic and differential topology.

44.1 Introduction

Riemannian geometry was first put forward in generality by Bernhard Riemann in the nineteenth century. It deals with a broad range of geometries whose metric properties vary from point to point, including the standard types of Non-Euclidean geometry.

Any smooth manifold admits a Riemannian metric, which often helps to solve problems of differential topology. It also serves as an entry level for the more complicated structure of pseudo-Riemannian manifolds, which (in four dimensions) are the main objects of the theory of general relativity. Other generalizations of Riemannian geometry include Finsler geometry.

There exists a close analogy of differential geometry with the mathematical structure of defects in regular crystals. Dislocations and Disclinations produce torsions and curvature.[1][2]

The following articles provide some useful introductory material:

- Metric tensor

- Riemannian manifold

- Levi-Civita connection

- Curvature

- Curvature tensor

- List of differential geometry topics

- Glossary of Riemannian and metric geometry

Bernhard Riemann

44.2 Classical theorems in Riemannian geometry

What follows is an incomplete list of the most classical theorems in Riemannian geometry. The choice is made depending on its importance, beauty, and simplicity of formulation. Most of the results can be found in the classic monograph by Jeff Cheeger and D. Ebin (see below).

The formulations given are far from being very exact or the most general. This list is oriented to those who already know the basic definitions and want to know what these definitions are about.

44.2.1 General theorems

1. **Gauss–Bonnet theorem** The integral of the Gauss curvature on a compact 2-dimensional Riemannian manifold is equal to $2\pi\chi(M)$ where $\chi(M)$ denotes the Euler characteristic of M. This theorem has a generalization to any compact even-dimensional Riemannian manifold, see generalized Gauss-Bonnet theorem.

2. **Nash embedding theorems** also called fundamental theorems of Riemannian geometry. They state that every

Riemannian manifold can be isometrically embedded in a Euclidean space $\mathbf{R}^n$.

44.2.2 Geometry in large

In all of the following theorems we assume some local behavior of the space (usually formulated using curvature assumption) to derive some information about the global structure of the space, including either some information on the topological type of the manifold or on the behavior of points at "sufficiently large" distances.

Pinched sectional curvature

1. **Sphere theorem.** If M is a simply connected compact n-dimensional Riemannian manifold with sectional curvature strictly pinched between 1/4 and 1 then M is diffeomorphic to a sphere.

2. **Cheeger's finiteness theorem.** Given constants C, D and V, there are only finitely many (up to diffeomorphism) compact n-dimensional Riemannian manifolds with sectional curvature $|K| \leq C$, diameter $\leq D$ and volume $\geq V$.

3. **Gromov's almost flat manifolds.** There is an $\varepsilon n > 0$ such that if an n-dimensional Riemannian manifold has a metric with sectional curvature $|K| \leq \varepsilon n$ and diameter ≤ 1 then its finite cover is diffeomorphic to a nil manifold.

Sectional curvature bounded below

1. **Cheeger-Gromoll's Soul theorem.** If M is a non-compact complete non-negatively curved n-dimensional Riemannian manifold, then M contains a compact, totally geodesic submanifold S such that M is diffeomorphic to the normal bundle of S (S is called the **soul** of M.) In particular, if M has strictly positive curvature everywhere, then it is diffeomorphic to $\mathbf{R}^n$. G. Perelman in 1994 gave an astonishingly elegant/short proof of the Soul Conjecture: M is diffeomorphic to $\mathbf{R}^n$ if it has positive curvature at only one point.

2. **Gromov's Betti number theorem.** There is a constant $C = C(n)$ such that if M is a compact connected n-dimensional Riemannian manifold with positive sectional curvature then the sum of its Betti numbers is at most C.

3. **Grove–Petersen's finiteness theorem.** Given constants C, D and V, there are only finitely many homotopy types of compact n-dimensional Riemannian manifolds with sectional curvature $K \geq C$, diameter $\leq D$ and volume $\geq V$.

Sectional curvature bounded above

1. The **Cartan–Hadamard theorem** states that a complete simply connected Riemannian manifold M with non-positive sectional curvature is diffeomorphic to the Euclidean space $\mathbf{R}^n$ with $n = \dim M$ via the exponential map at any point. It implies that any two points of a simply connected complete Riemannian manifold with nonpositive sectional curvature are joined by a unique geodesic.

2. The geodesic flow of any compact Riemannian manifold with negative sectional curvature is ergodic.

3. If M is a complete Riemannian manifold with sectional curvature bounded above by a strictly negative constant k then it is a CAT(k) space. Consequently, its fundamental group $\Gamma = \pi_1(M)$ is Gromov hyperbolic. This has many implications for the structure of the fundamental group:

    - it is finitely presented;
    - the word problem for Γ has a positive solution;
    - the group Γ has finite virtual cohomological dimension;
    - it contains only finitely many conjugacy classes of elements of finite order;
    - the abelian subgroups of Γ are virtually cyclic, so that it does not contain a subgroup isomorphic to $\mathbf{Z} \times \mathbf{Z}$.

Ricci curvature bounded below

1. **Myers theorem.** If a compact Riemannian manifold has positive Ricci curvature then its fundamental group is finite.

2. **Splitting theorem.** If a complete n-dimensional Riemannian manifold has nonnegative Ricci curvature and a straight line (i.e. a geodesic that minimizes distance on each interval) then it is isometric to a direct product of the real line and a complete $(n-1)$-dimensional Riemannian manifold that has nonnegative Ricci curvature.

3. **Bishop–Gromov inequality.** The volume of a metric ball of radius r in a complete n-dimensional Riemannian manifold with positive Ricci curvature has volume at most that of the volume of a ball of the same radius r in Euclidean space.

4. **Gromov's compactness theorem.** The set of all Riemannian manifolds with positive Ricci curvature and diameter at most D is pre-compact in the Gromov-Hausdorff metric.

Negative Ricci curvature

1. The isometry group of a compact Riemannian manifold with negative Ricci curvature is discrete.

2. Any smooth manifold of dimension $n \geq 3$ admits a Riemannian metric with negative Ricci curvature.[3] (*This is not true for surfaces.*)

Positive scalar curvature

1. The n-dimensional torus does not admit a metric with positive scalar curvature.

2. If the injectivity radius of a compact n-dimensional Riemannian manifold is $\geq \pi$ then the average scalar curvature is at most $n(n-1)$.

44.3 See also

- Shape of the universe
- Basic introduction to the mathematics of curved spacetime
- Normal coordinates
- Systolic geometry
- Riemann–Cartan geometry in Einstein–Cartan theory (motivation)
- Riemann's minimal surface

44.4 Literature

[1] Kleinert, Hagen (1989). "Gauge Fields in Condensed Matter Vol II". pp. 743–1440.

[2] Kleinert, Hagen (2008). "Multivalued Fields in Condensed Matter, Electromagnetism, and Gravitation". pp. 1–496.

[3] Joachim Lohkamp has shown (Annals of Mathematics, 1994) that any manifold of dimension greater than two admits a metric of negative Ricci curvature.

44.5 References

Books

- Berger, Marcel (2000), *Riemannian Geometry During the Second Half of the Twentieth Century*, University Lecture Series **17**, Rhode Island: American Mathematical Society, ISBN 0-8218-2052-4. *(Provides a historical review and survey, including hundreds of references.)*

- Cheeger, Jeff; Ebin, David G. (2008), *Comparison theorems in Riemannian geometry*, Providence, RI: AMS Chelsea Publishing; Revised reprint of the 1975 original.

- Gallot, Sylvestre; Hulin, Dominique; Lafontaine, Jacques (2004), *Riemannian geometry*, Universitext (3rd ed.), Berlin: Springer-Verlag.

- Jost, Jürgen (2002), *Riemannian Geometry and Geometric Analysis*, Berlin: Springer-Verlag, ISBN 3-540-42627-2.

- Petersen, Peter (2006), *Riemannian Geometry*, Berlin: Springer-Verlag, ISBN 0-387-98212-4

Papers

- Brendle, Simon; Schoen, Richard M. (2007), *Classification of manifolds with weakly 1/4-pinched curvatures*, arXiv:0705.3963

44.6 External links

- Riemannian geometry by V. A. Toponogov at the Encyclopedia of Mathematics
- Weisstein, Eric W., "Riemannian Geometry", *MathWorld*.

Chapter 45

Affine connection

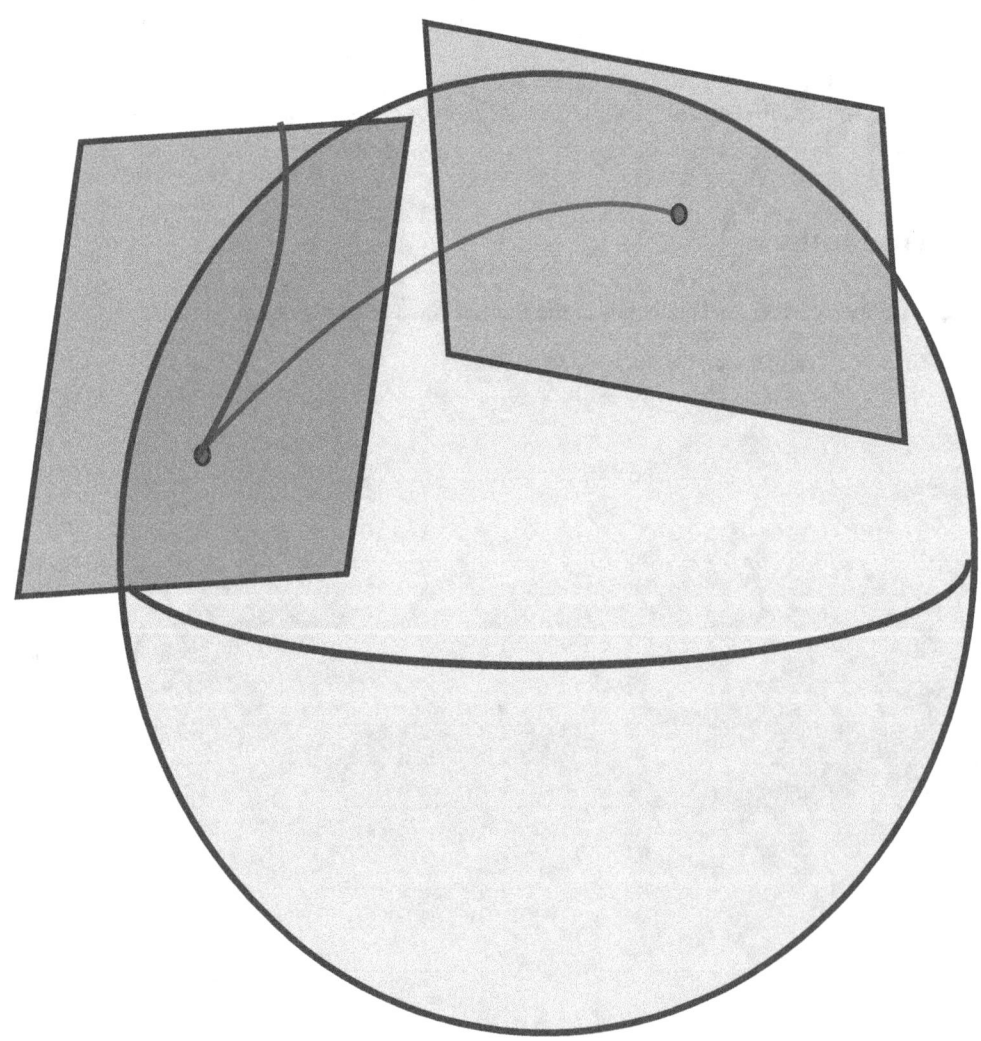

An affine connection on the sphere rolls the affine tangent plane from one point to another. As it does so, the point of contact traces out a curve in the plane: the development.

In the branch of mathematics called differential geometry, an **affine connection** is a geometric object on a smooth manifold which *connects* nearby tangent spaces, and so permits tangent vector fields to be differentiated as if they were functions on the manifold with values in a fixed vector space. The notion of an affine connection has its roots in 19th-century geometry and tensor calculus, but was not fully developed until the early 1920s, by Élie Cartan (as part of his general theory of connections) and Hermann Weyl (who used the notion as a part of his foundations for general relativity). The terminology is due to Cartan and has its origins in the identification of tangent spaces in Euclidean space $\mathbf{R}^n$ by translation: the idea is that a choice of affine connection makes a manifold look infinitesimally like Euclidean space not just smoothly, but as an affine space.

On any manifold of positive dimension there are infinitely many affine connections. If the manifold is further endowed with a Riemannian metric then there is a natural choice of affine connection, called the Levi-Civita connection. The choice of an affine connection is equivalent to prescribing a way of differentiating vector fields which satisfies several reasonable properties (linearity and the Leibniz rule). This yields a possible definition of an affine connection as a covariant derivative or (linear) connection on the tangent bundle. A choice of affine connection is also equivalent to a notion of parallel transport, which is a method for transporting tangent vectors along curves. This also defines a parallel transport on the frame bundle. Infinitesimal parallel transport in the frame bundle yields another description of an affine connection, either as a Cartan connection for the affine group or as a principal connection on the frame bundle.

The main invariants of an affine connection are its torsion and its curvature. The torsion measures how closely the Lie bracket of vector fields can be recovered from the affine connection. Affine connections may also be used to define (affine) geodesics on a manifold, generalizing the *straight lines* of Euclidean space, although the geometry of those straight lines can be very different from usual Euclidean geometry; the main differences are encapsulated in the curvature of the connection.

45.1 Motivation and history

A smooth manifold is a mathematical object which looks locally like a smooth deformation of Euclidean space $\mathbf{R}^n$: for example a smooth curve or surface looks locally like a smooth deformation of a line or a plane. Smooth functions and vector fields can be defined on manifolds, just as they can on Euclidean space, and scalar functions on manifolds can be differentiated in a natural way. However, differentiation of vector fields is less straightforward: this is a simple matter in Euclidean space, because the tangent space of based vectors at a point p can be identified naturally (by translation) with the tangent space at a nearby point q. On a general manifold, there is no such natural identification between nearby tangent spaces, and so tangent vectors at nearby points cannot be compared in a well-defined way. The notion of an affine connection was introduced to remedy this problem by *connecting* nearby tangent spaces. The origins of this idea can be traced back to two main sources: surface theory and tensor calculus.

45.1.1 Motivation from surface theory

See also: Cartan connection

Consider a smooth surface S in 3-dimensional Euclidean space. Near to any point, S can be approximated by its tangent plane at that point, which is an affine subspace of Euclidean space. Differential geometers in the 19th century were interested in the notion of development in which one surface was *rolled* along another, without *slipping* or *twisting*. In particular, the tangent plane to a point of S can be rolled on S: this should be easy to imagine when S is a surface like the 2-sphere, which is the smooth boundary of a convex region. As the tangent plane is rolled on S, the point of contact traces out a curve on S. Conversely, given a curve on S, the tangent plane can be rolled along that curve. This provides a way to identify the tangent planes at different points along the curve: in particular, a tangent vector in the tangent space at one point on the curve is identified with a unique tangent vector at any other point on the curve. These identifications are always given by affine transformations from one tangent plane to another.

This notion of parallel transport of tangent vectors, by affine transformations, along a curve has a characteristic feature: the point of contact of the tangent plane with the surface *always moves* with the curve under parallel translation (i.e., as the tangent plane is rolled along the surface, the point of contact moves). This generic condition is characteristic of Cartan connections. In more modern approaches, the point of contact is viewed as the *origin* in the tangent plane (which is then a vector space), and the movement of the origin is corrected by a translation, so that parallel transport is linear, rather than affine.

In the point of view of Cartan connections, however, the affine subspaces of Euclidean space are *model* surfaces —

they are the simplest surfaces in Euclidean 3-space, and are homogeneous under the affine group of the plane — and every smooth surface has a unique model surface tangent to it at each point. These model surfaces are *Klein geometries* in the sense of Felix Klein's Erlangen programme. More generally, an n-dimensional affine space is a Klein geometry for the affine group Aff(n), the stabilizer of a point being the general linear group GL(n). An affine n-manifold is then a manifold which looks infinitesimally like n-dimensional affine space.

45.1.2 Motivation from tensor calculus

See also: covariant derivative

The second motivation for affine connections comes from the notion of a covariant derivative of vector fields. Before the advent of coordinate-independent methods, it was necessary to work with vector fields using their components in coordinate charts. These components can be differentiated, but the derivatives do not transform in a manageable way under changes of coordinates. Correction terms were introduced by Elwin Bruno Christoffel (following ideas of Bernhard Riemann) in the 1870s so that the (corrected) derivative of one vector field along another transformed covariantly under coordinate transformations — these correction terms subsequently came to be known as Christoffel symbols. This idea was developed into the theory of the *absolute differential calculus* (now known as tensor calculus) by Gregorio Ricci-Curbastro and his student Tullio Levi-Civita between 1880 and the turn of the 20th century.

The tensor calculus really came to life, however, with the advent of Albert Einstein's theory of general relativity in 1915. A few years after this, Levi-Civita formalized the unique connection associated to a Riemannian metric, now known as the Levi-Civita connection. More general affine connections were then studied around 1920, by Hermann Weyl,[1] who developed a detailed mathematical foundation for general relativity, and Élie Cartan,[2] who made the link with the geometrical ideas coming from surface theory.

45.1.3 Approaches

The complex history has led to the development of widely varying approaches to and generalizations of the affine connection concept.

The most popular approach is probably the definition motivated by covariant derivatives. On the one hand, the ideas of Weyl were taken up by physicists in the form of gauge theory and gauge covariant derivatives. On the other hand, the notion of covariant differentiation was abstracted by Jean-Louis Koszul, who defined (linear or Koszul) connections on vector bundles. In this language, an affine connection is simply a covariant derivative or (linear) connection on the tangent bundle.

However, this approach does not explain the geometry behind affine connections nor how they acquired their name.[3] The term really has its origins in the identification of tangent spaces in Euclidean space by translation: this property means that Euclidean n-space is an affine space. (Alternatively, Euclidean space is a principal homogeneous space or torsor under the group of translations, which is a subgroup of the affine group.) As mentioned in the introduction, there are several ways to make this precise: one uses the fact that an affine connection defines a notion of parallel transport of vector fields along a curve. This also defines a parallel transport on the frame bundle. Infinitesimal parallel transport in the frame bundle yields another description of an affine connection, either as a Cartan connection for the affine group Aff(n) or as a principal GL(n) connection on the frame bundle.

45.2 Formal definition as a differential operator

See also: covariant derivative and connection (vector bundle)

Let M be a smooth manifold and let $C^\infty(M,TM)$ be the space of vector fields on M, that is, the space of smooth sections of the tangent bundle TM. Then an **affine connection** on M is a bilinear map

$$\begin{array}{rcl} C^\infty(M,TM) \times C^\infty(M,TM) & \to & C^\infty(M,TM) \\ (X,Y) & \mapsto & \nabla_X Y, \end{array}$$

such that for all smooth functions f in $C^\infty(M,\mathbf{R})$ and all vector fields X, Y on M:

1. $\nabla_{fX}Y = f\nabla_X Y$, that is, ∇ is $C^\infty(M,\mathbf{R})$-*linear* in the first variable;

2. $\nabla_X(fY) = \mathrm{d}f(X)Y + f\nabla_X Y$, that is, ∇ satisfies *Leibniz rule* in the second variable.

45.2.1 Elementary properties

- It follows from the property (1) above that the value of ∇XY at a point $x \in M$ depends only on the value of X at x and not on the value of X on $M-\{x\}$. It also follows from property (2) above that the value of ∇XY at a point $x \in M$ depends only on the value of Y on a neighbourhood of x.

- If ∇^1, ∇^2 are affine connections then the value at x of $\nabla^1 XY - \nabla^2 XY$ may be written $\Gamma x(Xx,Yx)$ where

  $$\Gamma x\colon \mathrm{T}xM \times \mathrm{T}xM \to \mathrm{T}xM$$

  is bilinear and depends smoothly on x (i.e., it defines a smooth bundle homomorphism). Conversely if ∇ is an affine connection and Γ is such a smooth bilinear bundle homomorphism (called a connection form on M) then $\nabla + \Gamma$ is an affine connection.

- If M is an open subset of $\mathbf{R}^n$, then the tangent bundle of M is the trivial bundle $M \times \mathbf{R}^n$. In this situation there is a canonical affine connection d on M: any vector field Y is given by a smooth function V from M to $\mathbf{R}^n$; then $\mathrm{d}XY$ is the vector field corresponding to the smooth function $\mathrm{d}V(X)=\partial XY$ from M to $\mathbf{R}^n$. Any other affine connection ∇ on M may therefore be written $\nabla = \mathrm{d} + \Gamma$, where Γ is a connection form on M.

- More generally, a local trivialization of the tangent bundle is a bundle isomorphism between the restriction of TM to an open subset U of M, and $U \times \mathbf{R}^n$. The restriction of an affine connection ∇ to U may then be written in the form $\mathrm{d} + \Gamma$ where Γ is a connection form on U.

45.3 Parallel transport for affine connections

See also: parallel transport

Comparison of tangent vectors at different points on a manifold is generally not a well-defined process. An affine connection provides one way to remedy this using the notion of parallel transport, and indeed this can be used to give a definition of an affine connection.

Let M be a manifold with an affine connection ∇. Then a vector field X is said to be **parallel** if $\nabla X = 0$ in the sense that for any vector field Y, $\nabla YX=0$. Intuitively speaking, parallel vectors have *all their derivatives equal to zero* and are therefore in some sense *constant*. By evaluating a parallel vector field at two points x and y, an identification between a tangent vector at x and one at y is obtained. Such tangent vectors are said to be **parallel transports** of each other.

Unfortunately, nonzero parallel vector fields do not, in general, exist, because the equation $\nabla X = 0$ is a partial differential equation which is overdetermined: the integrability condition for this equation is the vanishing of the **curvature** of ∇ (see below). However, if this equation is restricted to a curve from x to y it becomes an ordinary differential equation. There is then a unique solution for any initial value of X at x.

More precisely, if $\gamma : I \to M$ a smooth curve parametrized by an interval $[a,b]$ and $\xi \in \mathrm{T}xM$, where $x=\gamma(a)$, then a vector field X along γ (and in particular, the value of this vector field at $y=\gamma(b)$) is called the **parallel transport of ξ along γ** if

1. $\nabla_{\dot\gamma(t)} X = 0$, for all $t \in [a,b]$

2. $X_{\gamma(a)} = \xi$.

Formally, the first condition means that X is parallel with respect to the pullback connection on the pullback bundle γ^*TM. However, in a local trivialization it is a first-order system of linear ordinary differential equations, which has a unique solution for any initial condition given by the second condition (for instance, by the Picard–Lindelöf theorem).

Thus parallel transport provides a way of moving tangent vectors along a curve using the affine connection to keep them "pointing in the same direction" in an intuitive sense, and this provides a linear isomorphism between the tangent spaces at the two ends of the curve. The isomorphism obtained in this way will in general depend on the choice of

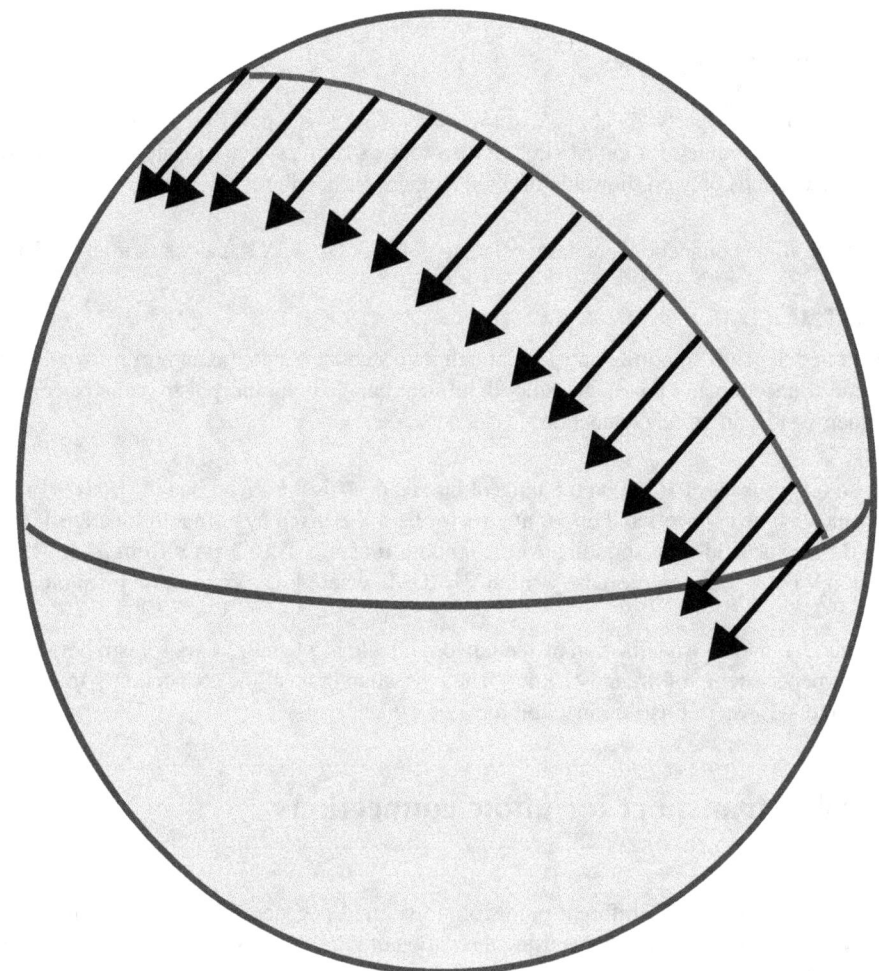

Parallel transport of a tangent vector along a curve in the sphere.

the curve: if it does not, then parallel transport along every curve can be used to define parallel vector fields on M, which can only happen if the curvature of ∇ is zero.

A linear isomorphism is determined by its action on an ordered basis or **frame**. Hence parallel transport can also be characterized as a way of transporting elements of the (tangent) frame bundle GL(M) along a curve. In other words, the affine connection provides a **lift** of any curve γ in M to a curve $\tilde{\gamma}$ in GL(M).

45.4 Formal definition on the frame bundle

See also: connection (principal bundle)

An affine connection may also be defined as a principal GL(n) connection ω on the frame bundle FM or GL(M) of a manifold M. In more detail, ω is a smooth map from the tangent bundle T(FM) of the frame bundle to the space of $n \times n$ matrices (which is the Lie algebra **gl**(n) of the Lie group GL(n) of invertible $n \times n$ matrices) satisfying two properties:

1. ω is equivariant with respect to the action of GL(n) on T(FM) and **gl**(n);

2. ω(Xξ) = ξ for any ξ in **gl**(n), where Xξ is the vector field on FM corresponding to ξ.

Such a connection ω immediately defines a covariant derivative not only on the tangent bundle, but on vector bundles associated to any group representation of GL(n), including bundles of tensors and tensor densities. Conversely, an affine connection on the tangent bundle determines an affine connection on the frame bundle, for instance, by requiring that ω vanishes on tangent vectors to the lifts of curves to the frame bundle defined by parallel transport.

The frame bundle also comes equipped with a solder form θ : T(FM) → **R**^{n} which is **horizontal** in the sense that it vanishes on vertical vectors such as the point values of the vector fields Xξ: indeed θ is defined first by projecting a tangent vector (to FM at a frame f) to M, then by taking the components of this tangent vector on M with respect to the frame f. Note that θ is also GL(n)-equivariant (where GL(n) acts on **R**^{n} by matrix multiplication).

The pair (θ,ω) define a bundle isomorphism of T(FM) with the trivial bundle FM × **aff**(n), where **aff**(n) is the cartesian product of **R**^{n} and **gl**(n) (viewed as the Lie algebra of the affine group, which is actually a semidirect product — see below).

45.5 Affine connections as Cartan connections

See also: Cartan connection

Affine connections can be defined within Cartan's general framework.[4] In the modern approach, this is closely related to the definition of affine connections on the frame bundle. Indeed, in one formulation, a Cartan connection is an absolute parallelism of a principal bundle satisfying suitable properties. From this point of view the **aff**(n)-valued 1-form (θ,ω): T(FM) → **aff**(n) on the frame bundle (of an affine manifold) is a Cartan connection. However, Cartan's original approach was different from this in a number of ways:

- the concept of frame bundles or principal bundles did not exist;
- a connection was viewed in terms of parallel transport between infinitesimally nearby points;[5]
- this parallel transport was affine, rather than linear;
- the objects being transported were not tangent vectors in the modern sense, but elements of an affine space with a marked point, which the Cartan connection ultimately *identifies* with the tangent space.

45.5.1 Explanations and historical intuition

The points just raised are easiest to explain in reverse, starting from the motivation provided by surface theory. In this situation, although the planes being rolled over the surface are tangent planes in a naive sense, the notion of a tangent space is really an infinitesimal notion,[6] whereas the planes, as affine subspaces of **R**^{3}, are infinite in extent. However these affine planes all have a marked point, the point of contact with the surface, and they are tangent to the surface at this point. The confusion therefore arises because an affine space with a marked point can be identified with its tangent space at that point. However, the parallel transport defined by rolling does not fix this "origin": it is affine rather than linear; the linear parallel transport can be recovered by applying a translation.

Abstracting this idea, an affine manifold should therefore be an n-manifold M with an affine space Ax, of dimension n, *attached* to each $x \in M$ at a marked point $ax \in Ax$, together with a method for transporting elements of these affine spaces along any curve C in M. This method is required to satisfy several properties:

1. for any two points x, y on C, parallel transport is an affine transformation from Ax to Ay;

2. parallel transport is defined infinitesimally in the sense that it is differentiable at any point on C and depends only on the tangent vector to C at that point;

3. the derivative of the parallel transport at x determines a linear isomorphism from TxM to $T_{a_x} A_x$.

These last two points are quite hard to make precise,[7] so affine connections are more often defined infinitesimally. To motivate this, it suffices to consider how affine frames of reference transform infinitesimally with respect to parallel transport. (This is the origin of Cartan's method of moving frames.) An affine frame at a point consists of a list $(p, \mathbf{e}_1, ..., \mathbf{e}_n)$, where $p \in A_x$[8] and the $\mathbf{e}_i$ form a basis of $T_p(A_x)$. The affine connection is then given symbolically by a first order differential system

$$(*)\begin{cases} dp & = \theta^1 \mathbf{e}_1 + \cdots + \theta^n \mathbf{e}_n \\ d\mathbf{e}_i & = \omega_i^1 \mathbf{e}_1 + \cdots + \omega_i^n \mathbf{e}_n, \qquad i = 1, 2, \ldots, n \end{cases}$$

defined by a collection of one-forms (θ^j, ω_i^j). Geometrically, an affine frame undergoes a displacement travelling along a curve γ from $\gamma(t)$ to $\gamma(t + \delta t)$ given (approximately, or infinitesimally) by

$$\begin{aligned} p(\gamma(t+\delta t)) - p(\gamma(t)) &= \bigl(\theta^1(\gamma'(t))\mathbf{e}_1 + \cdots + \theta^n(\gamma'(t))\mathbf{e}_n\bigr)\delta t \\ \mathbf{e}_i(\gamma(t+\delta t)) - \mathbf{e}_i(\gamma(t)) &= \bigl(\omega_i^1(\gamma'(t))\mathbf{e}_1 + \cdots + \omega_i^n(\gamma'(t))\mathbf{e}_n\bigr)\delta t. \end{aligned}$$

Furthermore, the affine spaces A_x are required to be *tangent* to M in the informal sense that the displacement of a_x along γ can be identified (approximately or infinitesimally) with the tangent vector $\gamma'(t)$ to γ at $x=\gamma(t)$ (which is the infinitesimal displacement of x). Since

$$a_x(\gamma(t+\delta t)) - a_x(\gamma(t)) = \theta(\gamma'(t))\delta t,$$

where θ is defined by $\theta(X) = \theta^1(X)\mathbf{e}_1 + ... + \theta^n(X)\mathbf{e}_n$, this identification is given by θ, so the requirement is that θ should be a linear isomorphism at each point.

The tangential affine space A_x is thus identified intuitively with an *infinitesimal affine neighborhood* of x.

The modern point of view makes all this intuition more precise using principal bundles (the essential idea is to replace a frame or a *variable* frame by the space of all frames and functions on this space). It also draws on the inspiration of Felix Klein's Erlangen programme,[9] in which a *geometry* is defined to be a homogeneous space. Affine space is a geometry in this sense, and is equipped with a *flat* Cartan connection. Thus a general affine manifold is viewed as *curved* deformation of the flat model geometry of affine space.

45.5.2 Affine space as the flat model geometry

Definition of an affine space

Informally, an **affine space** is a vector space without a fixed choice of origin. It describes the geometry of points and free vectors in space. As a consequence of the lack of origin, points in affine space cannot be added together as this requires a choice of origin with which to form the parallelogram law for vector addition. However, a vector v may be added to a point p by placing the initial point of the vector at p and then transporting p to the terminal point. The operation thus described $p \to p+v$ is the **translation** of p along v. In technical terms, affine n-space is a set $\mathbf{A}^n$ equipped with a free transitive action of the vector group $\mathbf{R}^n$ on it through this operation of translation of points: $\mathbf{A}^n$ is thus a principal homogeneous space for the vector group $\mathbf{R}^n$.

The general linear group $GL(n)$ is the group of transformations of $\mathbf{R}^n$ which preserve the *linear structure* of $\mathbf{R}^n$ in the sense that $T(av+bw) = aT(v) + bT(w)$. By analogy, the **affine group** $\mathrm{Aff}(n)$ is the group of transformations of $\mathbf{A}^n$ preserving the *affine structure*. Thus $\varphi \in \mathrm{Aff}(n)$ must *preserve translations* in the sense that

$$\phi(p+v) = \alpha(p) + T(v)$$

where T is a general linear transformation. The map sending $\varphi \in \mathrm{Aff}(n)$ to $T \in GL(n)$ is a group homomorphism. Its kernel is the group of translations $\mathbf{R}^n$. The stabilizer of any point p in A can thus be identified with $GL(n)$ using this projection: this realises the affine group as a semidirect product of $GL(n)$ and $\mathbf{R}^n$, and affine space as the homogeneous space $\mathrm{Aff}(n)/GL(n)$.

45.5. AFFINE CONNECTIONS AS CARTAN CONNECTIONS

Affine frames and the flat affine connection

An *affine frame* for A consists of a point $p \in A$ and a basis $(\mathbf{e}_1,...,\mathbf{e}_n)$ of the vector space $T_pA = \mathbf{R}^n$. The general linear group $GL(n)$ acts freely on the set FA of all affine frames by fixing p and transforming the basis $(\mathbf{e}_1,...,\mathbf{e}_n)$ in the usual way, and the map π sending an affine frame $(p;\mathbf{e}_1,...,\mathbf{e}n)$ to p is the quotient map. Thus FA is a principal $GL(n)$-bundle over A. The action of $GL(n)$ extends naturally to a free transitive action of the affine group $\mathrm{Aff}(n)$ on FA, so that FA is an $\mathrm{Aff}(n)$-torsor, and the choice of a reference frame identifies $FA \to A$ with the principal bundle $\mathrm{Aff}(n) \to \mathrm{Aff}(n)/GL(n)$.

On FA there is a collection of $n+1$ functions defined by

$$\pi(p;\mathbf{e}_1,\ldots,\mathbf{e}_n) = p$$
$$\epsilon_i(p;\mathbf{e}_1,\ldots,\mathbf{e}_n) = \mathbf{e}_i.$$

After choosing a basepoint for A, these are all functions with values in $\mathbf{R}^n$, so it is possible to take their exterior derivatives to obtain differential 1-forms with values in $\mathbf{R}^n$. Since the functions ϵi yield a basis for $\mathbf{R}^n$ at each point of FA, these 1-forms must be expressible as sums of the form

$$\begin{aligned} d\pi &= \theta^1 \epsilon_1 + \cdots + \theta^n \epsilon_n \\ d\epsilon_i &= \omega_i^1 \epsilon_1 + \cdots + \omega_i^n \epsilon_n \end{aligned}$$

for some collection $(\theta^i, \omega_j^k)_{1 \leq i,j,k \leq n}$ of real-valued one-forms on $\mathrm{Aff}(n)$. This system of one-forms on the principal bundle $FA \to A$ defines the affine connection on A.

Taking the exterior derivative a second time, and using the fact that $d^2=0$ as well as the linear independence of the ϵi, the following relations are obtained:

$$d\theta^j - \sum_i \omega_i^j \wedge \theta^i = 0$$
$$d\omega_i^j - \sum_k \omega_k^j \wedge \omega_i^k = 0.$$

These are the Maurer-Cartan equations for the Lie group $\mathrm{Aff}(n)$ (identified with FA by the choice of a reference frame). Furthermore:

- the Pfaffian system $\theta^j=0$ (for all j) is integrable, and its integral manifolds are the fibres of the principal bundle $\mathrm{Aff}(n) \to A$.

- the Pfaffian system $\omega i^j=0$ (for all i, j) is also integrable, and its integral manifolds define parallel transport in FA.

Thus the forms (ωi^j) define a flat principal connection on $FA \to A$.

For a strict comparison with the motivation, one should actually define parallel transport in a principal $\mathrm{Aff}(n)$-bundle over A. This can be done by pulling back FA by the smooth map $\varphi : \mathbf{R}^n \times A \to A$ defined by translation. Then the composite $\varphi'^*FA \to FA \to A$ is a principal $\mathrm{Aff}(n)$-bundle over A, and the forms (θ^i, ω_j^k) pull back to give a flat principal $\mathrm{Aff}(n)$-connection on this bundle.

45.5.3 General affine geometries: formal definitions

An affine space, as with essentially any smooth Klein geometry, is a manifold equipped with a flat Cartan connection. More general affine manifolds or affine geometries are obtained easily by dropping the flatness condition expressed by the Maurer-Cartan equations. There are several ways to approach the definition and two will be given. Both definitions are facilitated by the realisation that 1-forms (θ^i, ω_j^k) in the flat model fit together to give a 1-form with values in the Lie algebra $\mathbf{aff}(n)$ of the affine group $\mathrm{Aff}(n)$.

In these definitions, M is a smooth n-manifold and $A = \mathrm{Aff}(n)/GL(n)$ is an affine space of the same dimension.

Definition via absolute parallelism

Let M be a manifold, and P a principal GL(n)-bundle over M. Then an **affine connection** is a 1-form η on P with values in **aff**(n) satisfying the following properties

1. η is equivariant with respect to the action of GL(n) on P and **aff**(n);
2. $\eta(X\xi) = \xi$ for all ξ in the Lie algebra **gl**(n) of all $n \times n$ matrices;
3. η is a linear isomorphism of each tangent space of P with **aff**(n).

The last condition means that η is an **absolute parallelism** on P, i.e., it identifies the tangent bundle of P with a trivial bundle (in this case $P \times$ **aff**(n)). The pair (P,η) defines the structure of an **affine geometry** on M, making it into an **affine manifold**.

The affine Lie algebra **aff**(n) splits as a semidirect product of $\mathbf{R}^n$ and **gl**(n) and so η may be written as a pair (θ,ω) where θ takes values in $\mathbf{R}^n$ and ω takes values in **gl**(n). The conditions (1) and (2) are equivalent to ω being a principal GL(n)-connection and θ being a horizontal equivariant 1-form, which induces a bundle homomorphism from TM to the associated bundle $P \times_{\text{GL}(n)} \mathbf{R}^n$. The condition (3) is equivalent to the fact that this bundle homomorphism is an isomorphism. (However, this decomposition is a consequence of the rather special structure of the affine group.) Since P is the frame bundle of $P \times_{\text{GL}(n)} \mathbf{R}^n$, it follows that θ provides a bundle isomorphism between P and the frame bundle FM of M; this recovers the definition of an affine connection as a principal GL(n)-connection on FM.

The 1-forms arising in the flat model are just the components of θ and ω.

Definition as a principal affine connection

An **affine connection** on M is a principal Aff(n)-bundle Q over M, together with a principal GL(n)-subbundle P of Q and a principal Aff(n)-connection α (a 1-form on Q with values in **aff**(n)) which satisfies the following (generic) *Cartan condition*. The $\mathbf{R}^n$ component of pullback of α to P is a horizontal equivariant 1-form and so defines a bundle homomorphism from TM to $P \times_{\text{GL}(n)} \mathbf{R}^n$: this is required to be an isomorphism.

Relation to the motivation

Since Aff(n) acts on A, there is, associated to the principal bundle Q, a bundle $A = Q \times_{\text{Aff}(n)} A$, which is a fiber bundle over M whose fiber at x in M is an affine space Ax. A section a of A (defining a marked point ax in Ax for each $x \in M$) determines a principal GL(n)-subbundle P of Q (as the bundle of stabilizers of these marked points) and vice versa. The principal connection α defines an Ehresmann connection on this bundle, hence a notion of parallel transport. The Cartan condition ensures that the distinguished section a always moves under parallel transport.

45.6 Further properties

45.6.1 Curvature and torsion

Curvature and torsion are the main invariants of an affine connection. As there are many equivalent ways to define the notion of an affine connection, so there are many different ways to define curvature and torsion.

From the Cartan connection point of view, the curvature is the failure of the affine connection η to satisfy the Maurer-Cartan equation

$$d\eta + \tfrac{1}{2}[\eta \wedge \eta] = 0,$$

where the second term on the left hand side is the wedge product using the Lie bracket in **aff**(n) to contract the values. By expanding η into the pair (θ,ω) and using the structure of the Lie algebra **aff**(n), this left hand side can be expanded into the two formulae

$$d\theta + \omega \wedge \theta$$

$$d\omega + \omega \wedge \omega,$$

where the wedge products are evaluated using matrix multiplication. The first expression is called the torsion of the connection, and the second is also called the curvature.

These expressions are differential 2-forms on the total space of a frame bundle. However, they are horizontal and equivariant, and hence define tensorial objects. These can be defined directly from the induced covariant derivative ∇ on TM as follows.

The torsion is given by the formula

$$T^\nabla(X,Y) = \nabla_X Y - \nabla_Y X - [X,Y].$$

If the torsion vanishes, the connection is said to be *torsion-free* or *symmetric*.

The curvature is given by the formula

$$R^\nabla_{X,Y} Z = \nabla_X \nabla_Y Z - \nabla_Y \nabla_X Z - \nabla_{[X,Y]} Z.$$

When both curvature and torsion vanish, the connection defines a pre-Lie algebra structure on the space of global sections of the tangent bundle.

45.6.2 The Levi-Civita connection

If (M,g) is a Riemannian manifold then there is a unique affine connection ∇ on M with the following two properties:

- the connection is torsion-free, i.e., T^∇ is zero;
- parallel transport is an isometry, i.e., the inner products (defined using g) between tangent vectors are preserved.

This connection is called the **Levi-Civita connection**.

The second condition means that the connection is a **metric connection** in the sense that the Riemannian metric g is parallel: $\nabla g = 0$. In local coordinates the components of the connection form are called Christoffel symbols: because of the uniqueness of the Levi-Civita connection, there is a formula for these components in terms of the components of g.

45.6.3 Geodesics

Since straight lines are a concept in affine geometry, affine connections define a generalized notion of (parametrized) straight lines on any affine manifold, called affine geodesics. Abstractly, a parametric curve $\gamma : I \to M$ is a straight line if its tangent vector remains parallel and equipollent with itself when it is transported along γ. From the linear point of view, an affine connection M distinguishes the affine geodesics in the following way: a smooth curve $\gamma : I \to M$ is an **affine geodesic** if $\dot\gamma$ is parallel transported along γ, that is

$$\tau_t^s \dot\gamma(s) = \dot\gamma(t)$$

where $\tau_t^s : T_{\gamma s} M \to T_{\gamma t} M$ is the parallel transport map defining the connection.

In terms of the infinitesimal connection ∇, the derivative of this equation implies

$$\nabla_{\dot\gamma(t)} \dot\gamma(t) = 0 \text{ for all } t \in I.$$

Conversely, any solution of this differential equation yields a curve whose tangent vector is parallel transported along the curve. For every $x \in M$ and every $X \in T_xM$, there exists a unique affine geodesic $\gamma : I \to M$ with $\gamma(0) = x$ and $\dot{\gamma}(0) = X$ and where I is the maximal open interval in $\mathbf{R}$, containing 0, on which the geodesic is defined. This follows from the Picard–Lindelöf theorem, and allows for the definition of an exponential map associated to the affine connection.

In particular, when M is a (pseudo-)Riemannian manifold and ∇ is the Levi-Civita connection, then the affine geodesic are the usual geodesics of Riemannian geometry and are the locally distance minimizing curves.

The geodesics defined here are sometimes called **affinely parametrized**, since a given straight line in M determines a parametric curve γ through the line up to a choice of affine reparametrization $\gamma(t) \to \gamma(at+b)$, where a and b are constants. The tangent vector to an affine geodesic is parallel and equipollent along itself. An unparametrized geodesic, or one which is merely parallel along itself without necessarily being equipollent, need only satisfy

$$\nabla_{\dot{\gamma}} \dot{\gamma} = k \dot{\gamma}$$

for some function k defined along γ. Unparametrized geodesics are often studied from the point of view of projective connections.

45.6.4 Development

An affine connection defines a notion of **development** of curves. Intuitively, development captures the notion that if x_t is a curve in M, then the affine tangent space at x_0 may be *rolled* along the curve. As it does so, the marked point of contact between the tangent space and the manifold traces out a curve C_t in this affine space: the development of x_t.

In formal terms, let $\tau_t^0 : T_{x_t}M \to T_{x_0}M$ be the linear parallel transport map associated to the affine connection. Then the development C_t is the curve in $T_{x_0}M$ starts off at 0 and is parallel to the tangent of x_t for all time t:

$$\dot{C}_t = \tau_t^0 \dot{x}_t, \quad C_0 = 0.$$

In particular, x_t is a *geodesic* if and only if its development is an affinely parametrized straight line in $T_{x_0}M$.[10]

45.7 Surface theory revisited

If M is a surface in $\mathbf{R}^3$, it is easy to see that M has a natural affine connection. From the linear connection point of view, the covariant derivative of a vector field is defined by differentiating the vector field, viewed as a map from M to $\mathbf{R}^3$, and then projecting the result orthogonally back onto the tangent spaces of M. It is easy to see that this affine connection is torsion-free. Furthermore, it is a metric connection with respect to the Riemannian metric on M induced by the inner product on $\mathbf{R}^3$, hence it is the Levi-Civita connection of this metric.

45.7.1 Example: the unit sphere in Euclidean space

Let $\langle , \rangle$ be the usual scalar product on $\mathbf{R}^3$, and let $\mathbf{S}^2$ be the unit sphere. The tangent space to $\mathbf{S}^2$ at a point x is naturally identified with the vector sub-space of $\mathbf{R}^3$ consisting of all vectors orthogonal to x. It follows that a vector field Y on $\mathbf{S}^2$ can be seen as a map $Y : \mathbf{S}^2 \to \mathbf{R}^3$ which satisfies

$$\langle Y_x, x \rangle = 0, \quad \forall x \in \mathbf{S}^2.$$

Denote by dY the differential of such a map. Then we have:

> **Lemma**. The formula
> $$(\nabla_X Y)_x = dY_x(X) + \langle X_x, Y_x \rangle x$$
> defines an affine connection on $\mathbf{S}^2$ with vanishing torsion.

Proof. It is straightforward to prove that ∇ satisfies the Leibniz identity and is $C^\infty(\mathbf{S}^2)$ linear in the first variable. So all that needs to be proved here is that the map above does indeed define a tangent vector field. That is, we need to prove that for all x in $\mathbf{S}^2$

$$\langle (\nabla_X Y)_x, x \rangle = 0 \qquad (1).$$

Consider the map

$$\begin{cases} f : \mathbf{S}^2 \to \mathbf{R} \\ x \longmapsto \langle Y_x, x \rangle. \end{cases}$$

The map f is constant, hence its differential vanishes. In particular

$$\mathrm{d}f_x(X) = \langle (\mathrm{d}Y)_x(X), x \rangle + \langle Y_x, X_x \rangle = 0.$$

The equation (1) above follows. $\square$

45.8 See also

- Atlas (topology)
- Chart (topology)
- Connection (mathematics)
- Connection (fibred manifold)
- Connection (affine bundle)
- Differentiable manifold
- Differential geometry
- Introduction to mathematics of general relativity
- Levi-Civita connection
- List of formulas in Riemannian geometry
- Riemannian geometry

45.9 Notes

[1] Weyl 1918, 5 editions to 1922.

[2] Cartan 1923.

[3] As a result, many mathematicians use the term *linear connection* (instead of *affine connection*) for a connection on the tangent bundle, on the grounds that parallel transport is linear and not affine. However, the same property holds for any (Koszul or linear Ehresmann) connection on a vector bundle. Originally the term *affine connection* is short for an affine *connection* in the sense of Cartan, and this implies that the connection is defined on the tangent bundle, rather than an arbitrary vector bundle. The notion of a linear Cartan connection does not really make much sense, because linear representations are not transitive.

[4] Cartan 1926.

[5] It is difficult to make Cartan's intuition precise without invoking smooth infinitesimal analysis, but one way is to regard his points being *variable*, that is maps from some unseen parameter space into the manifold, which can then be differentiated.

[6] Classically, the tangent space was viewed as an infinitesimal approximation, while in modern differential geometry, tangent spaces are often defined in terms of differential objects such as derivations (see Kobayashi & Nomizu 1996, Volume 1, sections 1.1–1.2).

[7] For details, see Ü. Lumiste (2001b). The following intuitive treatment is that of Cartan (1923) and Cartan (1926).

[8] This can be viewed as a choice of origin: actually it suffices to consider only the case $p=ax$; Cartan implicitly identifies this with x in M.

[9] Cf. R. Hermann (1983), Appendix 1–3 to Cartan (1951), and also Sharpe (1997).

[10] This treatment of development is from Kobayashi & Nomizu (1996, Volume 1, Proposition III.3.1); see section III.3 for a more geometrical treatment. See also Sharpe (1997) for a thorough discussion of development in other geometrical situations.

45.10 References

45.10.1 Primary historical references

- Christoffel, Elwin Bruno (1869), *Über die Transformation der homogenen Differentialausdrücke zweiten Grades*, J. Für die Reine und Angew. Math. **70**: 46–70

- Levi-Civita, Tullio (1917), *Nozione di parallelismo in una varietà qualunque e conseguente specificazione geometrica della curvatura Riemanniana*, Rend. Circ. Mat. Palermo **42**: 173–205, doi:10.1007/bf03014898

- Cartan, Élie (1923), *Sur les variétés à connexion affine, et la théorie de la relativité généralisée (première partie)*, Annales Scientifiques de l'École Normale Supérieure **40**: 325–412

- Cartan, Élie (1924), *Sur les variétés à connexion affine, et la théorie de la relativité généralisée (première partie) (Suite)*, Annales Scientifiques de l'École Normale Supérieure **41**: 1–25

    Cartan's treatment of affine connections as motivated by the study of relativity theory. Includes a detailed discussion of the physics of reference frames, and how the connection reflects the physical notion of transport along a worldline.

- Cartan, Élie (1926), *Espaces à connexion affine, projective et conforme*, Acta Math. **48**: 1–42, doi:10.1007/BF02629755

    A more mathematically motivated account of affine connections.

- Cartan, Élie (1951), with appendices by Robert Hermann, ed., *Geometry of Riemannian Spaces* (translation by James Glazebrook of *Leçons sur la géométrie des espaces de Riemann*, 2nd ed.), Math Sci Press, Massachusetts (published 1983), ISBN 978-0-915692-34-7.

    Affine connections from the point of view of Riemannian geometry. Robert Hermann's appendices discuss the motivation from surface theory, as well as the notion of affine connections in the modern sense of Koszul. He develops the basic properties of the differential operator ∇, and relates them to the classical affine connections in the sense of Cartan.

- Weyl, Hermann (1918), *Raum, Zeit, Materie* (5 editions to 1922, with notes by Jürgen Ehlers (1980), translated 4th edition *Space, Time, Matter* by Henry Brose, 1922 (Methuen, reprinted 1952 by Dover) ed.), Springer, Berlin, ISBN 0-486-60267-2

45.10.2 Secondary references

- Kobayashi, Shoshichi; Nomizu, Katsumi (1996), *Foundations of Differential Geometry, Vols. 1 & 2* (New ed.), Wiley-Interscience, ISBN 0-471-15733-3.

45.10. REFERENCES

> This is the main reference for the technical details of the article. Volume 1, chapter III gives a detailed account of affine connections from the perspective of principal bundles on a manifold, parallel transport, development, geodesics, and associated differential operators. Volume 1 chapter VI gives an account of affine transformations, torsion, and the general theory of affine geodesy. Volume 2 gives a number of applications of affine connections to homogeneous spaces and complex manifolds, as well as to other assorted topics.

- Ü. Lumiste (2001a), "Affine connection", *Encyclopaedia of Mathematics*, Kluwer Academic Publishers, ISBN 978-1-55608-010-4 |first1= missing |last1= in Editors list (help).

- Ü. Lumiste (2001b), "Connections on a manifold", *Encyclopaedia of Mathematics*, Kluwer Academic Publishers, ISBN 978-1-55608-010-4 |first1= missing |last1= in Editors list (help).

> Two articles by Lumiste, giving precise conditions on parallel transport maps in order that they define affine connections. They also treat curvature, torsion, and other standard topics from a classical (non-principal bundle) perspective.

- Sharpe, R.W. (1997), *Differential Geometry: Cartan's Generalization of Klein's Erlangen Program*, Springer-Verlag, New York, ISBN 0-387-94732-9.

> This fills in some of the historical details, and provides a more reader-friendly elementary account of Cartan connections in general. Appendix A elucidates the relationship between the principal connection and absolute parallelism viewpoints. Appendix B bridges the gap between the classical "rolling" model of affine connections, and the modern one based on principal bundles and differential operators.

Chapter 46

De Sitter universe

A **de Sitter universe** is a cosmological solution to Einstein's field equations of General Relativity which is named after Willem de Sitter. It models the universe as spatially flat and neglects ordinary matter, so the dynamics of the universe are dominated by the cosmological constant, thought to correspond to dark energy in our universe or the inflaton field in the early universe. According to the models of inflation and current observations of the accelerating universe, the concordance models of physical cosmology are converging on a consistent model where our universe was best described as a de Sitter universe at about a time $t = 10^{-33}$ seconds after the fiducial Big Bang singularity, and far into the future.

46.1 Mathematical expression

A de Sitter universe has no ordinary matter content but with a positive cosmological constant (Λ) which sets the expansion rate, H . A larger cosmological constant leads to a larger expansion rate:

$$H \propto \sqrt{\Lambda},$$

where the constants of proportionality depend on conventions.

It is common to describe a patch of this solution as an expanding universe of the FLRW form where the scale factor is given by[1]

$$a(t) = e^{Ht},$$

where the constant H is the Hubble expansion rate and t is time. As in all FLRW spaces, $a(t)$, the scale factor, describes the expansion of physical spatial distances.

Unique to universes described by the FLRW metric, a de Sitter universe has a Hubble Law which is not only consistent through all space, but also through all time (since the deceleration parameter is equal to $q = -1$), thus satisfying the perfect cosmological principle that assumes isotropy and homogeneity throughout space and time. As a class of models with different values of the Hubble constant, the static universe that Einstein developed, and for which he invented the cosmological constant, can be considered a special case of the de Sitter universe where the expansion is finely tuned to just cancel out the collapse associated with the positive curvature associated with a non-zero matter density. There are ways to cast de Sitter space with static coordinates (see de Sitter space), so unlike other FLRW models, de Sitter space can be thought of as a static solution to Einstein's equations even though the geodesics followed by observers necessarily diverge in the normal way expected from the expansion of physical spatial dimensions. As a model for the universe, de Sitter's solution was not considered viable for the observed universe until models for inflation and dark energy were developed. Before then, it was assumed that the Big Bang implied only an acceptance of the weaker cosmological principle which holds isotropy true only for spatial extents but not temporal extents.[2]

46.2 Potential for the Universe

Because our Universe entered the Dark Energy Dominated Era a few billion years ago, our universe is probably approaching a de Sitter universe in the infinite future. If the current acceleration of our universe is due to a cosmological constant then as the universe continues to expand all of the matter and radiation will be diluted. Eventually there will be almost nothing left but the vacuum energy, tiny thermal fluctuations, quantum fluctuations and our universe will have become a de Sitter universe.

46.3 Relative expansion

The exponential expansion of the scale factor means that the physical distance between any two non-accelerating observers will eventually be growing faster than the speed of light. At this point those two observers will no longer be able to make contact. Therefore any observer in a de Sitter universe would see event horizons beyond which that observer can never see nor learn any information. If our universe is approaching a de Sitter universe then eventually we will not be able to observe any galaxies other than our own Milky Way (and any others in the gravitationally bound Local Group, assuming they were to somehow survive to that time without merging).

46.4 Modelling cosmic inflation

Another application of de Sitter space is in the early universe during cosmic inflation. Many inflationary models are approximately de Sitter space and can be modelled by giving the Hubble parameter a mild time dependence. For simplicity, some calculations involving inflation in the early universe can be performed in de Sitter space rather than a more realistic inflationary universe. By using the de Sitter universe instead, where the expansion is truly exponential, there are many simplifications.

46.5 See also

- Cosmic inflation
- De Sitter space for more mathematical properties
- Deceleration parameter
- Causal patch

46.6 References

[1] Adler, Ronald; Maurice Bazin; Menahem Schiffer (1965). *Introduction to General Relativity*. NY: McGraw-Hill. p. 468.

[2] Dodelson, Scott (2003). *Modern Cosmology* (4. [print.]. ed.). San Diego: Academic Press. ISBN 978-0-12-219141-1.

Chapter 47

Standard Model

This article is about the Standard Model of particle physics. For other uses, see Standard model (disambiguation). This article is a non-mathematical general overview of the Standard Model. For a mathematical description, see the article Standard Model (mathematical formulation).

The **Standard Model** of particle physics is a theory concerning the electromagnetic, weak, and strong nuclear

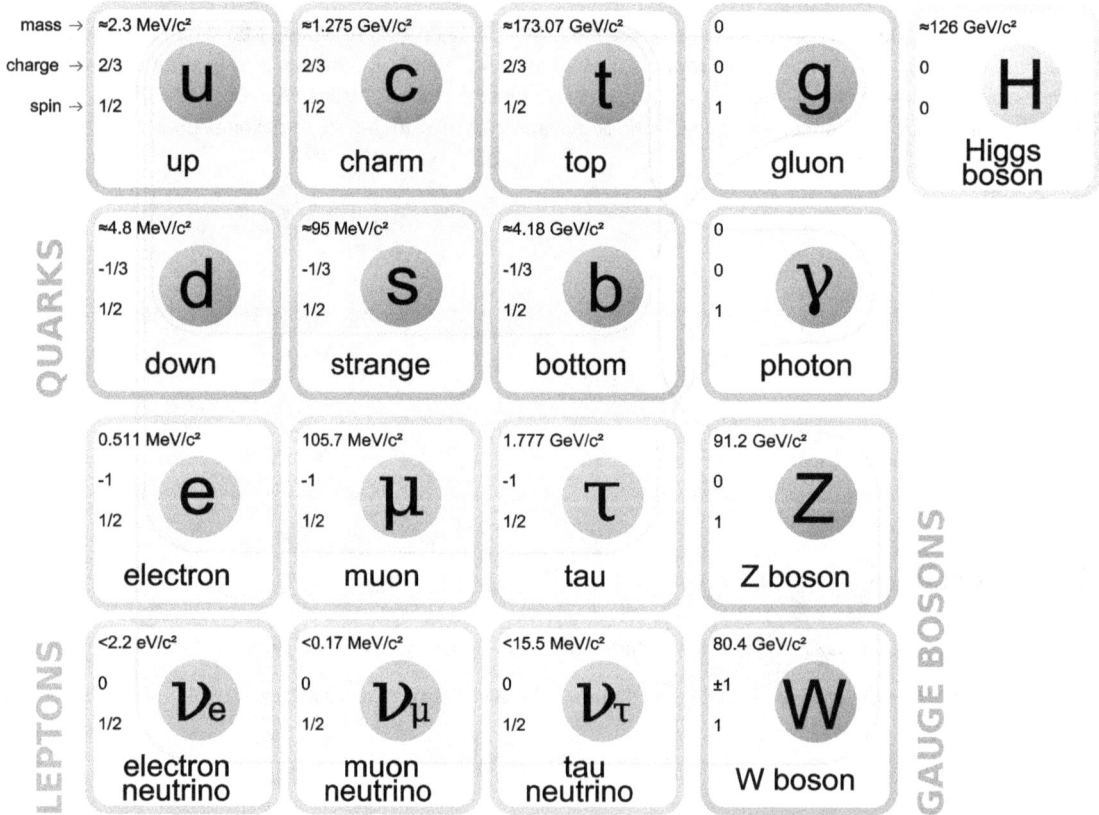

The Standard Model of elementary particles, with the three generations of matter, gauge bosons in the fourth column, and the Higgs boson in the fifth.

interactions, which mediate the dynamics of the known subatomic particles. It was developed throughout the latter half of the 20th century, as a collaborative effort of scientists around the world.[1] The current formulation was finalized in the mid-1970s upon experimental confirmation of the existence of quarks. Since then, discoveries of the top quark (1995), the tau neutrino (2000), and more recently the Higgs boson (2013), have given further credence to the Standard Model. Because of its success in explaining a wide variety of experimental results, the Standard Model is sometimes regarded as a "theory of almost everything".

The Standard Model falls short of being a complete theory of fundamental interactions. It does not incorporate the full theory of gravitation[2] as described by general relativity, or account for the accelerating expansion of the universe (as possibly described by dark energy). The model does not contain any viable dark matter particle that possesses all of the required properties deduced from observational cosmology. It also does not incorporate neutrino oscillations (and their non-zero masses). Although the Standard Model is believed to be theoretically self-consistent[3] and has demonstrated huge and continued successes in providing experimental predictions, it does leave some phenomena unexplained.

The development of the Standard Model was driven by theoretical and experimental particle physicists alike. For theorists, the Standard Model is a paradigm of a quantum field theory, which exhibits a wide range of physics including spontaneous symmetry breaking, anomalies, non-perturbative behavior, etc. It is used as a basis for building more exotic models that incorporate hypothetical particles, extra dimensions, and elaborate symmetries (such as supersymmetry) in an attempt to explain experimental results at variance with the Standard Model, such as the existence of dark matter and neutrino oscillations.

47.1 Historical background

The first step towards the Standard Model was Sheldon Glashow's discovery in 1961 of a way to combine the electromagnetic and weak interactions.[4] In 1967 Steven Weinberg[5] and Abdus Salam[6] incorporated the Higgs mechanism[7][8][9] into Glashow's electroweak theory, giving it its modern form.

The Higgs mechanism is believed to give rise to the masses of all the elementary particles in the Standard Model. This includes the masses of the W and Z bosons, and the masses of the fermions, i.e. the quarks and leptons.

After the neutral weak currents caused by Z boson exchange were discovered at CERN in 1973,[10][11][12][13] the electroweak theory became widely accepted and Glashow, Salam, and Weinberg shared the 1979 Nobel Prize in Physics for discovering it. The W and Z bosons were discovered experimentally in 1981, and their masses were found to be as the Standard Model predicted.

The theory of the strong interaction, to which many contributed, acquired its modern form around 1973–74, when experiments confirmed that the hadrons were composed of fractionally charged quarks.

47.2 Overview

At present, matter and energy are best understood in terms of the kinematics and interactions of elementary particles. To date, physics has reduced the laws governing the behavior and interaction of all known forms of matter and energy to a small set of fundamental laws and theories. A major goal of physics is to find the "common ground" that would unite all of these theories into one integrated theory of everything, of which all the other known laws would be special cases, and from which the behavior of all matter and energy could be derived (at least in principle).[14]

47.3 Particle content

The Standard Model includes members of several classes of elementary particles (fermions, gauge bosons, and the Higgs boson), which in turn can be distinguished by other characteristics, such as color charge.

47.3.1 Fermions

The Standard Model includes 12 elementary particles of spin-½ known as fermions. According to the spin-statistics theorem, fermions respect the Pauli exclusion principle. Each fermion has a corresponding antiparticle.

The fermions of the Standard Model are classified according to how they interact (or equivalently, by what charges they carry). There are six quarks (up, down, charm, strange, top, bottom), and six leptons (electron, electron neutrino, muon, muon neutrino, tau, tau neutrino). Pairs from each classification are grouped together to form a generation, with corresponding particles exhibiting similar physical behavior (see table).

The defining property of the quarks is that they carry color charge, and hence, interact via the strong interaction. A phenomenon called color confinement results in quarks being perpetually (or at least since very soon after the start of

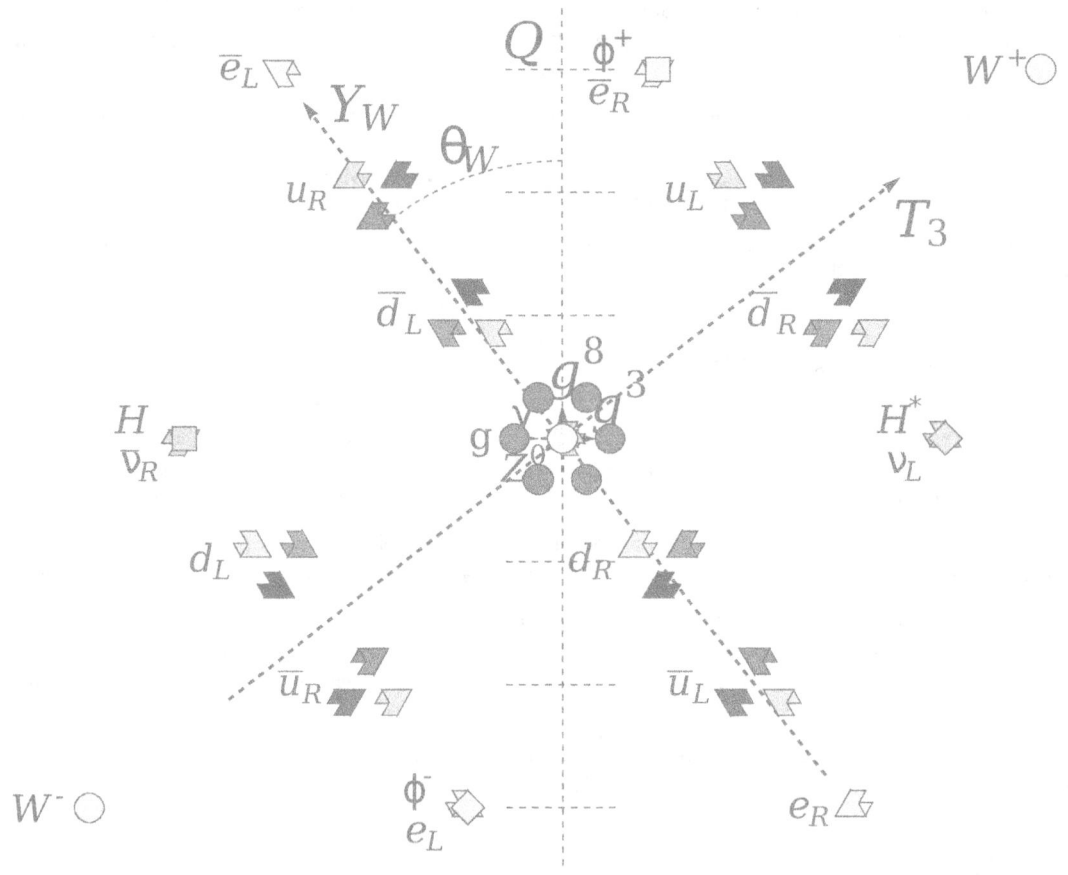

The pattern of weak isospin, T_3, weak hypercharge, Y_W, and color charge of all known elementary particles, rotated by the weak mixing angle to show electric charge, Q, roughly along the vertical. The neutral Higgs field (gray square) breaks the electroweak symmetry and interacts with other particles to give them mass.

the Big Bang) bound to one another, forming color-neutral composite particles (hadrons) containing either a quark and an antiquark (mesons) or three quarks (baryons). The familiar proton and the neutron are the two baryons having the smallest mass. Quarks also carry electric charge and weak isospin. Hence they interact with other fermions both electromagnetically and via the weak interaction.

The remaining six fermions do not carry colour charge and are called leptons. The three neutrinos do not carry electric charge either, so their motion is directly influenced only by the weak nuclear force, which makes them notoriously difficult to detect. However, by virtue of carrying an electric charge, the electron, muon, and tau all interact electromagnetically.

Each member of a generation has greater mass than the corresponding particles of lower generations. The first generation charged particles do not decay; hence all ordinary (baryonic) matter is made of such particles. Specifically, all atoms consist of electrons orbiting atomic nuclei ultimately constituted of up and down quarks. Second and third generations charged particles, on the other hand, decay with very short half lives, and are observed only in very high-energy environments. Neutrinos of all generations also do not decay, and pervade the universe, but rarely interact with baryonic matter.

47.3.2 Gauge bosons

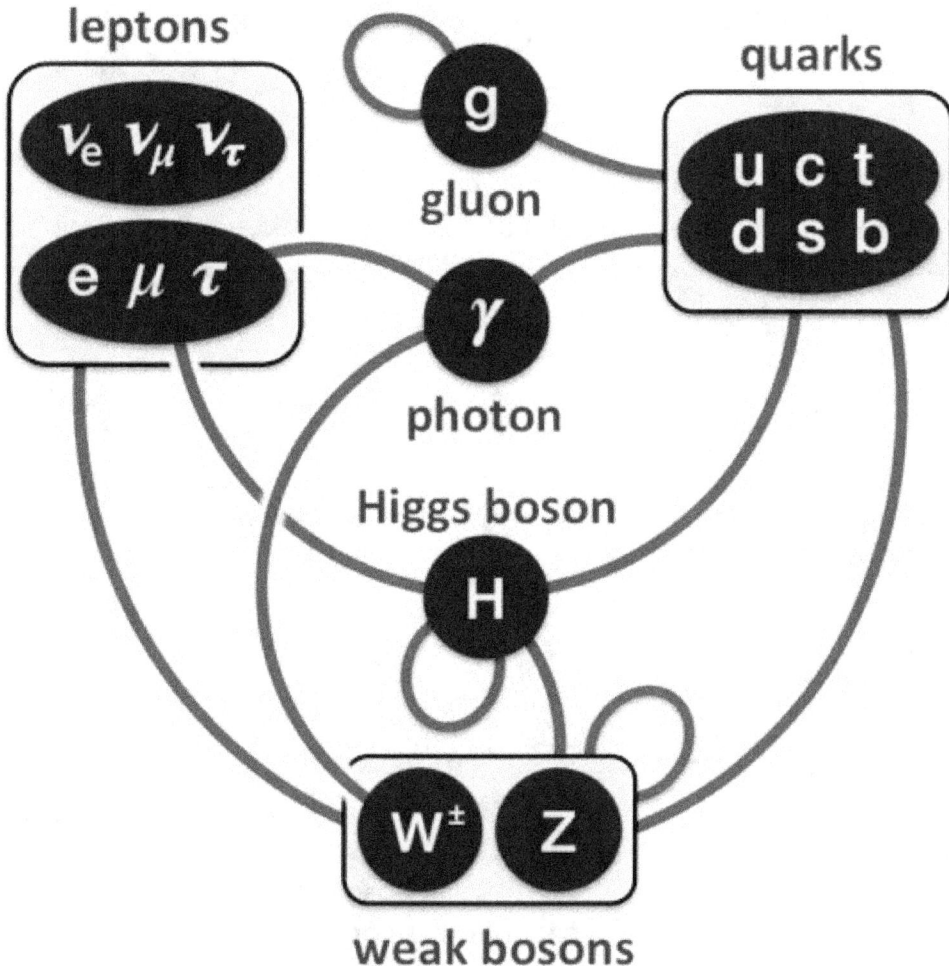

Summary of interactions between particles described by the Standard Model.

In the Standard Model, gauge bosons are defined as force carriers that mediate the strong, weak, and electromagnetic fundamental interactions.

Interactions in physics are the ways that particles influence other particles. At a macroscopic level, electromagnetism allows particles to interact with one another via electric and magnetic fields, and gravitation allows particles with mass to attract one another in accordance with Einstein's theory of general relativity. The Standard Model explains such forces as resulting from matter particles exchanging other particles, known as *force mediating particles* (strictly speaking, this is only so if interpreting literally what is actually an *approximation method* known as perturbation theory). When a force-mediating particle is exchanged, at a macroscopic level the effect is equivalent to a force influencing both of them, and the particle is therefore said to have *mediated* (i.e., been the agent of) that force. The Feynman diagram calculations, which are a graphical representation of the perturbation theory approximation, invoke "force mediating particles", and when applied to analyze high-energy scattering experiments are in reasonable agreement with the data. However, perturbation theory (and with it the concept of a "force-mediating particle") fails in other situations. These include low-energy quantum chromodynamics, bound states, and solitons.

The gauge bosons of the Standard Model all have spin (as do matter particles). The value of the spin is 1, making them bosons. As a result, they do not follow the Pauli exclusion principle that constrains fermions: thus bosons (e.g. photons) do not have a theoretical limit on their spatial density (number per volume). The different types of gauge bosons are described below.

Standard Model Interactions
(Forces Mediated by Gauge Bosons)

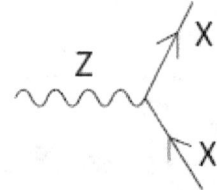

X is any fermion in the Standard Model.

X is electrically charged.

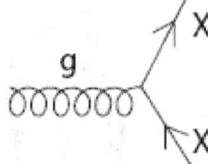

X is any quark.

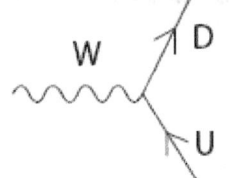

U is a up-type quark; D is a down-type quark.

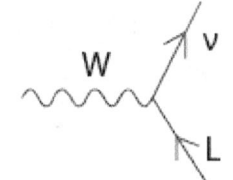

L is a lepton and ν is the corresponding neutrino.

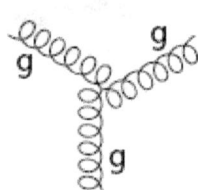

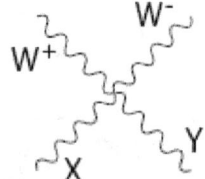

X is a photon or Z-boson.   X and Y are any two electroweak bosons such that charge is conserved.

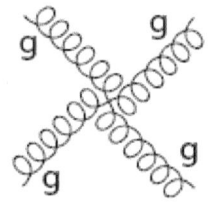

The above interactions form the basis of the standard model. Feynman diagrams in the standard model are built from these vertices. Modifications involving Higgs boson interactions and neutrino oscillations are omitted. The charge of the W bosons is dictated by the fermions they interact with; the conjugate of each listed vertex (i.e. reversing the direction of arrows) is also allowed.

- Photons mediate the electromagnetic force between electrically charged particles. The photon is massless and is well-described by the theory of quantum electrodynamics.

- The W+, W−, and Z gauge bosons mediate the weak interactions between particles of different flavors (all quarks and leptons). They are massive, with the Z being more massive than the W±. The weak interactions involving the W± exclusively act on *left-handed* particles and *right-handed* antiparticles. Furthermore, the W± carries an electric charge of +1 and −1 and couples to the electromagnetic interaction. The electrically neutral Z boson interacts with both left-handed particles and antiparticles. These three gauge bosons along with the photons are grouped together, as collectively mediating the electroweak interaction.

- The eight gluons mediate the strong interactions between color charged particles (the quarks). Gluons are massless. The eightfold multiplicity of gluons is labeled by a combination of color and anticolor charge (e.g. red–antigreen).[nb 1] Because the gluons have an effective color charge, they can also interact among themselves. The gluons and their interactions are described by the theory of quantum chromodynamics.

The interactions between all the particles described by the Standard Model are summarized by the diagrams on the right of this section.

47.3.3 Higgs boson

Main article: Higgs boson

The Higgs particle is a massive scalar elementary particle theorized by Robert Brout, François Englert, Peter Higgs, Gerald Guralnik, C. R. Hagen, and Tom Kibble in 1964 (see 1964 PRL symmetry breaking papers) and is a key building block in the Standard Model.[7][8][9][15] It has no intrinsic spin, and for that reason is classified as a boson (like the gauge bosons, which have integer spin).

The Higgs boson plays a unique role in the Standard Model, by explaining why the other elementary particles, except the photon and gluon, are massive. In particular, the Higgs boson explains why the photon has no mass, while the W and Z bosons are very heavy. Elementary particle masses, and the differences between electromagnetism (mediated by the photon) and the weak force (mediated by the W and Z bosons), are critical to many aspects of the structure of microscopic (and hence macroscopic) matter. In electroweak theory, the Higgs boson generates the masses of the leptons (electron, muon, and tau) and quarks. As the Higgs boson is massive, it must interact with itself.

Because the Higgs boson is a very massive particle and also decays almost immediately when created, only a very high-energy particle accelerator can observe and record it. Experiments to confirm and determine the nature of the Higgs boson using the Large Hadron Collider (LHC) at CERN began in early 2010, and were performed at Fermilab's Tevatron until its closure in late 2011. Mathematical consistency of the Standard Model requires that any mechanism capable of generating the masses of elementary particles become visible at energies above 1.4 TeV;[16] therefore, the LHC (designed to collide two 7 to 8 TeV proton beams) was built to answer the question of whether the Higgs boson actually exists.[17]

On 4 July 2012, the two main experiments at the LHC (ATLAS and CMS) both reported independently that they found a new particle with a mass of about 125 GeV/c^2 (about 133 proton masses, on the order of 10^{-25} kg), which is "consistent with the Higgs boson." Although it has several properties similar to the predicted "simplest" Higgs,[18] they acknowledged that further work would be needed to conclude that it is indeed the Higgs boson, and exactly which version of the Standard Model Higgs is best supported if confirmed.[19][20][21][22][23]

On 14 March 2013 the Higgs Boson was tentatively confirmed to exist.[24]

47.3.4 Full particle count

Counting particles by a rule that distinguishes between particles and their corresponding antiparticles, and among the many color states of quarks and gluons, gives a total of 61 elementary particles.[25]

47.4 Theoretical aspects

Main article: Standard Model (mathematical formulation)

47.4.1 Construction of the Standard Model Lagrangian

Technically, quantum field theory provides the mathematical framework for the Standard Model, in which a Lagrangian controls the dynamics and kinematics of the theory. Each kind of particle is described in terms of a dynamical field that pervades space-time. The construction of the Standard Model proceeds following the modern method of constructing most field theories: by first postulating a set of symmetries of the system, and then by writing down the most general renormalizable Lagrangian from its particle (field) content that observes these symmetries.

The global Poincaré symmetry is postulated for all relativistic quantum field theories. It consists of the familiar translational symmetry, rotational symmetry and the inertial reference frame invariance central to the theory of special relativity. The local SU(3)×SU(2)×U(1) gauge symmetry is an internal symmetry that essentially defines the Standard Model. Roughly, the three factors of the gauge symmetry give rise to the three fundamental interactions. The fields fall into different representations of the various symmetry groups of the Standard Model (see table). Upon writing the most general Lagrangian, one finds that the dynamics depend on 19 parameters, whose numerical values are established by experiment. The parameters are summarized in the table above (note: with the Higgs mass is at 125 GeV, the Higgs self-coupling strength $\lambda \sim 1/8$).

Quantum chromodynamics sector

Main article: Quantum chromodynamics

The quantum chromodynamics (QCD) sector defines the interactions between quarks and gluons, with SU(3) symmetry, generated by T^a. Since leptons do not interact with gluons, they are not affected by this sector. The Dirac Lagrangian of the quarks coupled to the gluon fields is given by

$$\mathcal{L}_{QCD} = i\overline{U}(\partial_\mu - ig_s G_\mu^a T^a)\gamma^\mu U + i\overline{D}(\partial_\mu - ig_s G_\mu^a T^a)\gamma^\mu D.$$

G_μ^a is the SU(3) gauge field containing the gluons, γ^μ are the Dirac matrices, D and U are the Dirac spinors associated with up- and down-type quarks, and g_s is the strong coupling constant.

Electroweak sector

Main article: Electroweak interaction

The electroweak sector is a Yang–Mills gauge theory with the simple symmetry group U(1)×SU(2)L,

$$\mathcal{L}_{EW} = \sum_\psi \bar{\psi}\gamma^\mu \left(i\partial_\mu - g'\frac{1}{2}Y_W B_\mu - g\frac{1}{2}\vec{\tau}_L \vec{W}_\mu\right)\psi$$

where $B\mu$ is the U(1) gauge field; Y_W is the weak hypercharge—the generator of the U(1) group; $\vec{W}_\mu$ is the three-component SU(2) gauge field; $\vec{\tau}_L$ are the Pauli matrices—infinitesimal generators of the SU(2) group. The subscript L indicates that they only act on left fermions; g' and g are coupling constants.

Higgs sector

Main article: Higgs mechanism

In the Standard Model, the Higgs field is a complex scalar of the group SU(2)L:

$$\varphi = \frac{1}{\sqrt{2}}\begin{pmatrix}\varphi^+\\ \varphi^0\end{pmatrix},$$

where the indices + and 0 indicate the electric charge (Q) of the components. The weak isospin (Y_W) of both components is 1.

Before symmetry breaking, the Higgs Lagrangian is:

$$\mathcal{L}_H = \varphi^\dagger \left(\partial^\mu - \frac{i}{2}\left(g'Y_W B^\mu + g\vec{\tau}\vec{W}^\mu\right)\right)\left(\partial_\mu + \frac{i}{2}\left(g'Y_W B_\mu + g\vec{\tau}\vec{W}_\mu\right)\right)\varphi - \frac{\lambda^2}{4}(\varphi^\dagger\varphi - v^2)^2,$$

which can also be written as:

$$\mathcal{L}_H = \left|\left(\partial_\mu + \frac{i}{2}\left(g'Y_W B_\mu + g\vec{\tau}\vec{W}_\mu\right)\right)\varphi\right|^2 - \frac{\lambda^2}{4}(\varphi^\dagger\varphi - v^2)^2.$$

47.5 Tests and predictions

The Standard Model (SM) predicted the existence of the W and Z bosons, gluon, and the top and charm quarks before these particles were observed. Their predicted properties were experimentally confirmed with good precision. To give an idea of the success of the SM, the following table compares the measured masses of the W and Z bosons with the masses predicted by the SM:

The SM also makes several predictions about the decay of Z bosons, which have been experimentally confirmed by the Large Electron-Positron Collider at CERN.

In May 2012 BaBar Collaboration reported that their recently analyzed data may suggest possible flaws in the Standard Model of particle physics.[26][27] These data show that a particular type of particle decay called "B to D-star-tau-nu" happens more often than the Standard Model says it should. In this type of decay, a particle called the B-bar meson decays into a D meson, an antineutrino and a tau-lepton. While the level of certainty of the excess (3.4 sigma) is not enough to claim a break from the Standard Model, the results are a potential sign of something amiss and are likely to impact existing theories, including those attempting to deduce the properties of Higgs bosons.[28]

On December 13, 2012, physicists reported the constancy, over space and time, of a basic physical constant of nature that supports the *standard model of physics*. The scientists, studying methanol molecules in a distant galaxy, found the change ($\Delta\mu/\mu$) in the proton-to-electron mass ratio μ to be equal to "$(0.0 \pm 1.0) \times 10^{-7}$ at redshift z = 0.89" and consistent with "a null result".[29][30]

47.6 Challenges

See also: Physics beyond the Standard Model

Self-consistency of the Standard Model (currently formulated as a non-abelian gauge theory quantized through path-integrals) has not been mathematically proven. While regularized versions useful for approximate computations (for example lattice gauge theory) exist, it is not known whether they converge (in the sense of S-matrix elements) in the limit that the regulator is removed. A key question related to the consistency is the Yang–Mills existence and mass gap problem.

Experiments indicate that neutrinos have mass, which the classic Standard Model did not allow.[31] To accommodate this finding, the classic Standard Model can be modified to include neutrino mass.

If one insists on using only Standard Model particles, this can be achieved by adding a non-renormalizable interaction of leptons with the Higgs boson.[32] On a fundamental level, such an interaction emerges in the seesaw mechanism where heavy right-handed neutrinos are added to the theory. This is natural in the left-right symmetric extension of the Standard Model [33][34] and in certain grand unified theories.[35] As long as new physics appears below or around 10^{14} GeV, the neutrino masses can be of the right order of magnitude.

Theoretical and experimental research has attempted to extend the Standard Model into a Unified field theory or a Theory of everything, a complete theory explaining all physical phenomena including constants. Inadequacies of the Standard Model that motivate such research include:

- It does not attempt to explain gravitation, although a theoretical particle known as a graviton would help explain it, and unlike for the strong and electroweak interactions of the Standard Model, there is no known way of describing general relativity, the canonical theory of gravitation, consistently in terms of quantum field theory. The reason for this is, among other things, that quantum field theories of gravity generally break down before reaching the Planck scale. As a consequence, we have no reliable theory for the very early universe;

- Some consider it to be *ad hoc* and inelegant, requiring 19 numerical constants whose values are unrelated and arbitrary. Although the Standard Model, as it now stands, can explain why neutrinos have masses, the specifics of neutrino mass are still unclear. It is believed that explaining neutrino mass will require an additional 7 or 8 constants, which are also arbitrary parameters;

- The Higgs mechanism gives rise to the hierarchy problem if some new physics (coupled to the Higgs) is present at high energy scales. In these cases in order for the weak scale to be much smaller than the Planck scale, severe fine tuning of the parameters is required; there are, however, other scenarios that include quantum gravity in which such fine tuning can be avoided. [36]

- It should be modified so as to be consistent with the emerging "Standard Model of cosmology." In particular, the Standard Model cannot explain the observed amount of cold dark matter (CDM) and gives contributions to dark energy which are many orders of magnitude too large. It is also difficult to accommodate the observed predominance of matter over antimatter (matter/antimatter asymmetry). The isotropy and homogeneity of the visible universe over large distances seems to require a mechanism like cosmic inflation, which would also constitute an extension of the Standard Model.

Currently, no proposed Theory of Everything has been widely accepted or verified.

47.7 See also

- Fundamental interaction:
    - Quantum electrodynamics
    - Strong interaction: Color charge, Quantum chromodynamics, Quark model
    - Weak interaction: Electroweak theory, Fermi theory of beta decay, Weak hypercharge, Weak isospin
- Gauge theory: Nontechnical introduction to gauge theory
- Generation
- Higgs mechanism: Higgs boson, Higgsless model
- J. C. Ward
- J. J. Sakurai Prize for Theoretical Particle Physics
- Lagrangian
- Open questions: BTeV experiment, CP violation, Neutrino masses, Quark matter
- Penguin diagram
- Quantum field theory
- Standard Model: Mathematical formulation of, Physics beyond the Standard Model

47.8 Notes and references

[1] Technically, there are nine such color–anticolor combinations. However, there is one color-symmetric combination that can be constructed out of a linear superposition of the nine combinations, reducing the count to eight.

47.9 References

[1] R. Oerter (2006). *The Theory of Almost Everything: The Standard Model, the Unsung Triumph of Modern Physics* (Kindle ed.). Penguin Group. p. 2. ISBN 0-13-236678-9.

[2] Sean Carroll, Ph.D., Cal Tech, 2007, The Teaching Company, *Dark Matter, Dark Energy: The Dark Side of the Universe*, Guidebook Part 2 page 59, Accessed Oct. 7, 2013, "...Standard Model of Particle Physics: The modern theory of elementary particles and their interactions ... It does not, strictly speaking, include gravity, although it's often convenient to include gravitons among the known particles of nature..."

[3] In fact, there are mathematical issues regarding quantum field theories still under debate (see e.g. Landau pole), but the predictions extracted from the Standard Model by current methods applicable to current experiments are all self-consistent. For a further discussion see e.g. Chapter 25 of R. Mann (2010). *An Introduction to Particle Physics and the Standard Model*. CRC Press. ISBN 978-1-4200-8298-2.

[4] S.L. Glashow (1961). "Partial-symmetries of weak interactions". *Nuclear Physics* **22** (4): 579–588. Bibcode:1961NucPh..22..579G. doi:10.1016/0029-5582(61)90469-2.

[5] S. Weinberg (1967). "A Model of Leptons". *Physical Review Letters* **19** (21): 1264–1266. Bibcode:1967PhRvL..19.1264W. doi:10.1103/PhysRevLett.19.1264.

[6] A. Salam (1968). N. Svartholm, ed. "Elementary Particle Physics: Relativistic Groups and Analyticity". Eighth Nobel Symposium. Stockholm: Almquvist and Wiksell. p. 367.

[7] F. Englert, R. Brout (1964). "Broken Symmetry and the Mass of Gauge Vector Mesons". *Physical Review Letters* **13** (9): 321–323. Bibcode:1964PhRvL..13..321E. doi:10.1103/PhysRevLett.13.321.

[8] P.W. Higgs (1964). "Broken Symmetries and the Masses of Gauge Bosons". *Physical Review Letters* **13** (16): 508–509. Bibcode:1964PhRvL..13..508H. doi:10.1103/PhysRevLett.13.508.

[9] G.S. Guralnik, C.R. Hagen, T.W.B. Kibble (1964). "Global Conservation Laws and Massless Particles". *Physical Review Letters* **13** (20): 585–587. Bibcode:1964PhRvL..13..585G. doi:10.1103/PhysRevLett.13.585.

[10] F.J. Hasert *et al.* (1973). "Search for elastic muon-neutrino electron scattering". *Physics Letters B* **46** (1): 121. Bibcode:1973PhLB...46..121H. doi:10.1016/0370-2693(73)90494-2.

[11] F.J. Hasert *et al.* (1973). "Observation of neutrino-like interactions without muon or electron in the Gargamelle neutrino experiment". *Physics Letters B* **46** (1): 138. Bibcode:1973PhLB...46..138H. doi:10.1016/0370-2693(73)90499-1.

[12] F.J. Hasert *et al.* (1974). "Observation of neutrino-like interactions without muon or electron in the Gargamelle neutrino experiment". *Nuclear Physics B* **73** (1): 1. Bibcode:1974NuPhB..73....1H. doi:10.1016/0550-3213(74)90038-8.

[13] D. Haidt (4 October 2004). "The discovery of the weak neutral currents". *CERN Courier*. Retrieved 8 May 2008.

[14] "Details can be worked out if the situation is simple enough for us to make an approximation, which is almost never, but often we can understand more or less what is happening." from *The Feynman Lectures on Physics*, Vol 1. pp. 2–7

[15] G.S. Guralnik (2009). "The History of the Guralnik, Hagen and Kibble development of the Theory of Spontaneous Symmetry Breaking and Gauge Particles". *International Journal of Modern Physics A* **24** (14): 2601–2627. arXiv:0907.3466. Bibcode:2009IJMPA..24.2601G. doi:10.1142/S0217751X09045431.

[16] B.W. Lee, C. Quigg, H.B. Thacker (1977). "Weak interactions at very high energies: The role of the Higgs-boson mass". *Physical Review D* **16** (5): 1519–1531. Bibcode:1977PhRvD..16.1519L. doi:10.1103/PhysRevD.16.1519.

[17] "Huge $10 billion collider resumes hunt for 'God particle'". CNN. 11 November 2009. Retrieved 2010-05-04.

[18] M. Strassler (10 July 2012). "Higgs Discovery: Is it a Higgs?". Retrieved 2013-08-06.

[19] "CERN experiments observe particle consistent with long-sought Higgs boson". CERN. 4 July 2012. Retrieved 2012-07-04.

[20] "Observation of a New Particle with a Mass of 125 GeV". CERN. 4 July 2012. Retrieved 2012-07-05.

[21] "ATLAS Experiment". ATLAS. 1 January 2006. Retrieved 2012-07-05.

[22] "Confirmed: CERN discovers new particle likely to be the Higgs boson". *YouTube*. Russia Today. 4 July 2012. Retrieved 2013-08-06.

[23] D. Overbye (4 July 2012). "A New Particle Could Be Physics' Holy Grail". *New York Times*. Retrieved 2012-07-04.

[24] "New results indicate that new particle is a Higgs boson". CERN. 14 March 2013. Retrieved 2013-08-06.

[25] S. Braibant, G. Giacomelli, M. Spurio (2009). *Particles and Fundamental Interactions: An Introduction to Particle Physics*. Springer. pp. 313–314. ISBN 978-94-007-2463-1.

[26] "BABAR Data in Tension with the Standard Model". SLAC. 31 May 2012. Retrieved 2013-08-06.

[27] BaBar Collaboration (2012). "Evidence for an excess of $B \to D^{(*)} \tau^- \nu_\tau$ decays". *Physical Review Letters* **109** (10): 101802. arXiv:1205.5442. Bibcode:2012PhRvL.109j1802L. doi:10.1103/PhysRevLett.109.101802.

[28] "BaBar data hint at cracks in the Standard Model". *e! Science News*. 18 June 2012. Retrieved 2013-08-06.

[29] J. Bagdonaite *et al.* (2012). "A Stringent Limit on a Drifting Proton-to-Electron Mass Ratio from Alcohol in the Early Universe". *Science* **339** (6115): 46. Bibcode:2013Sci...339...46B. doi:10.1126/science.1224898.

[30] C. Moskowitz (13 December 2012). "Phew! Universe's Constant Has Stayed Constant". Space.com. Retrieved 2012-12-14.

[31] "Particle chameleon caught in the act of changing". CERN. 31 May 2010. Retrieved 2012-07-05.

[32] S. Weinberg (1979). "Baryon and Lepton Nonconserving Processes". *Physical Review Letters* **43** (21): 1566. Bibcode:1979PhRvL..43.1566 doi:10.1103/PhysRevLett.43.1566.

[33] P. Minkowski (1977). "$\mu \to e\,\gamma$ at a Rate of One Out of 10^9 Muon Decays?". *Physics Letters B* **67** (4): 421. Bibcode:1977PhLB...67..421M. doi:10.1016/0370-2693(77)90435-X.

[34] R. N. Mohapatra, G. Senjanovic (1980). "Neutrino Mass and Spontaneous Parity Nonconservation". *Physical Review Letters* **44** (14): 912–915. Bibcode:1980PhRvL..44..912M. doi:10.1103/PhysRevLett.44.912.

[35] M. Gell-Mann, P. Ramond and R. Slansky (1979). F. van Nieuwenhuizen and D. Z. Freedman, ed. *Supergravity*. North Holland. pp. 315–321. ISBN 0-444-85438-X.

[36] Salvio, Strumia (2014-03-17). "Agravity". *JHEP 1406 (2014) 080*. arXiv:1403.4226. Bibcode:2014arXiv1403.4226S.

47.10 Further reading

- R. Oerter (2006). *The Theory of Almost Everything: The Standard Model, the Unsung Triumph of Modern Physics*. Plume.

- B.A. Schumm (2004). *Deep Down Things: The Breathtaking Beauty of Particle Physics*. Johns Hopkins University Press. ISBN 0-8018-7971-X.

Introductory textbooks

- I. Aitchison, A. Hey (2003). *Gauge Theories in Particle Physics: A Practical Introduction*. Institute of Physics. ISBN 978-0-585-44550-2.

- W. Greiner, B. Müller (2000). *Gauge Theory of Weak Interactions*. Springer. ISBN 3-540-67672-4.

- G.D. Coughlan, J.E. Dodd, B.M. Gripaios (2006). *The Ideas of Particle Physics: An Introduction for Scientists*. Cambridge University Press.

- D.J. Griffiths (1987). *Introduction to Elementary Particles*. John Wiley & Sons. ISBN 0-471-60386-4.

- G.L. Kane (1987). *Modern Elementary Particle Physics*. Perseus Books. ISBN 0-201-11749-5.

Advanced textbooks

- T.P. Cheng, L.F. Li (2006). *Gauge theory of elementary particle physics*. Oxford University Press. ISBN 0-19-851961-3. Highlights the gauge theory aspects of the Standard Model.

- J.F. Donoghue, E. Golowich, B.R. Holstein (1994). *Dynamics of the Standard Model*. Cambridge University Press. ISBN 978-0-521-47652-2. Highlights dynamical and phenomenological aspects of the Standard Model.

- L. O'Raifeartaigh (1988). *Group structure of gauge theories*. Cambridge University Press. ISBN 0-521-34785-8.

- Nagashima Y. Elementary Particle Physics: Foundations of the Standard Model, Volume 2. (Wiley 2013) 920 рапуы

- Schwartz, M.D. Quantum Field Theory and the Standard Model (Cambridge University Press 2013) 952 pages

- Langacker P. The standard model and beyond. (CRC Press, 2010) 670 pages Highlights group-theoretical aspects of the Standard Model.

Journal articles

- E.S. Abers, B.W. Lee (1973). "Gauge theories". *Physics Reports* **9**: 1–141. Bibcode:1973PhR.....9....1A. doi:10.1016/0370-1573(73)90027-6.

- M. Baak *et al.* (2012). "The Electroweak Fit of the Standard Model after the Discovery of a New Boson at the LHC". *The European Physical Journal C* **72** (11). arXiv:1209.2716. Bibcode:2012EPJC...72.2205B. doi:10.1140/epjc/s10052-012-2205-9.

- Y. Hayato *et al.* (1999). "Search for Proton Decay through $p \to \nu K^+$ in a Large Water Cherenkov Detector". *Physical Review Letters* **83** (8): 1529. arXiv:hep-ex/9904020. Bibcode:1999PhRvL..83.1529H. doi:10.1103/PhysRevLett.83.1529.

- S.F. Novaes (2000). "Standard Model: An Introduction". arXiv:hep-ph/0001283 [hep-ph].

- D.P. Roy (1999). "Basic Constituents of Matter and their Interactions — A Progress Report". arXiv:hep-ph/9912523 [hep-ph].

- F. Wilczek (2004). "The Universe Is A Strange Place". *Nuclear Physics B - Proceedings Supplements* **134**: 3. arXiv:astro-ph/0401347. Bibcode:2004NuPhS.134....3W. doi:10.1016/j.nuclphysbps.2004.08.001.

47.11 External links

- "The Standard Model explained in Detail by CERN's John Ellis" omega tau podcast.

- "LHC sees hint of lightweight Higgs boson" "New Scientist".

- "Standard Model may be found incomplete," *New Scientist*.

- "Observation of the Top Quark" at Fermilab.

- "The Standard Model Lagrangian." After electroweak symmetry breaking, with no explicit Higgs boson.

- "Standard Model Lagrangian" with explicit Higgs terms. PDF, PostScript, and LaTeX versions.

- "The particle adventure." Web tutorial.

- Nobes, Matthew (2002) "Introduction to the Standard Model of Particle Physics" on Kuro5hin: Part 1, Part 2, Part 3a, Part 3b.

- "The Standard Model" The Standard Model on the CERN web site explains how the basic building blocks of matter interact, governed by four fundamental forces.

Chapter 48

Gauge theory

For a more accessible and less technical introduction to this topic, see Introduction to gauge theory.

In physics, a **gauge theory** is a type of field theory in which the Lagrangian is invariant under a continuous group of local transformations.

The term *gauge* refers to redundant degrees of freedom in the Lagrangian. The transformations between possible gauges, called *gauge transformations*, form a Lie group—referred to as the *symmetry group* or the *gauge group* of the theory. Associated with any Lie group is the Lie algebra of group generators. For each group generator there necessarily arises a corresponding vector field called the *gauge field*. Gauge fields are included in the Lagrangian to ensure its invariance under the local group transformations (called *gauge invariance*). When such a theory is quantized, the quanta of the gauge fields are called *gauge bosons*. If the symmetry group is non-commutative, the gauge theory is referred to as *non-abelian*, the usual example being the Yang–Mills theory.

Many powerful theories in physics are described by Lagrangians that are invariant under some symmetry transformation groups. When they are invariant under a transformation identically performed at *every* point in the space in which the physical processes occur, they are said to have a global symmetry. The requirement of local symmetry, the cornerstone of gauge theories, is a stricter constraint. In fact, a global symmetry is just a local symmetry whose group's parameters are fixed in space-time.

Gauge theories are important as the successful field theories explaining the dynamics of elementary particles. Quantum electrodynamics is an abelian gauge theory with the symmetry group $U(1)$ and has one gauge field, the electromagnetic four-potential, with the photon being the gauge boson. The Standard Model is a non-abelian gauge theory with the symmetry group $U(1)\times SU(2)\times SU(3)$ and has a total of twelve gauge bosons: the photon, three weak bosons and eight gluons.

Gauge theories are also important in explaining gravitation in the theory of general relativity. Its case is somewhat unique in that the gauge field is a tensor, the Lanczos tensor. Theories of quantum gravity, beginning with gauge gravitation theory, also postulate the existence of a gauge boson known as the graviton. Gauge symmetries can be viewed as analogues of the principle of general covariance of general relativity in which the coordinate system can be chosen freely under arbitrary diffeomorphisms of spacetime. Both gauge invariance and diffeomorphism invariance reflect a redundancy in the description of the system. An alternative theory of gravitation, gauge theory gravity, replaces the principle of general covariance with a true gauge principle with new gauge fields.

Historically, these ideas were first stated in the context of classical electromagnetism and later in general relativity. However, the modern importance of gauge symmetries appeared first in the relativistic quantum mechanics of electrons – quantum electrodynamics, elaborated on below. Today, gauge theories are useful in condensed matter, nuclear and high energy physics among other subfields.

48.1 History and importance

The earliest field theory having a gauge symmetry was Maxwell's formulation of electrodynamics in 1864. The importance of this symmetry remained unnoticed in the earliest formulations. Similarly unnoticed, Hilbert had derived the Einstein field equations by postulating the invariance of the action under a general coordinate transformation.

Later Hermann Weyl, in an attempt to unify general relativity and electromagnetism, conjectured that *Eichinvarianz* or invariance under the change of scale (or "gauge") might also be a local symmetry of general relativity. After the development of quantum mechanics, Weyl, Vladimir Fock and Fritz London modified gauge by replacing the scale factor with a complex quantity and turned the scale transformation into a change of phase, which is a U(1) gauge symmetry. This explained the electromagnetic field effect on the wave function of a charged quantum mechanical particle. This was the first widely recognised gauge theory, popularised by Pauli in the 1940s.[1]

In 1954, attempting to resolve some of the great confusion in elementary particle physics, Chen Ning Yang and Robert Mills introduced **non-abelian gauge theories** as models to understand the strong interaction holding together nucleons in atomic nuclei. (Ronald Shaw, working under Abdus Salam, independently introduced the same notion in his doctoral thesis.) Generalizing the gauge invariance of electromagnetism, they attempted to construct a theory based on the action of the (non-abelian) SU(2) symmetry group on the isospin doublet of protons and neutrons. This is similar to the action of the U(1) group on the spinor fields of quantum electrodynamics. In particle physics the emphasis was on using **quantized gauge theories**.

This idea later found application in the quantum field theory of the weak force, and its unification with electromagnetism in the electroweak theory. Gauge theories became even more attractive when it was realized that non-abelian gauge theories reproduced a feature called asymptotic freedom. Asymptotic freedom was believed to be an important characteristic of strong interactions. This motivated searching for a strong force gauge theory. This theory, now known as quantum chromodynamics, is a gauge theory with the action of the SU(3) group on the color triplet of quarks. The Standard Model unifies the description of electromagnetism, weak interactions and strong interactions in the language of gauge theory.

In the 1970s, Sir Michael Atiyah began studying the mathematics of solutions to the classical Yang–Mills equations. In 1983, Atiyah's student Simon Donaldson built on this work to show that the differentiable classification of smooth 4-manifolds is very different from their classification up to homeomorphism. Michael Freedman used Donaldson's work to exhibit exotic $\mathbf{R}^4$s, that is, exotic differentiable structures on Euclidean 4-dimensional space. This led to an increasing interest in gauge theory for its own sake, independent of its successes in fundamental physics. In 1994, Edward Witten and Nathan Seiberg invented gauge-theoretic techniques based on supersymmetry that enabled the calculation of certain topological invariants (the Seiberg–Witten invariants). These contributions to mathematics from gauge theory have led to a renewed interest in this area.

The importance of gauge theories in physics is exemplified in the tremendous success of the mathematical formalism in providing a unified framework to describe the quantum field theories of electromagnetism, the weak force and the strong force. This theory, known as the Standard Model, accurately describes experimental predictions regarding three of the four fundamental forces of nature, and is a gauge theory with the gauge group SU(3) × SU(2) × U(1). Modern theories like string theory, as well as general relativity, are, in one way or another, gauge theories.

See Pickering[2] for more about the history of gauge and quantum field theories.

48.2 Description

48.2.1 Global and local symmetries

In physics, the mathematical description of any physical situation usually contains excess degrees of freedom; the same physical situation is equally well described by many equivalent mathematical configurations. For instance, in Newtonian dynamics, if two configurations are related by a Galilean transformation (an inertial change of reference frame) they represent the same physical situation. These transformations form a group of "symmetries" of the theory, and a physical situation corresponds not to an individual mathematical configuration but to a class of configurations related to one another by this symmetry group.

This idea can be generalized to include local as well as global symmetries, analogous to much more abstract "changes of coordinates" in a situation where there is no preferred "inertial" coordinate system that covers the entire physical system. A gauge theory is a mathematical model that has symmetries of this kind, together with a set of techniques for making physical predictions consistent with the symmetries of the model.

48.2.2 Example of global symmetry

When a quantity occurring in the mathematical configuration is not just a number but has some geometrical significance, such as a velocity or an axis of rotation, its representation as numbers arranged in a vector or matrix is also changed by a coordinate transformation. For instance, if one description of a pattern of fluid flow states that the fluid velocity in the neighborhood of (x=1, y=0) is 1 m/s in the positive x direction, then a description of the same situation in which the coordinate system has been rotated clockwise by 90 degrees states that the fluid velocity in the neighborhood of (x=0, y=1) is 1 m/s in the positive y direction. The coordinate transformation has affected both the coordinate system used to identify the *location* of the measurement and the basis in which its *value* is expressed. As long as this transformation is performed globally (affecting the coordinate basis in the same way at every point), the effect on values that represent the *rate of change* of some quantity along some path in space and time as it passes through point P is the same as the effect on values that are truly local to P.

48.2.3 Use of fiber bundles to describe local symmetries

In order to adequately describe physical situations in more complex theories, it is often necessary to introduce a "coordinate basis" for some of the objects of the theory that do not have this simple relationship to the coordinates used to label points in space and time. (In mathematical terms, the theory involves a fiber bundle in which the fiber at each point of the base space consists of possible coordinate bases for use when describing the values of objects at that point.) In order to spell out a mathematical configuration, one must choose a particular coordinate basis at each point (a *local section* of the fiber bundle) and express the values of the objects of the theory (usually "fields" in the physicist's sense) using this basis. Two such mathematical configurations are equivalent (describe the same physical situation) if they are related by a transformation of this abstract coordinate basis (a change of local section, or *gauge transformation*).

In most gauge theories, the set of possible transformations of the abstract gauge basis at an individual point in space and time is a finite-dimensional Lie group. The simplest such group is U(1), which appears in the modern formulation of quantum electrodynamics (QED) via its use of complex numbers. QED is generally regarded as the first, and simplest, physical gauge theory. The set of possible gauge transformations of the entire configuration of a given gauge theory also forms a group, the *gauge group* of the theory. An element of the gauge group can be parameterized by a smoothly varying function from the points of spacetime to the (finite-dimensional) Lie group, such that the value of the function and its derivatives at each point represents the action of the gauge transformation on the fiber over that point.

A gauge transformation with constant parameter at every point in space and time is analogous to a rigid rotation of the geometric coordinate system; it represents a global symmetry of the gauge representation. As in the case of a rigid rotation, this gauge transformation affects expressions that represent the rate of change along a path of some gauge-dependent quantity in the same way as those that represent a truly local quantity. A gauge transformation whose parameter is *not* a constant function is referred to as a local symmetry; its effect on expressions that involve a derivative is qualitatively different from that on expressions that don't. (This is analogous to a non-inertial change of reference frame, which can produce a Coriolis effect.)

48.2.4 Gauge fields

The "gauge covariant" version of a gauge theory accounts for this effect by introducing a gauge field (in mathematical language, an Ehresmann connection) and formulating all rates of change in terms of the covariant derivative with respect to this connection. The gauge field becomes an essential part of the description of a mathematical configuration. A configuration in which the gauge field can be eliminated by a gauge transformation has the property that its field strength (in mathematical language, its curvature) is zero everywhere; a gauge theory is *not* limited to these configurations. In other words, the distinguishing characteristic of a gauge theory is that the gauge field does not merely compensate for a poor choice of coordinate system; there is generally no gauge transformation that makes the gauge field vanish.

When analyzing the dynamics of a gauge theory, the gauge field must be treated as a dynamical variable, similarly to other objects in the description of a physical situation. In addition to its interaction with other objects via the covariant derivative, the gauge field typically contributes energy in the form of a "self-energy" term. One can obtain the equations for the gauge theory by:

- starting from a naïve ansatz without the gauge field (in which the derivatives appear in a "bare" form);

- listing those global symmetries of the theory that can be characterized by a continuous parameter (generally an abstract equivalent of a rotation angle);

- computing the correction terms that result from allowing the symmetry parameter to vary from place to place; and

- reinterpreting these correction terms as couplings to one or more gauge fields, and giving these fields appropriate self-energy terms and dynamical behavior.

This is the sense in which a gauge theory "extends" a global symmetry to a local symmetry, and closely resembles the historical development of the gauge theory of gravity known as general relativity.

48.2.5 Physical experiments

Gauge theories are used to model the results of physical experiments, essentially by:

- limiting the universe of possible configurations to those consistent with the information used to set up the experiment, and then

- computing the probability distribution of the possible outcomes that the experiment is designed to measure.

The mathematical descriptions of the "setup information" and the "possible measurement outcomes" (loosely speaking, the "boundary conditions" of the experiment) are generally not expressible without reference to a particular coordinate system, including a choice of gauge. (If nothing else, one assumes that the experiment has been adequately isolated from "external" influence, which is itself a gauge-dependent statement.) Mishandling gauge dependence in boundary conditions is a frequent source of anomalies in gauge theory calculations, and gauge theories can be broadly classified by their approaches to anomaly avoidance.

48.2.6 Continuum theories

The two gauge theories mentioned above (continuum electrodynamics and general relativity) are examples of continuum field theories. The techniques of calculation in a continuum theory implicitly assume that:

- given a completely fixed choice of gauge, the boundary conditions of an individual configuration can in principle be completely described;

- given a completely fixed gauge and a complete set of boundary conditions, the principle of least action determines a unique mathematical configuration (and therefore a unique physical situation) consistent with these bounds;

- the likelihood of possible measurement outcomes can be determined by:

  - establishing a probability distribution over all physical situations determined by boundary conditions that are consistent with the setup information,

  - establishing a probability distribution of measurement outcomes for each possible physical situation, and

  - convolving these two probability distributions to get a distribution of possible measurement outcomes consistent with the setup information; and

- fixing the gauge introduces no anomalies in the calculation, due either to gauge dependence in describing partial information about boundary conditions or to incompleteness of the theory.

These assumptions are close enough to be valid across a wide range of energy scales and experimental conditions, to allow these theories to make accurate predictions about almost all of the phenomena encountered in daily life, from light, heat, and electricity to eclipses and spaceflight. They fail only at the smallest and largest scales (due to omissions in the theories themselves) and when the mathematical techniques themselves break down (most notably in the case of turbulence and other chaotic phenomena).

48.2.7 Quantum field theories

Other than these "classical" continuum field theories, the most widely known gauge theories are quantum field theories, including quantum electrodynamics and the Standard Model of elementary particle physics. The starting point of a quantum field theory is much like that of its continuum analog: a gauge-covariant action integral that characterizes "allowable" physical situations according to the principle of least action. However, continuum and quantum theories differ significantly in how they handle the excess degrees of freedom represented by gauge transformations. Continuum theories, and most pedagogical treatments of the simplest quantum field theories, use a gauge fixing prescription to reduce the orbit of mathematical configurations that represent a given physical situation to a smaller orbit related by a smaller gauge group (the global symmetry group, or perhaps even the trivial group).

More sophisticated quantum field theories, in particular those that involve a non-abelian gauge group, break the gauge symmetry within the techniques of perturbation theory by introducing additional fields (the Faddeev–Popov ghosts) and counterterms motivated by anomaly cancellation, in an approach known as BRST quantization. While these concerns are in one sense highly technical, they are also closely related to the nature of measurement, the limits on knowledge of a physical situation, and the interactions between incompletely specified experimental conditions and incompletely understood physical theory . The mathematical techniques that have been developed in order to make gauge theories tractable have found many other applications, from solid-state physics and crystallography to low-dimensional topology.

48.3 Classical gauge theory

48.3.1 Classical electromagnetism

Historically, the first example of gauge symmetry discovered was classical electromagnetism. In electrostatics, one can either discuss the electric field, **E**, or its corresponding electric potential, V. Knowledge of one makes it possible to find the other, except that potentials differing by a constant, $V \to V + C$, correspond to the same electric field. This is because the electric field relates to *changes* in the potential from one point in space to another, and the constant C would cancel out when subtracting to find the change in potential. In terms of vector calculus, the electric field is the gradient of the potential, $\mathbf{E} = -\nabla V$. Generalizing from static electricity to electromagnetism, we have a second potential, the vector potential **A**, with

$$\mathbf{E} = -\nabla V - \frac{\partial \mathbf{A}}{\partial t}$$
$$\mathbf{B} = \nabla \times \mathbf{A}$$

The general gauge transformations now become not just $V \to V + C$ but

$$\mathbf{A} \to \mathbf{A} + \nabla f$$
$$V \to V - \frac{\partial f}{\partial t}$$

where f is any function that depends on position and time. The fields remain the same under the gauge transformation, and therefore Maxwell's equations are still satisfied. That is, Maxwell's equations have a gauge symmetry.

48.3.2 An example: Scalar O(n) gauge theory

The remainder of this section requires some familiarity with classical or quantum field theory, and the use of Lagrangians.

Definitions in this section: gauge group, gauge field, interaction Lagrangian, gauge boson.

The following illustrates how local gauge invariance can be "motivated" heuristically starting from global symmetry properties, and how it leads to an interaction between originally non-interacting fields.

48.3. CLASSICAL GAUGE THEORY

Consider a set of n non-interacting real scalar fields, with equal masses m. This system is described by an action that is the sum of the (usual) action for each scalar field φ_i

$$S = \int d^4x \sum_{i=1}^{n} \left[\frac{1}{2} \partial_\mu \varphi_i \partial^\mu \varphi_i - \frac{1}{2} m^2 \varphi_i^2 \right]$$

The Lagrangian (density) can be compactly written as

$$\mathcal{L} = \frac{1}{2} (\partial_\mu \Phi)^T \partial^\mu \Phi - \frac{1}{2} m^2 \Phi^T \Phi$$

by introducing a vector of fields

$$\Phi = (\varphi_1, \varphi_2, \ldots, \varphi_n)^T$$

The term ∂_μ is Einstein notation for the partial derivative of Φ in each of the four dimensions. It is now transparent that the Lagrangian is invariant under the transformation

$$\Phi \mapsto \Phi' = G\Phi$$

whenever G is a *constant* matrix belonging to the *n*-by-*n* orthogonal group O(*n*). This is seen to preserve the Lagrangian, since the derivative of Φ transforms identically to Φ and both quantities appear inside dot products in the Lagrangian (orthogonal transformations preserve the dot product).

$$(\partial_\mu \Phi) \mapsto (\partial_\mu \Phi)' = G \partial_\mu \Phi$$

This characterizes the *global* symmetry of this particular Lagrangian, and the symmetry group is often called the **gauge group**; the mathematical term is **structure group**, especially in the theory of G-structures. Incidentally, Noether's theorem implies that invariance under this group of transformations leads to the conservation of the *currents*

$$J_\mu^a = i \partial_\mu \Phi^T T^a \Phi$$

where the T^a matrices are generators of the SO(*n*) group. There is one conserved current for every generator.

Now, demanding that this Lagrangian should have *local* O(*n*)-invariance requires that the G matrices (which were earlier constant) should be allowed to become functions of the space-time coordinates x.

Unfortunately, the G matrices do not "pass through" the derivatives, when $G = G(x)$,

$$\partial_\mu (G\Phi) \neq G(\partial_\mu \Phi)$$

The failure of the derivative to commute with "G" introduces an additional term (in keeping with the product rule), which spoils the invariance of the Lagrangian. In order to rectify this we define a new derivative operator such that the derivative of Φ again transforms identically with Φ

$$(D_\mu \Phi)' = G D_\mu \Phi$$

This new "derivative" is called a (gauge) covariant derivative and takes the form

$$D_\mu = \partial_\mu + ig A_\mu$$

Where g is called the coupling constant; a quantity defining the strength of an interaction. After a simple calculation we can see that the **gauge field** $A(x)$ must transform as follows

$$A'_\mu = G A_\mu G^{-1} + \frac{i}{g}(\partial_\mu G) G^{-1}$$

The gauge field is an element of the Lie algebra, and can therefore be expanded as

$$A_\mu = \sum_a A_\mu^a T^a$$

There are therefore as many gauge fields as there are generators of the Lie algebra.

Finally, we now have a *locally gauge invariant* Lagrangian

$$\mathcal{L}_{\text{loc}} = \frac{1}{2}(D_\mu \Phi)^T D^\mu \Phi - \frac{1}{2} m^2 \Phi^T \Phi$$

Pauli uses the term *gauge transformation of the first type* to mean the transformation of Φ, while the compensating transformation in A is called a *gauge transformation of the second type*.

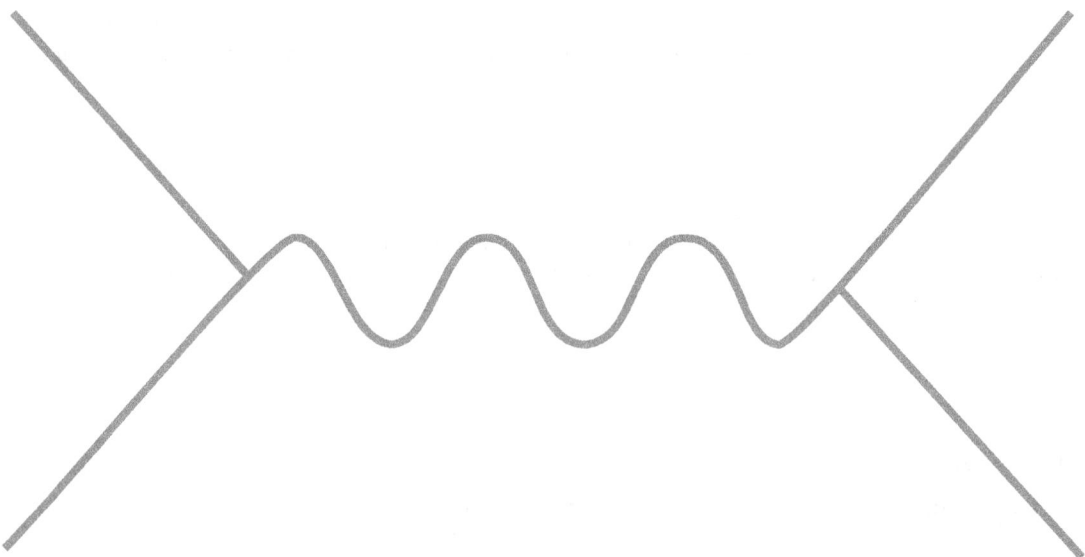

Feynman diagram of scalar bosons interacting via a gauge boson

The difference between this Lagrangian and the original *globally gauge-invariant* Lagrangian is seen to be the **interaction Lagrangian**

$$\mathcal{L}_{\text{int}} = i\frac{g}{2}\Phi^T A_\mu^T \partial^\mu \Phi + i\frac{g}{2}(\partial_\mu \Phi)^T A^\mu \Phi - \frac{g^2}{2}(A_\mu \Phi)^T A^\mu \Phi$$

This term introduces interactions between the n scalar fields just as a consequence of the demand for local gauge invariance. However, to make this interaction physical and not completely arbitrary, the mediator $A(x)$ needs to propagate in space. That is dealt with in the next section by adding yet another term, $\mathcal{L}_{\text{gf}}$, to the Lagrangian. In the quantized version of the obtained classical field theory, the quanta of the gauge field $A(x)$ are called gauge bosons. The interpretation of the interaction Lagrangian in quantum field theory is of scalar bosons interacting by the exchange of these gauge bosons.

48.3.3 The Yang–Mills Lagrangian for the gauge field

Main article: Yang–Mills theory

The picture of a classical gauge theory developed in the previous section is almost complete, except for the fact that to define the covariant derivatives D, one needs to know the value of the gauge field $A(x)$ at all space-time points. Instead of manually specifying the values of this field, it can be given as the solution to a field equation. Further requiring that the Lagrangian that generates this field equation is locally gauge invariant as well, one possible form for the gauge field Lagrangian is (conventionally) written as

$$\mathcal{L}_{\text{gf}} = -\frac{1}{2} \text{Tr}(F^{\mu\nu} F_{\mu\nu})$$

with

$$F_{\mu\nu} = \frac{1}{ig}[D_\mu, D_\nu]$$

and the trace being taken over the vector space of the fields. This is called the **Yang–Mills action**. Other gauge invariant actions also exist (e.g., nonlinear electrodynamics, Born–Infeld action, Chern–Simons model, theta term, etc.).

Note that in this Lagrangian term there is no field whose transformation counterweighs the one of A. Invariance of this term under gauge transformations is a particular case of *a priori* classical (geometrical) symmetry. This symmetry must be restricted in order to perform quantization, the procedure being denominated gauge fixing, but even after restriction, gauge transformations may be possible.[3]

The complete Lagrangian for the gauge theory is now

$$\mathcal{L} = \mathcal{L}_{\text{loc}} + \mathcal{L}_{\text{gf}} = \mathcal{L}_{\text{global}} + \mathcal{L}_{\text{int}} + \mathcal{L}_{\text{gf}}$$

48.3.4 An example: Electrodynamics

As a simple application of the formalism developed in the previous sections, consider the case of electrodynamics, with only the electron field. The bare-bones action that generates the electron field's Dirac equation is

$$S = \int \bar{\psi}(i\hbar c \gamma^\mu \partial_\mu - mc^2)\psi \, d^4x$$

The global symmetry for this system is

$$\psi \mapsto e^{i\theta}\psi$$

The gauge group here is U(1), just rotations of the phase angle of the field, with the particular rotation determined by the constant θ.

"Local"ising this symmetry implies the replacement of θ by $\theta(x)$. An appropriate covariant derivative is then

$$D_\mu = \partial_\mu - i\frac{e}{\hbar}A_\mu$$

Identifying the "charge" e (not to be confused with the mathematical constant e in the symmetry description) with the usual electric charge (this is the origin of the usage of the term in gauge theories), and the gauge field $A(x)$ with the four-vector potential of electromagnetic field results in an interaction Lagrangian

$$\mathcal{L}_{\text{int}} = \frac{e}{\hbar}\bar{\psi}(x)\gamma^\mu\psi(x)A_\mu(x) = J^\mu(x)A_\mu(x)$$

where $J^\mu(x)$ is the usual four vector electric current density. The gauge principle is therefore seen to naturally introduce the so-called minimal coupling of the electromagnetic field to the electron field.

Adding a Lagrangian for the gauge field $A_\mu(x)$ in terms of the field strength tensor exactly as in electrodynamics, one obtains the Lagrangian used as the starting point in quantum electrodynamics.

$$\mathcal{L}_{\text{QED}} = \bar{\psi}(i\hbar c\,\gamma^\mu D_\mu - mc^2)\psi - \frac{1}{4\mu_0}F_{\mu\nu}F^{\mu\nu}$$

See also: Dirac equation, Maxwell's equations, Quantum electrodynamics

48.4 Mathematical formalism

Gauge theories are usually discussed in the language of differential geometry. Mathematically, a *gauge* is just a choice of a (local) section of some principal bundle. A **gauge transformation** is just a transformation between two such sections.

Although gauge theory is dominated by the study of connections (primarily because it's mainly studied by high-energy physicists), the idea of a connection is not central to gauge theory in general. In fact, a result in general gauge theory shows that affine representations (i.e., affine modules) of the gauge transformations can be classified as sections of a jet bundle satisfying certain properties. There are representations that transform covariantly pointwise (called by physicists gauge transformations of the first kind), representations that transform as a connection form (called by physicists gauge transformations of the second kind, an affine representation)—and other more general representations, such as the B field in BF theory. There are more general nonlinear representations (realizations), but these are extremely complicated. Still, nonlinear sigma models transform nonlinearly, so there are applications.

If there is a principal bundle P whose base space is space or spacetime and structure group is a Lie group, then the sections of P form a principal homogeneous space of the group of gauge transformations.

Connections (gauge connection) define this principal bundle, yielding a covariant derivative ∇ in each associated vector bundle. If a local frame is chosen (a local basis of sections), then this covariant derivative is represented by the connection form A, a Lie algebra-valued 1-form, which is called the **gauge potential** in physics. This is evidently not an intrinsic but a frame-dependent quantity. The curvature form F, a Lie algebra-valued 2-form that is an intrinsic quantity, is constructed from a connection form by

$$\mathbf{F} = \mathrm{d}\mathbf{A} + \mathbf{A} \wedge \mathbf{A}$$

where d stands for the exterior derivative and $\wedge$ stands for the wedge product. ($\mathbf{A}$ is an element of the vector space spanned by the generators T^a, and so the components of $\mathbf{A}$ do not commute with one another. Hence the wedge product $\mathbf{A} \wedge \mathbf{A}$ does not vanish.)

Infinitesimal gauge transformations form a Lie algebra, which is characterized by a smooth Lie-algebra-valued scalar, ε. Under such an infinitesimal gauge transformation,

$$\delta_\varepsilon \mathbf{A} = [\varepsilon, \mathbf{A}] - \mathrm{d}\varepsilon$$

where $[\cdot,\cdot]$ is the Lie bracket.

One nice thing is that if $\delta_\varepsilon X = \varepsilon X$, then $\delta_\varepsilon DX = \varepsilon DX$ where D is the covariant derivative

$$DX \stackrel{\text{def}}{=} \mathrm{d}X + \mathbf{A}X$$

Also, $\delta_\varepsilon \mathbf{F} = \varepsilon \mathbf{F}$, which means $\mathbf{F}$ transforms covariantly.

Not all gauge transformations can be generated by infinitesimal gauge transformations in general. An example is when the base manifold is a compact manifold without boundary such that the homotopy class of mappings from that manifold to the Lie group is nontrivial. See instanton for an example.

The *Yang–Mills action* is now given by

$$\frac{1}{4g^2} \int \mathrm{Tr}[*F \wedge F]$$

where * stands for the Hodge dual and the integral is defined as in differential geometry.

A quantity which is **gauge-invariant** (i.e., invariant under gauge transformations) is the Wilson loop, which is defined over any closed path, γ, as follows:

$$\chi^{(\rho)}\left(\mathcal{P}\left\{e^{\int_\gamma A}\right\}\right)$$

where χ is the character of a complex representation ρ and $\mathcal{P}$ represents the path-ordered operator.

48.5 Quantization of gauge theories

Main article: Quantum gauge theory

Gauge theories may be quantized by specialization of methods which are applicable to any quantum field theory. However, because of the subtleties imposed by the gauge constraints (see section on Mathematical formalism, above) there are many technical problems to be solved which do not arise in other field theories. At the same time, the richer structure of gauge theories allows simplification of some computations: for example Ward identities connect different renormalization constants.

48.5.1 Methods and aims

The first gauge theory quantized was quantum electrodynamics (QED). The first methods developed for this involved gauge fixing and then applying canonical quantization. The Gupta–Bleuler method was also developed to handle this problem. Non-abelian gauge theories are now handled by a variety of means. Methods for quantization are covered in the article on quantization.

The main point to quantization is to be able to compute quantum amplitudes for various processes allowed by the theory. Technically, they reduce to the computations of certain correlation functions in the vacuum state. This involves a renormalization of the theory.

When the running coupling of the theory is small enough, then all required quantities may be computed in perturbation theory. Quantization schemes intended to simplify such computations (such as canonical quantization) may be called **perturbative quantization schemes**. At present some of these methods lead to the most precise experimental tests of gauge theories.

However, in most gauge theories, there are many interesting questions which are non-perturbative. Quantization schemes suited to these problems (such as lattice gauge theory) may be called **non-perturbative quantization schemes**. Precise computations in such schemes often require supercomputing, and are therefore less well-developed currently than other schemes.

48.5.2 Anomalies

Some of the symmetries of the classical theory are then seen not to hold in the quantum theory; a phenomenon called an **anomaly**. Among the most well known are:

- The scale anomaly, which gives rise to a *running coupling constant*. In QED this gives rise to the phenomenon of the Landau pole. In Quantum Chromodynamics (QCD) this leads to asymptotic freedom.

- The chiral anomaly in either chiral or vector field theories with fermions. This has close connection with topology through the notion of instantons. In QCD this anomaly causes the decay of a pion to two photons.

- The gauge anomaly, which must cancel in any consistent physical theory. In the electroweak theory this cancellation requires an equal number of quarks and leptons.

48.6 Pure gauge

A pure gauge is the set of field configurations obtained by a gauge transformation on the null-field configuration, i.e., a gauge-transform of zero. So it is a particular "gauge orbit" in the field configuration's space.

Thus, in the abelian case, where $A_\mu(x) \to A'_\mu(x) = A_\mu(x) + \partial_\mu f(x)$, the pure gauge is just the set of field configurations $A'_\mu(x) = \partial_\mu f(x)$ for all $f(x)$.

48.7 See also

48.8 References

[1] Wolfgang Pauli (1941) "Relativistic Field Theories of Elementary Particles," *Rev. Mod. Phys.* **13**: 203–32.

[2] Pickering, A. (1984). *Constructing Quarks*. University of Chicago Press. ISBN 0-226-66799-5.

[3] Sakurai, *Advanced Quantum Mechanics*, sect 1–4

48.9 Bibliography

General readers

- Schumm, Bruce (2004) *Deep Down Things*. Johns Hopkins University Press. Esp. chpt. 8. A serious attempt by a physicist to explain gauge theory and the Standard Model with little formal mathematics.

Texts

- Bromley, D.A. (2000). *Gauge Theory of Weak Interactions*. Springer. ISBN 3-540-67672-4.

- Cheng, T.-P.; Li, L.-F. (1983). *Gauge Theory of Elementary Particle Physics*. Oxford University Press. ISBN 0-19-851961-3.

- Frampton, P. (2008). *Gauge Field Theories* (3rd ed.). Wiley-VCH.

- Kane, G.L. (1987). *Modern Elementary Particle Physics*. Perseus Books. ISBN 0-201-11749-5.

Articles

- Becchi, C. (1997). "Introduction to Gauge Theories". p. 5211. arXiv:hep-ph/9705211. Bibcode:1997hep.ph....5211B.

- Gross, D. (1992). "Gauge theory – Past, Present and Future". Retrieved 2009-04-23.

- Jackson, J.D. (2002). "From Lorenz to Coulomb and other explicit gauge transformations". *Am.J.Phys* **70** (9): 917–928. arXiv:physics/0204034. Bibcode:2002AmJPh..70..917J. doi:10.1119/1.1491265.

- Svetlichny, George (1999). "Preparation for Gauge Theory". p. 2027. arXiv:math-ph/9902027. Bibcode:1999math.ph...2027

48.10 External links

- Hazewinkel, Michiel, ed. (2001), "Gauge transformation", *Encyclopedia of Mathematics*, Springer, ISBN 978-1-55608-010-4
- Yang–Mills equations on DispersiveWiki
- Gauge theories on Scholarpedia

Chapter 49

Quantum field theory

"Relativistic quantum field theory" redirects here. For other uses, see Relativity.

In theoretical physics, **quantum field theory** (**QFT**) is a theoretical framework for constructing quantum mechanical models of subatomic particles in particle physics and quasiparticles in condensed matter physics. A QFT treats particles as excited states of an underlying physical field, so these are called field quanta.

For example, quantum electrodynamics (QED) has one electron field and one photon field; quantum chromodynamics (QCD) has one field for each type of quark; and, in condensed matter, there is an atomic displacement field that gives rise to phonon particles. Edward Witten describes QFT as "by far" the most difficult theory in modern physics.[1]

In QFT, quantum mechanical interactions between particles are described by interaction terms between the corresponding underlying fields. QFT interaction terms are similar in spirit to those between charges with electric and magnetic fields in Maxwell's equations. However, unlike the classical fields of Maxwell's theory, fields in QFT generally exist in quantum superpositions of states and are subject to the laws of quantum mechanics.

Quantum mechanical systems have a fixed number of particles, with each particle having a finite number of degrees of freedom. In contrast, the excited states of a QFT can represent any number of particles. This makes quantum field theories especially useful for describing systems where the particle count/number may change over time, a crucial feature of relativistic dynamics.

Because the fields are continuous quantities over space, there exist excited states with arbitrarily large numbers of particles in them, providing QFT systems with an effectively infinite number of degrees of freedom. Infinite degrees of freedom can easily lead to divergences of calculated quantities (i.e., the quantities become infinite). Techniques such as renormalization of QFT parameters or discretization of spacetime, as in lattice QCD, are often used to avoid such infinities so as to yield physically meaningful results.

Most theories in standard particle physics are formulated as **relativistic quantum field theories**, such as QED, QCD, and the Standard Model. QED, the quantum field-theoretic description of the electromagnetic field, approximately reproduces Maxwell's theory of electrodynamics in the low-energy limit, with small non-linear corrections to the Maxwell equations required due to virtual electron–positron pairs.

In the perturbative approach to quantum field theory, the full field interaction terms are approximated as a perturbative expansion in the number of particles involved. Each term in the expansion can be thought of as forces between particles being mediated by other particles. In QED, the electromagnetic force between two electrons is caused by an exchange of photons. Similarly, intermediate vector bosons mediate the weak force and gluons mediate the strong force in QCD. The notion of a force-mediating particle comes from perturbation theory, and does not make sense in the context of non-perturbative approaches to QFT, such as with bound states.

The gravitational field and the electromagnetic field are the only two fundamental fields in nature that have infinite range and a corresponding classical low-energy limit, which greatly diminishes and hides their "particle-like" excitations. Albert Einstein in 1905, attributed "particle-like" and discrete exchanges of momenta and energy, characteristic of "field quanta", to the electromagnetic field. Originally, his principal motivation was to explain the thermodynamics of radiation. Although the photoelectric effect and Compton scattering strongly suggest the existence of the photon, it is now understood that they can be explained without invoking a quantum electromagnetic field; therefore, a more definitive proof of the quantum nature of radiation is now taken up into modern quantum optics as in the antibunching effect.[2]

There is currently no complete quantum theory of the remaining fundamental force, gravity. Many of the proposed theories to describe gravity as a QFT postulate the existence of a graviton particle that mediates the gravitational force. Presumably, the as yet unknown correct quantum field-theoretic treatment of the gravitational field will behave like Einstein's general theory of relativity in the low-energy limit. Quantum field theory of the fundamental forces itself has been postulated to be the low-energy effective field theory limit of a more fundamental theory such as superstring theory.

49.1 History

Main article: History of quantum field theory

49.1.1 Foundations

The early development of the field involved Dirac, Fock, Pauli, Heisenberg and Bogolyubov. This phase of development culminated with the construction of the theory of quantum electrodynamics in the 1950s.

49.1.2 Gauge theory

Gauge theory was formulated and quantized, leading to the **unification of forces** embodied in the standard model of particle physics. This effort started in the 1950s with the work of Yang and Mills, was carried on by Martinus Veltman and a host of others during the 1960s and completed by the 1970s through the work of Gerard 't Hooft, Frank Wilczek, David Gross and David Politzer.

49.1.3 Grand synthesis

Parallel developments in the understanding of phase transitions in condensed matter physics led to the study of the renormalization group. This in turn led to the **grand synthesis** of theoretical physics, which unified theories of particle and condensed matter physics through quantum field theory. This involved the work of Michael Fisher and Leo Kadanoff in the 1970s, which led to the seminal reformulation of quantum field theory by Kenneth G. Wilson.

49.2 Principles

49.2.1 Classical and quantum fields

Main article: Classical field theory

A classical field is a function defined over some region of space and time.[3] Two physical phenomena which are described by classical fields are Newtonian gravitation, described by Newtonian gravitational field $\mathbf{g}(\mathbf{x}, t)$, and classical electromagnetism, described by the electric and magnetic fields $\mathbf{E}(\mathbf{x}, t)$ and $\mathbf{B}(\mathbf{x}, t)$. Because such fields can in principle take on distinct values at each point in space, they are said to have infinite degrees of freedom.[3]

Classical field theory does not, however, account for the quantum-mechanical aspects of such physical phenomena. For instance, it is known from quantum mechanics that certain aspects of electromagnetism involve discrete particles—photons—rather than continuous fields. The business of *quantum* field theory is to write down a field that is, like a classical field, a function defined over space and time, but which also accommodates the observations of quantum mechanics. This is a *quantum field*.

It is not immediately clear *how* to write down such a quantum field, since quantum mechanics has a structure very unlike a field theory. In its most general formulation, quantum mechanics is a theory of abstract operators (observables) acting on an abstract state space (Hilbert space), where the observables represent physically observable quantities and the state space represents the possible states of the system under study.[4] For instance, the fundamental observables associated with the motion of a single quantum mechanical particle are the position and momentum operators $\hat{x}$ and $\hat{p}$. Field theory, in contrast, treats x as a way to index the field rather than as an operator.[5]

There are two common ways of developing a quantum field: the path integral formalism and canonical quantization.[6] The latter of these is pursued in this article.

Lagrangian formalism

Quantum field theory frequently makes use of the Lagrangian formalism from classical field theory. This formalism is analogous to the Lagrangian formalism used in classical mechanics to solve for the motion of a particle under the influence of a field. In classical field theory, one writes down a Lagrangian density, $\mathcal{L}$, involving a field, $\varphi(\mathbf{x},t)$, and possibly its first derivatives ($\partial\varphi/\partial t$ and $\nabla\varphi$), and then applies a field-theoretic form of the Euler–Lagrange equation. Writing coordinates $(t, \mathbf{x}) = (x^0, x^1, x^2, x^3) = x^\mu$, this form of the Euler–Lagrange equation is[3]

$$\frac{\partial}{\partial x^\mu}\left[\frac{\partial \mathcal{L}}{\partial(\partial\phi/\partial x^\mu)}\right] - \frac{\partial \mathcal{L}}{\partial\phi} = 0,$$

where a sum over μ is performed according to the rules of Einstein notation.

By solving this equation, one arrives at the "equations of motion" of the field.[3] For example, if one begins with the Lagrangian density

$$\mathcal{L}(\phi, \nabla\phi) = -\rho(t,\mathbf{x})\,\phi(t,\mathbf{x}) - \frac{1}{8\pi G}|\nabla\phi|^2,$$

and then applies the Euler–Lagrange equation, one obtains the equation of motion

$$4\pi G \rho(t,\mathbf{x}) = \nabla^2 \phi.$$

This equation is Newton's law of universal gravitation, expressed in differential form in terms of the gravitational potential $\varphi(t, \mathbf{x})$ and the mass density $\rho(t, \mathbf{x})$. Despite the nomenclature, the "field" under study is the gravitational potential, φ, rather than the gravitational field, **g**. Similarly, when classical field theory is used to study electromagnetism, the "field" of interest is the electromagnetic four-potential (V/c, **A**), rather than the electric and magnetic fields **E** and **B**.

Quantum field theory uses this same Lagrangian procedure to determine the equations of motion for quantum fields. These equations of motion are then supplemented by commutation relations derived from the canonical quantization procedure described below, thereby incorporating quantum mechanical effects into the behavior of the field.

49.2.2 Single- and many-particle quantum mechanics

Main articles: Quantum mechanics and First quantization

In quantum mechanics, a particle (such as an electron or proton) is described by a complex wavefunction, $\psi(x, t)$, whose time-evolution is governed by the Schrödinger equation:

$$-\frac{\hbar^2}{2m}\frac{\partial^2}{\partial x^2}\psi(x,t) + V(x)\psi(x,t) = i\hbar\frac{\partial}{\partial t}\psi(x,t).$$

Here m is the particle's mass and $V(x)$ is the applied potential. Physical information about the behavior of the particle is extracted from the wavefunction by constructing probability density functions for various quantities; for example, the p.d.f. for the particle's position is $\psi^*(x)\, x\, \psi(x)$, and the p.d.f. for the particle's momentum is $-i\hbar\psi^*(x)d\psi/dx$; integrating over x yields the expectation for the position, and the momentum respectively. This treatment of quantum mechanics, where a particle's wavefunction evolves against a classical background potential $V(x)$, is sometimes called *first quantization*.

This description of quantum mechanics can be extended to describe the behavior of multiple particles, so long as the number and the type of particles remain fixed. The particles are described by a wavefunction $\psi(x_1, x_2, ..., xN, t)$, which is governed by an extended version of the Schrödinger equation.

Often one is interested in the case where N particles are all of the same type (for example, the 18 electrons orbiting a neutral argon nucleus). As described in the article on identical particles, this implies that the state of the entire system must be either symmetric (bosons) or antisymmetric (fermions) when the coordinates of its constituent particles are exchanged. This is achieved by using a Slater determinant as the wavefunction of a fermionic system (and a Slater permanent for a bosonic system), which is equivalent to an element of the symmetric or antisymmetric subspace of a tensor product.

For example, the general quantum state of a system of N bosons is written as

$$|\phi_1 \cdots \phi_N\rangle = \sqrt{\frac{\prod_j N_j!}{N!}} \sum_{p \in S_N} |\phi_{p(1)}\rangle \otimes \cdots \otimes |\phi_{p(N)}\rangle,$$

where $|\phi_i\rangle$ are the single-particle states, N_j is the number of particles occupying state j, and the sum is taken over all possible permutations p acting on N elements. In general, this is a sum of $N!$ (N factorial) distinct terms. $\sqrt{\frac{\prod_j N_j!}{N!}}$ is a normalizing factor.

There are several shortcomings to the above description of quantum mechanics, which are addressed by quantum field theory. First, it is unclear how to extend quantum mechanics to include the effects of special relativity.[7] Attempted replacements for the Schrödinger equation, such as the Klein–Gordon equation or the Dirac equation, have many unsatisfactory qualities; for instance, they possess energy eigenvalues that extend to $-\infty$, so that there seems to be no easy definition of a ground state. It turns out that such inconsistencies arise from relativistic wavefunctions having a probabilistic interpretation in position space, as probability conservation is not a relativistically covariant concept. The second shortcoming, related to the first, is that in quantum mechanics there is no mechanism to describe particle creation and annihilation;[8] this is crucial for describing phenomena such as pair production, which result from the conversion between mass and energy according to the relativistic relation $E = mc^2$.

49.2.3 Second quantization

Main article: Second quantization

In this section, we will describe a method for constructing a quantum field theory called **second quantization**. This basically involves choosing a way to index the quantum mechanical degrees of freedom in the space of multiple identical-particle states. It is based on the Hamiltonian formulation of quantum mechanics.

Several other approaches exist, such as the Feynman path integral,[9] which uses a Lagrangian formulation. For an overview of some of these approaches, see the article on quantization.

Bosons

For simplicity, we will first discuss second quantization for bosons, which form perfectly symmetric quantum states. Let us denote the mutually orthogonal single-particle states which are possible in the system by $|\phi_1\rangle, |\phi_2\rangle, |\phi_3\rangle$, and so on. For example, the 3-particle state with one particle in state $|\phi_1\rangle$ and two in state $|\phi_2\rangle$ is

$$\frac{1}{\sqrt{3}} \left[|\phi_1\rangle|\phi_2\rangle|\phi_2\rangle + |\phi_2\rangle|\phi_1\rangle|\phi_2\rangle + |\phi_2\rangle|\phi_2\rangle|\phi_1\rangle \right].$$

The first step in second quantization is to express such quantum states in terms of **occupation numbers**, by listing the number of particles occupying each of the single-particle states $|\phi_1\rangle, |\phi_2\rangle$, etc. This is simply another way of labelling the states. For instance, the above 3-particle state is denoted as

$$|1, 2, 0, 0, 0, \ldots\rangle.$$

An N-particle state belongs to a space of states describing systems of N particles. The next step is to combine the individual N-particle state spaces into an extended state space, known as Fock space, which can describe systems of any number of particles. This is composed of the state space of a system with no particles (the so-called vacuum state, written as $|0\rangle$), plus the state space of a 1-particle system, plus the state space of a 2-particle system, and so forth. States describing a definite number of particles are known as Fock states: a general element of Fock space will be a linear combination of Fock states. There is a one-to-one correspondence between the occupation number representation and valid boson states in the Fock space.

At this point, the quantum mechanical system has become a quantum field in the sense we described above. The field's elementary degrees of freedom are the occupation numbers, and each occupation number is indexed by a number j indicating which of the single-particle states $|\phi_1\rangle, |\phi_2\rangle, \ldots, |\phi_j\rangle, \ldots$ it refers to:

$$|N_1, N_2, N_3, \ldots, N_j, \ldots\rangle.$$

The properties of this quantum field can be explored by defining creation and annihilation operators, which add and subtract particles. They are analogous to ladder operators in the quantum harmonic oscillator problem, which added and subtracted energy quanta. However, these operators literally create and annihilate particles of a given quantum state. The bosonic annihilation operator a_2 and creation operator $a_2^\dagger$ are easily defined in the occupation number representation as having the following effects:

$$a_2|N_1, N_2, N_3, \ldots\rangle = \sqrt{N_2}\,|\,N_1, (N_2-1), N_3, \ldots\rangle,$$

$$a_2^\dagger|N_1, N_2, N_3, \ldots\rangle = \sqrt{N_2+1}\,|\,N_1, (N_2+1), N_3, \ldots\rangle.$$

It can be shown that these are operators in the usual quantum mechanical sense, i.e. linear operators acting on the Fock space. Furthermore, they are indeed Hermitian conjugates, which justifies the way we have written them. They can be shown to obey the commutation relation

$$[a_i, a_j] = 0 \quad, \quad \left[a_i^\dagger, a_j^\dagger\right] = 0 \quad, \quad \left[a_i, a_j^\dagger\right] = \delta_{ij},$$

where δ stands for the Kronecker delta. These are precisely the relations obeyed by the ladder operators for an infinite set of independent quantum harmonic oscillators, one for each single-particle state. Adding or removing bosons from each state is therefore analogous to exciting or de-exciting a quantum of energy in a harmonic oscillator.

Applying an annihilation operator a_k followed by its corresponding creation operator $a_k^\dagger$ returns the number N_k of particles in the kth single-particle eigenstate:

$$a_k^\dagger a_k|\ldots, N_k, \ldots\rangle = N_k|\ldots, N_k, \ldots\rangle.$$

The combination of operators $a_k^\dagger a_k$ is known as the number operator for the kth eigenstate.

The Hamiltonian operator of the quantum field (which, through the Schrödinger equation, determines its dynamics) can be written in terms of creation and annihilation operators. For instance, for a field of free (non-interacting) bosons, the total energy of the field is found by summing the energies of the bosons in each energy eigenstate. If the kth single-particle energy eigenstate has energy E_k and there are N_k bosons in this state, then the total energy of these bosons is $E_k N_k$. The energy in the *entire* field is then a sum over k:

$$E_{\text{tot}} = \sum_k E_k N_k$$

This can be turned into the Hamiltonian operator of the field by replacing N_k with the corresponding number operator, $a_k^\dagger a_k$. This yields

$$H = \sum_k E_k\, a_k^\dagger a_k.$$

Fermions

It turns out that a different definition of creation and annihilation must be used for describing fermions. According to the Pauli exclusion principle, fermions cannot share quantum states, so their occupation numbers Ni can only take on the value 0 or 1. The fermionic annihilation operators c and creation operators $c^\dagger$ are defined by their actions on a Fock state thus

$$c_j|N_1, N_2, \ldots, N_j = 0, \ldots\rangle = 0$$

$$c_j|N_1, N_2, \ldots, N_j = 1, \ldots\rangle = (-1)^{(N_1+\cdots+N_{j-1})}|N_1, N_2, \ldots, N_j = 0, \ldots\rangle$$

$$c_j^\dagger|N_1, N_2, \ldots, N_j = 0, \ldots\rangle = (-1)^{(N_1+\cdots+N_{j-1})}|N_1, N_2, \ldots, N_j = 1, \ldots\rangle$$

$$c_j^\dagger|N_1, N_2, \ldots, N_j = 1, \ldots\rangle = 0.$$

These obey an anticommutation relation:

$$\{c_i, c_j\} = 0 \quad , \quad \{c_i^\dagger, c_j^\dagger\} = 0 \quad , \quad \{c_i, c_j^\dagger\} = \delta_{ij}.$$

One may notice from this that applying a fermionic creation operator twice gives zero, so it is impossible for the particles to share single-particle states, in accordance with the exclusion principle.

Field operators

We have previously mentioned that there can be more than one way of indexing the degrees of freedom in a quantum field. Second quantization indexes the field by enumerating the single-particle quantum states. However, as we have discussed, it is more natural to think about a "field", such as the electromagnetic field, as a set of degrees of freedom indexed by position.

To this end, we can define *field operators* that create or destroy a particle at a particular point in space. In particle physics, these operators turn out to be more convenient to work with, because they make it easier to formulate theories that satisfy the demands of relativity.

Single-particle states are usually enumerated in terms of their momenta (as in the particle in a box problem.) We can construct field operators by applying the Fourier transform to the creation and annihilation operators for these states. For example, the bosonic field annihilation operator $\phi(\mathbf{r})$ is

$$\phi(\mathbf{r}) \stackrel{\text{def}}{=} \sum_j e^{i\mathbf{k}_j \cdot \mathbf{r}} a_j.$$

The bosonic field operators obey the commutation relation

$$[\phi(\mathbf{r}), \phi(\mathbf{r}')] = 0 \quad , \quad [\phi^\dagger(\mathbf{r}), \phi^\dagger(\mathbf{r}')] = 0 \quad , \quad [\phi(\mathbf{r}), \phi^\dagger(\mathbf{r}')] = \delta^3(\mathbf{r} - \mathbf{r}')$$

where $\delta(x)$ stands for the Dirac delta function. As before, the fermionic relations are the same, with the commutators replaced by anticommutators.

The field operator is not the same thing as a single-particle wavefunction. The former is an operator acting on the Fock space, and the latter is a quantum-mechanical amplitude for finding a particle in some position. However, they are closely related, and are indeed commonly denoted with the same symbol. If we have a Hamiltonian with a space representation, say

$$H = -\frac{\hbar^2}{2m}\sum_i \nabla_i^2 + \sum_{i<j} U(|\mathbf{r}_i - \mathbf{r}_j|)$$

where the indices *i* and *j* run over all particles, then the field theory Hamiltonian (in the non-relativistic limit and for negligible self-interactions) is

$$H = -\frac{\hbar^2}{2m} \int d^3r \, \phi^\dagger(\mathbf{r}) \nabla^2 \phi(\mathbf{r}) + \frac{1}{2} \int d^3r \int d^3r' \, \phi^\dagger(\mathbf{r}) \phi^\dagger(\mathbf{r'}) U(|\mathbf{r} - \mathbf{r'}|) \phi(\mathbf{r'}) \phi(\mathbf{r}).$$

This looks remarkably like an expression for the expectation value of the energy, with ϕ playing the role of the wavefunction. This relationship between the field operators and wavefunctions makes it very easy to formulate field theories starting from space-projected Hamiltonians.

49.2.4 Dynamics

Once the Hamiltonian operator is obtained as part of the canonical quantization process, the time dependence of the state is described with the Schrödinger equation, just as with other quantum theories. Alternatively, the Heisenberg picture can be used where the time dependence is in the operators rather than in the states.

49.2.5 Implications

Unification of fields and particles

The "second quantization" procedure that we have outlined in the previous section takes a set of single-particle quantum states as a starting point. Sometimes, it is impossible to define such single-particle states, and one must proceed directly to quantum field theory. For example, a quantum theory of the electromagnetic field *must* be a quantum field theory, because it is impossible (for various reasons) to define a wavefunction for a single photon.[10] In such situations, the quantum field theory can be constructed by examining the mechanical properties of the classical field and guessing the corresponding quantum theory. For free (non-interacting) quantum fields, the quantum field theories obtained in this way have the same properties as those obtained using second quantization, such as well-defined creation and annihilation operators obeying commutation or anticommutation relations.

Quantum field theory thus provides a unified framework for describing "field-like" objects (such as the electromagnetic field, whose excitations are photons) and "particle-like" objects (such as electrons, which are treated as excitations of an underlying electron field), so long as one can treat interactions as "perturbations" of free fields. There are still unsolved problems relating to the more general case of interacting fields that may or may not be adequately described by perturbation theory. For more on this topic, see Haag's theorem.

Physical meaning of particle indistinguishability

The second quantization procedure relies crucially on the particles being identical. We would not have been able to construct a quantum field theory from a distinguishable many-particle system, because there would have been no way of separating and indexing the degrees of freedom.

Many physicists prefer to take the converse interpretation, which is that *quantum field theory explains what identical particles are*. In ordinary quantum mechanics, there is not much theoretical motivation for using symmetric (bosonic) or antisymmetric (fermionic) states, and the need for such states is simply regarded as an empirical fact. From the point of view of quantum field theory, particles are identical if and only if they are excitations of the same underlying quantum field. Thus, the question "why are all electrons identical?" arises from mistakenly regarding individual electrons as fundamental objects, when in fact it is only the electron field that is fundamental.

Particle conservation and non-conservation

During second quantization, we started with a Hamiltonian and state space describing a fixed number of particles (*N*), and ended with a Hamiltonian and state space for an arbitrary number of particles. Of course, in many common situations *N* is an important and perfectly well-defined quantity, e.g. if we are describing a gas of atoms sealed in a box. From the point of view of quantum field theory, such situations are described by quantum states that are eigenstates of the number operator $\hat{N}$, which measures the total number of particles present. As with any quantum mechanical observable, $\hat{N}$ is conserved if it commutes with the Hamiltonian. In that case, the quantum state is trapped in the

N-particle subspace of the total Fock space, and the situation could equally well be described by ordinary N-particle quantum mechanics. (Strictly speaking, this is only true in the noninteracting case or in the low energy density limit of renormalized quantum field theories)

For example, we can see that the free-boson Hamiltonian described above conserves particle number. Whenever the Hamiltonian operates on a state, each particle destroyed by an annihilation operator a_k is immediately put back by the creation operator $a_k^\dagger$.

On the other hand, it is possible, and indeed common, to encounter quantum states that are *not* eigenstates of $\hat{N}$, which do not have well-defined particle numbers. Such states are difficult or impossible to handle using ordinary quantum mechanics, but they can be easily described in quantum field theory as quantum superpositions of states having different values of N. For example, suppose we have a bosonic field whose particles can be created or destroyed by interactions with a fermionic field. The Hamiltonian of the combined system would be given by the Hamiltonians of the free boson and free fermion fields, plus a "potential energy" term such as

$$H_I = \sum_{k,q} V_q (a_q + a^\dagger_{-q}) c^\dagger_{k+q} c_k,$$

where $a_k^\dagger$ and a_k denotes the bosonic creation and annihilation operators, $c_k^\dagger$ and c_k denotes the fermionic creation and annihilation operators, and V_q is a parameter that describes the strength of the interaction. This "interaction term" describes processes in which a fermion in state k either absorbs or emits a boson, thereby being kicked into a different eigenstate $k+q$. (In fact, this type of Hamiltonian is used to describe interaction between conduction electrons and phonons in metals. The interaction between electrons and photons is treated in a similar way, but is a little more complicated because the role of spin must be taken into account.) One thing to notice here is that even if we start out with a fixed number of bosons, we will typically end up with a superposition of states with different numbers of bosons at later times. The number of fermions, however, is conserved in this case.

In condensed matter physics, states with ill-defined particle numbers are particularly important for describing the various superfluids. Many of the defining characteristics of a superfluid arise from the notion that its quantum state is a superposition of states with different particle numbers. In addition, the concept of a coherent state (used to model the laser and the BCS ground state) refers to a state with an ill-defined particle number but a well-defined phase.

49.2.6 Axiomatic approaches

The preceding description of quantum field theory follows the spirit in which most physicists approach the subject. However, it is not mathematically rigorous. Over the past several decades, there have been many attempts to put quantum field theory on a firm mathematical footing by formulating a set of axioms for it. These attempts fall into two broad classes.

The first class of axioms, first proposed during the 1950s, include the Wightman, Osterwalder–Schrader, and Haag–Kastler systems. They attempted to formalize the physicists' notion of an "operator-valued field" within the context of functional analysis, and enjoyed limited success. It was possible to prove that any quantum field theory satisfying these axioms satisfied certain general theorems, such as the spin-statistics theorem and the CPT theorem. Unfortunately, it proved extraordinarily difficult to show that any realistic field theory, including the Standard Model, satisfied these axioms. Most of the theories that could be treated with these analytic axioms were physically trivial, being restricted to low-dimensions and lacking interesting dynamics. The construction of theories satisfying one of these sets of axioms falls in the field of constructive quantum field theory. Important work was done in this area in the 1970s by Segal, Glimm, Jaffe and others.

During the 1980s, a second set of axioms based on geometric ideas was proposed. This line of investigation, which restricts its attention to a particular class of quantum field theories known as topological quantum field theories, is associated most closely with Michael Atiyah and Graeme Segal, and was notably expanded upon by Edward Witten, Richard Borcherds, and Maxim Kontsevich. However, most of the physically relevant quantum field theories, such as the Standard Model, are not topological quantum field theories; the quantum field theory of the fractional quantum Hall effect is a notable exception. The main impact of axiomatic topological quantum field theory has been on mathematics, with important applications in representation theory, algebraic topology, and differential geometry.

Finding the proper axioms for quantum field theory is still an open and difficult problem in mathematics. One of the Millennium Prize Problems—proving the existence of a mass gap in Yang–Mills theory—is linked to this issue.

49.3 Associated phenomena

In the previous part of the article, we described the most general properties of quantum field theories. Some of the quantum field theories studied in various fields of theoretical physics possess additional special properties, such as renormalizability, gauge symmetry, and supersymmetry. These are described in the following sections.

49.3.1 Renormalization

Main article: Renormalization

Early in the history of quantum field theory, it was found that many seemingly innocuous calculations, such as the perturbative shift in the energy of an electron due to the presence of the electromagnetic field, give infinite results. The reason is that the perturbation theory for the shift in an energy involves a sum over all other energy levels, and there are infinitely many levels at short distances that each give a finite contribution which results in a divergent series.

Many of these problems are related to failures in classical electrodynamics that were identified but unsolved in the 19th century, and they basically stem from the fact that many of the supposedly "intrinsic" properties of an electron are tied to the electromagnetic field that it carries around with it. The energy carried by a single electron—its self energy—is not simply the bare value, but also includes the energy contained in its electromagnetic field, its attendant cloud of photons. The energy in a field of a spherical source diverges in both classical and quantum mechanics, but as discovered by Weisskopf with help from Furry, in quantum mechanics the divergence is much milder, going only as the logarithm of the radius of the sphere.

The solution to the problem, presciently suggested by Stueckelberg, independently by Bethe after the crucial experiment by Lamb, implemented at one loop by Schwinger, and systematically extended to all loops by Feynman and Dyson, with converging work by Tomonaga in isolated postwar Japan, comes from recognizing that all the infinities in the interactions of photons and electrons can be isolated into redefining a finite number of quantities in the equations by replacing them with the observed values: specifically the electron's mass and charge: this is called renormalization. The technique of renormalization recognizes that the problem is essentially purely mathematical, that extremely short distances are at fault. In order to define a theory on a continuum, first place a cutoff on the fields, by postulating that quanta cannot have energies above some extremely high value. This has the effect of replacing continuous space by a structure where very short wavelengths do not exist, as on a lattice. Lattices break rotational symmetry, and one of the crucial contributions made by Feynman, Pauli and Villars, and modernized by 't Hooft and Veltman, is a symmetry-preserving cutoff for perturbation theory (this process is called regularization). There is no known symmetrical cutoff outside of perturbation theory, so for rigorous or numerical work people often use an actual lattice.

On a lattice, every quantity is finite but depends on the spacing. When taking the limit of zero spacing, we make sure that the physically observable quantities like the observed electron mass stay fixed, which means that the constants in the Lagrangian defining the theory depend on the spacing. Hopefully, by allowing the constants to vary with the lattice spacing, all the results at long distances become insensitive to the lattice, defining a continuum limit.

The renormalization procedure only works for a certain class of quantum field theories, called **renormalizable quantum field theories**. A theory is **perturbatively renormalizable** when the constants in the Lagrangian only diverge at worst as logarithms of the lattice spacing for very short spacings. The continuum limit is then well defined in perturbation theory, and even if it is not fully well defined non-perturbatively, the problems only show up at distance scales that are exponentially small in the inverse coupling for weak couplings. The Standard Model of particle physics is perturbatively renormalizable, and so are its component theories (quantum electrodynamics/electroweak theory and quantum chromodynamics). Of the three components, quantum electrodynamics is believed to not have a continuum limit, while the asymptotically free $SU(2)$ and $SU(3)$ weak hypercharge and strong color interactions are nonperturbatively well defined.

The renormalization group describes how renormalizable theories emerge as the long distance low-energy effective field theory for any given high-energy theory. Because of this, renormalizable theories are insensitive to the precise nature of the underlying high-energy short-distance phenomena. This is a blessing because it allows physicists to formulate low energy theories without knowing the details of high energy phenomenon. It is also a curse, because once a renormalizable theory like the standard model is found to work, it gives very few clues to higher energy processes. The only way high energy processes can be seen in the standard model is when they allow otherwise forbidden events, or if they predict quantitative relations between the coupling constants.

49.3.2 Haag's theorem

See also: Haag's theorem

From a mathematically rigorous perspective, there exists no interaction picture in a Lorentz-covariant quantum field theory. This implies that the perturbative approach of Feynman diagrams in QFT is not strictly justified, despite producing vastly precise predictions validated by experiment. This is called Haag's theorem, but most particle physicists relying on QFT largely shrug it off.

49.3.3 Gauge freedom

A gauge theory is a theory that admits a symmetry with a local parameter. For example, in every quantum theory the global phase of the wave function is arbitrary and does not represent something physical. Consequently, the theory is invariant under a global change of phases (adding a constant to the phase of all wave functions, everywhere); this is a global symmetry. In quantum electrodynamics, the theory is also invariant under a *local* change of phase, that is – one may shift the phase of all wave functions so that the shift may be different at every point in space-time. This is a *local* symmetry. However, in order for a well-defined derivative operator to exist, one must introduce a new field, the gauge field, which also transforms in order for the local change of variables (the phase in our example) not to affect the derivative. In quantum electrodynamics this gauge field is the electromagnetic field. The change of local gauge of variables is termed gauge transformation.

In quantum field theory the excitations of fields represent particles. The particle associated with excitations of the gauge field is the gauge boson, which is the photon in the case of quantum electrodynamics.

The degrees of freedom in quantum field theory are local fluctuations of the fields. The existence of a gauge symmetry reduces the number of degrees of freedom, simply because some fluctuations of the fields can be transformed to zero by gauge transformations, so they are equivalent to having no fluctuations at all, and they therefore have no physical meaning. Such fluctuations are usually called "non-physical degrees of freedom" or *gauge artifacts*; usually some of them have a negative norm, making them inadequate for a consistent theory. Therefore, if a classical field theory has a gauge symmetry, then its quantized version (i.e. the corresponding quantum field theory) will have this symmetry as well. In other words, a gauge symmetry cannot have a quantum anomaly. If a gauge symmetry is anomalous (i.e. not kept in the quantum theory) then the theory is non-consistent: for example, in quantum electrodynamics, had there been a gauge anomaly, this would require the appearance of photons with longitudinal polarization and polarization in the time direction, the latter having a negative norm, rendering the theory inconsistent; another possibility would be for these photons to appear only in intermediate processes but not in the final products of any interaction, making the theory non-unitary and again inconsistent (see optical theorem).

In general, the gauge transformations of a theory consist of several different transformations, which may not be commutative. These transformations are together described by a mathematical object known as a gauge group. Infinitesimal gauge transformations are the gauge group generators. Therefore the number of gauge bosons is the group dimension (i.e. number of generators forming a basis).

All the fundamental interactions in nature are described by gauge theories. These are:

- Quantum chromodynamics, whose gauge group is **SU**(3). The gauge bosons are eight gluons.

- The electroweak theory, whose gauge group is **U**(1) × **SU**(2), (a direct product of **U**(1) and **SU**(2)).

- Gravity, whose classical theory is general relativity, admits the equivalence principle, which is a form of gauge symmetry. However, it is explicitly non-renormalizable.

49.3.4 Multivalued gauge transformations

The gauge transformations which leave the theory invariant involve, by definition, only single-valued gauge functions $\Lambda(x_i)$ which satisfy the Schwarz integrability criterion

$$\partial_{x_i x_j} \Lambda = \partial_{x_j x_i} \Lambda.$$

An interesting extension of gauge transformations arises if the gauge functions $\Lambda(x_i)$ are allowed to be multivalued functions which violate the integrability criterion. These are capable of changing the physical field strengths and are therefore no proper symmetry transformations. Nevertheless, the transformed field equations describe correctly the physical laws in the presence of the newly generated field strengths. See the textbook by H. Kleinert cited below for the applications to phenomena in physics.

49.3.5 Supersymmetry

Main article: Supersymmetry

Supersymmetry assumes that every fundamental fermion has a superpartner that is a boson and vice versa. It was introduced in order to solve the so-called Hierarchy Problem, that is, to explain why particles not protected by any symmetry (like the Higgs boson) do not receive radiative corrections to its mass driving it to the larger scales (GUT, Planck...). It was soon realized that supersymmetry has other interesting properties: its gauged version is an extension of general relativity (Supergravity), and it is a key ingredient for the consistency of string theory.

The way supersymmetry protects the hierarchies is the following: since for every particle there is a superpartner with the same mass, any loop in a radiative correction is cancelled by the loop corresponding to its superpartner, rendering the theory UV finite.

Since no superpartners have yet been observed, if supersymmetry exists it must be broken (through a so-called soft term, which breaks supersymmetry without ruining its helpful features). The simplest models of this breaking require that the energy of the superpartners not be too high; in these cases, supersymmetry is expected to be observed by experiments at the Large Hadron Collider. The Higgs particle has been detected at the LHC, and no such superparticles have been discovered.

49.4 See also

- Abraham–Lorentz force
- Basic concepts of quantum mechanics
- Common integrals in quantum field theory
- Constructive quantum field theory
- Einstein–Maxwell–Dirac equations
- Feynman path integral
- Form factor (quantum field theory)
- Fundamental equation of unified field theory
- Green–Kubo relations
- Green's function (many-body theory)
- Invariance mechanics
- List of quantum field theories
- Pauli exclusion principle
- Photon polarization
- Pseudoscalar Field
- Quantum field theory in curved spacetime
- Quantum flavordynamics
- Quantum geometrodynamics

- Quantum hydrodynamics
- Quantum magnetodynamics
- Quantum triviality
- Relation between Schrödinger's equation and the path integral formulation of quantum mechanics
- Relationship between string theory and quantum field theory
- Schwinger–Dyson equation
- Static forces and virtual-particle exchange
- Symmetry in quantum mechanics
- Theoretical and experimental justification for the Schrödinger equation
- Ward–Takahashi identity
- Wheeler–Feynman absorber theory
- Wigner's classification
- Wigner's theorem

49.5 Notes

49.6 References

[1] "Beautiful Minds, Vol. 20: Ed Witten". la Repubblica. 2010. Retrieved 22 June 2012. See here.

[2] J. J. Thorn et al. (2004). Observing the quantum behavior of light in an undergraduate laboratory. . J. J. Thorn, M. S. Neel, V. W. Donato, G. S. Bergreen, R. E. Davies, and M. Beck. American Association of Physics Teachers, 2004.DOI: 10.1119/1.1737397.

[3] David Tong, *Lectures on Quantum Field Theory*, chapter 1.

[4] Srednicki, Mark. *Quantum Field Theory* (1st ed.). p. 19.

[5] Srednicki, Mark. *Quantum Field Theory* (1st ed.). pp. 25–6.

[6] Zee, Anthony. *Quantum Field Theory in a Nutshell* (2nd ed.). p. 61.

[7] David Tong, *Lectures on Quantum Field Theory*, Introduction.

[8] Zee, Anthony. *Quantum Field Theory in a Nutshell* (2nd ed.). p. 3.

[9] Abraham Pais, *Inward Bound: Of Matter and Forces in the Physical World* ISBN 0-19-851997-4. Pais recounts how his astonishment at the rapidity with which Feynman could calculate using his method. Feynman's method is now part of the standard methods for physicists.

[10] Newton, T.D.; Wigner, E.P. (1949). "Localized states for elementary systems". *Reviews of Modern Physics* **21** (3): 400–406. Bibcode:1949RvMP...21..400N. doi:10.1103/RevModPhys.21.400.

49.7 Further reading

General readers

- Feynman, R.P. (2001) [1964]. *The Character of Physical Law*. MIT Press. ISBN 0-262-56003-8.
- Feynman, R.P. (2006) [1985]. *QED: The Strange Theory of Light and Matter*. Princeton University Press. ISBN 0-691-12575-9.

- Gribbin, J. (1998). *Q is for Quantum: Particle Physics from A to Z*. Weidenfeld & Nicolson. ISBN 0-297-81752-3.

- Schumm, Bruce A. (2004) *Deep Down Things*. Johns Hopkins Univ. Press. Chpt. 4.

Introductory texts

- McMahon, D. (2008). *Quantum Field Theory*. McGraw-Hill. ISBN 978-0-07-154382-8.

- Bogoliubov, N.; Shirkov, D. (1982). *Quantum Fields*. Benjamin-Cummings. ISBN 0-8053-0983-7.

- Frampton, P.H. (2000). *Gauge Field Theories*. Frontiers in Physics (2nd ed.). Wiley.

- Greiner, W; Müller, B. (2000). *Gauge Theory of Weak Interactions*. Springer. ISBN 3-540-67672-4.

- Itzykson, C.; Zuber, J.-B. (1980). *Quantum Field Theory*. McGraw-Hill. ISBN 0-07-032071-3.

- Kane, G.L. (1987). *Modern Elementary Particle Physics*. Perseus Books. ISBN 0-201-11749-5.

- Kleinert, H.; Schulte-Frohlinde, Verena (2001). *Critical Properties of φ^4-Theories*. World Scientific. ISBN 981-02-4658-7.

- Kleinert, H. (2008). *Multivalued Fields in Condensed Matter, Electrodynamics, and Gravitation*. World Scientific. ISBN 978-981-279-170-2.

- Loudon, R (1983). *The Quantum Theory of Light*. Oxford University Press. ISBN 0-19-851155-8.

- Mandl, F.; Shaw, G. (1993). *Quantum Field Theory*. John Wiley & Sons. ISBN 978-0-471-94186-6.

- Peskin, M.; Schroeder, D. (1995). *An Introduction to Quantum Field Theory*. Westview Press. ISBN 0-201-50397-2.

- Ryder, L.H. (1985). *Quantum Field Theory*. Cambridge University Press. ISBN 0-521-33859-X.

- Schwartz, M.D. (2014). *Quantum Field Theory and the Standard Model*. Cambridge University Press. ISBN 978-1107034730.

- Srednicki, Mark (2007) *Quantum Field Theory*. Cambridge Univ. Press.

- Ynduráin, F.J. (1996). *Relativistic Quantum Mechanics and Introduction to Field Theory* (1st ed.). Springer. ISBN 978-3-540-60453-2.

- Zee, A. (2003). *Quantum Field Theory in a Nutshell*. Princeton University Press. ISBN 0-691-01019-6.

Advanced texts

- Brown, Lowell S. (1994). *Quantum Field Theory*. Cambridge University Press. ISBN 978-0-521-46946-3.

- Bogoliubov, N.; Logunov, A.A.; Oksak, A.I.; Todorov, I.T. (1990). *General Principles of Quantum Field Theory*. Kluwer Academic Publishers. ISBN 978-0-7923-0540-8.

- Weinberg, S. (1995). *The Quantum Theory of Fields* **1–3**. Cambridge University Press.

Articles:

- Gerard 't Hooft (2007) "The Conceptual Basis of Quantum Field Theory" in Butterfield, J., and John Earman, eds., *Philosophy of Physics, Part A*. Elsevier: 661–730.

- Frank Wilczek (1999) "Quantum field theory", *Reviews of Modern Physics* 71: S83–S95. Also doi=10.1103/Rev. Mod. Phys. 71.

49.8 External links

- Hazewinkel, Michiel, ed. (2001), "Quantum field theory", *Encyclopedia of Mathematics*, Springer, ISBN 978-1-55608-010-4
- Stanford Encyclopedia of Philosophy: "Quantum Field Theory", by Meinard Kuhlmann.
- Siegel, Warren, 2005. *Fields*. A free text, also available from arXiv:hep-th/9912205.
- Quantum Field Theory by P. J. Mulders

49.9 Text and image sources, contributors, and licenses

49.9.1 Text

- **Grand Unified Theory** *Source:* http://en.wikipedia.org/wiki/Grand%20Unified%20Theory?oldid=621604996 *Contributors:* AxelBoldt, Lee Daniel Crocker, Mav, AstroNomer, XJaM, Heron, Michael Hardy, Zocky, CesarB, Looxix, Emperorbma, Dysprosia, Phys, Omegatron, Northgrove, Robbot, Securiger, Lowellian, Meelar, Caknuck, Giftlite, Jmnbpt, Herbee, Fropuff, Xerxes314, Golbez, Ary29, Sam Hocevar, Lumidek, IcycleMort, M1ss1ontomars2k4, Mike Rosoft, Discospinster, 4pq1injbok, Pjacobi, Silence, JustinWick, Bobo192, Smalljim, John Vandenberg, Apyule, Foobaz, QTxVi4bEMRbrNqOorWBV, Jeodesic, Physicistjedi, Jérôme, Alansohn, Krischik, Sligocki, Mac Davis, GeorgeStepanek, RJFJR, Lee-Anne, DV8 2XL, Simetrical, Linas, Mindmatrix, FeanorStar7, GregorB, Ruziklan, Mekong Bluesman, Ashmoo, Rachel1, Rjwilmsi, Strait, Drrngrvy, FlaBot, Margosbot, DannyWilde, Rune.welsh, BradBeattie, Snailwalker, Phoenix2, Guanxi, DVdm, YurikBot, Ugha, Hairy Dude, Michael Slone, Gaius Cornelius, CambridgeBayWeather, NawlinWiki, Wiki alf, Joel7687, JocK, Zwobot, IceCreamAntisocial, Ms2ger, Noclip, CWenger, Smurrayinchester, Curpsbot-unicodify, Caco de vidro, Jaysbro, Sbyrnes321, SmackBot, Tom Lougheed, Eskimbot, Dauto, Silly rabbit, DHN-bot, Colonies Chris, Scwlong, QFT, Addshore, Robma, Dreadstar, Pwjb, Gbinal, Thorsen, Vina-iwbot, ArglebargleIV, Rory096, Ben Jos, Mr. Lefty, Ckatz, SirFozzie, Quaeler, Baderyp, Richwhite10, Cydebot, Peripitus, Michael C Price, Tawkerbot4, Thijs!bot, Headbomb, J.christianson, Luna Santin, Alphachimpbot, JAnDbot, Satarsa, Homy, Mbarbier, Durianking, Danmctaggart, Maliz, JCarlos, AstroHurricane001, Adavidb, Bogey97, Qatter, Jeepday, Econofire, Lseixas, Jaffar33, Eismc2, Alphanon, Praveen pillay, KabbalistPhysicist, PaddyLeahy, Hemadh, Will Scot 55, Datpol, Moffitma, ClueBot, DFRussia, James edmiston, Ordinaterr, Djr32, PixelBot, Weysheehai, Sun Creator, Subdolous, Dekisugi, AnonyScientist, TimothyRias, SilvonenBot, Bywater100, Truthnlove, Balungifrancis, Addbot, Micromaster, Favonian, Mohitsridhar, 84user, OlEnglish, WikiDan61, Amirobot, AnomieBOT, Girl Scout cookie, Theunify, Karanmohan, Materialscientist, Citation bot, Obersachsebot, Under22Entreprenuer, Dale Ritter, Senouf, Ernsts, FrescoBot, Paine Ellsworth, Steven Avraham Rosten, Ironboy11, Thamntamil, Sławomir Biały, GreenRoot, Ysyoon, John85, Gil987, Stupidsimple, Casimir9999, RobinK, Aknochel, Grandunifier, RjwilmsiBot, Afteread, EmausBot, Arbnos, L Kensington, ClueBot NG, ClaudeDes, Helpful Pixie Bot, Bibcode Bot, Bernard Rementilla, Kkumer, Wer900, Dilaton, Hilander316, Ryanr666, Davidyevgeny, Cjean42, Franzl aus tirol, Sagnac, GabeIglesia, Lmboyer04, Ovidiu cupsa, Jwratner1, Gilitejman1 and Anonymous: 127

- **Theory of everything** *Source:* http://en.wikipedia.org/wiki/Theory%20of%20everything?oldid=633835091 *Contributors:* AxelBoldt, Paul Drye, CYD, The Anome, Eclecticology, Roadrunner, Zippy, Stevertigo, Lorenzarius, Michael Hardy, Rojclague, Nixdorf, Takuya-Murata, Karada, Skysmith, Kosebamse, CesarB, Anders Feder, Angela, Julesd, Salsa Shark, Ugen64, Poor Yorick, Evercat, Schneelocke, Feedmecereal, Timwi, Dcoetzee, Dysprosia, Jitse Niesen, Wik, Jakenelson, Omegatron, Raul654, Nnh, Kevin M C Harkess, UninvitedCompany, Fredrik, Altenmann, Nurg, Naddy, Gandalf61, Mirv, Academic Challenger, Rursus, Blainster, Caknuck, Wereon, Diberri, Pengo, Tobias Bergemann, Hooloovoo, Ancheta Wis, Dbenbenn, Mporter, Jabra, Ferkelparade, Bfinn, Xerxes314, Curps, Alison, FeloniousMonk, McGravin, Behnam, Gzornenplatz, JRR Trollkien, Steuard, Andycjp, Sonjaaa, Antandrus, Kim54, Tomruen, Lumidek, Gscshoyru, WpZurp, TJSwoboda, Zondor, JimJast, Discospinster, Rich Farmbrough, H0riz0n, Pjacobi, Vsmith, Pluke, Autiger, Mal, Pavel Vozenilek, Floorsheim, El C, Lycurgus, Sourcecode, Oldsoul, PhilHibbs, Sietse Snel, Jpgordon, Atraxani, Smalljim, Slicky, LostLeviathan, Matpitka, Juesch, Danski14, Alansohn, Gary, DariuszT, ShardPhoenix, Kocio, Pion, Hdeasy, Bart133, Schaefer, BanyanTree, ClockworkSoul, Tycho, Suruena, Count Iblis, DV8 2XL, Gene Nygaard, Euphrosyne, Squidwina, Ott, Siafu, Roylee, Woohookitty, Mindmatrix, RHaworth, TigerShark, Savantnavas, MrDarcy, Mpatel, GregorB, Athletec64, Christopher Thomas, Ashmoo, BD2412, Drbogdan, Rjwilmsi, Kinu, Strait, Lordsatri, Dennis Estenson II, HappyCamper, LjL, Bubba73, The wub, Yamamoto Ichiro, JohnD-Buell, FayssalF, ColinJF, Wragge, Windchaser, Musical Linguist, Mindloss, RexNL, Gurch, Pete.Hurd, Lmatt, Diza, Zayani, Spencerk, Chobot, Sharkface217, DVdm, Hmonroe, Ptah, Ugha, Wavelength, Hillman, StuffOfInterest, Phantomsteve, John Smith's, Zigamorph, SpuriousQ, Jobe457, Stephenb, CambridgeBayWeather, Rsrikanth05, Vibritannia, Neilbeach, Salsb, Big Brother 1984, Anomalocaris, NawlinWiki, Joncolvin, ErkDemon, Trovatore, ETTan, Schrei, THB, Syrthiss, Wknight94, Richardcavell, FF2010, CWenger, Kevin, Caco de vidro, Katieh5584, Banus, Sbyrnes321, Btipling, SmackBot, R.E. Freak, Kurochka, DuoDeathscyther 02, Bayardo, McGeddon, Delldot, Kintetsubuffalo, Portillo, Rmosler2100, Bluebot, Jjalexand, 7777777s, Silly rabbit, George Church, Colonies Chris, A. B., Calc rulz, Nicknitro71, Zsinj, TallyJoe, John Hyams, Jamse, Scott3, Jefffire, Serenity-Fr, Bilgrau, Avb, Rrburke, Addshore, DrL, Mr.LMNOP, Rassisi, Spanyard, Byelf2007, Nishkid64, Giovanni33, Soap, Cronholm144, Loadmaster, Stupid Corn, Benjaminlobato, FredrickS, SirFozzie, Waggers, Alexander Gieg, Gcavep, Abel Cavaşi, Newone, Courcelles, Tubezone, Esn, Dave Runger, Valoem, JRSpriggs, Kurtan, 0-8, Duduong, Friendly Neighbour, CRGreathouse, Geremia, Tkoeppe, Ken Gallager, DepartedUser2, Cydebot, Vanished user 2340rujowierfj08234irjwfw4, Ninguém, Steel, Peterdjones, Hebrides, David edwards, Michael C Price, Raoul NK, Wortzman, Ulnevets, Konradek, Mojo Hand, Raymond Feilner, Headbomb, Marek69, Inve40, Twcjr, Duncan McB, KrakatoaKatie, Luna Santin, Gdo01, Byrgenwulf, Myanw, Knotwork, Len Raymond, JAnDbot, Barek, MER-C, Txomin, Matthew Fennell, Instinct, MoralMajority, Promking, Bongwarrior, VoABot II, JamesBWatson, JBKramer, DAGwyn, Theroadislong, Lenschulwitz, 28421u2232nfenfcenc, Peatbog, Allstarecho, Fang 23, Spellmaster, Philg88, Peter J Schoen, Denis tarasov, MartinBot, R'n'B, JCarlos, J.delanoy, Pharaoh of the Wizards, Maurice Carbonaro, LordAnubisBOT, Pyrospirit, AntiSpamBot, NewEnglandYankee, Cometstyles, WJBscribe, Foofighter20x, Econofire, Squids and Chips, Germanium, Reelrt, ChaosCon343, Danwills, RingtailedFox, Jeff G., TXiKiBoT, Nxavar, Rei-bot, Vishal144, Pouya sh, Corvus cornix, Michael H 34, Martin451, Cheffoxx, Betanon, BotKung, Everything counts, Popopp, MrMelonhead, Stephenmolesey, James McBride, Deanlsinclair, Pageman, Monty845, Logan, Kpa4941, PaddyLeahy, Dogah, SieBot, Tiddly Tom, Robdunst, Wing gundam, Gammanon, Likebox, Tiptoety, SteakNShake, Momo san, BartekChom, Monkeyspangler, Lightmouse, Anakin101, Divinestuff, Carbogen, Ayleuss, Soporaeternus, ArepoEn, ClueBot, LAX, Cliff, Ian the Aussie, Monomath1, Boing! said Zebedee, Heldbacktheband, LonelyBeacon, Neverquick, Excirial, WikiZorro, Tamaratrouts, Wndl42, Brews ohare, PhySusie, Morel, Mastertek, Mikaey, 7, Crowsnest, Thinking Stone, TimothyRias, PatDunphey, JKeck, XLinkBot, Bvssvni, Ougner, Truthnlove, YeAaMsLtA, Thatguyflint, Tayste, Balungifrancis, Addbot, Proofreader77, Some jerk on the Internet, Uruk2008, DOI bot, Couchie, Johnchang6868, Discrepancy, Mjamja, Bobtron5000, Fluffernutter, KaityJoe, MrOllie, Favonian, Barak Sh, F Notebook, Tide rolls, Scientryst, WikiDreamer Bot, Meisam, Blah28948, Yobot, Finiter, Ptbotgourou, Ezequiels.90, Jgmoxness, Amble, Mirandamir, RDemelo, AnomieBOT, ^musaz, Girl Scout cookie, 9258fahsflkh917fas, Theunify, Anxfisa, Kanat Abildinov, Materialscientist, Citation bot, Subhajit Ganguly, Fleaman5000, Amareto2, Addihockey10, Smk65536, Mlpearc, GrouchoBot, Rwmeo, Omnipaedista, Shirik, RibotBOT, Fa.alt3r3g0, Fsdjfsdfk, Chaseroads, FrescoBot, Paine Ellsworth, Ribashka, Steven Avraham Rosten, PhysicsExplorer, Ottokar, Tank hasmukh Khimjibhai, Tank theorist of everything, Hasmukh Khimjibhai Tank, DivineAlpha, Citation bot 1, Gil987, Three887, A8UDI, NarSakSasLee, Casimir9999, AndrewGrieder, Aknochel, IVAN3MAN, SchreyP, Noel Edward, Natwatchmaker, Weedwhacker128, Suffusion of Yellow, Koozedine, RjwilmsiBot, Specal ops, Afteread, DASHBot, Golumbo, EmausBot, Ikerus, Katherine, RA0808, K6ka, Zero939, Thecheesykid, Hhhippo, Traxs7, Arbnos, SporkBot, DanielBurnstein, FinalRapture, Aatu Koskensilta, Staszek Lem, Sridattadev, M00se1989, Wiggles007, Vedoder, Donner60, GIAN PHIL, Davidaedwards, WHF Christie, Terra Novus, Matevz91, Isocliff, Sanno89, Cgtdk,

Will Beback Auto, ClueBot NG, Stein Sivertsen, ClaudeDes, Lord God Almighty, Hindustanilanguage, Helpful Pixie Bot, Nightingale.zj, B21O303V3941W42371, Bibcode Bot, Wiki13, Neutral current, Aranea Mortem, Stimulieconomy, Steven.w.kowalski, MathewTownsend, Flyerbri, GroupT, Megajakeroo, La marts boys, Zofo, LightandDark2000, Josepht404, Nickhwee, Davidyevgeny, Kingcircle, Vith Nix, Illuusio, Davidyevgenyroven, QuantumNico, Vladimir Leonov, HesterShaw, Sol1, Phaedrx, Jwratner1, Jmassion, HeymynamesJon, Bigfootrobert, Elitousson, Mdsheraj, JaconaFrere, Somecdnguy4, Monkbot, LollyBear12, StacyPoyPie, Mujii loving, Mayojohns, Yoyosami, Hakan tomaşoğlu, Cirksena, Svm sudhan and Anonymous: 497

- **Technicolor (physics)** *Source:* http://en.wikipedia.org/wiki/Technicolor%20(physics)?oldid=633167974 *Contributors:* Maury Markowitz, Michael Hardy, IMSoP, Timwi, Grendelkhan, Phys, Xerxes314, CryptoDerk, Pjacobi, David Schaich, Dbachmann, Bender235, RJHall, Jag123, Boredzo, Guy Harris, Rjwilmsi, Brendan Moody, Chobot, Bgwhite, Conscious, SCZenz, Pyrotec, SmackBot, Melchoir, Chris the speller, Njerseyguy, Vichka, Valoem, JRSpriggs, Michael C Price, Headbomb, OrenBochman, Maliz, Laager, Lseixas, Miztli, BotKung, Snideology, ClueBot, Mild Bill Hiccup, Wikisannino, WikiMSM, Addbot, WikiMSSM, Luckas-bot, Wireader, AnomieBOT, Rjanag, Citation bot, Paine Ellsworth, Citation bot 2, Citation bot 1, RjwilmsiBot, DacodaNelson, Arbnos, Suslindisambiguator, Bibcode Bot, ChrisGualtieri, Жаворонок, Drscientific, Doctor Dashiki and Anonymous: 22

- **Kaluza–Klein theory** *Source:* http://en.wikipedia.org/wiki/Kaluza%E2%80%93Klein%20theory?oldid=632716746 *Contributors:* Sodium, The Anome, XJaM, Roadrunner, Rlee0001, Stevertigo, JohnOwens, Michael Hardy, Looxix, Ahoerstemeier, Susurrus, Smack, Charles Matthews, Timwi, Reddi, Wik, Phys, Carbuncle, Ancheta Wis, Kim Bruning, Fropuff, Mdob, Iantresman, Rauyran, Rich Farmbrough, Roo72, Ponder, Paul August, Bender235, Szquirrel, John Vandenberg, QTxVi4bEMRbrNqOorWBV, Geschichte, RJFJR, Notjim, Linas, Lgallindo, Trapolator, Mpatel, Joke137, MarSch, YurikBot, Rt66lt, Hillman, Geologician, Buster79, Gillis, Kewp, Petri Krohn, Pred, Caco de vidro, KasugaHuang, Jodarom, Kurochka, Nihonjoe, Zazaban, Unyoyega, Hmains, Chris the speller, MalafayaBot, Colonies Chris, QFT, Legaleagle86, Jon Awbrey, Beetstra, DabMachine, Rschwieb, FrEd 00, Tawkerbot2, CmdrObot, Mattbr, Wfdavis, Moyerjax, Epbr123, Mojo Hand, Headbomb, JustAGal, Isilanes, Golf Bravo, Magioladitis, JoseAntonioOrtegaRuiz, Jpod2, Mbc362, Maliz, Cpiral, TomyDuby, Quantling, Lseixas, Sheliak, Cuzkatzimhut, Red Act, Impunv, AlleborgoBot, ArdClose, EoGuy, Masterpiece2000, Canis Lupus, EverettYou, AnonyScientist, Albambot, Addbot, Gravitophoton, Lightbot, Yobot, Turul2, Jo3sampl, Citation bot, Omnipaedista, RibotBOT, Paine Ellsworth, Quiden711, RockSolidCosmo, Crabhiggins, Bj norge, David.c.stone, Arbnos, Quondum, TonyMath, Helpful Pixie Bot, Bibcode Bot, Zerothat, Ownedroad9, Metsfreak2121, Mogism, Lianatajo, MuonRay, Orderofmagnitudeapproximation, Frinthruit, Monkbot, ManitouLance and Anonymous: 80

- **String theory** *Source:* http://en.wikipedia.org/wiki/String%20theory?oldid=630688453 *Contributors:* AxelBoldt, Sodium, Mav, Bryan Derksen, Zundark, The Anome, Tarquin, Taw, Eean, Malcolm Farmer, Hephaestos, Olivier, Drseudo, Stevertigo, Spiff, Edward, PhilipMW, Michael Hardy, Bewildebeast, Dante Alighieri, Gabbe, Graue, Tgeorgescu, Mcarling, CesarB, Looxix, Ahoerstemeier, Theresa knott, Suisui, Angela, Den fjättrade ankan, Jdforrester, Julesd, Salsa Shark, Schneelocke, Charles Matthews, Timwi, Bemoeial, Jitse Niesen, 4lex, Greenrd, ErikStewart, Furrykef, Saltine, Phys, Omegatron, Bevo, Trent, Nufy8, Robbot, Craig Stuntz, Fredrik, Chris 73, R3m0t, COGDEN, Mirv, Wjhonson, Sverdrup, Academic Challenger, DHN, Hadal, Khlo, ElBenevolente, HaeB, Tobias Bergemann, Giftlite, DocWatson42, Christopher Parham, Awolf002, Mporter, Amorim Parga, Mikez, Harp, Kim Bruning, Tom harrison, Ferkelparade, Leflyman, Fropuff, No Guru, Anville, Moyogo, Curps, Pashute, Nomad, Mboverload, Solipsist, SWAdair, DemonThing, Wmahan, Btphelps, MSTCrow, Decoy, Chowbok, Gadfium, Steuard, Pgan002, Quadell, Carandol, Antandrus, Beland, JoJan, Khaosworks, Tothebarricades.tk, Thincat, Tomruen, Shidobu, Icairns, Lumidek, NoPetrol, Avihu, Fanghong, Trevor MacInnis, Lacrimosus, Zro, D6, Urvabara, Felix Wan, Jkl, Discospinster, ElTyrant, Rich Farmbrough, Rhobite, Pjacobi, Alien life form, Vapour, Silence, Kzzl, LindsayH, Mani1, Pavel Vozenilek, Paul August, Bender235, Kjoonlee, Mashford, Kelvinc, Perlman10s, Panu, Brian0918, Dpotter, Livajo, El C, Laurascudder, Shanes, Zegoma beach, RoyBoy, Causa sui, Bobo192, Directorstratton, Janna Isabot, John Vandenberg, Flxmghvgvk, QTxVi4bEMRbrNqOorWBV, Physicistjedi, Bongoo, 4v4l0n42, Merope, Geschichte, Linuxlad, Phils, Merenta, Alansohn, Gary, JYolkowski, Enirac Sum, Ryanmcdaniel, Arthena, Borisblue, Rd232, Plumbago, Axl, R Calvete, Lightdarkness, Kocio, Bart133, Wtmitchell, Isaac, Tycho, Cal 1234, Fadereu, CloudNine, Sciurinæ, Computerjoe, Kusma, DV8 2XL, Pwqn, Gene Nygaard, Ringbang, Ceyockey, Falcorian, Bobrayner, Joriki, Mel Etitis, Linas, BillC, Jacobolus, HFarmer, Before My Ken, Netdragon, MONGO, GeorgeOrr, Mpatel, Bbatsell, GregorB, מוטיבציה, Joke137, Christopher Thomas, Dysepsion, GSlicer, Jan.bannister, Graham87, Magister Mathematicae, Hillbrand, BD2412, Elvey, Galwhaa, Raymond Hill, JIP, RxS, Athelwulf, Edison, Sjakkalle, Rjwilmsi, Xgamer4, Jake Wartenberg, Arabani, MarSch, TheRingess, Jmcc150, Aero66, Crazynas, Juan Marquez, R.e.b., DoubleBlue, Zelos, AlisonW, Asafavi, Lionelbrits, Conorific, Zunz, Mathbot, Crazycomputers, RexNL, Gurch, Algri, TeaDrinker, Zifnabxar, XAXISx, Erik4, Phoenix2, Antimatter15, Ggb667, Chobot, Visor, DVdm, Mhking, VolatileChemical, Bgwhite, Algebraist, Ben Tibbetts, YurikBot, Ugha, Wavelength, Borgx, NuclearFusion, Angus Lepper, Hairy Dude, Jimp, Hillman, Cyferx, Wolfmankurd, Pip2andahalf, RussBot, Moronoman, Crazytales, Pippo2001, Bhny, Pigman, SpuriousQ, Branman515, Stephenb, Gaius Cornelius, Eleassar, Bovineone, Cheesus, Shanel, NawlinWiki, Tong, Mike18xx, SCZenz, Cleared as filed, Bdiah, Pym98, SColombo, Haemo, FF2010, Closedmouth, Reyk, Brina700, Chris Brennan, Vicarious, Brianlucas, Geoffrey.landis, Hitchhiker89, Spliffy, Pred, ArielGold, Roy Fultun, Ilmari Karonen, Katieh5584, Pentasyllabic, Lunch, DVD R W, WikiFew, That Guy, From That Show!, Street Scholar, AndrewWTaylor, QSquared, Sardanaphalus, Vanka5, MacsBug, Hvitlys, SmackBot, Kurochka, Zazaban, Tom Lougheed, Prodego, KnowledgeOfSelf, Hydrogen Iodide, Melchoir, Vald, Skrewtape, Atomota, Canthusus, GaeusOctavius, Cool3, Andyvn22, Skizzik, RobertM525, Dauto, Bluebot, SSJ 5, Keegan, Aidan Croft, Thumperward, Oli Filth, Silly rabbit, Timneu22, SchfiftyThree, Moshe Constantine Hassan Al-Silverburg, Complexica, Rediahs, RayAYang, Aero77, Adamstevenson, Ikiroid, Epastore, Baronnet, Ned Scott, Sbharris, Colonies Chris, Konstable, Sct72, Scwlong, Can't sleep, clown will eat me, Timothy Clemans, Onorem, Neilanderson, EvelinaB, TKD, KerathFreeman, Addshore, UU, The tooth, Pepsidrinka, Somebody2292, --=The Doctor=--, Fughettaboutit, Cybercobra, Irish Souffle, Nakon, Jdlambert, James McNally, MichaelBillington, Lostart, Insineratehymn, Drphilharmonic, SpiderJon, DMacks, Ihatetoregister, Where, Michael IFA, Yevgeny Kats, Vasiliy Faronov, Byelf2007, Angela26, Visium, Rory096, Zymurgy, Harryboyles, Mdl53711, T-dot, Titus III, Ergative rlt, MagnaMopus, UberCryxic, Linnell, Mgiganteus1, Nonsuch, IronGargoyle, Ckatz, DoItAgain, AstroGod, Kirbytime, Jimbo Mahoney, FredrickS, Invisifan, Ryulong, Ryanjunk, MathStuf, Mike Doughney, Norm mit, Hindol, Dan Gluck, Huntscorpio, Iridescent, K, Sunoco, You? Me? Us?, CzarB, Rabinzkaman, JoeBot, Lottamiata, Tony Fox, Vrkaul, Torrazzo, Gil Gamesh, Areldyb, Courcelles, Tawkerbot2, Gebrah, Shamvil, DKqwerty, Lbr123, Harold f, Heqs, Devourer09, Duduong, Sarvagnya, Dewayne76, JForget, Cg-realms, InvisibleK, CRGreathouse, CmdrObot, Earthlyreason, Van helsing, Olaf Davis, CBM, Rawling, Jibal, Witten Is God, Nunquam Dormio, Giko, KnightLago, Thubsch, Leujohn, SlashDot, TheTito, Karenjc, Myasuda, Emarv, Cydebot, Gmusser, Gogo Dodo, Kahananite, Quajafrie, Michael C Price, Dougweller, DumbBOT, Narayanese, AlphaNumeric, SRoughsedge, Jguard18, Woland37, Zalgo, Daniel Olsen, UberScienceNerd, Bkazaz, DJBullfish, Thijs!bot, Epbr123, Rwmnau, Babemachine, Pimpin101, Mbell, O, Faigl.ladislav, Ucanlookitup, Andyjsmith, Headbomb, Tcturner2002, Marek69, Brahmajnani, Arthurcprado, Y.t., D3gtrd, Babemonkey, Dark dude, Duncan McB, EdJohnston, MichaelMaggs, Ancientanubis, Natalie Erin, Hempfel, Jomoal99, Mmortal03, Mentifisto, Geekdom04, AntiVandalBot, Luna Santin, Seaphoto, Ed270791, Opelio, Doc Tropics, David136a, NithinBekal, Dotdotdotdash, Helicoptr, Poszombie, MontanNito, Dylan Lake, Maximilian77, Shlomi Hillel, Db63376, SamIAmNot, Knotwork, Res2216firestar, Superior IQ Genius, MER-C,

Andonic, Sitethief, 100110100, TallulahBelle, Nestamachine, Savant13, Daynightrader, Goldenglove, Charibdis, Acroterion, Ophion, Aigisthos, Editmyhandman, Aruben537, Magioladitis, WolfmanSF, Bongwarrior, VoABot II, Yandman, JamesBWatson, باسم, Qutt, Jespinos, Kevinmon, Aka042, Froid, DAGwyn, Catgut, Panser Born, Ensign beedrill, Perspectival, JJ Harrison, Dirac66, Justanother, Aziz1005, Cpl Syx, ChazBeckett, Teardrop onthefire, WLU, Stephen shenker, Robin S, SkepticVK, Joshua Davis, Mkroh, B9 hummingbird hovering, S3000, Hdt83, MartinBot, FlieGerFaUstMe262, Ytomem, Shimwell, Arjun01, KrishSundaresan, Anaxial, Jay Litman, Alexcalamaro, Andrej.westermann, Smokizzy, LedgendGamer, Cyrus Andiron, Peteryoung144, Tgeairn, Artaxiad, HEL, AlphaEta, J.delanoy, AstroHurricane001, Maurice Carbonaro, Yonidebot, Morris729, M C Y 1008, 69gangsta420, It Is Me Here, Shawn in Montreal, Janus Shadowsong, Bailo26, Fredsie, Madagaskar07, Duchesserin, AntiSpamBot, CHIAGEHYANG, Watsup1313, Belovedfreak, HaloInverse, NewEnglandYankee, Scott1329m, Thesis4Eva, Policron, Jrcla2, WJBscribe, Rnricklefs, Jamesofur, Eyelidlessness, Jonnyk aus, Kvdveer, JavierMC, Izno, Xiahou, CardinalDan, Sheliak, HamatoKameko, Malik Shabazz, Concertmusic, JohnBlackburne, JustinHagstrom, Fences and windows, Wooba doob, Philip Trueman, DoorsAjar, HowardFrampton, TXiKiBoT, Zidonuke, Red Act, Kriak, Calwiki, Technopat, Hqb, Andrius.v, Anonymous Dissident, Crohnie, AlysTarr, Qxz, Vanished user ikijeirw34iuaeolaseriffic, Impunv, Seraphim, Martin451, Don4of4, ABigGreenHippo, Huperphuff, LeaveSleaves, Kaenneth, StringyGuy, Maxim, Erth64net, Meters, Rickstauduhar, Enviroboy, Turgan, Anna512, PhysPhD, Northfox, NPguy, Matthew Sanders, Luke Walkerson, Newbyguesses, MissMJ, SieBot, Escher26, J.A.Ireland, BA (IHPST), 4wajzkd02, Robdunst, Dreamafter, Pallab1234, Dbelange, MTHarden, Lemonflash, Kylemew, Yintan, GlassCobra, Wpegden, Likebox, Flyer22, Exert, ProGeek314, Arbor to SJ, Babawhitemoose, Caidh, Dhatfield, Audree, Oxymoron83, Pretty Green, Weaselstomp, Manway, Alex.muller, Taco Manipulator, Tschach, Manheat84, Anchor Link Bot, Mikebernstein, ImperialismGo, Nergaal, Ionfield, Ayleuss, Sh4wz0r, Naturespace, Martarius, Phyte, ClueBot, The Thing That Should Not Be, String4d, Illusion96, Polyamorph, Mpd1989, Alexdeburca18, Wiggl3sLimited, Excirial, Kjramesh, Jusdafax, Resoru, WikiZorro, Eeekster, Verum, Tamaratrouts, Gtstricky, Humanino, Brews ohare, NuclearWarfare, Cenarium, Razorflame, Scoobey, BOTarate, Sideswiper, Thingg, Capudo, BVBede, Versus22, Introductory adverb clause, MelonBot, SoxBot III, Egmontaz, Notpayingthepsychiatrist, DumZiBoT, BahTab, TimothyRias, Aj00200, Reaperfromhell, Dunkaroo207, XLinkBot, AlexGWU, Impshum, Saeed.Veradi, Little Mountain 5, Guy392, David424, Truthnlove, Qweeveen, Tayste, addbot, Steven66s, Denali134, Elemented9, Varrey280303, Eric Drexler, Some jerk on the Internet, Fizzycyst, Uruk2008, DOI bot, Jojhutton, AngryBacon, Captain-tucker, Auspex1729, Kongr43gpen, Fgnievinski, Rhetoric Of A Sophist, Ronhjones, CanadianLinuxUser, Cst17, Download, Glane23, Bassbonerocks, Chzz, Favonian, Kronix35, LinkFA-Bot, Udugunit, Aktsu, Tassedethe, Numbo3-bot, Anpecota, Tide rolls, HerpesVirus, SDJ, OlEnglish, Scourge of God, Davidmedlar, Couldbenoway66, Yobot, Maxdamantus, Terrisknickers, Kartano, TaBOT-zerem, Julia W, Unique and proud of it, FireMouseHQ, Terrifictriffid, ArchonMagnus, CinchBug, Synchronism, AnomieBOT, Cleeseheb, 1exec1, Charlesvi, Bigdaddy4x4, Gitman4, Jim1138, IRP, Mintrick, Drweetmola, Ornamentalone, M00npirate, Gautam10, Csigabi, Poli-Psy, Materialscientist, 90 Auto, Citation bot, Teleprinter Sleuth, Vuerqex, Twri, Frankenpuppy, Fuzzy Bob Saget, DirlBot, Georgepowell2008, Heidisql, Cureden, Ekwos, Capricorn42, Gensanders, NFD9001, Anna Frodesiak, Tomwsulcer, A23649, Pra1998, Coretheapple, Ruy Pugliesi, Jagbag2, Vandalism destroyer, Ab1, Omnipaedista, Bandit5005, Shirik, RibotBOT, Waleswatcher, Saalstin, Amaury, Aaron35510, Caz34, Doulos Christos, Sewblon, Born Gay, Capricorn24, SchnitzelMannGreek, A. di M., SpacePyjamas, Kierkkadon, A.amitkumar, Dougofborg, StringLove, Nobelprizewinner, Astiburg, FrescoBot, Fortdj33, Paine Ellsworth, Goodbye Galaxy, HJ Mitchell, Steve Quinn, Vhann, Kwiki, Xhaoz, Citation bot 1, Batong, Gil987, Pinethicket, I dream of horses, Tallboyhoops1991, Three887, Steveo27five, RedBot, Sardinita, Vhsatheeshkumar, Swisstingle, DeletionUK, Reconsider the static, IVAN3MAN, Remingtonhill1, Orenburg1, Coltonhs, Willy Weazley, Smamaret, Bethovenn, Dinamik-bot, Dc987, Oswaldo Zapata, Egemont, Syebo, Alaithiran, Reaper Eternal, Seahorseruler, Ybungalobill, Quaker phil, Specs112, Dr. Aakash Patel, Tbhotch, StormbringerUK, Minimac, Mathgenius3141592, Keegscee, Omgwaffels, Mick le pick, Solancel, Aznhero3793, Dwielark, Afteread, Enauspeaker, EmausBot, MaoooaM, Immunize, Az29, Milkocookie, Fotoni, RA0808, RenamedUser01302013, 8digits, Yukieaside, Slightsmile, Tommy2010, Winner 42, Wikipelli, JonezyKiDx, Joe Gazz84, ZéroBot, Timeitsways, John Cline, Cogiati, Quaqa, Chrispaps2413, Nasulikid, Vollrath2323, Benjamin1414141414141414, Arbnos, Green Lane, A930913, Azeraphale, H3llBot, Encyclopadia, Danga1988, Ollainen, PoisonGM, Wayne Slam, OnePt618, Knome335, L Kensington, Lulzprotuns, Kranix, Rpcappello, Vastly, Donner60, CatFiggy, CountMacula, Orange Suede Sofa, Etov, M1k3 101, Bill william compton, Wakabaloola, TERBAFAN, Nickslspride34, NeuralLotus, Isocliff, Brechbill123, Xanchester, ClueBot NG, Martti Muukkonen, KagakuKyouju, Jeff Song, This lousy T-shirt, Satellizer, Name Omitted, Marcdean123, Wiki incorp, O.Koslowski, Alexdamaino9, Dream of Nyx, Blackhall616, Widr, Sashhere, WikiPuppies, Stu181, T00g00d96, Pluma, Storm.sarup, Helpful Pixie Bot, Manzeet, Waffleboy36, HMSSolent, Mikeshelton1, Bibcode Bot, 2001:db8, Phillip.phillipson, Hoaxinator, Lowercase sigmabot, Thor cherubim, Mrshabam, Nishch, Flowerhat15, AvocatoBot, Housegeek224, MahRanch, Benzband, Altaïr, Benhenchdickthomas, Shreyakstring, Sweaty maori sphincter, DaFalk, Dsabo74, Ratanmaitra, MM4EVAH, Steven.w.kowalski, Minsbot, JGallardo2600, Dylanlatham, Myfriendganesha, OCCullens, Likeaboss189, Sean271293, LinusE8, BattyBot, Several Pending, Aldrich2122, CommanderMoka, The Illusive Man, ChrisGualtieri, KoalamaN2, Trevorkid45, Catsloveit07, Alex Modzz, Rustyjamsen, Goh ryangoh, Dexbot, Exolius, Hilander316, Alman1234321, SuperCalzer, LightandDark2000, MeekMelange, BQND, Cdarra1, TheMonkeyboy524, Michael Anon, Mattfat8, Lugia2453, Anruy, Rachel weld, Jamesx12345, AHusain314, BossEditors, Hillbillyholiday, Mattninja, Theshadow444, Asaa82, Jakemarz197, Kzhang1025, Epicgenius, Spongbob456789, ☺, TestMaster, Ianreisterariola, GrapperJ, Makeitnasty, Moemajdi, I am One of Many, NualaIvy, BAZINGASS, St3fanPC, Eyesnore, Isaac grozd, Jordanissexyaf1999, Baruch6525, Mosbruckercj, Ihatedirac2k13, Jonamithy121314, 123physicsquantum, Jt198, HeyJude70, AParker628, DimReg, A.k.blaze1, Joshuk, Zenibus, Nianoobasik, Ihelpapplen, Gamo To Apoel, SacredLabyrinth, Ginsuloft, Vampre1122, Dimension10, Howard Wolowitz, AddWittyNameHere, Polytope24, Tutun12$, Longerboats5, SimonWombat8, Konveyor Belt, Vtank54, Micheal545, Hck24, Hexafish, Simpick, TheRealTheKoi, Bballbro62, Monkbot, ArmyPath, TheQ Editor, Jtsmith098, Joshmiller1, Hanseer360, XXvPIEvXx, Dbennett 24, Ghikpenos, Nick65633, Saundra03, Thehippothatknows, Sewwgers, Teelaskeletor, Cirksena, Balockaye1234, PloppyDoo, Yesufu29, Lumpy2k14, Podayeruma, Abstract92, Sbenfiel, Monkman2k4, Swegwegdgfyetkfoffkkfkfkv, John95541234, Poopman224 and Anonymous: 1524

- **Superfluid vacuum theory** *Source:* http://en.wikipedia.org/wiki/Superfluid%20vacuum%20theory?oldid=632213894 *Contributors:* Rich Farmbrough, Cedders, Mu301, DeadlyAssassin, Serendipodous, Tom Morris, Colonies Chris, Thijs!bot, R'n'B, Uncle Milty, Schreiber-Bike, TimothyRias, Addbot, Yobot, Edstamos, AnomieBOT, FrescoBot, Paine Ellsworth, Tom.Reding, RjwilmsiBot, Arbnos, Baseball Watcher, Bibcode Bot, Machoota, Brainssturm, Uioplk, Account12098, Intogain891, Andyhowlett, Rolf h nelson, Gravytacky67 and Anonymous: 13

- **Supersymmetry** *Source:* http://en.wikipedia.org/wiki/Supersymmetry?oldid=634585777 *Contributors:* Bryan Derksen, Taw, Andre Engels, Roadrunner, Maury Markowitz, Ewen, Stevertigo, Edward, Michael Hardy, Arpingstone, Theresa knott, IMSoP, Jeandré du Toit, Samw, Smack, Charles Matthews, Maximus Rex, Phys, Raul654, BenRG, Rursus, Mor, Ancheta Wis, Giftlite, Mporter, Ferkelparade, Monedula, Fropuff, Xerxes314, Anville, Gus Polly, Moyogo, Unconcerned, DO'Neil, Maarten van Vliet, Pharotic, LiDaobing, Sam Hocevar, Lumidek, Deglr6328, Arivero, Rich Farmbrough, Roybb95, Bender235, El C, Nornagon, Duk, Tweet Tweet, LostLeviathan, Pearle, Gary, Francescog, Wtmitchell, RJFJR, Reaverdrop, Blaxthos, Killing Vector, Jordan14, Ted BJ, MONGO, Mpatel, MFH, SeventyThree, Bodera, VermillionBird, Rjwilmsi, Josiah Rowe, R.e.b., Bubba73, Maxim Razin, Drrngrvy, FlaBot, Cless Alvein,

Nowhither, Itinerant1, Gparker, KFP, Lmatt, Chobot, Vyroglyph, YurikBot, Wavelength, RussBot, Ohwilleke, Bhny, Epolk, Maxim Leyenson, Chaos, Romanc19s, Bota47, Mgnbar, Closedmouth, Arthur Rubin, RG2, That Guy, From That Show!, A bit iffy, SmackBot, Mira, Kurochka, Wangjiaji, Gilliam, Bluebot, Cadmasteradam, Complexica, Bazonka, Colonies Chris, Can't sleep, clown will eat me, QFT, Ruff ilb, Robma, Solarapex, Radagast83, Jgwacker, TheMaster42, Ligulembot, Acjohnson55, Yevgeny Kats, Charleswestbrook, TriTertButoxy, Lambiam, Tktktk, Xiaphias, JarahE, Mdanziger, Dan Gluck, Marysunshine, Tawkerbot2, Cydebot, David edwards, Michael C Price, Crum375, Koeplinger, Headbomb, J.christianson, Escarbot, Salgueiro, Kborland, Jpod2, Cgingold, Maliz, TimidGuy, C9, Kostisl, R'n'B, Zentropa77, Natsirtguy, Maurice Carbonaro, Shawn in Montreal, Idioma-bot, Sheliak, Cuzkatzimhut, Nxavar, Kawakameha, Cuboidal, Ptrslv72, PhysPhD, SieBot, Nn123645, ClueBot, Jcpilman, Chessmaster7m, Rhododendrites, Mastertek, Mishas42, Scrabby, TimothyRias, WikHead, MystBot, Addbot, DOI bot, Zahd, Barak Sh, F Notebook, Lightbot, Luckasbot, Yobot, Ibayn, TaBOT-zerem, Amirobot, Nonnormalizable, AnomieBOT, Girl Scout cookie, Citation bot, ArthurBot, Plumpurple, Tomwsulcer, Omnipaedista, Gsard, CES1596, FrescoBot, HaloStereo1, Paine Ellsworth, Citation bot 1, Gil987, Kikeku, Jonesey95, Eddie Nixon, MondalorBot, Aknochel, Gagoga ju, TobeBot, Puzl bustr, EmausBot, Djloststylez, Ddimensões, Arbnos, Susy is it, ChuispastonBot, Isocliff, ClueBot NG, KagakuKyouju, IJVin, Frietjes, Helpful Pixie Bot, Bibcode Bot, BG19bot, Teika kazura, JayBeeEye, Ninmacer20, ChrisGualtieri, Logosun, AHusain314, NA48, Katherine Pendleton, Lioinnisfree, Liquidityinsta, TaiSakuma, Kdmeaney, Qxxxxxq, Almaionescu, Monkbot, Janhaithabu, Mammoth2011, Stacie Croquet, Cuttlas1 and Anonymous: 160

- **Quantum gravity** Source: http://en.wikipedia.org/wiki/Quantum%20gravity?oldid=633515209 Contributors: AstroNomer, Matusz, Miguel, Roadrunner, Stevertigo, Ubiquity, Bobby D. Bryant, Mcarling, NuclearWinner, Anders Feder, Susurrus, Coren, Charles Matthews, Timwi, Reddi, Tpbradbury, Phys, Bevo, Raul654, BenRG, Frazzydee, Jeffq, Sdedeo, Rholton, Wereon, Ilya (usurped), Seth Ilys, Ancheta Wis, Giftlite, Herbee, Fropuff, Endlessnameless, Malyctenar, Jason Quinn, Finn-Zoltan, YapaTi, Lumidek, Marcus2, Joyous!, TJSwoboda, Vitaleyes, Davidclifford, JimJast, Guanabot, FT2, Masudr, Pjacobi, Pie4all88, David Schaich, Bender235, Clement Cherlin, El C, PhilHibbs, Army1987, Apyule, VBGFscJUn3, PWilkinson, Daniel Arteaga, Keenan Pepper, Cjthellama, DonJStevens, Velella, Dabbler, Tycho, Cal 1234, RJFJR, Count Iblis, ThomasWinwood, Anarchimede, Scarykitty, Woohookitty, Igny, ToddFincannon, Mpatel, GregorB, Joke137, Christopher Thomas, Marudubshinki, Graham87, Yurik, Kroggz, Rjwilmsi, Eoghanacht, Jrasowsky, JHMM13, Smithfarm, Ems57fcva, FayssalF, Itinerant1, Lmatt, Chobot, Hmonroe, YurikBot, Hillman, ErkDemon, JocK, SCZenz, Roy Brumback, Bota47, Zunaid, JonathanD, 2over0, Arthur Rubin, Modify, LeonardoRob0t, Caco de vidro, RG2, KasugaHuang, Resolute, SmackBot, Samdutton, Vald, Eskimbot, Hbackman, Onebravemonkey, Chris the speller, Ben.c.roberts, Cthuljew, Silly rabbit, Complexica, Colonies Chris, QFT, Soosed, Theanphibian, Shushruth, Ck lostsword, Yevgeny Kats, DJIndica, Lambiam, Vampus, Vincenzo.romano, Jaganath, JorisvS, RoboDick, IronGargoyle, Dicklyon, SirFozzie, Treyp, Twunchy, Piccor, Kurtan, Harold f, CalebNoble, Duduong, Paulmlieberman, TVC 15, UncleBubba, TAz69x, Sam Staton, ST47, B, Patrick O'Leary, Epbr123, Koeplinger, Klasovsky, Markus Pössel, Keraunos, Headbomb, Marek69, MichaelMaggs, Tim Shuba, MER-C, ParadiZio, Perlygatekeeper, VoABot II, Alvatros, Bdalevin, SHCarter, Jpod2, DAGwyn, Nucleophilic, LorenzoB, Rickard Vogelberg, DancingPenguin, Rettetast, Victor Blacus, AstroHurricane001, Yonidebot, Acalamari, Mstuomel, Fullmetal2887, NewEnglandYankee, DorganBot, CardinalDan, Idioma-bot, Sheliak, VolkovBot, Pleasantville, Seattle Skier, AlnoktaBOT, TXiKiBoT, Dllahr, Rdekleer, Saibod, Cyberchip, Wikiwikimoore, Carlorovelli, StevenJohnston, SieBot, LeadSongDog, ReluctantPhilosopher, StaticG, GarbagEcol, ClueBot, The Thing That Should Not Be, EoGuy, Polyamorph, Andwor9, Notburnt, Tms9, Alexbot, Resoru, Eeekster, Tamaratrouts, Brews ohare, SchreiberBike, Askahrc, BOTarate, Lambtron, DumZiBoT, XLinkBot, Rror, Facts707, SilvonenBot, Theonlydavewilliams, Mhsb, Truthnlove, Ttimespan, Trifonov, Addbot, Mortense, Grayfell, Eric Drexler, Gravitophoton, DOI bot, AkhtaBot, CanadianLinuxUser, Frosty726, LaaknorBot, Delaszk, Tassedethe, Tide rolls, Taketa, Titan1129, Avono, Luckas-bot, Yobot, WikiDan61, Pigetrational, Wireader, Allowgolf, Wiki Roxor, Jim1138, IRP, Sz-iwbot, Quantity, Materialscientist, Citation bot, ArthurBot, LilHelpa, Amareto2, Ekwos, KrisBogdanov, Rolfguthmann, StealthCopyEditor, Dan6hell66, Rabsmith, Hep thinker, Paine Ellsworth, DrArthurRubinPHD, Lagelspeil, Nunc aut numquam, Vacuunaut, Van Speijk, Knowandgive, Craig Pemberton, Udifuchs, Citation bot 2, Citation bot 1, Citation bot 4, Hirvenkürpa, Pmokeefe, Casimir9999, Dac04, Dude1818, Valeriy Pischenko, Follyland, TrueTeargem, N0814444, Earthandmoon, Korepin, DARTH SIDIOUS 2, Musictime4me, RjwilmsiBot, EmausBot, Francophile124, Octaazacubane, Fotoni, Slightsmile, Garfield Salazar, Hhhippo, JSquish, John Cline, Fæ, Brazmyth, Throwmeaway, Arbnos, Ebrambot, Kusername, DanielBurnstein, TonyMath, L Kensington, Maschen, Donner60, Parusaro, Apratim07, Terra Novus, Isocliff, Googledin!, ClueBot NG, SpikeTorontoRCP, Science writer, Preon, Raidr, Jhmmok, 336, Widr, Helpful Pixie Bot, Bibcode Bot, Bardsley Rides a Segway, Apelikedawg, FiveColourMap, Mr.viktor.stepanov, Brainssturm, BattyBot, Jimw338, Ryanr666, Kryomaxim, Garuda0001, Saehry, Sanathdevalapurkar, GabeIglesia, Sanathlab, Roiwallace, Spencer.mccormick, Spencerfjase, MrShlongNo1, Marc D. Garrett, D00d00ballz, Gigantmozg, Polytope24, Frinthruit, Anrnusna, Dfyytj, Monkbot, Umut Alihan Dikel, Amortias, Klj1234, Pfpguy and Anonymous: 279

- **Simple group** Source: http://en.wikipedia.org/wiki/Simple%20group?oldid=633643782 Contributors: AxelBoldt, Zundark, Patrick, Michael Hardy, TakuyaMurata, Silverfish, Schneelocke, Jitse Niesen, Tobias Bergemann, Giftlite, Lethe, Beland, Pyrop, Gadykozma, ArnoldReinhold, Vipul, Axeman89, Galaxiaad, Oleg Alexandrov, Linas, Magister Mathematicae, Rjwilmsi, HappyCamper, R.e.b., Chobot, Roboto de Ajvol, YurikBot, DYLAN LENNON, Cullinane, Pred, Eskimbot, MalafayaBot, Nbarth, Hgrosser, Weregerbil, Jim.belk, Gandalfxviv, Schildt.a, Noleander, Bruno321, Cydebot, Mato, Thijs!bot, Headbomb, Lifthrasir1, .anacondabot, Albmont, Jakob.scholbach, JoergenB, BigrTex, Maproom, TXiKiBoT, Random Hippopotamus, JackSchmidt, Razimantv, He7d3r, Xodarap00, DumZiBoT, XLinkBot, Addbot, Mrchapel0203, Legobot, Luckas-bot, Yobot, Kilom691, JackieBot, Citation bot, Citation bot 1, Trappist the monk, WikitanvirBot, Chricho, Helpful Pixie Bot, Pariefracture, CsDix and Anonymous: 17

- **Georgi–Glashow model** Source: http://en.wikipedia.org/wiki/Georgi%E2%80%93Glashow%20model?oldid=630420125 Contributors: Roadrunner, Michael Hardy, TakuyaMurata, Phys, Fropuff, Alison, JeffBobFrank, David Schaich, MBisanz, Keenan Pepper, Mpatel, BradBeattie, Bgwhite, Fram, RodVance, Sbyrnes321, SmackBot, Colonies Chris, Dan Gluck, Peterdjones, Michael C Price, Headbomb, Wasell, Jpod2, Faizhaider, Businessman332211, Cyanolinguophile, Legoktm, ArdClose, Addbot, Yobot, AnomieBOT, 777sms, EmausBot, AvicBot, Frietjes, BattyBot, Mogism, Cjean42, AHusain314 and Anonymous: 6

- **Simple Lie group** Source: http://en.wikipedia.org/wiki/Simple%20Lie%20group?oldid=612063018 Contributors: Zundark, Nonenmac, Michael Hardy, TakuyaMurata, Charles Matthews, Phys, Giftlite, Fropuff, Cambyses, Sigfpe, Tomruen, MIT Trekkie, Oleg Alexandrov, GregorB, Rjwilmsi, Salix alba, R.e.b., Buster79, Kier07, SmackBot, Bluebot, Nbarth, Jim.belk, CRGreathouse, Myasuda, Secular mind, Thijs!bot, Escarbot, Ludvikus, R'n'B, TomyDuby, Gill110951, Red Act, Michael H 34, Arcfrk, Shadrack-dva, Addbot, Discrepancy, Niout, AnomieBOT, Erik9bot, GoingBatty, Helpful Pixie Bot, CsDix and Anonymous: 15

- **Group representation** Source: http://en.wikipedia.org/wiki/Group%20representation?oldid=622431952 Contributors: AxelBoldt, Zundark, Youandme, Chas zzz brown, Michael Hardy, GTBacchus, Looxix, Stevenj, Loren Rosen, Revolver, Charles Matthews, Dysprosia, Michael Larsen, Phys, Aleph4, Robbot, Josh Cherry, Huppybanny, Rvollmert, Mattblack82, Mohan ravichandran, MathMartin, Weialawaga, Giftlite, Fropuff, Zteitler, DefLog, Almit39, Gauge, Cmdrjameson, Crust, Msh210, Pion, Mlm42, Oleg Alexandrov, Linas, Mpatel, Graham87, MarSch, Salix alba, HappyCamper, Mathbot, Chobot, Ashsong, Michael Slone, KSmrq, Grafen, Bluebot, MalafayaBot,

Cícero, Halio, JamieVicary, Anthos, Echocampfire, Paul Matthews, Thijs!bot, Atmd, RobHar, Serpent's Choice, Jakob.scholbach, Andre.holzner, Felixbecker2, Huzzlet the bot, STBotD, Clarince63, Geometry guy, StefanKarlsson, YonaBot, JackSchmidt, Sjn28, Emesee, Anchor Link Bot, Schrodu, Rhubbarb, Cenarium, Marc van Leeuwen, Addbot, Luckas-bot, Ht686rg90, Kiefer.Wolfowitz, Quondum, Maschen, Kodiologist, Jose Brox, Hamoudafg and Anonymous: 41

- **SO(10) (physics)** *Source:* http://en.wikipedia.org/wiki/SO(10)%20(physics)?oldid=598893904 *Contributors:* Charles Matthews, Phys, Bkell, Alison, Mpatel, Ugha, Noname301, Dauto, Colonies Chris, Doug Bell, Harryboyles, Dan Gluck, DangerousPanda, Michael C Price, David Eppstein, C9, 1ForTheMoney, LilHelpa, Omnipaedista, FrescoBot, GreenRoot, PigFlu Oink, DrilBot, Afteread, Howardgeorgi, Cjean42 and Anonymous: 6

- **Lie superalgebra** *Source:* http://en.wikipedia.org/wiki/Lie%20superalgebra?oldid=599227684 *Contributors:* Schneelocke, Charles Matthews, Wik, Phys, Giftlite, Fropuff, Sam Hocevar, Nabla, Woohookitty, Mpatel, Marudubshinki, R.e.b., Bgwhite, Wavelength, KnightRider, Silly rabbit, Nbarth, Colonies Chris, MTd2, RandomP, Vyznev Xnebara, Headbomb, Kyle the bot, Hesam7, Geometry guy, Sun Creator, Addbot, Goykhman, Luckas-bot, Justpasha, FrescoBot, ChrisGualtieri, Ruhland, Immusson and Anonymous: 21

- **Yang–Mills theory** *Source:* http://en.wikipedia.org/wiki/Yang%E2%80%93Mills%20theory?oldid=626134896 *Contributors:* The Anome, William Avery, Michael Hardy, TakuyaMurata, Arpingstone, Timwi, Kbk, Giftlite, Gadfium, HorsePunchKid, Lumidek, Rich Farmbrough, ArnoldReinhold, MuDavid, Bender235, Grutness, Kocio, Falcorian, LoopZilla, Rjwilmsi, Lmatt, Chobot, Wavelength, Huw Powell, SmackBot, YellowMonkey, Scwlong, Berland, MTd2, Henning Makholm, Esrever, Atoll, Cydebot, Michael C Price, Bajo, Headbomb, Gamebm, Yill577, JaGa, Salsa man, Tikiwont, Policron, Sheliak, Cuzkatzimhut, Alexandria, YuryKirienko, Axiomsofchoice, Lejarrag, StevenJohnston, Henry Delforn (old), StewartMH, SchreiberBike, Qwfp, Addbot, Oberflaechenelement, CarsracBot, ??, Skippy le Grand Gourou, Yobot, Citation bot, Pra1998, Omnipaedista, Tabarr, Peterwoit, Ajtolland, Aliotra, Ysyoon, Pmokeefe, EmausBot, Quondum, Asi013, Helpful Pixie Bot, Bibcode Bot, BG19bot, ChrisGualtieri, Zatrp, Makecat-bot, ?, Zanpan, Airwoz, Nationalfannz, ASCarretero and Anonymous: 54

- **Minimal Supersymmetric Standard Model** *Source:* http://en.wikipedia.org/wiki/Minimal%20Supersymmetric%20Standard%20Model?oldid=633972076 *Contributors:* Phys, Dmytro, Gandalf61, Rursus, Connelly, Marcika, Waltpohl, Pharotic, Carandol, HorsePunchKid, Grunt, Pjacobi, Jensbn, El C, Jag123, JohnyDog, RJFJR, DV8 2XL, Woohookitty, Mpatel, VermillionBird, Rjwilmsi, Goudzovski, Bhny, JabberWok, Shawn81, SCZenz, Closedmouth, Caco de vidro, Tom Lougheed, Stepa, Dauto, Chris the speller, Bluebot, Colonies Chris, Sl1982, QFT, MBlume, Jgwacker, Pulu, CenozoicEra, NNemec, Waggers, Dan Gluck, Iridescent, Antonio Prates, Lottamiata, CmdrObot, Michael C Price, Dchristle, RoadMap, Headbomb, CannedhamX, Knotwork, Yill577, Paulnilsson, Maliz, Dr. Morbius, Andre.holzner, Wilsonge, Red Act, Pjoef, StewartMH, PipepBot, ArdClose, Mastertek, Rreagan007, SkyLined, Addbot, DOI bot, Mjamja, Tokikake, Luckas-bot, Yobot, Wireader, AnomieBOT, Citation bot, GenQuest, GrouchoBot, Omnipaedista, Ernsts, Paine Ellsworth, Identitaamore, Citation bot 1, PigFlu Oink, Puzl bustr, RjwilmsiBot, Akrose, EmausBot, WCEngineer, Arbnos, Suslindisambiguator, AManWithNoPlan, Isocliff, Zukertort, Bibcode Bot, BG19bot, ElphiBot, Physlad, ChrisGualtieri, Cinaro and Anonymous: 46

- **Grand unification energy** *Source:* http://en.wikipedia.org/wiki/Grand%20unification%20energy?oldid=581875171 *Contributors:* Bryan Derksen, Michael Hardy, Emperorbma, Phys, Raul654, Fropuff, LucasVB, Linas, Nowa, Caco de vidro, Rotiro, Headbomb, AstroHurricane001, Wilsonge, TXiKiBoT, Ergo leu, Fratrep, Addbot, FrescoBot, LucienBOT, Kirota, Aknochel, Helpful Pixie Bot and Anonymous: 4

- **Hierarchy problem** *Source:* http://en.wikipedia.org/wiki/Hierarchy%20problem?oldid=611672367 *Contributors:* The Anome, WhisperToMe, Phys, AnonMoos, Jni, Giftlite, Xerxes314, Thincat, Lumidek, Rich Farmbrough, FT2, Pt, Jag123, QTxVi4bEMRbrNqOorWBV, GregorB, VermillionBird, Coemgenus, Mattmartin, Strait, Salix alba, UkPaolo, Ugha, Bhny, Netrapt, Ephraim33, QFT, Jgwacker, NNemec, Ninjakannon, Shambolic Entity, Dr. Morbius, Drgnrave, X!, James Banogon, Megalekaitrane, D.scain.farenzena, Alexbot, Lalegria, Addbot, Mixen Dixon, TutterMouse, Debresser, Topquark22, Yobot, AnomieBOT, Yemibedu, Materialscientist, Citation bot, Neurolysis, ArthurBot, Pra1998, Omnipaedista, A. di M., Erik9bot, FrescoBot, Paine Ellsworth, Puzl bustr, Bj norge, Hauntedpz, RjwilmsiBot, EmausBot, Arbnos, Suslindisambiguator, Quondum, Jbackroyd, Bibcode Bot, Ervin Goldfain, Drcooljoe, IluvatarBot, Ownedroad9 and Anonymous: 35

- **Left–right symmetry** *Source:* http://en.wikipedia.org/wiki/Left%E2%80%93right%20symmetry?oldid=598894957 *Contributors:* Charles Matthews, Phys, Bkell, Giftlite, Christopherlin, Paul August, Jag123, Woohookitty, Mpatel, Conscious, SmackBot, Tom Lougheed, Colonies Chris, Doug Bell, Braddodson, MessedRobot, Michael C Price, Headbomb, WVhybrid, Adavidb, TXiKiBoT, 1ForTheMoney, Addbot, Mjamja, Lightbot, Legobot, Yobot, Erik9bot, Argumzio, Thinking of England, TBrandley, Dark Silver Crow, Cinaro and Anonymous: 5

- **Trinification** *Source:* http://en.wikipedia.org/wiki/Trinification?oldid=598865212 *Contributors:* Charles Matthews, Phys, Pjacobi, Oleg Alexandrov, Mpatel, Ugha, RL0919, SmackBot, Colonies Chris, R'n'B, Arakunem, 1ForTheMoney, Mjamja, Erik9bot, Wiggles007, BattyBot and Anonymous: 3

- **SU(6) (physics)** *Source:* http://en.wikipedia.org/wiki/SU(6)%20(physics)?oldid=598894484 *Contributors:* Phys, Mpatel, Conscious, RL0919, SmackBot, Colonies Chris, 1ForTheMoney, Mifalco, Yobot, GoingBatty, AvicAWB, Delusion23, BattyBot and Anonymous: 7

- **E6 (mathematics)** *Source:* http://en.wikipedia.org/wiki/E6%20(mathematics)?oldid=618289083 *Contributors:* Zundark, Michael Hardy, Charles Matthews, Phys, Giftlite, Fropuff, Gro-Tsen, Tomruen, Almit39, Lumidek, Ukexpat, Giraffedata, Arthena, Oleg Alexandrov, Rjwilmsi, Koavf, Salix alba, R.e.b., John Baez, Algebraist, YurikBot, Gaius Cornelius, Nbarth, Vanished User 0001, Cronholm144, Jim.belk, Phuzion, Dan Gluck, Headbomb, .anacondabot, Magioladitis, Exceptq, Ludvikus, Rocchini, Remember the dot, Drschawrz, Ioverka, Nilradical, Addbot, DOI bot, Eall Ân Ûle, Apaul00, Luckas-bot, Yobot, Jgmoxness, Citation bot, The tree stump, WildBot, GoingBatty, Joel B. Lewis, Helpful Pixie Bot, Bilingsley, Mark L MacDonald, Cjean42, CsDix, Monkbot and Anonymous: 9

- **331 model** *Source:* http://en.wikipedia.org/wiki/331%20model?oldid=621613183 *Contributors:* Mpatel, Rjwilmsi, Fram, SmackBot, Chris the speller, Headbomb, Avicennasis, Maliz, Natsirtguy, Lseixas, HowardFrampton, 1ForTheMoney, DOI bot, JohnHarold, Citation bot, Omnipaedista, Citation bot 1, Obankston, Bibcode Bot, BattyBot, Mfb and Anonymous: 5

- **Chiral color** *Source:* http://en.wikipedia.org/wiki/Chiral%20color?oldid=464933925 *Contributors:* Mpatel, SmackBot, Kintetsubuffalo, Headbomb, Maliz, Ryan Postlethwaite, HowardFrampton, Toddst1, 1ForTheMoney, TimothyRias, JohnHarold, AnomieBOT, Citation bot, Citation bot 1, Bibcode Bot and Anonymous: 3

- **Flipped SU(5)** *Source:* http://en.wikipedia.org/wiki/Flipped%20SU(5)?oldid=613116742 *Contributors:* Charles Matthews, Dysprosia, Phys, Herbee, RJFJR, DV8 2XL, Oleg Alexandrov, Mpatel, Mathbot, SmackBot, GaeusOctavius, Chris the speller, Colonies Chris, Will Beback, Beetstra, Foice, Kupirijo, Headbomb, TimidGuy, R'n'B, Littleolive oil, Auntof6, Fladrif, Debresser, Yobot, Erik9bot, GreenRoot, Hickorybark, Helpful Pixie Bot, ChrisGualtieri and Anonymous: 9

- **Pati–Salam model** *Source:* http://en.wikipedia.org/wiki/Pati%E2%80%93Salam%20model?oldid=613092645 *Contributors:* Michael Hardy, Phys, Bkell, Paul August, Velella, Woohookitty, Mpatel, Rjwilmsi, John Baez, RussBot, RL0919, Fram, SmackBot, Nberger, Baronnet, Colonies Chris, Doug Bell, Michael C Price, Headbomb, R'n'B, Rakeshsumit, DragonBot, 1ForTheMoney, Addbot, Mjamja, OlEnglish, Legobot, Yobot, Xqbot, FrescoBot, D'ohBot, Jschnur, Thinking of England, 777sms, AvicBot, Jaycee55, BattyBot, GroupT, Benastephens and Anonymous: 7

- **Flipped SO(10)** *Source:* http://en.wikipedia.org/wiki/Flipped%20SO(10)?oldid=598894453 *Contributors:* Phys, Mpatel, Ohwilleke, Fram, SmackBot, Colonies Chris, Headbomb, Stepshep, 1ForTheMoney, LilHelpa, AvicBot, BattyBot and Anonymous: 1

- **Little Higgs** *Source:* http://en.wikipedia.org/wiki/Little%20Higgs?oldid=628125530 *Contributors:* Michael Hardy, Phys, Lumidek, David Schaich, BD2412, Rjwilmsi, Erkcan, Conscious, Crasshopper, SmackBot, Jgwacker, Jim.belk, Headbomb, Maliz, Mjamja, Dreamer08, AnomieBOT, SassoBot, Jesse V., EmausBot and Anonymous: 15

- **Preon** *Source:* http://en.wikipedia.org/wiki/Preon?oldid=632388001 *Contributors:* Maury Markowitz, Heron, Ewen, Edward, Kickaha, Dcljr, Timwi, David Latapie, Chrisjj, BenRG, Altenmann, Merovingian, Xanzzibar, David Gerard, Giftlite, Graeme Bartlett, Herbee, Monedula, Semorrison, Mboverload, Eequor, Icairns, Urhixidur, Rich Farmbrough, Pjacobi, Drhex, John Vandenberg, Jag123, Calton, Alai, Uncle G, GregorB, BD2412, Ketiltrout, Rjwilmsi, Fragglet, Mathrick, Smithbrenon, CJLL Wright, Chobot, ScottAlanHill, Jp-kotta, YurikBot, Ugha, Bambaiah, Phmer, Ohwilleke, Merick, Gcapp1959, Dialectric, Buster79, Trovatore, Długosz, Closedmouth, Iell-wood, Paul D. Anderson, Lserni, SmackBot, RockMaestro, Bayardo, Stepa, GwydionM, Kmarinas86, DocKrin, JesseStone, Trekphiler, V1adis1av, Fatla00, SilverStar, Jaganath, Md2perpe, Will314159, Friendly Neighbour, CmdrObot, Doc W, AlphaNumeric, LactoseTI, Mglg, Keraunos, Headbomb, Thadius856, Joe Schmedley, Ph.eyes, GurchBot, Yill577, Randyfurlong, Dr. Morbius, Lexivore, Experien-tial, ChauriCh, Tanaats, OliverHarris, VolkovBot, Fences and windows, Calwiki, Anonymous Dissident, Thrawn562, Synthebot, Antixt, Pegasus1965, AHMartin, PlanetStar, Work permit, WereSpielChequers, 1ForTheMoney, SkyLined, Addbot, Lightbot, OlEnglish, Zor-robot, The Bushranger, Luckas-bot, Yobot, AnomieBOT, Icalanise, Materialscientist, Citation bot, Jsharpminor, GrouchoBot, FrescoBot, Goodbye Galaxy, Citation bot 1, MastiBot, Bj norge, RjwilmsiBot, Ofercomay, Detogain, WikitanvirBot, Slightsmile, The Mysterious El Willstro, Hhhippo, Wyvern Rex., Suslindisambiguator, Gilderien, Preon, Helpful Pixie Bot, Bibcode Bot, BG19bot, Marioedesouza, Andyhowlett, I am One of Many, FrigidNinja, Draconnis caput, Delbert7 and Anonymous: 99

- **M-theory** *Source:* http://en.wikipedia.org/wiki/M-theory?oldid=634444258 *Contributors:* AxelBoldt, CYD, Eloquence, BF, Bryan Derk-sen, Zundark, The Anome, Ap, Tim Chambers, Hari, Stevertigo, Michael Hardy, Tim Starling, Gabbe, Tompagenet, Ixfd64, CesarB, Looxix, JWSchmidt, Darkwind, Marco Krohn, Jeandré du Toit, Evercat, Schneelocke, Charles Matthews, Timwi, Reddi, Malcohol, Bevo, Jusjih, Slawojarek, Sander123, Fredrik, R3m0t, RedWolf, Blainster, DHN, Hadal, HaeB, Tobias Bergemann, David Gerard, Giftlite, DocWatson42, Jmnbpt, Barbara Shack, Fropuff, Moyogo, Sigfpe, Daen, Antandrus, Lumidek, ChrisCostello, Mike Rosoft, Spiffy sperry, Urvabara, Noisy, Discospinster, H0riz0n, Vsmith, Loren36, El C, Momotaro, RoyBoy, Triona, Constantine, QTxVi4bEMRbrNqOorWBV, Giraffedata, Wolfrider, Physicistjedi, MPerel, Gsklee, ShardPhoenix, Axl, Mac Davis, Kocio, Burn, Hu, Wtmitchell, SidP, DV8 2XL, Ringbang, Kazvorpal, Omnist, Sharkie, Joelpt, Angr, Firsfron, FeanorStar7, Pol098, WadeSimMiser, Mpatel, GregorB, Jugger90, Paxsim-ius, Mandarax, Chun-hian, Grammarbot, Rjwilmsi, Nightscream, Zbxgscqf, Oblivious, Yug, Lionelbrits, Ruidlopes, The ARK, Latka, Mathbot, Diza, Phoenix2, DVdm, Eric B, Loom91, Zafiroblue05, Bhny, Stephenb, KSchutte, Bovineone, Salsb, Erielhonan, Bobak, Asarelah, Dna-webmaster, Sandstein, Superdude99, Zzuuzz, Imaninjapirate, Arthur Rubin, Ilmari Karonen, Caco de vidro, DVD R W, Hide&Reason, Jmeden2000, Teo64x, Sardanaphalus, MartinGugino, RupertMillard, SmackBot, Kurochka, K-UNIT, Rwp, Rlbates99, Ajt, Gilliam, Wlmg, DividedByNegativeZero, Bluebot, Cush, SMP, Ben.c.roberts, MalafayaBot, Nbarth, DHN-bot, Joemah, N.MacInnes, Xiner, Nunocordeiro, Mbertsch, Addshore, EPM, Nakon, Kiplantt, Bigmantonyd, Martijn Hoekstra, Kabain52, Brdforallseasons, Say-den, Doug Bell, Jaganath, Shadowlynk, IronGargoyle, Jochietoch, Hu12, Jxh2154, Tawkerbot2, Valoem, Gebrah, Albertod4, Kurtan, Harold f, Devourer09, Cyrusc, CRGreathouse, Olaf Davis, Lambertian, Friendlystar, Rowellcf, Bmk, Myasuda, DepartedUser2, Ekajati, Cydebot, Fluence, Meno25, Gagueci, Kahananite, Michael C Price, Alexnye, IComputerSaysNo, Lord Satorious, Krowe, Mrockman, Thijs!bot, Epbr123, Daniel, Headbomb, NeilHalfway, James086, KrakatoaKatie, AntiVandalBot, Blue Tie, Alphachimpbot, J rowley, Shambolic Entity, SuperLuigi31, Buchhemi, Fetchcomms, 100110100, VoABot II, Madevin314, SHCarter, Rami R, Jqshenker, Just H, Rickard Vogelberg, Stephen Shenker, Theoretic, MartinBot, Kostisl, R'n'B, Euku, Numbo3, Maurice Carbonaro, Nly8nchz, Thucy-dides411, LordAnubisBOT, Janus Shadowsong, Peskydan, Isoko, Belovedfreak, Antony-22, Wesino, WJBscribe, Thomas795135, Blood Oath Bot, Idioma-bot, Sheliak, Gogobera, Jeff G., Rei-bot, Ask123, Pennstatephil, JhsBot, Mazarin07, Peace keeper II, Antixt, Why Not A Duck, PhysPhD, Rknasc, Guystout, Drschawrz, SieBot, Robdunst, Paradoctor, Wing gundam, Holt27, Astroboyretro, Caidh, OKBot, Divinestuff, Wpac5, Ayleuss, Beofluff, Loren.wilton, ClueBot, Master Shake 9, The Thing That Should Not Be, Haemor-rhage, Arakunem, Drmies, IMNTU, Yupjohnny, Huntthetroll, Patrik Andersson, Gardv, DumZiBoT, Jfosc, Truthnlove, Autocoast, Al-bambot, Addbot, Uruk2008, Cuaxdon, CanadianLinuxUser, WikiUserPedia, Barak Sh, Tassedethe, Carapheonix, Togekiss101, Tide rolls, OlEnglish, Snaily, Legobot, Luckas-bot, Yobot, Fraggle81, Pcap, Foolo, CinchBug, Tempodivalse, AnomieBOT, KDS4444, Götz, Charlesvi, Dalton h, Marcka, Alexzabbey, Jim1138, IRP, AdjustShift, Materialscientist, Citation bot, Quebec99, Ruike, Tinucherian-Bot II, Ekwos, Techwiz2000, Omnipaedista, Peanuts4life, Pinethicket, Vicenarian, EDG161, Jusses2, ActivExpression, SkyMachine, Tkachyk, 122589423KM, சஞ்சீவி சிவகுமாரி், Reaper Eternal, Apb91781, 786 zikhar, LcawteHuggle, Adam1217, EmausBot, Going-Batty, Pyschobbens, StringTheory11, Smiwi, Suslindisambiguator, SporkBot, PoisonGM, Besneatte, SBaker43, Denholm Reynholm, RockMagnetist, ClueBot NG, Blueshift333, Rgwkenyon, Helpful Pixie Bot, Bibcode Bot, SharkinthePool, Msaunier, Copernicus01, El-ginfball10, Qed3, ShotmanMaslo, Zujua, Kooky2, Mediran, Chris5631, FEYKATD, Ecila3, Lugia2453, Frosty, AHusain314, Among Men, Faizan, Epicgenius, Diekilldie, Beakr, DavidLeighEllis, Vampre1122, Polytope24, Evandas, Oneidiotsavant, Pretickle, TheRe-alTheKoi, Shantsforeverandalways, QuantumMatt101 and Anonymous: 437

- **Loop quantum gravity** *Source:* http://en.wikipedia.org/wiki/Loop%20quantum%20gravity?oldid=634568081 *Contributors:* Bryan Derk-sen, The Anome, AstroNomer, RK, Toby Bartels, Miguel, Schewek, Ewen, Michael Hardy, TakuyaMurata, Islandboy99, GTBacchus, Mcarling, Looxix, Ahoerstemeier, Cyp, Kimiko, Palfrey, Jordi Burguet Castell, Mxn, Charles Matthews, Sanxiyn, Maximus Rex, Phys, Omegatron, Finlay McWalter, Dmytro, Sdedeo, Astronautics, Peak, Chris Roy, Mirv, Sverdrup, Kn1kda, Hadal, Jheise, Clementi, Con-nelly, Giftlite, Sj, Fastfission, Herbee, Anville, Dratman, Curps, JeffBobFrank, Jason Quinn, Gzornenplatz, C17GMaster, DÅ,ugosz, PhiloVivero, DefLog, Gadfium, HorsePunchKid, Sam Hocevar, Lumidek, Tdent, Joyous!, M1ss1ontomars2k4, Eep², Poccil, Rich Farm-brough, Avriette, Pjacobi, MuDavid, Pavel Vozenilek, Bender235, ESkog, Clement Cherlin, Peter M Gerdes, Drhex, John Vandenberg, C S, Cmdrjameson, GTubio, Tweet Tweet, Slicky, Ral315, Lysdexia, Arthena, Xaphan9966, Greg Kuperberg, Count Iblis, Egg, Lee-Anne, Kazvorpal, Killing Vector, Linas, Merlinme, HFarmer, Sympleko, Hfarmer, Mpatel, GregorB, J M Rice, Ae7flux, Tjbk tjb, Alienus, Fleisher, Sjö, Rjwilmsi, Nightscream, Zbxgscqf, Bubba73, FlaBot, John Baez, Don Gosiewski, Smithbrenon, Chobot, Bgwhite, Roboto de Ajvol, YurikBot, Wavelength, RobotE, Rt66lt, Hillman, DanMS, Chaos, Salsb, Welsh, Schmock, Crasshopper, Beanyk, Akashmitra, Bota47, JonathanD, Endomion, Modify, Petri Krohn, Ilmari Karonen, Caco de vidro, Benandorsqueaks, SmackBot, Bayardo, FlashSh-eridan, Unyoyega, Vald, JMiall, Chris the speller, IvanAndreevich, DHN-bot, Colonies Chris, Chlewbot, Pepsidrinka, Chrylis, Mega-

Hasher, TriTertButoxy, Lambiam, Vincenzo.romano, Loadmaster, Konklone, K, G-W, Kurtan, Harold f, Will314159, Friendly Neighbour, Vyznev Xnebara, Ian Beynon, Myasuda, Gmusser, Rjm656s, Fournax, Headbomb, Nick Number, MichaelMaggs, Edokter, Byrgenwulf, Knotwork, Arch dude, Igodard, Yill577, WolfmanSF, Tonyfaull, Skylights76, Rickard Vogelberg, Gwern, AltiusBimm, Melamed katz, Vanished user 47736712, WJBscribe, Izno, KittyHawker, Sheliak, AlnoktaBOT, Nxavar, Jackfork, Carlorovelli, Anotherak, SieBot, Keskival, AS, Robdunst, Hugh16, Senderista, Bnsreenath, Caidh, Oxymoron83, Dcattell, Swiebodzice, Sk8hack, Danthewhale, Martarius, Sfan00 IMG, Shaded0, Djr32, CohesionBot, JavierReynaldo, Arjayay, SchreiberBike, Pqnelson, Mjaniec, DumZiBoT, Ianbay, Neuralwarp, XLinkBot, Fastily, Tenner47, Arthur chos, Avoided, Tenderbuttons, Benplusnumber, Balungifrancis, Addbot, DOI bot, 15lsoucy, Tarosic, Debresser, SamatBot, Yobot, Ibayn, 4th-otaku, AnomieBOT, Decora, Archon 2488, Francois33, Citation bot, Xqbot, Imushfiq, MIRROR, Pra1998, Dumontierc, Franco3450, Rr2000, FrescoBot, Paine Ellsworth, Martlet1215, Citation bot 1, ROMVLVS, Casimir9999, RobinK, Meier99, Dinamik-bot, Bj norge, ElPeste, Afteread, EmausBot, Detogain, John of Reading, Racerx11, Going-Batty, XinaNicole, Ensabah6, Uploadvirus, ZéroBot, Arbnos, Zueignung, WaterCrane, Crown Prince, LaurentRDC, Isocliff, Vodkacannon, Raidr, Helpful Pixie Bot, Titodutta, Bibcode Bot, BG19bot, Spaligo, KateWishing, PhnomPencil, Kecchina, Halfb1t, Brad7777, Fylbecatulous, Jimw338, Mogism, LTWoods, Andyhowlett, Jawa0, &reasNink, SomeFreakOnTheInternet, Tentinator, EvergreenFir, DimReg, Pedarkwa, Db9199 24, Anrnusna, Notspelly, Ntomlin1996, Monkbot, Isbromberg, Dsprc and Anonymous: 314

- **Causal dynamical triangulation** Source: http://en.wikipedia.org/wiki/Causal%20dynamical%20triangulation?oldid=623055825 Contributors: The Anome, Charles Matthews, Jeffq, Tobias Bergemann, Pjacobi, Alamino, Anthony Appleyard, Tabletop, Tlroche, MarSch, Itinerant1, JocK, JonathanD, SmackBot, HTeutsch, Schmiteye, Colonies Chris, Fuhghettaboutit, Vincenzo.romano, Gmusser, Tonyfaull, Gibimi, Melamed katz, Lantonov, Sigmundur, Sheliak, Foresyte, AlleborgoBot, Pallab1234, Thehotelambush, Ideal gas equation, Greennature2, Shamanchill, Addbot, Eric Drexler, AnomieBOT, Prari, Paine Ellsworth, Steve Quinn, ZéroBot, Arbnos, Maschen, Raidr, DavidRideout, Fraulein451, &reasNink, Dimension10 and Anonymous: 21

- **Lie algebra** Source: http://en.wikipedia.org/wiki/Lie%20algebra?oldid=633784515 Contributors: AxelBoldt, Zundark, Miguel, Michael Hardy, Wshun, Joel Koerwer, TakuyaMurata, Suisui, Kragen, Rossami, Iorsh, Loren Rosen, Charles Matthews, Dysprosia, Michael Larsen, Grendelkhan, Phys, Tobias Bergemann, David Gerard, Weialawaga, Tosha, Giftlite, BenFrantzDale, Lethe, Fropuff, Curps, Jeremy Henty, Jason Quinn, Python eggs, Chameleon, DefLog, CryptoDerk, CSTAR, Pyrop, Guanabot, Pj.de.bruin, Vsmith, Gauge, Pt, Kwamikagami, Wood Thrush, Reinyday, Foobaz, Msh210, Arthena, Spangineer, Dirac1933, Drbreznjev, Oleg Alexandrov, Linas, Isnow, BD2412, NatusRoma, MarSch, Mathbot, Margosbot, RexNL, Masnevets, YurikBot, Wavelength, Hairy Dude, Michael Slone, Lenthe, Stephenb, Grubber, Trovatore, Asimy, Crasshopper, Curpsbot-unicodify, Sbyrnes321, SmackBot, Incnis Mrsi, Grokmoo, Kmarinas86, Bluebot, Silly rabbit, Nbarth, Thomas Bliem, Chlewbot, BlackFingolfin, Noegenesis, Rschwieb, AlainD, Harold f, CmdrObot, Shirulashem, Headbomb, Second Quantization, Dachande, RobHar, B-80, Jrw@pobox.com, Deflective, Englebert, Vanish2, R'n'B, Bogey97, Maurice Carbonaro, Supermanifold, Policron, Freiddie, Cuzkatzimhut, VolkovBot, JohnBlackburne, LokiClock, Ndbrian1, Hesam7, Geometry guy, Drorata, Arcfrk, StevenJohnston, YohanN7, SieBot, Stca74, Jenny Lam, Paolo.dL, JackSchmidt, Fatchat, Veromies, JP.Martin-Flatin, Count Truthstein, Addbot, Roentgenium111, Lightbot, Legobot, Luckas-bot, Yobot, Niout, Jason Recliner, Esq., Delilahblue, AnomieBOT, Twri, SassoBot, Kaoru Itou, D'ohBot, Darij, Juniuswikiae, Prtmrz, Rausch, Jkock, Adam cohenus, Tobe-Bot, Lotje, Doctor Zook, Slawekb, Quondum, Mikhail Ryazanov, ClueBot NG, Dd314, Teika kazura, Walterpfeifer, Pfeiferwalter, IkamusumeFan, Flbsimas, Deltahedron, Saung Tadashi, Mark L MacDonald, Danielbrice, Enyokoyama, CsDix, 314Username, Forgetfulfunctor00, CaptainLama and Anonymous: 87

- **Lie group** Source: http://en.wikipedia.org/wiki/Lie%20group?oldid=629571422 Contributors: AxelBoldt, Zundark, Josh Grosse, XJaM, Miguel, Stevertigo, Xavic69, Michael Hardy, TakuyaMurata, GTBacchus, Looxix, Barak, Charles Matthews, Dysprosia, Jitse Niesen, Zoicon5, David Shay, Itai, Phys, Josh Cherry, Saaska, Tobias Bergemann, Weialawaga, Tosha, Giftlite, JamesMLane, BenFrantzDale, Lethe, Fropuff, Wgmccallum, Jason Quinn, Bobblewik, DefLog, Lockeownzj00, Beland, Pmanderson, Abdull, Dablaze, MuDavid, Paul August, ChrisJ, Bender235, Tompw, Rgdboer, Kwamikagami, Shanes, Cherlin, Msh210, PAR, Alex Varghese, Oleg Alexandrov, Zntrip, Joriki, Linas, Dzordzm, Isnow, SDC, AnmaFinotera, Frankie1969, Graham87, Porcher, Rjwilmsi, NatusRoma, MarSch, Salix alba, HappyCamper, R.e.b., VKokielov, BMF81, Masnevets, Chobot, Algebraist, Wavelength, Hillman, RussBot, Michael Slone, KSmrq, Archelon, Buster79, Arkapravo, Smaines, Orthografer, Ekeb, Kier07, Pred, RodVance, JDspeeder1, SmackBot, Incnis Mrsi, Tom Lougheed, FlashSheridan, Davewild, Mhss, Kmarinas86, Bluebot, Badger014, Silly rabbit, DHN-bot, Bears16, Akriasas, KeithB, Ninte, Siva1979, John, Ulner, Jim.belk, Michael Kinyon, Inquisitus, Mathchem271828, Rschwieb, Krasnoludek, Yggdrasil014, CRGreathouse, CBM, Logical2u, Myasuda, Kupirijo, MotherFunctor, Dr.enh, Xantharius, Thijs!bot, Headbomb, JustAGal, RichardVeryard, RobHar, Salgueiro, Dougher, Len Raymond, JAnDbot, Deflective, Unifey, Homeworlds, Magioladitis, Bongwarrior, Cmelby, WhatamIdoing, Sullivan.t.j, David Eppstein, The Real Marauder, Benjamin.friedrich, David J Wilson, Jesper Carlstrom, Maproom, TomyDuby, Rocket71048576, Pidara, Freiddie, Dorftrottel, Lseixas, Borat fan, Trevorgoodchild, JohnBlackburne, Ndbrian1, James.r.a.gray, Hesam7, Geometry guy, Jmath666, Eubulides, Brian Huffman, Genuine0legend, Drorata, Arcfrk, Smylei, Oscarbaltazar, YohanN7, JackSchmidt, S2000magician, Beastinwith, Deciwill, Sidiropo, Leontios, Heckledpie, Cacadril, SchreiberBike, Marc van Leeuwen, MystBot, Addbot, Topology Expert, LaaknorBot, Ozob, Tanath, Tide rolls, Luckas-bot, Ht686rg90, Niout, Amirobot, AnomieBOT, Citation bot, ArthurBot, Br77rino, Kaoru Itou, FrescoBot, Anterior1, Sławomir Biały, RedBot, Tinfoilcat, EmausBot, KbReZiE 12, Darkfight, Slawekb, Suslindisambiguator, Maschen, Zueignung, ClueBot NG, Mgvongoeden, Kasirbot, Helpful Pixie Bot, Daviddwd, CitationCleanerBot, Fraisière, NotWith, MathKnight-at-TAU, Suhagja, Brirush, CsDix, Sol1, Blackbombchu, Abitslow, Cbartondock, Victoryhuy and Anonymous: 107

- **Heterotic string theory** Source: http://en.wikipedia.org/wiki/Heterotic%20string%20theory?oldid=629000979 Contributors: Charles Matthews, Denni, Hugo, Giftlite, Fropuff, Lumidek, Rich Farmbrough, Pearle, NTK, Mpatel, Chobot, YurikBot, Bhny, Hwasungmars, 2over0, Sardanaphalus, KnightRider, SmackBot, Schmiteye, Fplay, QFT, Jmnbatista, Headbomb, Lamontacranston, STBot, MarkJefferys, Sheliak, JohnBlackburne, Spiral5800, Legoktm, TimothyRias, MystBot, Addbot, Debresser, Tassedethe, Lightbot, OlEnglish, Jack who built the house, AnomieBOT, Citation bot, Omnipaedista, Erik9bot, Steve Quinn, Orenburg1, EmausBot, WaterfordPBR, Staszek Lem, ClueBot NG, BG19bot, AHusain314, I am One of Many, Dimension10, Polytope24 and Anonymous: 31

- **Topological defect** Source: http://en.wikipedia.org/wiki/Topological%20defect?oldid=634480721 Contributors: Bryan Derksen, Michael Hardy, EddEdmondson, Breakpoint, CesarB, Mkweise, Cyp, Wnissen, Phys, Owen, Merovingian, Jimpaz, Giftlite, Barbara Shack, Brianhe, MuDavid, Euyyn, QTxVi4bEMRbrNqOorWBV, Guy Harris, Ceyockey, Linas, Rjwilmsi, Lebha, McGinnis, Wavelength, Borgx, Pelago, Light current, 2over0, Poulpy, Erik J, Japhet, Cydebot, Wikid77, Headbomb, I do not exist, Ste4k, 55david, Shambolic Entity, Serpent's Choice, B9 hummingbird hovering, Sketchjoy, R'n'B, Adavidb, Tarotcards, EoGuy, Niceguyedc, Aaroncorey, Alexbot, Addbot, DOI bot, SPat, Yobot, Vectorsoliton, AnomieBOT, Charvest, Pierlumba, FrescoBot, Pratik.mallya, Citation bot 1, Xiaoshan Math, Kenchikuben, Pavithransiyer, ZéroBot, Primergrey, Helpful Pixie Bot, Bibcode Bot, Enyokoyama, Monkbot, A200b and Anonymous: 22

- **Magnetic monopole** Source: http://en.wikipedia.org/wiki/Magnetic%20monopole?oldid=634104829 Contributors: Bryan Derksen, The Anome, Ap, Andre Engels, Roadrunner, Maury Markowitz, Heron, Camembert, Patrick, Michael Hardy, Tim Starling, EddEdmondson,

Dominus, Ixfd64, Skysmith, Looxix, Mkweise, Ahoerstemeier, Stevenj, Aarchiba, Cyan, HolIgor, Charles Matthews, Timwi, Phys, Jerzy, BenRG, Jeffq, Henrygb, Rasmus Faber, Pengo, Cutler, Enochlau, Giftlite, Mintleaf, Xerxes314, Rapjo, Waltpohl, Pharotic, Peter Ellis, Nova77, ConradPino, Gzuckier, Beland, MFNickster, Anythingyouwant, Elektron, Icairns, Lumidek, Karl Dickman, Mike Rosoft, Urvabara, Jkl, Rich Farmbrough, TedPavlic, Pjacobi, ArnoldReinhold, MuDavid, Bender235, ESkog, Kjoonlee, El C, Sasquatch, Thuktun, Congruence, Alansohn, Anthony Appleyard, Cmprince, Pauli133, Nick Mks, Falcorian, Linas, JarlaxleArtemis, Ruud Koot, Mpatel, Tabletop, GregorB, CharlesC, TheAlphaWolf, Emerson7, Mandarax, Aarghdvaark, BD2412, Rjwilmsi, HonoluluMan, MarSch, Eyu100, Seraphimblade, DonSiano, Gareth McCaughan, R.e.b., Erkcan, Drrngrvy, Mathbot, Tardis, Adarsh116098, Chobot, DVdm, Amaurea, YurikBot, Phmer, RussBot, Xihr, JabberWok, Gaius Cornelius, PoorLeno, DragonHawk, Wiki alf, Welsh, Długosz, Gillis, Dchoulette, Jstrater, Crasshopper, Tony1, Crumley, SamuelRiv, 2over0, Reyk, KingCarrot, Sbyrnes321, Mhardcastle, SmackBot, Michaelliv, Melchoir, Jonathan Karlsson, Octahedron80, Skatche, V1adis1av, QFT, Alex Fix, Ianmacm, Khukri, Mohseng, Skiminki, Yevgeny Kats, Nat2, JorisvS, Loadmaster, Rock4arolla, Stephen B Streater, Norm mit, Gorog, Courcelles, Piccor, Achoo5000, Chetvorno, Disambiguator, Randall Nortman, GRB, Capefeather, Moyerjax, Michael C Price, Quibik, Dougweller, DumbBOT, Karl-H, Difty, Wikid77, Headbomb, Luna Santin, Thranduil, Fru1tbat, Spartaz, JAnDbot, Igodard, Catslash, Bakken, Stevvers, WLU, 2bithacker, C.R.Selvakumar, Nsande01, Nlalic, NerdyNSK, JA.Davidson, Rod57, Dawright12, Aoosten, Tarotcards, Plasticup, Loohcsnuf, Barraki, Ross Fraser, Ratfox, Dorftrottel, Trmatthe, VolkovBot, FDominec, Rei-bot, Lixo2, Mathfreak11235, Wingedsubmariner, Antixt, RaseaC, Stigin, SieBot, CatherS, Likebox, Pit-trout, Henke37, Lisatwo, Dickontoo, Skeptical scientist, Maxime.Debosschere, Martarius, Balashpersia, Unbuttered Parsnip, Razimantv, Mild Bill Hiccup, LonelyBeacon, Wrsh11, SchreiberBike, DumZiBoT, XLinkBot, Oldnoah, Avoided, Addbot, Jacopo Werther, DOI bot, Мыша, Barak Sh, 84user, Qaswqaswgd, Skippy le Grand Gourou, Luckas-bot, Munkel Davidson, KamikazeBot, AnomieBOT, Floquenbeam, Citation bot, Flying hazard, Xqbot, Renaissancee, Kbodouhi, Charvest, Nagualdesign, FrescoBot, Goodbye Galaxy, Citation bot 1, Relke, DrilBot, Cwedhrin, Q0k, MarcelB612, RedBot, YURi-21century, Morphotomy, Splartmaggot, FKLS, Deanmullen09, Wrotesolid, Waylah, Giscard2, John of Reading, WikitanvirBot, Wikipelli, ZéroBot, Prayerfortheworld, Cogiati, N0RND123, StringTheory11, Quondum, Maschen, Particle hep, Zooooooooooaa, Isocliff, David Thorne, ClueBot NG, Andrija radovic, MerlIwBot, Bibcode Bot, BG19bot, PearlSt82, Gorthian, F=q(E+v^B), Niqomi, ChrisGualtieri, Khazar2, MaxwellDecoherence, Enyokoyama, JRYon, Jaxcp3, GabeIglesia, Razibot, Consecutor, Monkbot and Anonymous: 224

- **Cosmic string** *Source:* http://en.wikipedia.org/wiki/Cosmic%20string?oldid=612566436 *Contributors:* Bryan Derksen, Tim Starling, Cyan, Wnissen, BenRG, Northgrove, Giftlite, Lethe, Geni, Lumidek, Rich Farmbrough, Pjacobi, C1k3, Hot pastrami, Jumbuck, Paradiso, Kocio, Alai, Axeman89, Joke137, Eteq, Amaxson, Bgwhite, Wavelength, Klazuka, SmackBot, Can't sleep, clown will eat me, Mrwuggs, Kukini, Lambiam, Me.johnnyb, Dgw, Myasuda, Ericagol, Nebarnix, Dr.enh, Headbomb, AnAj, Balloonguy, Dr. Morbius, Trusilver, Adavidb, STBotD, Cenarium, Kakofonous, Kekule123, XLinkBot, Addbot, DOI bot, Numbo3-bot, Lightbot, OlEnglish, Luckas-bot, Yobot, Ptbotgourou, Wireader, AnomieBOT, Citation bot, ArthurBot, GrouchoBot, Homyakchik, Omnipaedista, Mailsewer, NoRad, FrescoBot, Molitorppd22, Citation bot 1, Citation bot 4, RedBot, WikitanvirBot, Serketan, Wakabaloola, ClueBot NG, Sashhere, Ramaksoud2000, Bibcode Bot, Elauminri, BG19bot, Joydeep, Bsaranga, Kogge and Anonymous: 58

- **Domain wall (string theory)** *Source:* http://en.wikipedia.org/wiki/Domain%20wall%20(string%20theory)?oldid=634473066 *Contributors:* Michael Hardy, Paul A, Lord Roem, BML0309, Andyhowlett, Journalbug, Mark viking, Nixie9 and Polytope24

- **Inflation (cosmology)** *Source:* http://en.wikipedia.org/wiki/Inflation%20(cosmology)?oldid=633025392 *Contributors:* Bryan Derksen, The Anome, Diatarn iv, Roadrunner, David spector, Hephaestos, Stevertigo, Edward, Nealmcb, Boud, Michael Hardy, Tim Starling, Dcljr, Cyde, Ellywa, William M. Connolley, Theresa knott, Jeff Relf, Mxn, Timwi, Rednblu, Bartosz, Pierre Boreal, Raul654, Chuunen Baka, Robbot, Gandalf61, Rursus, Ancheta Wis, Giftlite, Barbara Shack, Mikez, Lethe, Dratman, Curps, Jcobb, Just Another Dan, Andycjp, HorsePunchKid, Beland, Elroch, JDoolin, Burschik, Shadypalm88, Eep[2], Mike Rosoft, DanielCD, Noisy, Rich Farmbrough, FT2, Pjacobi, Luxdormiens, Bender235, AdamSolomon, Pt, Worldtraveller, Art LaPella, Orlady, Drhex, Guettarda, QTxVi4bEMRbrNqOorWBV, Jeodesic, Rsholmes, Anthony Appleyard, Plumbago, JHG, Schaefer, EmmetCaulfield, Cgmusselman, Dirac1933, Oleg Alexandrov, Matevzk, Yeastbeast, StradivariusTV, BillC, Bluemoose, Wdanwatts, Joke137, Rnt20, Malangthon, Ketiltrout, Drbogdan, Rjwilmsi, Zbxgscqf, Mattmartin, Strait, Eyu100, Jehochman, Ems57fcva, Bubba73, FlaBot, Nihiltres, Itinerant1, Phoenix2, Chobot, Hermitage, Bgwhite, YurikBot, Wavelength, Supasheep, Ytrottier, Gaius Cornelius, Anomalocaris, NawlinWiki, LiamE, JonathanD, Enormousdude, 2over0, Arthur Rubin, Argo Navis, Physicsdavid, Profero, Luk, SmackBot, Haza-w, KnowledgeOfSelf, Lawrencekhoo, Onsly, Jdthood, Salmar, Jefffire, Hve, QFT, Vanished User 0001, Stevenmitchell, BIL, Lostart, Ligulembot, Yevgeny Kats, Byelf2007, Lambiam, Rcapone, JorisvS, Heliogabulus, Dan Gluck, Spebudmak, JoeBot, UncleDouggie, Fsotrain09, Oshah, JRSpriggs, Chetvorno, Friendly Neighbour, Drinibot, Vanished user 2345, Brownlee, SuperMidget, Cydebot, BobQQ, Mortus Est, Cyhawk, Ttiotsw, Julian Mendez, Dr.enh, Michael C Price, Kozuch, LilDice, Thijs!bot, Headbomb, Z10x, Jklumker, Alfredr, Rico402, Gmarsden, JAnDbot, Olaf, Linkin-Park, GurchBot, Magioladitis, Jpod2, Vanished user ty12kl89jq10, Rickard Vogelberg, Dr. Morbius, Bhenderson, TomS TDotO, Tarotcards, Wesino, Student7, Potatoswatter, Ollie 9045, Ja 62, Useight, Idioma-bot, Sheliak, Tokenhost, VolkovBot, ABF, ColdCase, Philip Trueman, TXiKiBoT, Calwiki, Thrawn562, Gobofro, SwordSmurf, Northfox, PaddyLeahy, SieBot, Wing gundam, OpenLoop, Likebox, Mimihitam, Hockeyboi34, Lightmouse, Sunrise, Southtown, Hamiltondaniel, Epistemion, ClueBot, ChandlerMapBot, Jusdafax, ResidueOfDesign, Ploft, Scog, SchreiberBike, TimothyRias, Katsushi, MidwestGeek, Addbot, Roentgenium111, DOI bot, Blethering Scot, Ronhjones, Glane23, Deamon138, TStein, Barak Sh, Tassedethe, Zorrobot, Ben Ben, Legobot, Yinweichen, Luckas-bot, Amirobot, Aldebaran66, Isotelesis, Magog the Ogre, AnomieBOT, Pyrrhon8, Rubinbot, Piano non troppo, Collieuk, Ulric1313, Citation bot, Xqbot, Plastadity, Capricorn42, P14nic997, False vacuum, Waleswatcher, Ignoranteconomist, Bigger digger, Chatul, ??, CES1596, FrescoBot, Mesterhd, Paine Ellsworth, Schnufflus, Charles Edwin Shipp, Bbhustles, Ahnoneemoos, Pinethicket, Σ, Aknochel, Mercy11, Trappist the monk, Jordgette, Wdanbae, Michael9422, CobraBot, Deathflyer, Mathewsyriac, EmausBot, Thucyd, GoingBatty, Wikipelli, Kiatdd, Italia2006, Werieth, ZéroBot, Chasrob, Wackywace, Bamyers99, Suslindisambiguator, RaptureBot, Maschen, HCPotter, Crux007, Rock-Magnetist, ClueBot NG, J kay831, Law of Entropy, Supermint, Helpful Pixie Bot, Bibcode Bot, Lowercase sigmabot, Negativecharge, MSgtpotter, Badon, BML0309, Hamish59, Minsbot, BattyBot, SupernovaExplosion, ChrisGualtieri, JYBot, Rfassbind, Astroali, Lepton01, Chwon, Rolf h nelson, Comp.arch, Kogge, Hilmer B, Anrrusna, Epaminondas of Thebes, Abitslow, Monkbot, Accnln, BradNorton1979 and Anonymous: 212

- **Doublet–triplet splitting problem** *Source:* http://en.wikipedia.org/wiki/Doublet%E2%80%93triplet%20splitting%20problem?oldid=598893865 *Contributors:* Edward, Charles Matthews, Phys, Lumidek, Jag123, Conscious, Colonies Chris, Scwlong, QFT, Jgwacker, Dan Gluck, Headbomb, TheMindsEye, Mild Bill Hiccup, Mjamja, Legobot, Yobot, Omnipaedista, GreenRoot, ChrisGualtieri and Anonymous: 2

- **Quantum chromodynamics** *Source:* http://en.wikipedia.org/wiki/Quantum%20chromodynamics?oldid=632387873 *Contributors:* AxelBoldt, CYD, Zundark, Youandme, Ewen, Stevertigo, Michael Hardy, Ahoerstemeier, Whkoh, Emperorbma, Jitse Niesen, Phys, Robbot, Fredrik, Ojigiri, Seth Ilys, Alan Liefting, Giftlite, JamesMLane, Monedula, Xerxes314, JeffBobFrank, Jason Quinn, Elroch, Icairns, Sam Hocevar, Lumidek, Sctfn, Eep[2], David Schaich, JonL, Goplat, AdamSolomon, Pt, El C, CDN99, Robotje, Slicky, Physicistjedi, Azn king28, Fwb22, Guy Harris, Ricky81682, TenOfAllTrades, Skyring, Kusma, Alai, Mpatel, Betsythedevine, Mendaliv, VermillionBird, Rjwilmsi, Coemgenus, FlaBot, Thenewdeal87, Adoniscik, Algebraist, YurikBot, Wavelength, Bambaiah, Hairy Dude, Moto Perpetuo, Ohwilleke, JabberWok, Kirill Lokshin, Spike Wilbury, BlackAndy, Thiseye, CecilWard, Voidxor, Zzuuzz, Banus, Finell, SmackBot, Henriok, Vald, ProveIt, GaeusOctavius, Chris the speller, Bluebot, TimBentley, Complexica, Colonies Chris, Modest Genius, Berland, Grover cleveland, Garry Denke, TriTertButoxy, DJIndica, Jaganath, RoboDick, NNemec, Slakr, Ryulong, Tawkerbot2, Memetics, Capefeather, Runningonbrains, Cydebot, DavidMcCabe, Headbomb, WVhybrid, Noclevername, Escarbot, Salgueiro, Shambolic Entity, Andonic, Hut 8.5, Pkoppenb, .anacondabot, Robomojo, Corvidaecorvus, Maliz, Connor Behan, TechnoFaye, R'n'B, HEL, DrKiernan, Acalamari, Shomroni, Lseixas, Skullfunk, GrahamHardy, Idioma-bot, Sheliak, VolkovBot, TXiKiBoT, Calwiki, Rei-bot, Saibod, KP-Adhikari, Ptrslv72, SieBot, Dawn Bard, Likebox, Anchor Link Bot, ClueBot, WDavis1911, Pechmerle, PixelBot, Brews ohare, Chrisarnesen, XLinkBot, SilvonenBot, SkyLined, Truthnlove, Addbot, DOI bot, AnnaFrance, SpBot, Lightbot, Zorrobot, Legobot, Luckas-bot, Yobot, Tamtamar, Nallimbot, Citation bot, LilHelpa, Info21, Chrisfox8, Pra1998, Petros000, FrescoBot, Ecuqkindler, Timmeken, Ganondolf, Meier99, Tarsilia, McSaks, Autumnalmonk, EmausBot, Mnkyman, Wikipelli, Brazmyth, Quondum, Aschwole, Rcsprinter123, Maschen, Fwilczek, RolteVolte, Neduard, QuantumSquirrel, ClueBot NG, Helpful Pixie Bot, Bibcode Bot, Dalit Llama, PhnomPencil, Vkpd11, Snow Blizzard, Cjean42, Trompedo and Anonymous: 135

- **Classical unified field theories** *Source:* http://en.wikipedia.org/wiki/Classical%20unified%20field%20theories?oldid=621896407 *Contributors:* Michael Hardy, William M. Connolley, Charles Matthews, Reddi, SJRubenstein, Diberri, Giftlite, Lethe, Alison, Brockert, Rich Farmbrough, Sfahey, Sjoerd visscher, Pearle, Daniel Arteaga, Yuckfoo, Linas, Duncan.france, Mpatel, Mandarax, YurikBot, Hillman, Gaius Cornelius, Salsb, Shanel, SEWilcoBot, Petri Krohn, SmackBot, Harald88, Hmains, Colonies Chris, Ligulembot, Lambiam, RHB, CmdrObot, Roger Anderton, Headbomb, Big Bird, RainbowCrane, VoABot II, DAGwyn, Catgut, Nikopopl, STBot, J.delanoy, Squids and Chips, Neparis, Oxymoron83, UserDoe, Alexbot, MelonBot, Truthnlove, Addbot, DOI bot, Bte99, Bathambaba, Yobot, Gsard, Citation bot 1, Hhhippo, Davidaedwards, Ggonzalm, Bibcode Bot, Teelaskeletor and Anonymous: 31

- **Riemannian geometry** *Source:* http://en.wikipedia.org/wiki/Riemannian%20geometry?oldid=628392826 *Contributors:* The Anome, Michael Hardy, Cyp, LittleDan, Kevin Baas, AugPi, EdH, Raven in Orbit, Charles Matthews, Terse, Dysprosia, Phys, Owen, Rvollmert, JensG, Tobias Bergemann, Tosha, Giftlite, Mikez, Jason Quinn, Just Another Dan, DemonThing, Vadmium, Klemen Kocjancic, Wmcjunkin, Rich Farmbrough, ESkog, Gauge, Ntmatter, Msh210, Linas, Oliphaunt, Ryan Reich, Mike Peel, Mathbot, Rob*, Chobot, YurikBot, Hakeem.gadi, JRawle, Finell, Sardanaphalus, SmackBot, Hbackman, Srnec, Silly rabbit, Complexica, Nbarth, Tamfang, Pwjb, Quaeler, Rschwieb, MightyWarrior, Headbomb, JAnDbot, Ttwo, Tarotcards, Haseldon, Policron, Squids and Chips, Sheliak, JohnBlackburne, Natural Philosopher, Turgan, Arcfrk, Katzmik, KoenDelaere, ClueBot, Jjauregui, Enrico Dirac, Masterpiece2000, WikHead, MystBot, Addbot, DOI bot, Dabsent, Pmod, Lightbot, Yobot, KamikazeBot, Götz, Maxis ftw, Xqbot, Ataleh, RibotBOT, Theshz, Nagualdesign, Sławomir Biały, Lochieisawsome, Baddykid13, Earthandmoon, Francisco Quiumento, Adamcheasley, Slawekb, JSquish, Rcsprinter123, Rostz, MerlIwBot, Turfy87, Brad7777, Frinthruit and Anonymous: 60

- **Affine connection** *Source:* http://en.wikipedia.org/wiki/Affine%20connection?oldid=627200220 *Contributors:* Edward, Michael Hardy, Silverfish, Charles Matthews, Phys, Robbot, Tosha, Giftlite, BenFrantzDale, Fropuff, DemonThing, Mdd, Keenan Pepper, Oleg Alexandrov, Linas, Mpatel, Isnow, Ryan Reich, Rjwilmsi, Salix alba, R.e.b., BradBeattie, Chobot, Evilbu, SmackBot, Paxse, Silly rabbit, Myasuda, WillowW, Synergy, Thijs!bot, Shambolic Entity, JAnDbot, Felix116, Meredyth, JaGa, R'n'B, Maurice Carbonaro, Typometer, Alan U. Kennington, XCelam, S.racaniere, Ylebru, Geometry guy, JemGage, YohanN7, Paolo.dL, Mild Bill Hiccup, Brews ohare, Addbot, DOI bot, Yobot, Ht686rg90, Amirobot, Kilom691, Point-set topologist, Gsard, Citation bot 1, SUL, Rausch, RjwilmsiBot, EmausBot, Slawekb, ZéroBot, Maschen, Mgvongoeden, Frietjes, Helpful Pixie Bot, Joydeep, Rauindia, Mark viking and Anonymous: 16

- **De Sitter universe** *Source:* http://en.wikipedia.org/wiki/De%20Sitter%20universe?oldid=621604571 *Contributors:* Stevertigo, Boud, Palfrey, Timwi, Robbot, Barbara Shack, ShaunMacPherson, LeYaYa, Fropuff, Eequor, SWAdair, Icairns, Lumidek, Cacycle, Pjacobi, Pt, Jag123, QTxVi4bEMRbrNqOorWBV, Falcorian, Mpatel, Christopher Thomas, Mattmartin, Carrionluggage, Hillman, Salsb, Larsobrien, SmackBot, Rentier, Eskimbot, Bluebot, Silly rabbit, Scwlong, QFT, Jmnbatista, Lpgeffen, Mindnumbed, Harold f, Cydebot, Michael C Price, Thijs!bot, Pervect, R'n'B, Ontarioboy, Dmcq, Vanished User 8902317830, Bobathon71, Addbot, SpBot, AnomieBOT, Dogbert66, RedBot, Lotje, EmausBot, Raidr, Helpful Pixie Bot, Monkbot, Thundergodz and Anonymous: 15

- **Standard Model** *Source:* http://en.wikipedia.org/wiki/Standard%20Model?oldid=633812376 *Contributors:* AxelBoldt, Derek Ross, CYD, Bryan Derksen, The Anome, Ed Poor, Andre Engels, Roadrunner, David spector, Isis, Youandme, Ram-Man, Stevertigo, Edward, Patrick, Boud, Michael Hardy, SebastianHelm, Looxix, Julesd, Glenn, AugPi, Mxn, Raven in Orbit, Reddi, Phr, Tpbradbury, Populus, Haoherb428, Phys, Floydian, Bevo, Pierre Boreal, AnonMoos, BenRG, Jeffq, Dmytro, Drxenocide, Robbot, Nurg, Securiger, Texture, Roscoe x, Fuelbottle, Superm401, Tobias Bergemann, Alan Liefting, Ancheta Wis, Giftlite, Dbenbenn, Harp, Herbee, Monedula, LeYaYa, Xerxes314, Dratman, Alison, JeffBobFrank, Dmmaus, Pharotic, Brockert, Bodhitha, Andycjp, Sonjaaa, HorsePunchKid, APH, Icairns, AmarChandra, Gscshoyru, Kate, Arivero, FT2, Rama, David Schaich, Xezbeth, D-Notice, Dfan, Bender235, Pt, El C, Laurascudder, Shanes, Drhex, Fogger, Brim, Rbj, Jeodesic, Jumbuck, Alansohn, Gary, ChristopherWillis, Guy Harris, Axl, Sligocki, Kocio, Stillnotelf, Alinor, Wtmitchell, Egg, TenOfAllTrades, H2g2bob, Killing Vector, Linas, Mindmatrix, Benbest, Dodiad, Mpatel, Faethon, TPickup, Faethon34, Palica, Dysepsion, Faethon36, Qwertyca, Drbogdan, Rjwilmsi, Zbxgscqf, Macumba, Strangethingintheland, Dstudent, R.e.b., Bubba73, Drrngrvy, Agasicles, FlaBot, Naraht, Agasides, DannyWilde, Dave1g, Itinerant1, Gparker, Jrtayloriv, Goudzovski, Chobot, Bgwhite, FrankTobia, YurikBot, Bambaiah, Ohwilleke, VoxMoose, Bhny, JabberWok, Bovineone, Krbabu, SCZenz, JulesH, Davemck, Lomn, E2mb0t, Dna-webmaster, Jrf, Dv82matt, Tetracube, Hirak 99, Netrapt, JLaTondre, Caco de vidro, RG2, GrinBot, That Guy, From That Show!, Hal peridol, SmackBot, YellowMonkey, Tom Lougheed, Melchoir, Bazza 7, KocjoBot, Jagged 85, Thunderboltz, Setanta747 (locked), Skizzik, Dauto, Chris the speller, Bluebot, TimBentley, Sirex98, Silly rabbit, Complexica, Metacomet, DHN-bot, MovGP0, QFT, Kittybrewster, Addshore, Jmnbatista, Cybercobra, Jgwacker, Soarhead77, Daniel.Cardenas, Yevgeny Kats, Byelf2007, TriTertButoxy, Craig Bolon, Ajnosek, Ekjon Lok, Bjankuloski06, Tarcieri, Waggers, JarahE, Michaelbusch, Lottamiata, Twas Now, IanOfNorwich, Srain, Patrickwooldridge, J Milburn, Mosaffa, Gatortpk, Vessels42, Geremia, Van helsing, Harrigan, Phatom87, Cydebot, David edwards, Verdy p, Michael C Price, Xantharius, Crum375, JamesAM, Thijs!bot, Epbr123, Headbomb, Phy1729, Stannered, Tariqhada, Seaphoto, Orionus, Gnixon, Jbaranao, Jrw@pobox.com, Len Raymond, Narssarssuaq, Bakken, CattleGirl, Davidoaf, Vanished user ty12kl89jq10, Lvwarren, Taborgate, HEL, J.delanoy, Hans Dunkelberg, Stephanwehner, Wbellido, Aoosten, Jacksonwalters, The Transliterator, DadaNeem, Student7, Joshmt, WJBscribe, Jozwolf, Hexane2000, BernardZ, Awren, Sheliak, Physicist brazuca, Schucker, Goop Goop, Fences and windows, Dextrose, Mcewan, Swamy g, TXiKiBoT, Sharikkamur, Thrawn562, Voorlandt, Escalona, Setreset, PDFbot, Pleroma, UnitedStatesian, Piyush Sriva, Kacser, Billinghurst, Francis Flinch, Moose-32, Ptrslv72, David Barnard, SieBot, ShiftFn, Robdunst, Jim E. Black, SheepNotGoats, Gerakibot, Nozzer42, Mr swordfish, Wing gundam, Bamkin, Likebox, Arthur Smart, HungarianBarbarian, Commutator, KathrynLybarger, Iomesus, C0nanPayne, Crazz bug 5, ClueBot, Superwj5, Wwheaton, Garyx, Elsweyn, Maldmac, DragonBot, Djr32, Diagramma Della Verita, Eeekster, Brews ohare, NuclearWarfare, PhySusie, Ordovico, Mastertek, DumZiBoT, BodhisattvaBot, Guarracino, Mitch Ames, Truthnlove, Stephen Poppitt, Tayste, Addbot, Deepmath,

Eric Drexler, DWHalliday, Mjamja, Leszek Jańczuk, NjardarBot, Mwoldin, Bassbonerocks, Barak Sh, AgadaUrbanit, Lightbot, Smeagol 17, Abjiklam, Luckas-bot, Yobot, Orion11M87, AnomieBOT, JackieBot, Icalanise, Citation bot, ArthurBot, Northryde, LilHelpa, Xqbot, Professor J Lawrence, Tomwsulcer, Edsegal, GrouchoBot, Trongphu, QMarion II, Ernsts, A. di M., Bytbox, FrescoBot, Paine Ellsworth, Aliotra, Steve Quinn, Citation bot 1, Rameshngbot, MJ94, RedBot, MastiBot, Aknochel, Sijothankam, Puzl bustr, Beta Orionis, Physics therapist, Bj norge, Innotata, Jesse V., RjwilmsiBot, Mathewsyriac, Afteread, EmausBot, Bookalign, WikitanvirBot, Wilhelm-physiker, Bdijkstra, DerNeedle, Kenmint, Dbraize, Tanner Swett, HeptishHotik, مهنشین بهار, Suslindisambiguator, Quondum, Webbeh, UniversumExNihilo, Vanished user fijw983kjaslkekfhj45, RockMagnetist, Stormymountain, Ζετα ζ, Whoop whoop pull up, Isocliff, ClueBot NG, Smtchahal, Snotbot, Tonypak, O.Koslowski, CharleyQuinton, Dsperlich, Theopolisme, ZakMarksbury, Helpful Pixie Bot, Bibcode Bot, BG19bot, Tirebiter78, AvocatoBot, Lukys, Stapletongrey, Ownedroad9, Chip123456, ChrisGualtieri, Khazar2, Billyfesh399, Rhlozier, JYBot, Dexbot, Doom636, Rongended, Cerabot, Cjean42, Jayanta mallick, Kowtje, JPaestpreornJeolhlna, Eyesnore, Euan Richard, Nigstomper, Particle physicist, Jernahthern, Ginsuloft, Dimension10, JNrgbKLM, Krabaey, Delbert7, BradNorton1979 and Anonymous: 343

- **Gauge theory** *Source:* http://en.wikipedia.org/wiki/Gauge%20theory?oldid=633276552 *Contributors:* The Anome, Michael Hardy, Tobias Bergemann, Ancheta Wis, TedPavlic, Xezbeth, MuDavid, Bender235, Pt, Phils, BD2412, Rjwilmsi, JocK, Modify, Teply, SmackBot, RDBury, Henning Makholm, Byelf2007, Michael C Price, Biblbroks, Headbomb, Nick Number, Fashionslide, VectorPosse, Magioladitis, Bakken, Email4mobile, JaGa, Policron, Squids and Chips, Cuzkatzimhut, VolkovBot, Red Act, Michael H 34, Setreset, Jwpitts, Tcamps42, Moonriddengirl, ClueBot, Mastertek, TimothyRias, XLinkBot, Addbot, Mortense, Eric Drexler, Bte99, Zorrobot, Luckas-bot, AnomieBOT, Christopher.Gordon3, Citation bot, Northryde, Xqbot, Pra1998, Gsard, A. di M., Erik9bot, FrescoBot, Fortdj33, Citation bot 1, Ganondolf, RedBot, RobinK, Mary at CERN, EmausBot, Brent Perreault, Slawekb, Cogiati, Maschen, Isocliff, ClueBot NG, Helpful Pixie Bot, Bibcode Bot, Dzustin, Brendan.Oz, ChrisGualtieri, SD5bot, Enyokoyama, Dath Thou Even Lift, Dhm4444 and Anonymous: 37

- **Quantum field theory** *Source:* http://en.wikipedia.org/wiki/Quantum%20field%20theory?oldid=632510599 *Contributors:* AxelBoldt, CYD, Mav, The Anome, XJaM, Roadrunner, Stevertigo, Michael Hardy, Tim Starling, IZAK, TakuyaMurata, SebastianHelm, Looxix, Ahoerstemeier, Cyp, Glenn, Rotem Dan, Stupidmoron, Charles Matthews, Timwi, Jitse Niesen, Kbk, Rudminjd, Wik, Phys, Bevo, BenRG, Northgrove, Robbot, Bkalafut, Gandalf61, Rursus, Fuelbottle, Tobias Bergemann, Ancheta Wis, Giftlite, Lethe, Dratman, Alison, St3vo, Mboverload, DefLog, ConradPino, Amarvc, Pcarbonn, Karol Langner, APH, AmarChandra, D6, CALR, Urvabara, Discospinster, Guanabot, Igorivanov, Masudr, Pjacobi, Vsmith, Nvj, MuDavid, Bender235, Pt, El C, Shanes, Sietse Snel, Physicistjedi, KarlHallowell, PWilkinson, Helix84, Thialfi, Varuna, Gcbirzan, Docboat, Count Iblis, Egg, Mpatel, Marudubshinki, Graham87, Opie, Vanderdecken, Rjwilmsi, MarSch, Earin, R.e.b., RE, Strobilomyces, Arnero, Itinerant1, Alfred Centauri, Srleffler, Chobot, UkPaolo, Wavelength, Bambaiah, Hairy Dude, RussBot, TimNelson, Archelon, CambridgeBayWeather, SCZenz, Odddmonster, E2mb0t, Semperf, Tetracube, Garion96, Erik J, Robert L, Banus, RG2, SmackBot, Stephan Schneider, Tom Lougheed, Melchoir, KocjoBot, Mcld, Dauto, Chris the speller, Complexica, Threepounds, RuudVisser, QFT, Jmnbatista, Cybercobra, Rebooted, Victor Eremita, DJIndica, Lambiam, Mgiganteus1, Zarniwoot, Jim.belk, Stwalkerster, SirFozzie, Hu12, Dan Gluck, Iridescent, Joseph Solis in Australia, Albertod4, Van helsing, GeorgeLouis, Witten Is God, Cydebot, Jamie Lokier, Meno25, Michael C Price, The 80s chick, Mendicus, AstroPig7, Msebast, Mbell, Headbomb, Nick Number, Mentifisto, AntiVandalBot, Bt414, Bananan, Martin Kostner, Moltrix, Kasimann, Kromatol, Puksik, Lerman, LLHolm, RogueNinja, Tlabshier, JEH, Nikolas Karalis, Storkk, JAnDbot, Igodard, Four Dog Night, N shaji, Bongwarrior, Andrea Allais, Soulbot, Etale, Maliz, Custos0, HEL, J.delanoy, Acalamari, Jeepday, Policron, Blckavnger, Juliancolton, Skou, Telecomtom, GrahamHardy, Sheliak, Cuzkatzimhut, VolkovBot, Bktennis2006, Marksr, HowardFrampton, The Original Wildbear, Dj thegreat, Markisgreen, Lejarrag, Moose-32, Raphtee, Sue Rangell, Neparis, Drschawrz, YohanN7, SieBot, TCO, Yintan, Likebox, Paolo.dL, Tugjob, Henry Delforn (old), Jecht (Final Fantasy X), OKBot, StewartMH, ClueBot, EoGuy, Wwheaton, The Wild West guy, Shvav, Bob108, Brews ohare, Thingg, Count Truthstein, XLinkBot, PSimeon, SilvonenBot, Truthnlove, HexaChord, Addbot, ConCompS, Pinkgoanna, Leapold, Dmhowarth26, Glane23, Hanish.polavarapu, Lightbot, R.ductor, Ettrig, Yndurain, Legobot, Luckas-bot, Yobot, Ht686rg90, Niout, Tamtamar, AnomieBOT, Ciphers, Palpher, IRP, Gjsreejith, Materialscientist, Citation bot, Bci2, ArthurBot, Northryde, LilHelpa, Caracolillo, Amareto2, MIRROR, Professor J Lawrence, Plasmon1248, Omnipaedista, RibotBOT, Spellage, JayJay, FrescoBot, Kenneth Dawson, D'ohBot, Knowandgive, N4tur4le, Hyqeom, Newt Winkler, Hickorybark, Lotje, Dinamik-bot, LilyKitty, Fortesque666, Reaper Eternal, Minimac, Marie Poise, Yaush, Dylan1946, EmausBot, Racerx11, GoingBatty, Carbosi, Thecheesykid, ZéroBot, Cogiati, Jjspinorfield1, Suslindisambiguator, Quondum, Maschen, Zueignung, Davidaedwards, Lom Konkreta, ClueBot NG, Gilderien, Iloveandrea, Vacation9, Heyheyheyhohoho, Fortune432, The ubik, Zak.estrada, Helpful Pixie Bot, Evanescent7, Ykentluo, Martin.uecker, Walterpfeifer, Pfeiferwalter, Klilidiplomus, W.D., CarrieVS, Khazar2, Momo1381, Dexbot, Cerabot, Garuda0001, AHusain314, Thepalerider2012, Mark viking, Faizan, Aj7s6, संजीव कुमार, Lemnaminor, BerFinelli, Axel.P.Hedstrom, Kclongstocking, Mutley1989, I art a troler, Liquidityinsta, DemonThuum, Dingdong2680, Asherkirschbaum, Monkbot, Gjbayes, Thedarkcheese, BradNorton1979, UareNumber6, Teelaskeletor, Mret81 and Anonymous: 285

49.9.2 Images

- **File:AdS3_(new).png** *Source:* http://upload.wikimedia.org/wikipedia/commons/f/fe/AdS3_%28new%29.png *License:* CC-BY-SA-3.0 *Contributors:* Own work *Original artist:* Polytope24

- **File:Ambox_content.png** *Source:* http://upload.wikimedia.org/wikipedia/en/f/f4/Ambox_content.png *License:* ? *Contributors:*
  Derived from Image:Information icon.svg *Original artist:*
  El T (original icon); David Levy (modified design); Penubag (modified color)

- **File:Biaxial.png** *Source:* http://upload.wikimedia.org/wikipedia/commons/7/7a/Biaxial.png *License:* Public domain *Contributors:* (1979). "The topological theory of defects in ordered media". *Reviews of Modern Physics* **51** (3). *Original artist:* N. D. Mermin

- **File:Black_Hole_Merger.jpg** *Source:* http://upload.wikimedia.org/wikipedia/commons/d/d1/Black_Hole_Merger.jpg *License:* Public domain *Contributors:* Taken from http://www.space.com/imageoftheday/image_of_day_060203.html credit is listed to NASA. *Original artist:* NASA

- **File:CDel_3.png** *Source:* http://upload.wikimedia.org/wikipedia/commons/c/c3/CDel_3.png *License:* Public domain *Contributors:* Own work *Original artist:* User:Tomruen

- **File:CDel_3ab.png** *Source:* http://upload.wikimedia.org/wikipedia/commons/d/d5/CDel_3ab.png *License:* Public domain *Contributors:* Own work *Original artist:* User:Tomruen

- **File:CDel_node.png** *Source:* http://upload.wikimedia.org/wikipedia/commons/5/5e/CDel_node.png *License:* Public domain *Contributors:* Own work *Original artist:* User:Tomruen
- **File:CDel_node_1.png** *Source:* http://upload.wikimedia.org/wikipedia/commons/b/bd/CDel_node_1.png *License:* Public domain *Contributors:* Own work *Original artist:* User:Tomruen
- **File:CDel_nodes.png** *Source:* http://upload.wikimedia.org/wikipedia/commons/1/1f/CDel_nodes.png *License:* Public domain *Contributors:* Own work *Original artist:* User:Tomruen
- **File:CDel_split1.png** *Source:* http://upload.wikimedia.org/wikipedia/commons/a/a1/CDel_split1.png *License:* Public domain *Contributors:* Own work *Original artist:* User:Tomruen
- **File:CERN_LHC_Tunnel1.jpg** *Source:* http://upload.wikimedia.org/wikipedia/commons/f/fc/CERN_LHC_Tunnel1.jpg *License:* CC-BY-SA-3.0 *Contributors:* Own work *Original artist:* Julian Herzog • [more photography on my website]
- **File:Calabi-Yau.png** *Source:* http://upload.wikimedia.org/wikipedia/commons/d/d4/Calabi-Yau.png *License:* CC-BY-SA-2.5 *Contributors:* own work by Lunch http://en.wikipedia.org/wiki/Image:Calabi-Yau.png (english Wikipedia) *Original artist:* en:User:Lunch
- **File:Calabi_yau.jpg** *Source:* http://upload.wikimedia.org/wikipedia/commons/f/f3/Calabi_yau.jpg *License:* Public domain *Contributors:* Mathematica output, created by author *Original artist:* Jbourjai
- **File:Circle_as_Lie_group.svg** *Source:* http://upload.wikimedia.org/wikipedia/commons/8/82/Circle_as_Lie_group.svg *License:* Public domain *Contributors:* self-made with en:Inkscape *Original artist:* Oleg Alexandrov
- **File:Commons-logo.svg** *Source:* http://upload.wikimedia.org/wikipedia/en/4/4a/Commons-logo.svg *License:* ? *Contributors:* ? *Original artist:* ?
- **File:Compactification_on_a_circle.png** *Source:* http://upload.wikimedia.org/wikipedia/commons/3/33/Compactification_on_a_circle.png *License:* CC-BY-SA-3.0 *Contributors:* Own work *Original artist:* Polytope24
- **File:Crab_Nebula.jpg** *Source:* http://upload.wikimedia.org/wikipedia/commons/0/00/Crab_Nebula.jpg *License:* Public domain *Contributors:* HubbleSite: gallery, release. *Original artist:* NASA, ESA, J. Hester and A. Loll (Arizona State University)
- **File:CuttingABarMagnet.svg** *Source:* http://upload.wikimedia.org/wikipedia/commons/4/43/CuttingABarMagnet.svg *License:* CC0 *Contributors:* Own work *Original artist:* Sbyrnes321
- **File:DoubleWellSoliton.png** *Source:* http://upload.wikimedia.org/wikipedia/commons/f/ff/DoubleWellSoliton.png *License:* CC-BY-SA-3.0 *Contributors:* Own work *Original artist:* Cyp
- **File:DoubleWellSolitonAntisoliton.gif** *Source:* http://upload.wikimedia.org/wikipedia/commons/8/8e/DoubleWellSolitonAntisoliton.gif *License:* CC-BY-SA-3.0 *Contributors:* Own work by uploader (source code below, click "show" to see) *Original artist:* Cyp
- **File:Dualities_of_string_and_M-theory.jpg** *Source:* http://upload.wikimedia.org/wikipedia/commons/b/bd/Dualities_of_string_and_M-theory.jpg *License:* CC-BY-SA-3.0 *Contributors:* Own work *Original artist:* Polytope24
- **File:Dyn-3.png** *Source:* http://upload.wikimedia.org/wikipedia/commons/b/b3/Dyn-3.png *License:* Public domain *Contributors:* Own work *Original artist:* self
- **File:Dyn-3s.png** *Source:* http://upload.wikimedia.org/wikipedia/commons/9/96/Dyn-3s.png *License:* Public domain *Contributors:* Own work *Original artist:* self
- **File:Dyn-loop2.png** *Source:* http://upload.wikimedia.org/wikipedia/commons/f/ff/Dyn-loop2.png *License:* Public domain *Contributors:* Own work *Original artist:* self
- **File:Dyn-node.png** *Source:* http://upload.wikimedia.org/wikipedia/commons/8/8b/Dyn-node.png *License:* Public domain *Contributors:* Own work *Original artist:* self
- **File:Dyn-nodes.png** *Source:* http://upload.wikimedia.org/wikipedia/commons/5/53/Dyn-nodes.png *License:* Public domain *Contributors:* Own work *Original artist:* self
- **File:Dyn2-3.png** *Source:* http://upload.wikimedia.org/wikipedia/commons/c/c8/Dyn2-3.png *License:* Public domain *Contributors:* Own work *Original artist:* self
- **File:Dyn2-branch.png** *Source:* http://upload.wikimedia.org/wikipedia/commons/7/7b/Dyn2-branch.png *License:* Public domain *Contributors:* Own work *Original artist:* self
- **File:Dyn2-node.png** *Source:* http://upload.wikimedia.org/wikipedia/commons/f/fb/Dyn2-node.png *License:* Public domain *Contributors:* Own work *Original artist:* self
- **File:DynkinE6.svg** *Source:* http://upload.wikimedia.org/wikipedia/commons/6/65/DynkinE6.svg *License:* CC-BY-SA-3.0 *Contributors:* Own work *Original artist:* Jgmoxness
- **File:Dynkin_diagram_E6.png** *Source:* http://upload.wikimedia.org/wikipedia/en/5/5b/Dynkin_diagram_E6.png *License:* ? *Contributors:*

Own work

Original artist:

Fropuff (talk) (Uploads)
- **File:E6Coxeter.svg** *Source:* http://upload.wikimedia.org/wikipedia/commons/a/a5/E6Coxeter.svg *License:* CC-BY-SA-3.0 *Contributors:* Own work *Original artist:* Jgmoxness
- **File:E6GUT.svg** *Source:* http://upload.wikimedia.org/wikipedia/commons/9/9c/E6GUT.svg *License:* CC-BY-SA-3.0 *Contributors:* Own work, Created from Garret Lisi's Elementary Particle Explorer *Original artist:* Cjean42
- **File:E6HassePoset.svg** *Source:* http://upload.wikimedia.org/wikipedia/commons/4/41/E6HassePoset.svg *License:* CC-BY-SA-3.0 *Contributors:* Own work *Original artist:* Jgmoxness

- **File:Edit-clear.svg** *Source:* http://upload.wikimedia.org/wikipedia/en/f/f2/Edit-clear.svg *License:* ? *Contributors:* The *Tango! Desktop Project*. *Original artist:*

    The people from the Tango! project. And according to the meta-data in the file, specifically: "Andreas Nilsson, and Jakub Steiner (although minimally)."

- **File:Edward_Witten_at_Harvard.jpg** *Source:* http://upload.wikimedia.org/wikipedia/commons/d/dd/Edward_Witten_at_Harvard.jpg *License:* CC-BY-3.0 *Contributors:* Own work by the original uploader *Original artist:* Lumidek at English Wikipedia

- **File:Elementary_particle_interactions_in_the_Standard_Model.png** *Source:* http://upload.wikimedia.org/wikipedia/commons/a/a7/Elementary_particle_interactions_in_the_Standard_Model.png *License:* CC0 *Contributors:* Own work *Original artist:* Eric Drexler

- **File:Em_dipoles.svg** *Source:* http://upload.wikimedia.org/wikipedia/commons/f/f0/Em_dipoles.svg *License:* CC0 *Contributors:* Own work *Original artist:* Maschen

- **File:Em_monopoles.svg** *Source:* http://upload.wikimedia.org/wikipedia/commons/2/2f/Em_monopoles.svg *License:* CC0 *Contributors:* Own work *Original artist:* Maschen

- **File:ExponentialMap-01.png** *Source:* http://upload.wikimedia.org/wikipedia/commons/0/06/ExponentialMap-01.png *License:* Public domain *Contributors:* ? *Original artist:* ?

- **File:FeynRulesEN.jpg** *Source:* http://upload.wikimedia.org/wikipedia/commons/7/7f/FeynRulesEN.jpg *License:* Public domain *Contributors:* Own work *Original artist:* Pra1998

- **File:Feynman-Diagram.svg** *Source:* http://upload.wikimedia.org/wikipedia/commons/e/e3/Feynman-Diagram.svg *License:* Public domain *Contributors:* own work, based on Image:Feynman-Diagram.jpg *Original artist:* helix84

- **File:Finite_Dynkin_diagrams.svg** *Source:* http://upload.wikimedia.org/wikipedia/commons/0/0c/Finite_Dynkin_diagrams.svg *License:* CC-BY-SA-3.0 *Contributors:* Created by me by copying File:Connected_Dynkin_Diagrams.svg *Original artist:* Tomruen

- **File:Fluxtube_meson.png** *Source:* http://upload.wikimedia.org/wikipedia/commons/9/93/Fluxtube_meson.png *License:* Public domain *Contributors:* http://inspirehep.net/record/840296 , arXiv:0912.3181 [hep-lat] and M. Cardoso et al., *Lattice QCD computation of the colour fields for the static hybrid quark-gluon-antiquark system, and microscopic study of the Casimir scaling*, Physical Review D, **81** (2010)) *Original artist:* RolteVolte (<a href='//commons.wikimedia.org/wiki/User_talk:RolteVolte' title='User talk:RolteVolte'>talk</a>) 16:42, 23 October 2011 (UTC)

- **File:Folder_Hexagonal_Icon.svg** *Source:* http://upload.wikimedia.org/wikipedia/en/4/48/Folder_Hexagonal_Icon.svg *License:* ? *Contributors:* ? *Original artist:* ?

- **File:Georg_Friedrich_Bernhard_Riemann.jpeg** *Source:* http://upload.wikimedia.org/wikipedia/commons/8/82/Georg_Friedrich_Bernhard_Riemann.jpeg *License:* Public domain *Contributors:* http://www.sil.si.edu/digitalcollections/hst/scientific-identity/explore.htm according to the German Wikipedia. *Original artist:*

- **File:Georgi-Glashow_charges.svg** *Source:* http://upload.wikimedia.org/wikipedia/commons/2/24/Georgi-Glashow_charges.svg *License:* CC-BY-SA-3.0 *Contributors:* Own work, Created from Garret Lisi's Elementary Particle Explorer *Original artist:* Cjean42

- **File:Gosset_1_22_polytope.svg** *Source:* http://upload.wikimedia.org/wikipedia/commons/a/a4/Gosset_1_22_polytope.svg *License:* CC-BY-3.0 *Contributors:* Own work *Original artist:* Claudio Rocchini

- **File:Gravity_Probe_B.jpg** *Source:* http://upload.wikimedia.org/wikipedia/commons/5/51/Gravity_Probe_B.jpg *License:* Public domain *Contributors:* ? *Original artist:* ?

- **File:History_of_the_Universe.svg** *Source:* http://upload.wikimedia.org/wikipedia/commons/d/db/History_of_the_Universe.svg *License:* CC-BY-SA-3.0 *Contributors:* Own work *Original artist:* Yinweichen

- **File:Horizonte_inflacionario.svg** *Source:* http://upload.wikimedia.org/wikipedia/commons/b/b4/Horizonte_inflacionario.svg *License:* CC-BY-SA-3.0 *Contributors:* Transferred from en.wikipedia; original: *I created this work in Adobe Illustrator*. *Original artist:* Original uploader was Joke137 at en.wikipedia

- **File:Hqmc-vector.svg** *Source:* http://upload.wikimedia.org/wikipedia/commons/6/68/Hqmc-vector.svg *License:* CC-BY-3.0 *Contributors:* Own work *Original artist:* VermillionBird

- **File:Hydrogen300.png** *Source:* http://upload.wikimedia.org/wikipedia/commons/a/ad/Hydrogen300.png *License:* Public domain *Contributors:* Transferred from en.wikipedia; transferred to Commons by User:OverlordQ using CommonsHelper. *Original artist:* PoorLeno (talk) Original uploader was PoorLeno at en.wikipedia

- **File:Infrared_gluon_propagator_of_Yang-Mills_theory.jpg** *Source:* http://upload.wikimedia.org/wikipedia/commons/6/64/Infrared_gluon_propagator_of_Yang-Mills_theory.jpg *License:* CC-BY-SA-3.0 *Contributors:* Own work *Original artist:* Pra1998

- **File:Kaluza_Klein_compactification.svg** *Source:* http://upload.wikimedia.org/wikipedia/commons/d/dd/Kaluza_Klein_compactification.svg *License:* Public domain *Contributors:* Own work *Original artist:* Isilanes

- **File:Knot_table-blank_unknot.svg** *Source:* http://upload.wikimedia.org/wikipedia/commons/f/f0/Knot_table-blank_unknot.svg *License:* Public domain *Contributors:* File:Knot table.svg *Original artist:* Jkasd (original), 84user (removed Unknot label)

- **File:LQG_black_hole_Horizon.jpg** *Source:* http://upload.wikimedia.org/wikipedia/en/9/9d/LQG_black_hole_Horizon.jpg *License:* CC-BY-SA-3.0 *Contributors:*

    created on xfig
    Previously published: 2007-09-01

    Original artist:

    Ibayn

- **File:Liealgebra.png** *Source:* http://upload.wikimedia.org/wikipedia/commons/d/d2/Liealgebra.png *License:* Public domain *Contributors:* http://en.wikipedia.org/wiki/File:Liealgebra.png *Original artist:* Phys

- **File:Limits_of_M-theory.png** *Source:* http://upload.wikimedia.org/wikipedia/commons/7/7b/Limits_of_M-theory.png *License:* CC-BY-SA-3.0 *Contributors:* Own work *Original artist:* Polytope24

- **File:MSSM_Flavor_Changing.svg** *Source:* http://upload.wikimedia.org/wikipedia/en/c/cb/MSSM_Flavor_Changing.svg *License:* CC-BY-SA-3.0 *Contributors:*
  I created this work entirely by myself.
  Original artist:
  JabberWok (talk)

- **File:MichaelDuff.JPG** *Source:* http://upload.wikimedia.org/wikipedia/en/5/5e/MichaelDuff.JPG *License:* PD *Contributors:*
  I created this work entirely by myself. I took the picture at Harvard in 2003 and release it to the public domain.
  Original artist:
  Lubos Motl

- **File:Nuvola_apps_edu_mathematics_blue-p.svg** *Source:* http://upload.wikimedia.org/wikipedia/commons/3/3e/Nuvola_apps_edu_mathematics_blue-p.svg *License:* GPL *Contributors:* Derivative work from Image:Nuvola apps edu mathematics.png and Image:Nuvola apps edu mathematics-p.svg *Original artist:* David Vignoni (original icon); Flamurai (SVG convertion); bayo (color)

- **File:Nuvola_apps_katomic.png** *Source:* http://upload.wikimedia.org/wikipedia/commons/7/73/Nuvola_apps_katomic.png *License:* LGPL *Contributors:* http://icon-king.com *Original artist:* David Vignoni / ICON KING

- **File:Office-book.svg** *Source:* http://upload.wikimedia.org/wikipedia/commons/a/a8/Office-book.svg *License:* Public domain *Contributors:* This and myself. *Original artist:* Chris Down/Tango project

- **File:Open_and_closed_strings.svg** *Source:* http://upload.wikimedia.org/wikipedia/commons/5/56/Open_and_closed_strings.svg *License:* CC0 *Contributors:* Own work *Original artist:* Xoneca

- **File:PIA17993-DetectorsForInfantUniverseStudies-20140317.jpg** *Source:* http://upload.wikimedia.org/wikipedia/commons/1/1a/PIA17993-DetectorsForInfantUniverseStudies-20140317.jpg *License:* Public domain *Contributors:* http://photojournal.jpl.nasa.gov/jpeg/PIA17993.jpg *Original artist:* NASA/JPL-Caltech

- **File:Padlock-silver.svg** *Source:* http://upload.wikimedia.org/wikipedia/commons/f/fc/Padlock-silver.svg *License:* CC0 *Contributors:* http://openclipart.org/people/Anonymous/padlock_aj_ashton_01.svg *Original artist:* This image file was created by AJ Ashton. Uploaded from English WP by User:Eleassar. Converted by User:AzaToth to a silver color.

- **File:Parallel_transport_sphere.svg** *Source:* http://upload.wikimedia.org/wikipedia/commons/f/f7/Parallel_transport_sphere.svg *License:* CC-BY-SA-3.0 *Contributors:* Originally from en.wikipedia; description page is/was here. *Original artist:* Original uploader was Silly rabbit at en.wikipedia

- **File:Parallel_transport_sphere2.svg** *Source:* http://upload.wikimedia.org/wikipedia/commons/2/24/Parallel_transport_sphere2.svg *License:* CC-BY-SA-3.0 *Contributors:* Own work *Original artist:* Silly rabbit at English Wikipedia

- **File:Point&string.png** *Source:* http://upload.wikimedia.org/wikipedia/commons/4/47/Point%26string.png *License:* Public domain *Contributors:* ? *Original artist:* ?

- **File:Portal-puzzle.svg** *Source:* http://upload.wikimedia.org/wikipedia/en/f/fd/Portal-puzzle.svg *License:* ? *Contributors:* ? *Original artist:* ?

- **File:Proton_decay3.svg** *Source:* http://upload.wikimedia.org/wikipedia/commons/b/bc/Proton_decay3.svg *License:* Public domain *Contributors:* Transferred from en.wikipedia; transferred to Commons by User:Ras67 using CommonsHelper.
  Original artist: GreenRoot (talk). Original uploader was GreenRoot at en.wikipedia

- **File:Proton_decay4.svg** *Source:* http://upload.wikimedia.org/wikipedia/en/5/51/Proton_decay4.svg *License:* PD *Contributors:*
  I drew this diagram using Inkscape.
  Original artist:
  GreenRoot (talk)

- **File:QCD.svg** *Source:* http://upload.wikimedia.org/wikipedia/commons/2/2b/QCD.svg *License:* CC-BY-SA-3.0 *Contributors:* Own work *Original artist:* Cjean42

- **File:Quantum_gravity.png** *Source:* http://upload.wikimedia.org/wikipedia/commons/6/6a/Quantum_gravity.png *License:* Public domain *Contributors:* Own work *Original artist:* Raidr

- **File:Question_book-new.svg** *Source:* http://upload.wikimedia.org/wikipedia/en/9/99/Question_book-new.svg *License:* ? *Contributors:*
  Created from scratch in Adobe Illustrator. Based on Image:Question book.png created by User:Equazcion *Original artist:*
  Tkgd2007

- **File:SO(10)_-_16_Weight_Diagram.svg** *Source:* http://upload.wikimedia.org/wikipedia/commons/c/c8/SO%2810%29_-_16_Weight_Diagram.svg *License:* CC-BY-SA-3.0 *Contributors:* Own work *Original artist:* Dauto

- **File:SO10.svg** *Source:* http://upload.wikimedia.org/wikipedia/commons/f/f3/SO10.svg *License:* CC-BY-SA-3.0 *Contributors:* Own work, Created from Garret Lisi's Elementary Particle Explorer *Original artist:* Cjean42

- **File:SU(5)_representation_of_fermions.png** *Source:* http://upload.wikimedia.org/wikipedia/commons/9/94/SU%285%29_representation_of_fermions.png *License:* CC-BY-SA-3.0 *Contributors:* Own work *Original artist:* Paul Bird

- **File:Spacetime_curvature.png** *Source:* http://upload.wikimedia.org/wikipedia/commons/2/22/Spacetime_curvature.png *License:* CC-BY-SA-3.0 *Contributors:* ? *Original artist:* ?

- **File:Spin_foam_from_Hamiltonian_constraint.jpg** *Source:* http://upload.wikimedia.org/wikipedia/en/3/34/Spin_foam_from_Hamiltonian_constraint.jpg *License:* CC-BY-SA-3.0 *Contributors:*
  On a graphics tool.
  Previously published: 2013-04-13
  Original artist:
  Ibayn

- **File:Spinnetwork.jpg** *Source:* http://upload.wikimedia.org/wikipedia/commons/d/d2/Spinnetwork.jpg *License:* CC-BY-SA-3.0-2.5-2.0-1.0 *Contributors:* Own work *Original artist:* Skoglund S
- **File:Standard_Model.svg** *Source:* http://upload.wikimedia.org/wikipedia/commons/f/f4/Standard_Model.svg *License:* CC-BY-SA-3.0 *Contributors:* Own work *Original artist:* Cjean42
- **File:Standard_Model_Feynman_Diagram_Vertices.png** *Source:* http://upload.wikimedia.org/wikipedia/commons/7/75/Standard_Model_Feynman_Diagram_Vertices.png *License:* CC-BY-SA-3.0 *Contributors:* I made it in Adobe Illustrator *Original artist:* Garyzx
- **File:Standard_Model_of_Elementary_Particles.svg** *Source:* http://upload.wikimedia.org/wikipedia/commons/0/00/Standard_Model_of_Elementary_Particles.svg *License:* CC-BY-3.0 *Contributors:* Own work by uploader, PBS NOVA [1], Fermilab, Office of Science, United States Department of Energy, Particle Data Group *Original artist:* MissMJ
- **File:String_theory.svg** *Source:* http://upload.wikimedia.org/wikipedia/commons/8/8a/String_theory.svg *License:* CC-BY-3.0 *Contributors:*
- Levels of magnification from [1] PBS: NOVA *Original artist:* MissMJ
- **File:Stylised_Lithium_Atom.svg** *Source:* http://upload.wikimedia.org/wikipedia/commons/e/e1/Stylised_Lithium_Atom.svg *License:* CC-BY-SA-3.0 *Contributors:* ? *Original artist:* ?
- **File:The_Mandelstam_identity.jpg** *Source:* http://upload.wikimedia.org/wikipedia/en/5/5f/The_Mandelstam_identity.jpg *License:* CC-BY-SA-3.0 *Contributors:*

On graphics package
Previously published: 2005-06-25

Original artist:

Ibayn

- **File:Uniform_tiling_433-t0.png** *Source:* http://upload.wikimedia.org/wikipedia/commons/1/13/Uniform_tiling_433-t0.png *License:* Public domain *Contributors:* Transferred from en.wikipedia *Original artist:* Tomruen at en.wikipedia
- **File:Wiki_letter_w.svg** *Source:* http://upload.wikimedia.org/wikipedia/en/6/6c/Wiki_letter_w.svg *License:* ? *Contributors:* ? *Original artist:* ?
- **File:Wiki_letter_w_cropped.svg** *Source:* http://upload.wikimedia.org/wikipedia/commons/1/1c/Wiki_letter_w_cropped.svg *License:* CC-BY-SA-3.0 *Contributors:*
- Wiki_letter_w.svg *Original artist:* Wiki_letter_w.svg: Jarkko Piiroinen
- **File:Wiktionary-logo-en.svg** *Source:* http://upload.wikimedia.org/wikipedia/commons/f/f8/Wiktionary-logo-en.svg *License:* Public domain *Contributors:* Vector version of Image:Wiktionary-logo-en.png. *Original artist:* Vectorized by Fvasconcellos (talk · contribs), based on original logo tossed together by Brion Vibber
- **File:World_lines_and_world_sheet.svg** *Source:* http://upload.wikimedia.org/wikipedia/commons/2/25/World_lines_and_world_sheet.svg *License:* Public domain *Contributors:* Point&string.png *Original artist:* Kurochka, svg version by Actam

49.9.3 Content license

- Creative Commons Attribution-Share Alike 3.0

www.ingramcontent.com/pod-product-compliance
Lightning Source LLC
Chambersburg PA
CBHW081612200526
45167CB00019B/2255